Migraine Neuroscience: From Experimental Models to Target Therapy

Migraine Neuroscience: From Experimental Models to Target Therapy

Editors

László Vécsei
Bernadett Tuka
Masaru Tanaka

MDPI

Basel • Beijing • Wuhan • Barcelona • Belgrade • Novi Sad • Cluj • Manchester

Editors

László Vécsei
University of Szeged
Szeged
Hungary

Bernadett Tuka
University of Szeged
Szeged
Hungary

Masaru Tanaka
University of Szeged
(HUN-REN-SZTE)
Szeged
Hungary

Editorial Office
MDPI AG
Grosspeteranlage 5
4052 Basel, Switzerland

This is a reprint of articles from the Special Issue published online in the open access journal *Cells* (ISSN 2073-4409) (available at: https://www.mdpi.com/journal/cells/special_issues/Migraine_Neuroscience).

For citation purposes, cite each article independently as indicated on the article page online and as indicated below:

Lastname, A.A.; Lastname, B.B. Article Title. *Journal Name* **Year**, *Volume Number*, Page Range.

ISBN 978-3-7258-1725-2 (Hbk)
ISBN 978-3-7258-1726-9 (PDF)
doi.org/10.3390/books978-3-7258-1726-9

Contents

About the Editors

László Vécsei

László Vécsei is the Professor of Neurology at the University of Szeged and head of the HUN-REN-SZTE Neuroscience Research Group. He received his MD from the Albert Szent-Györgyi Medical University in 1979, PhDs from the University of Szeged and the Hungarian Academy of Sciences in 1986, Lund University in 1989, and a DSc from the Hungarian Academy of Sciences in 1992. He served as Professor and Director of Neurology at the University of Szeged from 1993 to 2019, Dean of the Medical Faculty from 2010 to 2014, and President of the Medical Class of the Hungarian Academy of Sciences from 2011 to 2014. His research focuses on neurological disorders and kynurenines. Vécsei has published around 700 articles with a citation count of 27,200/20,500 (H-index: 62) on MTMT, 26,200 citations (H-index: 76) on Google Scholar, and research.com D-index: 70. He has authored 17 books and holds 9 patents. He has held leadership roles in the Society of Hungarian Neurologists and Psychiatrists, the Hungarian Medical Association, the European Society for Clinical Neuropharmacology, and the European Federation of Neurological Societies (EFNS). Currently, he is the Hungarian Delegate to the European Pain Federation and European Brain Council, the Hungarian Delegate to the World Federation of Neurology (WFN), and the President of the Standards and Evaluation Committee of the WFN.

Bernadett Tuka

Bernadett Tuka, Ph.D., is a Research Fellow and a Clinical Study Manager in the Neuroimaging Research Group in the Department of Radiology, University of Szeged. Earlier, she was an Assistant Research Fellow in the Neuroscience Research Group of the Hungarian Academy of Sciences and University of Szeged for 10 years, where she obtained a PhD in Neurobiology in 2014. Her scientific interest includes the pathomechanism of primary headache disorders (e.g., migraine and cluster headache) and the investigation of other neurological diseases. Her scientific focus is also the metabolic alterations of the tryptophan pathway, the expression changes in certain neuropeptides (e.g., PACAP and CGRP), and the associations of brain functional networks. Translational research between preclinical and clinical studies to identify potential therapeutic solutions is an important part of her research. She is the author of 74 publications and 37 papers in peer-reviewed journals. The number of her independent citations is 1077, and her metrics are 22 (h-index) and 37 (g-index). She is the owner of several Hungarian Youth Awards, and she is a member of the Hungarian Headache Society.

Masaru Tanaka

Masaru Tanaka, M.D., Ph.D. is a Senior Research Fellow in the Danube Neuroscience Research Laboratory, HUN-REN-SZTE Neuroscience Research Group, Hungarian Research Network, University of Szeged (HUN-REN-SZTE). His scientific interests include depression anxiety, dementia pain, their comorbid nature, and translational research in neurological diseases and psychiatric disorders. He is the author of six books and over 95 papers in peer-reviewed journals and ranked D1 in the global publication rankings. His articles have been ranked in the top 99.9 citation percentile in various fields. Not only did he win the Best Paper Award, but he was also named Best Reviewer of the Month. He is an editorial board member of Biomedicines, Biomolecules and Biomedicine, and Frontiers in Neuroscience, among others. He obtained a PhD in Medicine, an MD in general Medicine from the University of Szeged, and a bachelor's degree in Biophysics from the University of Illinois, Urbana-Champaign.

cells MDPI

Editorial

Navigating the Neurobiology of Migraine: From Pathways to Potential Therapies

Masaru Tanaka [1], Bernadett Tuka [2] and László Vécsei [1,3,*]

1 HUN-REN-SZTE Neuroscience Research Group, Danube Neuroscience Research Laboratory, Hungarian Research Network, University of Szeged (HUN-REN-SZTE), Tisza Lajos krt. 113, H-6725 Szeged, Hungary; tanaka.massaru.1@med.u-szeged.hu
2 Department of Radiology, Albert Szent-Györgyi Medical School, University of Szeged, Semmelweis u. 6, H-6725 Szeged, Hungary; tuka.bernadett@med.u-szeged.hu
3 Department of Neurology, Albert Szent-Györgyi Medical School, University of Szeged, Semmelweis u. 6, H-6725 Szeged, Hungary
* Correspondence: vecsei.laszlo@med.u-szeged.hu

Citation: Tanaka, M.; Tuka, B.; Vécsei, L. Navigating the Neurobiology of Migraine: From Pathways to Potential Therapies. *Cells* **2024**, *13*, 1098. https://doi.org/10.3390/cells13131098

Received: 14 June 2024
Accepted: 21 June 2024
Published: 25 June 2024

1. Introduction

Migraine is a debilitating neurological disorder characterized by recurring episodes of throbbing headaches that are frequently accompanied by sensory disturbances, nausea, and sensitivity to light and sound [1–4]. With a global prevalence of approximately 15% and a significant impact on individuals' quality of life, migraine represents a major public health concern [5–7]. Despite its high prevalence and impact, the underlying mechanisms of migraine remain complex and multifaceted, involving a combination of genetic, environmental, and neurobiological factors [8–10]. Understanding the pathophysiology of migraine is essential for developing targeted therapies that can effectively manage symptoms and improve patient outcomes [11–13].

Recent advances in migraine research have emphasized the importance of experimental models in understanding the neurobiological mechanisms that cause migraine attacks [14–18]. Experimental models, particularly murine models, have provided valuable insights into the molecular and cellular pathways involved in migraine pathophysiology, allowing researchers to investigate the neurotransmission, inflammation, and sensitization processes that underpin migraine attacks [19–22]. Researchers have identified key molecules, pathways, and neural circuits that contribute to the initiation and progression of migraine attacks by simulating migraine conditions in controlled experimental settings [23–25]. These experimental models not only enhance our understanding of migraine pathophysiology but also serve as valuable tools for testing potential therapeutic targets and interventions.

The pathophysiology of migraines is significantly influenced by neuroplasticity, which includes modifications to brain excitability, biochemistry, and functional connectivity [26–28]. Chronic migraines are linked to persistent changes in neural plasticity, such as central sensitization and impaired pain modulation mechanisms [29–33]. These changes can set off a vicious cycle of pain chronification, in which the brain's reaction to pain becomes maladaptive and pathological. Several receptors, including calcitonin gene-related peptide (CGRP), transient receptor potential vanilloid subtype 1 (TRPV1), and purinergic receptor P2X subtype 3 (P2X3), have also been implicated in migraine mechanisms. For example, CGRP has been shown to stimulate trigeminal afferent activity and sensitize nociceptive neurons, whereas TRPV1 and P2X3 receptors are involved in mechanical pain modulation and nociceptive signal generation, respectively. Understanding the interplay between neuroplasticity and receptor activity is essential for developing targeted therapies for neurological disorders, including migraine [34–37].

The following Special Issue, entitled "Migraine Neuroscience: From Experimental Models to Target Therapy", aims to showcase the latest research findings and advancements in the field of migraine neuroscience, with a specific focus on experimental models and therapeutic targets for migraine treatment. This collection of articles brings together cutting-edge studies that explore the intricate interplay of the neurobiological processes involved in migraine, from neurotransmitter signaling to neuroinflammation and genetic predisposition. By bridging the gap between basic science research and clinical applications, this Special Issue seeks to provide a comprehensive overview of the current state of migraine research and offer insights into novel therapeutic strategies for managing this complex neurological disorder.

2. Topic Articles

2.1. Calcitonin Gene-Related Peptide (CGRP)-Related Mechanisms and Therapies

CGRP plays a pivotal role in migraine pathophysiology [38–40]. During both migraine attacks (ictal phase) and headache-free periods between attacks (interictal phase), CGRP levels are elevated [41–43]. However, after abortive and prophylactic treatment, these levels tend to decrease [44,45]. Notably, CGRP can induce migraine-like headaches in patients, and studies have demonstrated its significant release during acute migraine and cluster headache attacks [46–48]. Chronic migraine may be associated with chronically elevated CGRP levels [49–51]. Beyond headache mechanisms, CGRP also contributes to maintaining a normal resting tone in brain circulation [52–54]. Perivascular CGRP appears to mediate a protective vasodilatory reflex triggered in response to vasoconstriction [55–57]. Additionally, trigeminal pathway activation leads to the release of both CGRP and substance P [58–60]. A literature review highlights the variability in study design, determination methods, and results for CGRP measurements in migraine patients [61]. Overall, the above findings collectively emphasize the importance of CGRP in migraine and highlight the need for standardized measurement methods, potentially informing therapeutic strategies targeting CGRP for migraine treatment.

2.1.1. Interaction between Calcitonin Gene-Related Peptide (CGRP) and Nitric Oxide in Migraine

Benedicter et al. investigated the complex relationship between glycerol trinitrate (GTN) and CGRP in the context of trigeminal nociception, a key mechanism in migraine pathogenesis [62]. The researchers used rodent experiments and the anti-CGRP antibody Fremanezumab to better understand GTN's downstream effects on CGRP signaling. Their findings provide compelling evidence that GTN acts downstream of CGRP in the trigeminal nociceptive system, suggesting that the modulation of CGRP is a critical factor in the complex pathways that drive migraine pain. By unraveling this interplay between two key players in migraine, the authors contribute valuable insights that may inform the development of targeted therapeutic strategies for migraine management. The findings of their study underscore the importance of understanding the intricate mechanisms underlying migraine in order to develop more effective and personalized treatments for this debilitating condition.

Greco et al. investigated the interaction between CGRP and pain mediators involved in neuronal sensitization in an animal model of chronic migraine [63]. The study authors aimed to investigate the central and peripheral mechanisms of CGRP receptor antagonism. The authors revealed that olcegepant reduced the incidence of trigeminal hyperalgesia by lowering the expression of CGRP in the trigeminal nucleus and proinflammatory markers such as cytokines, microRNA-132, and TRPA1. The results suggest that CGRP receptor antagonism is important in both the central and peripheral mechanisms of chronic migraine. The results of the above study have the potential to contribute to the development of novel therapeutic strategies for chronic migraine, potentially leading to improved treatment outcomes and enhanced patient quality of life.

Dux et al. investigated the effects of Fremanezumab, an anti-CGRP monoclonal antibody, on CGRP release from the rat dura mater and meningeal blood flow [64]. In the rat model, the authors found that administering Fremanezumab reduced CGRP release from the dura mater and, as a result, meningeal blood flow. These findings shed light on the mechanisms by which anti-CGRP antibodies, such as Fremanezumab, can modulate CGRP-mediated processes critical to migraine pathogenesis. The above study advances our understanding of the therapeutic potential of CGRP-targeted interventions in migraine management by elucidating Fremanezumab's effect on CGRP release and meningeal blood flow (Table 1).

Table 1. Major subtopics covering the Special Issue "Migraine Neuroscience: From Experimental Models to Target Therapy".

	Subtopics	Ref.
1.	CGRP-related mechanisms and therapies	
	a. Interaction between CGRP and nitric oxide in migraine	[62–64]
	b. Real-world outcomes and new therapeutic targets for migraine	[65,66]
2.	Metabolic pathways and migraine	
	a. Altered tryptophan metabolism and migraine susceptibility	[67]
	b. Tryptophan metabolism pathways in migraine: therapeutic implications	[68]
3.	Experimental Models and Therapeutic Targets	
	a. Dual FAAH/MAGL inhibitor in a migraine model	[69]
	b. SFK activity and CGRP–cytokine crosstalk	[70]
	c. KATP channels in migraine pathophysiology	[71]
4.	Inflammation and pathophysiology in migraine	
	a. Neurogenic neuroinflammation in migraine	[72]
	b. Complex symptomatology of migraine	[73]

CGRP: calcitonin gene-related peptide; FAAH: fatty acid amide hydrolase; KATP: ATP-sensitive potassium; MAGL: monoacylglycerol lipase; SFKs: Src family kinases.

2.1.2. Real-World Outcomes and New Therapeutic Targets for Migraine

Pavelic et al. conducted a systematic review to assess the real-world outcomes of monoclonal antibodies targeting CGRP for migraine prophylaxis [65]. CGRP has been recognized as a key player in migraine pathophysiology, and the development of CGRP-targeted therapies, including monoclonal antibodies, has revolutionized migraine management. In their review, the authors analyzed data from 134 publications, including retrospective and clinic-based studies, case reports, and other articles. The study findings suggest that treatment with anti-CGRP monoclonal antibodies is associated with lower healthcare utilization, better treatment adherence, and comparable efficacy to randomized controlled trials. The authors do, however, acknowledge that the retrospective study designs, small patient populations, and short follow-up periods used in the examined studies limit the availability of real-world data. They emphasize the need for large prospective studies with long-term follow-up to fully understand the real-world impact of these novel therapies. By synthesizing the available evidence on the efficacy and safety of anti-CGRP monoclonal antibodies in clinical practice, the authors contribute to the growing body of knowledge on the practical implications of these therapies in migraine management.

Tanaka et al. addressed the potential of other neuropeptides, such as pituitary adenylate cyclase-activating polypeptide (PACAP) and vasoactive intestinal peptide (VIP), to better understand the evolving landscape of migraine treatment beyond CGRP [66]. These neuropeptides have shown promise in novel migraine management as a potential therapeutic target in addition to CGRP. Their review delves into the transition from CGRP to PACAP, VIP, and beyond, shedding light on the next generation of migraine treatments. By unraveling the roles of these neuropeptides in migraine pathophysiology, the authors contribute to expanding the understanding of potential targets for migraine therapy beyond the traditional focus on CGRP. Their research paves the way for exploring novel treatment avenues and advancing the field of migraine management (Table 1).

2.2. Metabolic Pathways and Migraine

Altered tryptophan (Trp) metabolism plays a crucial role in migraine susceptibility through its main metabolites, serotonin and kynurenine (KYN), which affect pain processing, stress response, neural hypersensitivity, and inflammatory processes [74–82]. These pathways have been extensively studied in the context of migraine, with a focus on their influence on vascular and inflammatory mechanisms [68,83–85]. The involvement of Trp metabolism, particularly through the serotonin and KYN pathways, has been a key area of research in understanding migraine pathophysiology [86,87]. The authors of recent studies have identified potential therapeutic targets within these pathways for future drug development, emphasizing the importance of regulating Trp-KYN metabolism in migraine treatment [88,89]. The intricate interplay between neurotransmitters, neuropeptides, and inflammatory mediators underscores the complexity of migraine pathogenesis and the potential for targeted interventions to improve treatment outcomes [90,91].

2.2.1. Altered Tryptophan Metabolism and Migraine Susceptibility

Gecse et al. investigated the neuroendocrine response to citalopram in patients with migraine [67]. The authors used a neuroendocrine challenge to assess the metabolism of Trp and KYN, a key amino acid and its metabolite involved in migraine pathophysiology. The findings of their study indicate that patients with migraine exhibit altered Trp-KYN metabolism compared to healthy controls. Specifically, they observed increased Trp levels and decreased KYN levels in migraine patients. These changes suggest that the metabolism of these amino acids may play a crucial role in the development and continuation of migraine. By elucidating the neuroendocrine response to citalopram in migraine patients, the results of the above study advance our understanding of the complex mechanisms that underpin migraine, highlighting the potential therapeutic implications of targeting Trp-KYN metabolism in migraine management (Table 1).

2.2.2. Tryptophan Metabolism Pathways in Migraine: Therapeutic Implications

Körtési et al. present a narrative review that explores the role of Trp metabolism in migraine pathogenesis [68]. The authors looked at the complex interactions of Trp metabolic pathways and how they affect migraine-related mechanisms such as pain processing, stress response, neural and brain hypersensitivity, and vascular and inflammatory processes. Their review highlights the importance of Trp metabolism in migraine susceptibility and the potential therapeutic implications of targeting these pathways. By synthesizing the current knowledge of Trp metabolism in migraine, they provide a comprehensive overview of the underlying mechanisms and identify promising avenues for future drug development and migraine management strategies (Table 1).

2.3. Experimental Models and Therapeutic Targets

Experimental models and therapeutic targets are critical for better understanding migraine pathophysiology and developing effective treatments [92–94]. Experimental models, particularly murine models, provide invaluable insights into the biological mechanisms that underpin migraines, allowing researchers to test various pharmacological interventions [16,95]. These models simulate migraine conditions, allowing researchers to study neurotransmission pathways and identify key molecules involved in migraine attacks. Therapeutic targets identified by the authors of such studies provide promising treatment options, focusing on molecules and pathways that can be modulated to alleviate migraine symptoms [12,96–98]. The authors of recent studies have identified several potential targets, including enzymes, ion channels, and signaling pathways, each of which contribute uniquely to our understanding of migraine and offer new opportunities for therapeutic intervention [88,99–102]. Three papers delve into the specifics of these experimental models and therapeutic targets, providing a thorough overview of current progress and future directions in migraine research [69–71].

2.3.1. Dual Fatty Acid Amide Hydrolase (FAAH)/Monoacylglycerol Lipase (MAGL) Inhibitor in the Migraine Model

Greco et al. investigated the potential therapeutic benefits of targeting the endocannabinoid system in the context of migraine pathophysiology [69]. Utilizing a male rat model, the authors examined the effects of trigeminal hyperalgesia using AKU-005, a dual fatty acid amide hydrolase (FAAH) and monoacylglycerol lipase (MAGL) inhibitor. The study reported that administering AKU-005 significantly reduced trigeminal hyperalgesia in the rat model, implying that dual inhibition of FAAH and MAGL could be a useful strategy for managing migraine-related pain. By elucidating the effects of AKU-005 on trigeminal hyperalgesia, the findings of the above study contribute to the growing body of research on the role of the endocannabinoid system in migraine and highlight the potential therapeutic implications of this novel pharmacological intervention for migraine management.

2.3.2. Src Family Kinase (SFK) Activity and Calcitonin Gene-Related Peptide (CGRP)–Cytokine Crosstalk

Nie et al. investigated the role of Src family kinases (SFKs) in the interaction between CGRP and cytokines in sensitizing the trigeminal ganglion [70]. CGRP and cytokines are key players in migraine pathophysiology, and their interaction is thought to contribute to the sensitization of the trigeminal system, a crucial mechanism underlying migraine-related pain. The study authors revealed that SFKs play a crucial role in mediating the crosstalk between CGRP and cytokines in the trigeminal ganglion. Specifically, they demonstrated that SFKs facilitate the transmission of CGRP receptor signaling via the protein kinase A pathway, leading to the sensitization of trigeminal ganglion neurons. By elucidating the mechanisms by which SFKs regulate the interaction between CGRP and cytokines in the trigeminal system, the authors contribute to our understanding of the complex pathways involved in migraine pain and highlight the potential therapeutic implications of targeting SFKs in migraine management.

2.3.3. ATP-Sensitive Potassium (KATP) Channels in Migraine Pathophysiology

Clement et al. explored the potential role of ATP-sensitive potassium (KATP) channels in migraine pathophysiology and their therapeutic implications [71]. KATP channels are known to be involved in various physiological processes, including vascular regulation and pain perception, making them an intriguing target for migraine research. The researchers synthesized findings from both preclinical and clinical studies to provide a comprehensive overview of the current understanding of KATP channels in migraine. The above review emphasizes the translational potential of KATP channel modulation in migraine management, with the results of animal studies indicating that KATP channel openers may have anti-nociceptive effects and reduce the incidence of migraine-related behaviors. Additionally, genetic studies involving humans have identified associations between KATP channel subunits and migraine susceptibility. By integrating these findings, Clement et al. propose that targeting KATP channels could be a promising avenue for the development of novel migraine therapies. Their review underscores the importance of translational research in bridging the gap between preclinical discoveries and clinical applications in the field of migraine treatment (Table 1).

2.4. *Inflammation in the Pathophysiology of Migraine*

Inflammation plays a pivotal role in the pathophysiology of migraine [1,103–106]. The complex mechanisms underlying migraine include neurogenic inflammation, which contributes to peripheral and central sensitization, resulting in the condition's chronicity [107–110]. Two papers included in the present Special Issue examine the current understanding of neurogenic neuroinflammation and its role in migraine pathophysiology, focusing on how inflammatory processes in the meninges and trigeminal nerve pathways cause and sustain migraine pain [72,73]. Additionally, the authors of the second paper explore the wider symptomatology linked to migraines, highlighting the significance of recognizing and managing symptoms other than headache, such as photophobia and nausea, which have a major influence on patients' quality of life [73]. The following section provides a brief overview of the inflammatory processes involved in migraine, as well as potential strategies for mitigating these effects in order to improve patient outcomes.

2.4.1. Neurogenic Neuroinflammation in Migraine

Reducha et al. explored the potential of using experimental inflammation models to gain a better understanding of migraine pathophysiology [72]. Their review highlights how, in addition to other mechanisms, neurogenic neuroinflammation has been proposed to play a role in the chronification of migraine, which involves peripheral and central sensitization. Extensive data indicate that CGRP and intracranial meningeal inflammation may play a major role in triggering the sensitivity of trigeminal meningeal nociceptors. The authors discuss how several studies have utilized inflammatory animal models to investigate this concept, with a focus on the sensitization of trigeminovascular afferent nerve terminals. By applying a range of pharmacological interventions, these studies provide insights into the pathways involved in the inflammatory processes underlying migraine. However, the authors emphasize the importance of using animal models with care and carefully evaluating the outcomes in the context of migraine pathophysiology, as the invasive procedures used in these models may have implications for data interpretation. Overall, the above review underscores the potential of experimental inflammation models in enhancing our understanding of the complex mechanisms driving migraine while also highlighting the need for cautious interpretation and translation of the findings into a clinical setting.

2.4.2. Complex Symptomatology of Migraine

Villar-Martinez and Goadsby conducted a comprehensive review of the pathophysiology and treatment of migraine-associated symptoms [73]. Their review highlights that migraine is a complex and heterogeneous disorder that goes beyond the core symptom of headache. The authors discuss the various associated features, including photophobia, vomiting, and other symptoms that are often overlooked but can significantly impact patients' quality of life. The review authors emphasize the importance of understanding these associated features in the context of migraine pathophysiology as they can provide valuable insights into the underlying mechanisms driving the disorder. Additionally, the authors discuss the therapeutic implications of these associated features, including the potential for targeted interventions to alleviate symptoms and improve patient outcomes. By synthesizing the current understanding of migraine-associated features, the authors of the above review provide a comprehensive overview of the complex interplay between migraine pathophysiology and therapy, highlighting the need for a more holistic approach to managing this debilitating disorder.

3. Discussion

Migraine research has undergone significant advancements in recent years, shedding light on the intricate neurobiological mechanisms that underlie this debilitating neurological disorder [111–113]. The present Special Issue on migraine neuroscience serves as a comprehensive compilation of cutting-edge research in the field, with a specific emphasis on experimental models and therapeutic targets aimed at enhancing our understanding of migraine pathophysiology and improving patient outcomes. The ultimate goal of this line of research is to develop targeted therapies that can effectively manage migraine symptoms and alleviate the burden of this complex disorder on individuals' quality of life. By delving into the molecular and cellular pathways involved in migraine attacks, researchers strive to identify novel therapeutic strategies that can address the diverse array of symptoms experienced by migraine sufferers, ranging from throbbing headaches to sensory disturbances, nausea, and sensitivity to light and sound [66].

However, a significant challenge in the field of migraine research lies in bridging the gap between basic science discoveries and their translation into clinical applications. To overcome this challenge, researchers must collaborate across disciplines, integrating knowledge from neuroscience, genetics, pharmacology, and other fields to develop innovative treatment modalities. Leveraging cutting-edge technologies such as advanced imaging techniques, genetic sequencing, and computational modeling is crucial in unraveling the complexities of migraine pathophysiology and identifying potential therapeutic targets. For example, the functional interplay between error-related brain activity and the autonomic nervous system in migraine emphasizes the significance of coordinated interactions between the central and autonomic nervous systems [114,115]. These interactions are critical for survival because they aid in the detection and correction of errors that could be fatal [116–118].

Advancements in migraine research have not only deepened our understanding of the underlying mechanisms driving migraine attacks but have also paved the way for personalized and targeted treatment approaches. These studies expand on previous research on CGRP-related mechanisms, Trp metabolic pathways, and experimental models, providing new insights and having implications for future research directions in migraine neuroscience [119–121]. Traditional medicine, particularly those derived from herbal compounds, has proven to be effective in the treatment of a variety of neurological and mental health disorders [122–127]. These remedies have played an important role in the field, and ongoing research is being conducted to identify potential antimigraine agents [128–130]. Herbs and other traditional forms of treatment have been linked to the treatment of conditions such as migraine [131–136]. By investigating the properties of these herbs, researchers hope to develop more effective medicines and targeted treatments for various conditions, including migraine. Furthermore, neurophysiological procedures have emerged as promising noninvasive interventions for migraine treatment, such as transcranial magnetic stimulation (TMS), with the potential to reduce migraine frequency and severity by modulating cortical excitability and inhibiting cortical spreading depression. [137–140]. TMS has a lower side effect profile than traditional pharmacological treatments, making it an appealing option for patients who cannot tolerate medication [141,142].

The significance of this line of research extends beyond the realm of academia, with profound implications for clinical practice and patient care. By elucidating the neurobiological underpinnings of migraine and exploring novel therapeutic avenues, researchers aim to revolutionize the management of this complex neurological disorder, providing tailored treatment options that address the individual needs and symptoms of migraine sufferers. Furthermore, the collaborative efforts of basic science researchers and clinicians are essential in driving forward the field of migraine research and translating scientific discoveries into tangible benefits for patients. By embracing multidisciplinary approaches and innovative research methodologies, the study of migraine neuroscience holds enormous promise for future breakthroughs and improved outcomes for individuals affected by this debilitating condition.

4. Conclusions

The following Special Issue on migraine neuroscience has shed light on the intricate mechanisms underlying this debilitating neurological disorder. Through experimental models and therapeutic targets, researchers have made significant strides in understanding migraine pathophysiology, from genetic predisposition to neurotransmitter signaling and neuroinflammation. The emphasis on CGRP-related mechanisms and the exploration of Trp metabolic pathways have provided valuable insights into potential treatment avenues. Moving forward, future research directions could focus on further unraveling the complex interplay of neurobiological processes involved in migraine, including investigating novel therapeutic strategies and exploring the impact of environmental factors on migraine development. Collaborative efforts between basic science researchers and clinicians will be crucial in filling the gap between research findings and clinical applications, ultimately leading to improved management of this complex neurological disorder. By continuing to push the boundaries of our understanding and embracing multidisciplinary approaches, the field of migraine research holds enormous promise for innovative breakthroughs and improved outcomes for migraine sufferers.

Author Contributions: Conceptualization, L.V.; writing—original draft preparation, M.T.; writing—review and editing, M.T., B.T., and L.V.; supervision, L.V.; project administration, L.V.; funding acquisition, L.V. All authors have read and agreed to the published version of the manuscript.

Funding: This work was supported by the National Research, Development, and Innovation Office—NKFIH K138125, SZTE SZAOK-KKA No: 2022/5S729, and the HUN-REN Hungarian Research Network.

Conflicts of Interest: The authors declare no conflicts of interest.

Abbreviations

CGRP	Calcitonin gene-related peptide
FAAH	Fatty acid amide hydrolase
GTN	Glycerol trinitrate
KATP	ATP-sensitive potassium
KYN	Kynurenine
MAGL	Monoacylglycerol lipase
PACAP	Adenylate cyclase-activating polypeptide
P2X3	Purinergic receptor P2X 3
SFKs	Src family kinases
TMS	Transcranial magnetic stimulation
Trp	Tryptophan
TRPV1	Transient receptor potential vanilloid subtype 1
VIP	Vasoactive intestinal peptide

References

1. Edvinsson, L.; Haanes, K.A. Views on migraine pathophysiology: Where does it start? *Neurol. Clin. Neurosci.* **2020**, *8*, 120–127. [CrossRef]
2. Goadsby, P.J. Primary headache disorders: Five new things. *Neurol. Clin. Pract.* **2019**, *9*, 233–240. [CrossRef]
3. Ashina, M.; Hansen, J.M.; Do, T.P.; Melo-Carrillo, A.; Burstein, R.; Moskowitz, M.A. Migraine and the trigeminovascular system—40 years and counting. *Lancet Neurol.* **2019**, *18*, 795–804. [CrossRef] [PubMed]
4. Olesen, J. The International Classification of Headache Disorders: History and future perspectives. *Cephalalgia* **2024**, *44*, 03331024231214731. [CrossRef]
5. Steiner, T.J.; Stovner, L.J. Global epidemiology of migraine and its implications for public health and health policy. *Nat. Rev. Neurol.* **2023**, *19*, 109–117. [CrossRef] [PubMed]
6. Shimizu, T.; Sakai, F.; Miyake, H.; Sone, T.; Sato, M.; Tanabe, S.; Azuma, Y.; Dodick, D.W. Disability, quality of life, productivity impairment and employer costs of migraine in the workplace. *J. Headache Pain* **2021**, *22*, 1–11. [CrossRef]
7. Leonardi, M.; Raggi, A. A narrative review on the burden of migraine: When the burden is the impact on people's life. *J. Headache Pain* **2019**, *20*, 1–11. [CrossRef]
8. Sutherland, H.G.; Albury, C.L.; Griffiths, L.R. Advances in genetics of migraine. *J. Headache Pain* **2019**, *20*, 1–20. [CrossRef] [PubMed]
9. Liu, T.-H.; Wang, Z.; Xie, F.; Liu, Y.-Q.; Lin, Q. Contributions of aversive environmental stress to migraine chronification: Research update of migraine pathophysiology. *World J. Clin. Cases* **2021**, *9*, 2136. [CrossRef]
10. Qubty, W.; Patniyot, I. Migraine pathophysiology. *Pediatr. Neurol.* **2020**, *107*, 1–6. [CrossRef]
11. Rapoport, A.M.; Edvinsson, L. Some aspects on the pathophysiology of migraine and a review of device therapies for migraine and cluster headache. *Neurol. Sci.* **2019**, *40*, 75–80. [CrossRef] [PubMed]
12. Haanes, K.A.; Edvinsson, L. Pathophysiological Mechanisms in Migraine and the Identification of New Therapeutic Targets. *CNS Drugs* **2019**, *33*, 525–537. [CrossRef] [PubMed]
13. Fraser, C.L.; Hepschke, J.L.; Jenkins, B.; Prasad, S. Migraine Aura: Pathophysiology, Mimics, and Treatment Options. *Semin. Neurol.* **2019**, *39*, 739–748. [CrossRef] [PubMed]
14. Harriott, A.M.; Strother, L.C.; Vila-Pueyo, M.; Holland, P.R. Animal models of migraine and experimental techniques used to examine trigeminal sensory processing. *J. Headache Pain* **2019**, *20*, 91. [CrossRef] [PubMed]

15. Greco, R.; Demartini, C.; De Icco, R.; Martinelli, D.; Putortì, A.; Tassorelli, C. Migraine neuroscience: From experimental models to target therapy. *Neurol. Sci.* **2020**, *41*, 351–361. [CrossRef] [PubMed]
16. Tardiolo, G.; Bramanti, P.; Mazzon, E. Migraine: Experimental Models and Novel Therapeutic Approaches. *Int. J. Mol. Sci.* **2019**, *20*, 2932. [CrossRef] [PubMed]
17. Veréb, D.; Szabó, N.; Tuka, B.; Tajti, J.; Király, A.; Faragó, P.; Kocsis, K.; Tóth, E.; Kincses, B.; Bagoly, T. Correlation of neurochemical and imaging markers in migraine: PACAP38 and DTI measures. *Neurology* **2018**, *91*, e1166–e1174. [CrossRef]
18. Szabó, N.; Faragó, P.; Király, A.; Veréb, D.; Csete, G.; Tóth, E.; Kocsis, K.; Kincses, B.; Tuka, B.; Párdutz, Á. Evidence for plastic processes in migraine with aura: A diffusion weighted MRI study. *Front. Neuroanat.* **2018**, *11*, 138. [CrossRef]
19. Olesen, J. Personal view: Modelling pain mechanisms of migraine without aura. *Cephalalgia* **2022**, *42*, 1425–1435. [CrossRef]
20. Bertels, Z.; Pradhan, A.A.A. Emerging Treatment Targets for Migraine and Other Headaches. *Headache* **2019**, *59* (Suppl. 2), 50–65. [CrossRef]
21. Edvinsson, L.; Haanes, K.A.; Warfvinge, K. Does inflammation have a role in migraine? *Nat. Rev. Neurol.* **2019**, *15*, 483–490. [CrossRef] [PubMed]
22. Vuralli, D.; Wattiez, A.S.; Russo, A.F.; Bolay, H. Behavioral and cognitive animal models in headache research. *J. Headache Pain* **2019**, *20*, 11. [CrossRef] [PubMed]
23. Borkum, J.M. Brain Energy Deficit as a Source of Oxidative Stress in Migraine: A Molecular Basis for Migraine Susceptibility. *Neurochem. Res.* **2021**, *46*, 1913–1932. [CrossRef] [PubMed]
24. Andreou, A.P.; Edvinsson, L. Mechanisms of migraine as a chronic evolutive condition. *J. Headache Pain* **2019**, *20*, 117. [CrossRef] [PubMed]
25. Mercer Lindsay, N.; Chen, C.; Gilam, G.; Mackey, S.; Scherrer, G. Brain circuits for pain and its treatment. *Sci. Transl. Med.* **2021**, *13*, eabj7360. [CrossRef] [PubMed]
26. Viganò, A.; Toscano, M.; Puledda, F.; Di Piero, V. Treating chronic migraine with neuromodulation: The role of neurophysiological abnormalities and maladaptive plasticity. *Front. Pharmacol.* **2019**, *10*, 416813. [CrossRef]
27. Pozo-Rosich, P.; Coppola, G.; Pascual, J.; Schwedt, T.J. How does the brain change in chronic migraine? Developing disease biomarkers. *Cephalalgia* **2021**, *41*, 613–630. [CrossRef] [PubMed]
28. Veréb, D.; Szabó, N.; Tuka, B.; Tajti, J.; Király, A.; Faragó, P.; Kocsis, K.; Tóth, E.; Bozsik, B.; Kincses, B. Temporal instability of salience network activity in migraine with aura. *Pain* **2020**, *161*, 856–864. [CrossRef] [PubMed]
29. Battaglia, S.; Avenanti, A.; Vécsei, L.; Tanaka, M. Neural Correlates and Molecular Mechanisms of Memory and Learning. *Int. J. Mol. Sci.* **2024**, *25*, 2724. [CrossRef]
30. Su, M.; Yu, S. Chronic migraine: A process of dysmodulation and sensitization. *Mol. Pain* **2018**, *14*, 1744806918767697. [CrossRef]
31. Battaglia, S.; Avenanti, A.; Vécsei, L.; Tanaka, M. Neurodegeneration in cognitive impairment and mood disorders for experimental, clinical and translational neuropsychiatry. *Biomedicines* **2024**, *12*, 574. [CrossRef]
32. Puledda, F.; Viganò, A.; Sebastianelli, G.; Parisi, V.; Hsiao, F.-J.; Wang, S.-J.; Chen, W.-T.; Massimini, M.; Coppola, G. Electrophysiological findings in migraine may reflect abnormal synaptic plasticity mechanisms: A narrative review. *Cephalalgia* **2023**, *43*, 03331024231195780. [CrossRef] [PubMed]
33. Gregorio, F.D.; Battaglia, S. The intricate brain–body interaction in psychiatric and neurological diseases. *Adv. Clin. Exp. Med.* **2024**, *33*, 321–326. [CrossRef] [PubMed]
34. Tanaka, M.; Battaglia, S.; Giménez-Llort, L.; Chen, C.; Hepsomali, P.; Avenanti, A.; Vécsei, L. Innovation at the Intersection: Emerging Translational Research in Neurology and Psychiatry. *Cells* **2024**, *13*, 790. [CrossRef] [PubMed]
35. Battaglia, S.; Schmidt, A.; Hassel, S.; Tanaka, M. Case reports in neuroimaging and stimulation. *Front. Psychiatry* **2023**, *14*, 1264669. [CrossRef] [PubMed]
36. Tortora, F.; Hadipour, A.L.; Battaglia, S.; Falzone, A.; Avenanti, A.; Vicario, C.M. The role of serotonin in fear learning and memory: A systematic review of human studies. *Brain Sci.* **2023**, *13*, 1197. [CrossRef]
37. Battaglia, S.; Di Fazio, C.; Mazzà, M.; Tamietto, M.; Avenanti, A. Targeting Human Glucocorticoid Receptors in Fear Learning: A Multiscale Integrated Approach to Study Functional Connectivity. *Int. J. Mol. Sci.* **2024**, *25*, 864. [CrossRef] [PubMed]
38. Wattiez, A.S.; Sowers, L.P.; Russo, A.F. Calcitonin gene-related peptide (CGRP): Role in migraine pathophysiology and therapeutic targeting. *Expert. Opin. Ther. Targets* **2020**, *24*, 91–100. [CrossRef]
39. Russo, A.F.; Hay, D.L. CGRP physiology, pharmacology, and therapeutic targets: Migraine and beyond. *Physiol. Rev.* **2023**, *103*, 1565–1644. [CrossRef]
40. Edvinsson, L.; Warfvinge, K. Recognizing the role of CGRP and CGRP receptors in migraine and its treatment. *Cephalalgia* **2019**, *39*, 366–373. [CrossRef]
41. Kamm, K. CGRP and Migraine: What Have We Learned From Measuring CGRP in Migraine Patients So Far? *Front. Neurol.* **2022**, *13*, 930383. [CrossRef] [PubMed]
42. Dodick, D.W. A Phase-by-Phase Review of Migraine Pathophysiology. *Headache* **2018**, *58* (Suppl. 1), 4–16. [CrossRef] [PubMed]
43. Vincent, M.; Viktrup, L.; Nicholson, R.A.; Ossipov, M.H.; Vargas, B.B. The not so hidden impact of interictal burden in migraine: A narrative review. *Front. Neurol.* **2022**, *13*, 1032103. [CrossRef] [PubMed]
44. Argyriou, A.A.; Mantovani, E.; Mitsikostas, D.D.; Vikelis, M.; Tamburin, S. A systematic review with expert opinion on the role of gepants for the preventive and abortive treatment of migraine. *Expert. Rev. Neurother.* **2022**, *22*, 469–488. [CrossRef] [PubMed]

45. Rivera-Mancilla, E.; Villalón, C.M.; MaassenVanDenBrink, A. CGRP inhibitors for migraine prophylaxis: A safety review. *Expert. Opin. Drug Saf.* **2020**, *19*, 1237–1250. [CrossRef] [PubMed]
46. Ho, T.W.; Edvinsson, L.; Goadsby, P.J. CGRP and its receptors provide new insights into migraine pathophysiology. *Nat. Rev. Neurol.* **2010**, *6*, 573–582. [CrossRef]
47. Carmine Belin, A.; Ran, C.; Edvinsson, L. Calcitonin gene-related peptide (CGRP) and cluster headache. *Brain Sci.* **2020**, *10*, 30. [CrossRef] [PubMed]
48. Russo, A.F. CGRP as a neuropeptide in migraine: Lessons from mice. *Br. J. Clin. Pharmacol.* **2015**, *80*, 403–414. [CrossRef] [PubMed]
49. Greco, R.; De Icco, R.; Demartini, C.; Zanaboni, A.M.; Tumelero, E.; Sances, G.; Allena, M.; Tassorelli, C. Plasma levels of CGRP and expression of specific microRNAs in blood cells of episodic and chronic migraine subjects: Towards the identification of a panel of peripheral biomarkers of migraine? *J. Headache Pain* **2020**, *21*, 1–12. [CrossRef]
50. Deen, M.; Correnti, E.; Kamm, K.; Kelderman, T.; Papetti, L.; Rubio-Beltrán, E.; Vigneri, S.; Edvinsson, L.; Maassen Van Den Brink, A.; European Headache Federation School of Advanced Studies (EHF-SAS). Blocking CGRP in migraine patients—A review of pros and cons. *J. Headache Pain* **2017**, *18*, 1–9. [CrossRef]
51. Edvinsson, L. T he T rigeminovascular P athway: R ole of CGRP and CGRP R eceptors in M igraine. *Headache J. Head. Face Pain* **2017**, *57*, 47–55. [CrossRef] [PubMed]
52. Brian, J.E., Jr.; Faraci, F.M.; Heistad, D.D. Recent insights into the regulation of cerebral circulation. *Clin. Exp. Pharmacol. Physiol.* **1996**, *23*, 449–457. [CrossRef] [PubMed]
53. Kee, Z.; Kodji, X.; Brain, S.D. The role of calcitonin gene related peptide (CGRP) in neurogenic vasodilation and its cardioprotective effects. *Front. Physiol.* **2018**, *9*, 405886. [CrossRef] [PubMed]
54. Gulbenkian, S.; Uddman, R.; Edvinsson, L. Neuronal messengers in the human cerebral circulation. *Peptides* **2001**, *22*, 995–1007. [CrossRef]
55. Thengchaisri, N.; Rivers, R.J. Remote arteriolar dilations caused by methacholine: A role for CGRP sensory nerves? *Am. J. Physiol. Heart Circ. Physiol.* **2005**, *289*, H608–H613. [CrossRef] [PubMed]
56. Johansson, S.E.; Abdolalizadeh, B.; Sheykhzade, M.; Edvinsson, L.; Sams, A. Vascular pathology of large cerebral arteries in experimental subarachnoid hemorrhage: Vasoconstriction, functional CGRP depletion and maintained CGRP sensitivity. *Eur. J. Pharmacol.* **2019**, *846*, 109–118. [CrossRef] [PubMed]
57. Fromy, B.; Merzeau, S.; Abraham, P.; Saumet, J.L. Mechanisms of the cutaneous vasodilator response to local external pressure application in rats: Involvement of CGRP, neurokinins, prostaglandins and NO. *Br. J. Pharmacol.* **2000**, *131*, 1161–1171. [CrossRef] [PubMed]
58. Smith, D.; Hill, R.; Edvinsson, L.; Longmore, J. An immunocytochemical investigation of human trigeminal nucleus caudalis: CGRP, substance P and 5-HT1D-receptor immunoreactivities are expressed by trigeminal sensory fibres. *Cephalalgia* **2002**, *22*, 424–431. [CrossRef] [PubMed]
59. Edvinsson, J.C.; Reducha, P.V.; Sheykhzade, M.; Warfvinge, K.; Haanes, K.A.; Edvinsson, L. Neurokinins and their receptors in the rat trigeminal system: Differential localization and release with implications for migraine pain. *Mol. Pain* **2021**, *17*, 17448069211059400. [CrossRef]
60. Messlinger, K.; Balcziak, L.K.; Russo, A.F. Cross-talk signaling in the trigeminal ganglion: Role of neuropeptides and other mediators. *J. Neural Transm.* **2020**, *127*, 431–444. [CrossRef]
61. Gárate, G.; Pascual, J.; Pascual-Mato, M.; Madera, J.; Martín, M.M.-S.; González-Quintanilla, V. Untangling the mess of CGRP levels as a migraine biomarker: An in-depth literature review and analysis of our experimental experience. *J. Headache Pain* **2024**, *25*, 69. [CrossRef] [PubMed]
62. Benedicter, N.; Vogler, B.; Kuhn, A.; Schramm, J.; Mackenzie, K.D.; Stratton, J.; Dux, M.; Messlinger, K. Glycerol Trinitrate Acts Downstream of Calcitonin Gene-Related Peptide in Trigeminal Nociception—Evidence from Rodent Experiments with Anti-CGRP Antibody Fremanezumab. *Cells* **2024**, *13*, 572. [CrossRef] [PubMed]
63. Greco, R.; Demartini, C.; Francavilla, M.; Zanaboni, A.M.; Tassorelli, C. Antagonism of CGRP receptor: Central and peripheral mechanisms and mediators in an animal model of chronic migraine. *Cells* **2022**, *11*, 3092. [CrossRef] [PubMed]
64. Dux, M.; Vogler, B.; Kuhn, A.; Mackenzie, K.D.; Stratton, J.; Messlinger, K. The anti-CGRP antibody fremanezumab lowers CGRP release from rat dura mater and meningeal blood flow. *Cells* **2022**, *11*, 1768. [CrossRef] [PubMed]
65. Pavelic, A.R.; Wöber, C.; Riederer, F.; Zebenholzer, K. Monoclonal antibodies against calcitonin gene-related peptide for migraine prophylaxis: A systematic review of real-world data. *Cells* **2022**, *12*, 143. [CrossRef] [PubMed]
66. Tanaka, M.; Szabó, Á.; Körtési, T.; Szok, D.; Tajti, J.; Vécsei, L. From CGRP to PACAP, VIP, and Beyond: Unraveling the Next Chapters in Migraine Treatment. *Cells* **2023**, *12*, 2649. [CrossRef] [PubMed]
67. Gecse, K.; Édes, A.E.; Nagy, T.; Demeter, A.K.; Virág, D.; Király, M.; Dalmadi Kiss, B.; Ludányi, K.; Környei, Z.; Denes, A. Citalopram Neuroendocrine Challenge Shows Altered Tryptophan and Kynurenine Metabolism in Migraine. *Cells* **2022**, *11*, 2258. [CrossRef] [PubMed]
68. Körtési, T.; Spekker, E.; Vécsei, L. Exploring the tryptophan metabolic pathways in migraine-related mechanisms. *Cells* **2022**, *11*, 3795. [CrossRef]
69. Greco, R.; Demartini, C.; Francavilla, M.; Zanaboni, A.M.; Facchetti, S.; Palmisani, M.; Franco, V.; Tassorelli, C. Effects of the Dual FAAH/MAGL Inhibitor AKU-005 on Trigeminal Hyperalgesia in Male Rats. *Cells* **2024**, *13*, 830. [CrossRef]

70. Nie, L.; Sun, K.; Gong, Z.; Li, H.; Quinn, J.P.; Wang, M. Src Family Kinases Facilitate the Crosstalk between CGRP and Cytokines in Sensitizing Trigeminal Ganglion via Transmitting CGRP Receptor/PKA Pathway. *Cells* **2022**, *11*, 3498. [CrossRef]
71. Clement, A.; Guo, S.; Jansen-Olesen, I.; Christensen, S.L. ATP-sensitive potassium channels in migraine: Translational findings and therapeutic potential. *Cells* **2022**, *11*, 2406. [CrossRef]
72. Reducha, P.V.; Edvinsson, L.; Haanes, K.A. Could experimental inflammation provide better understanding of migraines? *Cells* **2022**, *11*, 2444. [CrossRef] [PubMed]
73. Villar-Martinez, M.D.; Goadsby, P.J. Pathophysiology and therapy of associated features of migraine. *Cells* **2022**, *11*, 2767. [CrossRef] [PubMed]
74. Staud, R.; Spaeth, M. Psychophysical and neurochemical abnormalities of pain processing in fibromyalgia. *CNS Spectr.* **2008**, *13*, 12–17. [CrossRef] [PubMed]
75. Guo, S.; Vecsei, L.; Ashina, M. The L-kynurenine signalling pathway in trigeminal pain processing: A potential therapeutic target in migraine? *Cephalalgia* **2011**, *31*, 1029–1038. [CrossRef] [PubMed]
76. Peyrot, F.; Ducrocq, C. Potential role of tryptophan derivatives in stress responses characterized by the generation of reactive oxygen and nitrogen species. *J. Pineal Res.* **2008**, *45*, 235–246. [CrossRef]
77. Won, E.; Kim, Y.-K. Stress, the autonomic nervous system, and the immune-kynurenine pathway in the etiology of depression. *Curr. Neuropharmacol.* **2016**, *14*, 665–673. [CrossRef] [PubMed]
78. Staats Pires, A.; Tan, V.X.; Heng, B.; Guillemin, G.J.; Latini, A. Kynurenine and tetrahydrobiopterin pathways crosstalk in pain hypersensitivity. *Front. Neurosci.* **2020**, *14*, 518534. [CrossRef] [PubMed]
79. Singh, R. Role of tryptophan in health and disease: Systematic review of the anti-oxidant, anti-inflammation, and nutritional aspects of tryptophan and its metabolites. *World Heart J.* **2019**, *11*, 161–178.
80. Wang, Q.; Liu, D.; Song, P.; Zou, M.-H. Tryptophan-kynurenine pathway is dysregulated in inflammation, and immune activation. *Front. Biosci. (Landmark Ed.)* **2015**, *20*, 1116–1143.
81. Tuka, B.; Nyári, A.; Cseh, E.K.; Körtési, T.; Veréb, D.; Tömösi, F.; Kecskeméti, G.; Janáky, T.; Tajti, J.; Vécsei, L. Clinical relevance of depressed kynurenine pathway in episodic migraine patients: Potential prognostic markers in the peripheral plasma during the interictal period. *J. Headache Pain* **2021**, *22*, 60. [CrossRef] [PubMed]
82. Király, A.; Szabó, N.; Párdutz, Á.; Tóth, E.; Tajti, J.; Csete, G.; Faragó, P.; Bodnár, P.; Szok, D.; Tuka, B. Macro-and microstructural alterations of the subcortical structures in episodic cluster headache. *Cephalalgia* **2018**, *38*, 662–673. [CrossRef] [PubMed]
83. Paolucci, M.; Altamura, C.; Vernieri, F. The role of endothelial dysfunction in the pathophysiology and cerebrovascular effects of migraine: A narrative review. *J. Clin. Neurol.* **2021**, *17*, 164. [CrossRef] [PubMed]
84. Fila, M.; Chojnacki, J.; Pawlowska, E.; Szczepanska, J.; Chojnacki, C.; Blasiak, J. Kynurenine pathway of tryptophan metabolism in migraine and functional gastrointestinal disorders. *Int. J. Mol. Sci.* **2021**, *22*, 10134. [CrossRef] [PubMed]
85. Jovanovic, F.; Candido, K.D.; Knezevic, N.N. The role of the kynurenine signaling pathway in different chronic pain conditions and potential use of therapeutic agents. *Int. J. Mol. Sci.* **2020**, *21*, 6045. [CrossRef]
86. Fila, M.; Chojnacki, C.; Chojnacki, J.; Blasiak, J. The kynurenine pathway of tryptophan metabolism in abdominal migraine in children–A therapeutic potential? *Eur. J. Paediatr. Neurol.* **2023**. [CrossRef] [PubMed]
87. Tuka, B.; Körtési, T.; Nánási, N.; Tömösi, F.; Janáky, T.; Veréb, D.; Szok, D.; Tajti, J.; Vécsei, L. Cluster headache and kynurenines. *J. Headache Pain* **2023**, *24*, 35. [CrossRef] [PubMed]
88. Tajti, J.; Szok, D.; Csáti, A.; Szabó, Á.; Tanaka, M.; Vécsei, L. Exploring novel therapeutic targets in the common pathogenic factors in migraine and neuropathic pain. *Int. J. Mol. Sci.* **2023**, *24*, 4114. [CrossRef] [PubMed]
89. Martos, D.; Lőrinczi, B.; Szatmári, I.; Vécsei, L.; Tanaka, M. The Impact of C-3 Side Chain Modifications on Kynurenic Acid: A Behavioral Analysis of Its Analogs in the Motor Domain. *Int. J. Mol. Sci.* **2024**, *25*, 3394. [CrossRef] [PubMed]
90. Jászberényi, M.; Thurzó, B.; Bagosi, Z.; Vécsei, L.; Tanaka, M. The Orexin/Hypocretin System, the Peptidergic Regulator of Vigilance, Orchestrates Adaptation to Stress. *Biomedicines* **2024**, *12*, 448. [CrossRef]
91. Tanaka, M.; Bohár, Z.; Vécsei, L. Are kynurenines accomplices or principal villains in dementia? Maintenance of kynurenine metabolism. *Molecules* **2020**, *25*, 564. [CrossRef]
92. Tanaka, M.; Vécsei, L. A Decade of Dedication: Pioneering Perspectives on Neurological Diseases and Mental Illnesses. *Biomedicines* **2024**, *12*, 1083. [CrossRef] [PubMed]
93. Tanaka, M.; Vécsei, L. From Lab to Life: Exploring Cutting-Edge Models for Neurological and Psychiatric Disorders. *Biomedicines* **2024**, *12*, 613. [CrossRef] [PubMed]
94. Tanaka, M.; Szabó, Á.; Vécsei, L.; Giménez-Llort, L. Emerging Translational Research in Neurological and Psychiatric Diseases: From In Vitro to In Vivo Models. *Int. J. Mol. Sci.* **2023**, *24*, 15739. [CrossRef] [PubMed]
95. Spekker, E.; Tanaka, M.; Szabó, Á.; Vécsei, L. Neurogenic inflammation: The participant in migraine and recent advancements in translational research. *Biomedicines* **2021**, *10*, 76. [CrossRef] [PubMed]
96. Khan, J.; Al Asoom, L.I.; Al Sunni, A.; Rafique, N.; Latif, R.; Al Saif, S.; Almandil, N.B.; Almohazey, D.; AbdulAzeez, S.; Borgio, J.F. Genetics, pathophysiology, diagnosis, treatment, management, and prevention of migraine. *Biomed. Pharmacother.* **2021**, *139*, 111557. [CrossRef]
97. Rainero, I.; Roveta, F.; Vacca, A.; Noviello, C.; Rubino, E. Migraine pathways and the identification of novel therapeutic targets. *Expert. Opin. Ther. Targets* **2020**, *24*, 245–253. [CrossRef]

98. Edvinsson, L.; Haanes, K.A. Identifying new antimigraine targets: Lessons from molecular biology. *Trends Pharmacol. Sci.* **2021**, *42*, 217–225. [CrossRef]
99. Lisicki, M.; Schoenen, J. Metabolic treatments of migraine. *Expert. Rev. Neurother.* **2020**, *20*, 295–302. [CrossRef]
100. Iannone, L.F.; De Logu, F.; Geppetti, P.; De Cesaris, F. The role of TRP ion channels in migraine and headache. *Neurosci. Lett.* **2022**, *768*, 136380. [CrossRef]
101. Thuraiaiyah, J.; Kokoti, L.; Al-Karagholi, M.A.-M.; Ashina, M. Involvement of adenosine signaling pathway in migraine pathophysiology: A systematic review of preclinical studies. *J. Headache Pain* **2022**, *23*, 43. [CrossRef] [PubMed]
102. Nisar, A.; Ahmed, Z.; Yuan, H. Novel therapeutic targets for migraine. *Biomedicines* **2023**, *11*, 569. [CrossRef] [PubMed]
103. Moskowitz, M.A. Neurogenic inflammation in the pathophysiology and treatment of migraine. *Neurol. Minneap.* **1993**, *43*, S16–S20.
104. Biscetti, L.; Cresta, E.; Cupini, L.M.; Calabresi, P.; Sarchielli, P. The putative role of neuroinflammation in the complex pathophysiology of migraine: From bench to bedside. *Neurobiol. Dis.* **2023**, 106072. [CrossRef] [PubMed]
105. Aczél, T.; Benczik, B.; Ágg, B.; Körtési, T.; Urbán, P.; Bauer, W.; Gyenesei, A.; Tuka, B.; Tajti, J.; Ferdinandy, P. Disease-and headache-specific microRNA signatures and their predicted mRNA targets in peripheral blood mononuclear cells in migraineurs: Role of inflammatory signalling and oxidative stress. *J. Headache Pain* **2022**, *23*, 113. [CrossRef] [PubMed]
106. Aczél, T.; Körtési, T.; Kun, J.; Urbán, P.; Bauer, W.; Herczeg, R.; Farkas, R.; Kovács, K.; Vásárhelyi, B.; Karvaly, G.B. Identification of disease-and headache-specific mediators and pathways in migraine using blood transcriptomic and metabolomic analysis. *J. Headache Pain* **2021**, *22*, 1–18. [CrossRef]
107. Suzuki, K.; Suzuki, S.; Shiina, T.; Kobayashi, S.; Hirata, K. Central sensitization in migraine: A narrative review. *J. Pain Res.* **2022**, 2673–2682. [CrossRef] [PubMed]
108. Ramachandran, R. Neurogenic inflammation and its role in migraine. *Semin. Immunopathol.* **2018**, *40*, 301–314. [CrossRef] [PubMed]
109. Mungoven, T.J.; Henderson, L.A.; Meylakh, N. Chronic migraine pathophysiology and treatment: A review of current perspectives. *Front. Pain Res.* **2021**, *2*, 705276. [CrossRef]
110. Bertels, Z.; Mangutov, E.; Conway, C.; Siegersma, K.; Asif, S.; Shah, P.; Huck, N.; Tawfik, V.L.; Pradhan, A.A. Migraine and peripheral pain models show differential alterations in neuronal complexity. *Headache* **2022**, *62*, 780–791. [CrossRef]
111. Karsan, N.; Goadsby, P.J. Migraine: Beyond pain. *Pract. Neurol.* **2021**, *21*, 475–480. [CrossRef] [PubMed]
112. Tuttolomondo, A.; Simonetta, I. Molecular Research on Migraine: From Pathogenesis to Treatment. *Int. J. Mol. Sci.* **2023**, *24*, 8681. [CrossRef] [PubMed]
113. Rattanawong, W.; Rapoport, A.; Srikiatkhachorn, A. Neurobiology of migraine progression. *Neurobiol. Pain* **2022**, *12*, 100094. [CrossRef]
114. Borsook, D.; Aasted, C.M.; Burstein, R.; Becerra, L. Migraine mistakes: Error awareness. *Neuroscientist* **2014**, *20*, 291–304. [CrossRef]
115. Feller, L.; Feller, G.; Ballyram, T.; Chandran, R.; Lemmer, J.; Khammissa, R.A.G. Interrelations between pain, stress and executive functioning. *Br. J. Pain* **2020**, *14*, 188–194. [CrossRef]
116. Battaglia, S.; Nazzi, C.; Thayer, J. Heart's tale of trauma: Fear-conditioned heart rate changes in post-traumatic stress disorder. *Acta Psychiatr. Scand.* **2023**, *148*, 463–466. [CrossRef]
117. Di Gregorio, F.; Steinhauser, M.; Maier, M.E.; Thayer, J.F.; Battaglia, S. Error-related cardiac deceleration: Functional interplay between error-related brain activity and autonomic nervous system in performance monitoring. *Neurosci. Biobehav. Rev.* **2024**, 105542. [CrossRef]
118. Battaglia, S.; Nazzi, C.; Thayer, J.F. Genetic differences associated with dopamine and serotonin release mediate fear-induced bradycardia in the human brain. *Transl. Psychiatry* **2024**, *14*, 24. [CrossRef] [PubMed]
119. Karsan, N.; Goadsby, P.J. CGRP mechanism antagonists and migraine management. *Curr. Neurol. Neurosci. Rep.* **2015**, *15*, 1–9. [CrossRef]
120. Gross, E.C.; Lisicki, M.; Fischer, D.; Sándor, P.S.; Schoenen, J. The metabolic face of migraine—From pathophysiology to treatment. *Nat. Rev. Neurol.* **2019**, *15*, 627–643. [CrossRef]
121. Arulmani, U.; Gupta, S.; VanDenBrink, A.M.; Centurión, D.; Villalón, C.; Saxena, P.R. Experimental migraine models and their relevance in migraine therapy. *Cephalalgia* **2006**, *26*, 642–659. [CrossRef] [PubMed]
122. Pagotto, G.L.d.O.; Santos, L.M.O.d.; Osman, N.; Lamas, C.B.; Laurindo, L.F.; Pomini, K.T.; Guissoni, L.M.; Lima, E.P.d.; Goulart, R.d.A.; Catharin, V.M.S. Ginkgo biloba: A Leaf of Hope in the Fight against Alzheimer's Dementia: Clinical Trial Systematic Review. *Antioxidants* **2024**, *13*, 651. [CrossRef]
123. Valotto Neto, L.J.; Reverete de Araujo, M.; Moretti Junior, R.C.; Mendes Machado, N.; Joshi, R.K.; dos Santos Buglio, D.; Barbalho Lamas, C.; Direito, R.; Fornari Laurindo, L.; Tanaka, M. Investigating the Neuroprotective and Cognitive-Enhancing Effects of Bacopa monnieri: A Systematic Review Focused on Inflammation, Oxidative Stress, Mitochondrial Dysfunction, and Apoptosis. *Antioxidants* **2024**, *13*, 393. [CrossRef] [PubMed]
124. Laurindo, L.F.; Rodrigues, V.D.; Minniti, G.; de Carvalho, A.C.A.; Zutin, T.L.M.; DeLiberto, L.K.; Bishayee, A.; Barbalho, S.M. Pomegranate (*Punica granatum* L.) phytochemicals target the components of metabolic syndrome. *J. Nutr. Biochem.* **2024**, *131*, 109670.
125. Direito, R.; Barbalho, S.M.; Sepodes, B.; Figueira, M.E. Plant-Derived Bioactive Compounds: Exploring Neuroprotective, Metabolic, and Hepatoprotective Effects for Health Promotion and Disease Prevention. *Pharmaceutics* **2024**, *16*, 577. [CrossRef] [PubMed]
126. Bássoli, R.; Audi, D.; Ramalho, B.; Audi, M.; Quesada, K.; Barbalho, S. The Effects of Curcumin on Neurodegenerative Diseases: A Systematic Review. *J. Herb. Med.* **2023**, *42*, 100771. [CrossRef]

127. Pasierski, M.; Szulczyk, B. Beneficial effects of capsaicin in disorders of the central nervous system. *Molecules* **2022**, *27*, 2484. [CrossRef] [PubMed]
128. Chen, Y.; Wang, S.; Wang, Y. Role of herbal medicine for prevention and treatment of migraine. *Phytother. Res.* **2022**, *36*, 730–760. [CrossRef]
129. Tauchen, J. Natural products and their (semi-) synthetic forms in the treatment of migraine: History and current status. *Curr. Med. Chem.* **2020**, *27*, 3784–3808. [CrossRef] [PubMed]
130. Wei, T.-H.; Hsieh, C.-L. Headaches, migraine, and herbal medicine. In *Treatments, Nutraceuticals, Supplements, and Herbal Medicine in Neurological Disorders*; Elsevier: Amsterdam, The Netherlands, 2023; pp. 401–419.
131. Minniti, G.; Laurindo, L.F.; Machado, N.M.; Duarte, L.G.; Guiguer, E.L.; Araujo, A.C.; Dias, J.A.; Lamas, C.B.; Nunes, Y.C.; Bechara, M.D. Mangifera indica L., By-Products, and Mangiferin on Cardio-Metabolic and Other Health Conditions: A Systematic Review. *Life* **2023**, *13*, 2270. [CrossRef]
132. Yarnell, E. Herbal medicine and migraine. *Altern. Complement. Ther.* **2017**, *23*, 192–201. [CrossRef]
133. Takeda, L.N.; Laurindo, L.F.; Guiguer, E.L.; Bishayee, A.; Araújo, A.C.; Ubeda, L.C.C.; Goulart, R.d.A.; Barbalho, S.M. Psidium guajava L.: A systematic review of the multifaceted health benefits and economic importance. *Food Rev. Int.* **2023**, *39*, 4333–4363. [CrossRef]
134. Laurindo, L.F.; Barbalho, S.M.; Araújo, A.C.; Guiguer, E.L.; Mondal, A.; Bachtel, G.; Bishayee, A. Açaí (*Euterpe oleracea* Mart.) in health and disease: A critical review. *Nutrients* **2023**, *15*, 989. [CrossRef]
135. Nishikito, D.F.; Borges, A.C.A.; Laurindo, L.F.; Otoboni, A.M.B.; Direito, R.; Goulart, R.d.A.; Nicolau, C.C.; Fiorini, A.M.; Sinatora, R.V.; Barbalho, S.M. Anti-inflammatory, antioxidant, and other health effects of dragon fruit and potential delivery systems for its bioactive compounds. *Pharmaceutics* **2023**, *15*, 159. [CrossRef]
136. Buglio, D.S.; Marton, L.T.; Laurindo, L.F.; Guiguer, E.L.; Araújo, A.C.; Buchaim, R.L.; Goulart, R.d.A.; Rubira, C.J.; Barbalho, S.M. The role of resveratrol in mild cognitive impairment and Alzheimer's disease: A systematic review. *J. Med. Food* **2022**, *25*, 797–806. [CrossRef] [PubMed]
137. Zhong, J.; Lan, W.; Feng, Y.; Yu, L.; Xiao, R.; Shen, Y.; Zou, Z.; Hou, X. Efficacy of repetitive transcranial magnetic stimulation on chronic migraine: A meta-analysis. *Front. Neurol.* **2022**, *13*, 1050090. [CrossRef] [PubMed]
138. Battaglia, S.; Nazzi, C.; Di Fazio, C.; Borgomaneri, S. The role of pre-supplementary motor cortex in action control with emotional stimuli: A repetitive transcranial magnetic stimulation study. *Ann. N. Y. Acad. Sci.* **2024**, *1536*, 151–166. [CrossRef]
139. Saltychev, M.; Juhola, J. Effectiveness of high-frequency repetitive transcranial magnetic stimulation in migraine: A systematic review and meta-analysis. *Am. J. Phys. Med. Rehabil.* **2022**, *101*, 1001–1006. [CrossRef]
140. Battaglia, S.; Nazzi, C.; Fullana, M.A.; di Pellegrino, G.; Borgomaneri, S. 'Nip it in the bud': Low-frequency rTMS of the prefrontal cortex disrupts threat memory consolidation in humans. *Behav. Res. Ther.* **2024**, *178*, 104548. [CrossRef] [PubMed]
141. Miron, J.-P.; Voetterl, H.; Fox, L.; Hyde, M.; Mansouri, F.; Dees, S.; Zhou, R.; Sheen, J.; Jodoin, V.D.; Mir-Moghtadaei, A. Optimized repetitive transcranial magnetic stimulation techniques for the treatment of major depression: A proof of concept study. *Psychiatry Res.* **2021**, *298*, 113790. [CrossRef]
142. Zhang, J.-H.; Liang, J.; Yang, Z.-W. Non-invasive brain stimulation for fibromyalgia: Current trends and future perspectives. *Front. Neurosci.* **2023**, *17*, 1288765. [CrossRef]

cells MDPI

Article

The Anti-CGRP Antibody Fremanezumab Lowers CGRP Release from Rat Dura Mater and Meningeal Blood Flow

Mária Dux [1], Birgit Vogler [2], Annette Kuhn [2], Kimberly D. Mackenzie [3], Jennifer Stratton [3] and Karl Messlinger [2,*]

[1] Department of Physiology, University of Szeged, H-6720 Szeged, Hungary; dux.maria@med.u-szeged.hu
[2] Institute of Physiology and Pathophysiology, Friedrich-Alexander-University, D-91054 Erlangen-Nürnberg, Germany; birgit.vogler@fau.de (B.V.); annette.kuhn@fau.de (A.K.)
[3] Teva Pharmaceuticals, Redwood City, CA 94063, USA; kimberly.mackenzie01@tevapharm.com (K.D.M.); jennifer.stratton@tevapharm.com (J.S.)
* Correspondence: karl.messlinger@fau.de; Tel.: +49-160-5521215

Abstract: Monoclonal antibodies directed against the neuropeptide calcitonin gene-related peptide (CGRP) belong to a new generation of therapeutics that are effective in the prevention of migraine. CGRP, a potent vasodilator, is strongly implicated in the pathophysiology of migraine, but its role remains to be fully elucidated. The hemisected rat head preparation and laser Doppler flowmetry were used to examine the effects on CGRP release from the dura mater and meningeal blood flow of the subcutaneously injected anti-CGRP monoclonal antibody fremanezumab at 30 mg/kg, when compared to an isotype control antibody. Some rats were administered glycerol trinitrate (GTN) intraperitoneally to produce a migraine-like sensitized state. When compared to the control antibody, the fremanezumab injection was followed by reduced basal and capsaicin-evoked CGRP release from day 3 up to 30 days. The difference was enhanced after 4 h of GTN application. The samples from the female rats showed a higher CGRP release compared to that of the males. The increases in meningeal blood flow induced by acrolein (100 µM) and capsaicin (100 nM) were reduced 13–20 days after the fremanezumab injection, and the direct vasoconstrictor effect of high capsaicin (10 µM) was intensified. In conclusion, fremanezumab lowers the CGRP release and lasts up to four weeks, thereby lowering the CGRP-dependent meningeal blood flow. The antibody may not only prevent the released CGRP from binding but may also influence the CGRP release stimulated by noxious agents relevant for the generation of migraine pain.

Keywords: fremanezumab; monoclonal antibody; calcitonin gene-related peptide; glycerol trinitrate; CGRP release; meningeal blood flow; rat; migraine pain

Citation: Dux, M.; Vogler, B.; Kuhn, A.; Mackenzie, K.D.; Stratton, J.; Messlinger, K. The Anti-CGRP Antibody Fremanezumab Lowers CGRP Release from Rat Dura Mater and Meningeal Blood Flow. *Cells* **2022**, *11*, 1768. https://doi.org/10.3390/cells11111768

Academic Editors: László Vécsei and Bernadett Tuka

Received: 15 April 2022
Accepted: 25 May 2022
Published: 28 May 2022

1. Introduction

The sensory neuropeptide calcitonin gene-related peptide (CGRP) is considered to be crucially involved in the generation and aggravation of migraine and trigemino-autonomic headaches [1–3]. Increased levels of CGRP have been found during migraine and cluster headache attacks in the venous outflow from the head and in peripheral blood [4–7]. Conversely, the infusion of CGRP caused delayed migraine-like headaches in migraineurs and cluster-like attacks in patients suffering from cluster headaches, respectively [8,9]. Targeting CGRP signalling has long been proven to be effective in the treatment of migraine. Triptans reduce CGRP release, and new small-molecule CGRP receptor antagonists have beneficial effects in migraine therapy [10–13]. Consistently, in preclinical models of migraine, CGRP-targeting antibodies or CGRP receptor antagonists have proven to be effective in reducing elevated trigeminal activity [14–17].

For more than five years now, monoclonal antibodies targeting CGRP or its receptor have shown their efficacy in preventing chronic, frequent, and episodic migraine [18–20]. Thus, it is quite evident that the attenuation of the CGRP signalling system is effective in reducing the occurrence and severity of migraine attacks; however, neither the sites nor the mechanisms of migraine inhibition or prevention are sufficiently elucidated. There are good reasons to assume that the main effects are peripheral as monoclonal antibodies cannot readily cross the blood-brain barrier (BBB) [21,22]. Much is still to be discovered regarding the mechanisms of CGRP action and the prevention of its signalling on trigeminal afferents, which causes sensitization in both migraine-related preclinical studies [14,23–27] and in patients suffering from primary headaches, triggering their specific types of headaches [9,28–31].

CGRP is present in a major proportion of spinal and trigeminal nociceptive afferents [11,32–35] and is released from activated peripheral and central terminals and possibly also from cell bodies within the sensory ganglia [35–37]. Most of the CGRP-containing trigeminal neurons are chemosensitive in nature, expressing the nociceptive cation channels transient receptor potential vanilloid (TRPV1) and the transient receptor potential ankyrin 1 (TRPA1) receptors. The basal and stimulated release of CGRP and substance P from rodent meningeal tissues has been used as an approved method that reflects the state of activity of trigeminal afferents [37–44]. In these studies, inflammatory mediators, such as high concentrations of potassium chloride (depolarizing agent), capsaicin (TRPV1 agonist), or acrolein (TRPA1 agonist), have been applied for stimulation. In clinical experiments, the infusion of CGRP or the nitrovasodilator glycerol trinitrate (GTN, nitroglycerin) induced immediate headaches and increased CGRP plasma levels [45,46]. In addition, these substances triggered delayed migraine-like headaches nearly exclusively in migraine patients [28,47]. Therefore, GTN infusion or injection has frequently been used to model meningeal nociception and migraine-like states in animal experiments [48–50]. Under experimental conditions, the injection of GTN resulted in an increased proportion of the trigeminal neurons that are immunoreactive for CGRP and neuronal NO synthase [51].

As a read-out of the functional condition of the trigeminal nociceptors and released CGRP, the vasodilatation of meningeal arteries and the changes in meningeal blood flow have been recorded in rodent in vivo models [52–54]. CGRP is the most effective vasodilator substance acting on arterial meningeal vessels [55] and is therefore mainly responsible for increased meningeal blood flow when the trigeminal afferents are activated to release their neuropeptides. Under experimental conditions, trigeminal afferents have been stimulated electrically and chemically, using substances such as acrolein and capsaicin [43,56–61]. These substances are well-known agonists at the TRPA1 and TRPV1 receptor channels, respectively, releasing neuropeptides upon calcium influx. From previous experiments, we determined that the meningeal blood flow response to capsaicin depends on the capsaicin concentration; at nanomolar concentrations, capsaicin causes arterial dilatation and increased blood flow, but at micromolar concentrations, it decreases meningeal perfusion [62]. Thus, capsaicin exerts a dual function on meningeal blood vessels; by releasing CGRP from trigeminal afferents, it increases meningeal blood flow, while capsaicin acting on vascular TRPV1 receptors constricts smooth muscle cells, reducing meningeal blood flow.

Further exploration of the mechanisms of the CGRP effect in trigeminal pathophysiology guided the rationale of the present preclinical study using fremanezumab [63–65]. We do not know exactly where the anti-CGRP antibodies exert their effect within the trigeminal tissues but assume that the cranial dura mater is an important site of action [21,66,67]. Systemically applied anti-CGRP antibodies have a long-lasting migraine prevention effect [68,69]. We therefore speculate that fremanezumab not only neutralizes the CGRP released into the tissues but may also reduce the amount of CGRP released in the trigeminovascular system upon stimulation. We used in vivo and ex vivo rat dura mater preparations to measure the CGRP release from the trigeminal afferents and the consequent changes in meningeal perfusion following the pre-administration of fremanezumab, when compared to an isotype control antibody.

2. Materials and Methods

The animal housing and all the experiments were carried out according to the German guidelines and regulations of the care and treatment of laboratory animals and the European Communities Council Directive of 24 November 1986 (86/609/EEC), amended 22 September 2010 (2010/63/EU). The experimental protocols were reviewed by an ethics committee and approved by the District Government of Middle Franconia (54-2532.1-21/12).

2.1. Animals

Adult Wistar rats of both sexes (body weight of females: 230–370 g; males: 230–450 g), bred and housed in the animal facility of the Institute of Physiology and Pathophysiology of the FAU Erlangen-Nürnberg, were used. They were kept in a 12 h light/dark cycle in standard cages in groups of 3–4 and fed with standard food pellets and water ad libitum. The animals were matched and distributed according to their gender and weight, as equally as possible, for the different experiments. We also matched the two antibodies used and the observation time after administration over the whole experimental period as far as possible. The oestrus state of the females was not assessed.

2.2. Administration of Antibodies

The rats were anaesthetized around 9 a.m. in a plastic box with a concentration of isoflurane increasing up to 4% (Forene, Abott, Wiesbaden, Germany), applied with an evaporator (Forane Vapor 19.3, Dräger AG, Lübeck, Germany). The animals were weighed, and the neck region was shaved and disinfected with 70% ethanol. Then, 30 mg/kg anti-CGRP antibody fremanezumab or isotype control antibody (Teva Pharmaceuticals, Redwood City, CA, USA) diluted in saline (10 mg/mL) was subcutaneously injected in an even distribution 2 cm left and right from the midline and 5 cm from the caudal of the occiput, using a syringe with a 27-gauge needle. A human IgG2 antibody-targeting keyhole limpet hemocyanin (KLH) was used as the isotype control antibody to assess the specific effect of the targeting CGRP. The examiners were blinded as to the identity of the antibodies. The rats were marked at their tail for identification and placed back in their cage, where they recovered from the anaesthesia usually within 2–3 min. The animals were inspected two times on every following day with regard to any unusual behaviour.

2.3. Preparation for CGRP Release Measurements

On day 1, 3, or 10 after the antibody injection at around 9 a.m., the rats were again shortly anaesthetized by isoflurane to receive an intraperitoneal (i.p.) injection of 5 mg/kg glycerol trinitrate (GTN, 1 mg/mL in saline) or the same volume of saline as a vehicle, equal to the GTN solution, using a 23 G needle. Four hours later, the rats were deeply anaesthetized and killed in an atmosphere of an increasing concentration of CO_2. The head was separated, skinned, and divided in the midline, and the two skull halves with adhering dura mater were prepared for the measurement of the CGRP release according to a standard protocol [39]. The skull halves were washed for 30 min with synthetic interstitial fluid (SIF) and mounted in a water bath above warm water (37 °C), holding the temperature constant. The SIF was composed of (in mM): 107.8 NaCl, 3.5 KCl, 0.69 $MgSO_4$ 7 H_2O, 26.2 $NaHCO_3$, 1.67 NaH_2PO_4 2 H_2O, 9.64 Na-gluconate, 5.55 glucose, 7.6 sucrose, and 1.53 $CaCl_2 \cdot 2\,H_2O$ buffered to pH 7.4 with carbogen gas (95% O_2, 5% CO_2). The skull halves were filled twice with 300 μL of SIF, followed by a solution of 500 nM capsaicin (dissolved in saline with 1% ethanol and further diluted with SIF) and another SIF; all the applications were at intervals of 5 min. The chosen capsaicin concentration exerts a robust CGRP release [70]. At the end of each interval, the fluid was carefully collected using a pipette without touching the tissue.

In some experiments, the long-term effect of the fremanezumab treatment on the stimulated CGRP release was measured. In these experiments, the meningeal afferents were stimulated by capsaicin (500 nM) application 28–30 days after antibody treatment and 4 h after GTN injection.

2.4. Analysis of Released CGRP Concentration

From the collected fluid samples, 100 µL was separated; immediately, 25 µL of enzyme-immunoassay (EIA) buffer containing peptidase inhibitors (Bertin Pharma/SPIbio, Montigny le Bretonneux, France) was added. The samples were deep-frozen and stored at −20 °C until their analysis, together with the samples of further experiments. After thawing, the samples were processed using an EIA kit for CGRP according to the instructions of the manufacturer (Bertin Pharma/SPIbio, Montigny le Bretonneux, France). The EIA is based on a double-antibody sandwich technique, with monoclonal capture and tracer antibodies binding the CGRP molecule; the tracer antibody is conjugated with acetylcholine esterase converting Ellman's reagent to a yellow substance, the absorbance of which is measured by a photo-spectrometer (Opsys MR, Dynex Technologies, Denkendorf, Germany). The assay has 100% reactivity for rat CGRP but <0.01% cross-reactivity with other proteins of the calcitonin family and detects both α- and β-CGRP with the same sensitivity. The lower limit of detection is 2 pg/mL according to the manufacturer's information. The CGRP concentrations in the original samples were calculated in pg/mL, considering the added volume of EIA buffer.

2.5. Preparation for Meningeal Blood Flow Recordings

Between day 13 and day 20 after the antibody injection, the rats were again anaesthetized by 4% isoflurane, followed by an application of 2% isoflurane through a tight mask. Atropine sulfate (B. Braun Melsungen AG, Melsungen, Germany, 0.5 mg/mL 1:10 with sodium chloride 0.9%) was injected subcutaneously to prevent salivation. The animals were tracheotomized in order to be artificially ventilated with a mixture of oxygen-enriched room air and 2% isoflurane. The depth of anaesthesia was routinely assessed and held at a level in which noxious stimuli (pinching of earlobes) failed to elicit motor reflexes or changes in systemic arterial pressure. The body temperature of the animals was recorded by a thermoprobe inserted into the rectum and was kept at 37–37.5 °C with a feedback-controlled heating pad. Systemic blood pressure was recorded with a pressure transducer connected to the catheter inserted into the right femoral artery. The expiratory CO_2 was continuously monitored (Artema MM 200, Karl Heyer, Bad Ems, Germany) and maintained at 3–3.5%. The head of the animal was fixed in a stereotaxic frame and held by ear bars and a snout clamp. The eyes were covered with dexpanthenol ointment (Bepanthen®, Bayer Vital GmbH, Leverkusen, Germany) to prevent dehydration of the cornea.

2.6. Meningeal Blood Flow Recordings

A median incision was made along the midline of the scalp, the periosteum was moved aside, and a cranial window of about 8 × 6 mm was drilled into the parietal bone under saline rinsing to expose the cranial dura mater. The exposed dura mater in the parietal cranial window was covered with SIF. Two needle type probes of a laser Doppler flowmeter (DRT4, Moor Instruments, Axminster, UK) were positioned over the branches of the middle meningeal artery supplying the dura mater. Blood flow was recorded at a sampling rate of 10 Hz and expressed in arbitrary perfusion units (AU), which apply to the output voltage (mV) of the flowmeter. The systemic blood pressure was recorded simultaneously. The data were stored and processed with the MoorSoft program for Windows. For chemical stimulation, the SIF was replaced by 40 µL of solutions, which were washed off after 5 min, 3 times with SIF, and replaced by the next solution 10 min after the last SIF application. The solutions were applied in a fixed order: SIF, acrolein 100 µM, SIF, capsaicin 100 nM, SIF, and capsaicin 10 µM. The basal blood flow was the mean flow value measured during a 3 min period prior to the stimulation of the dura mater. The blood flow values during stimulation were assessed as mean values, measured during 5 consecutive 1 min periods and during the whole 5 min application period, and were compared to the respective basal flow measured prior to stimulation.

2.7. Data Processing and Statistics

Statistical analysis was performed on non-normalized values using Statistica software (StatSoft, Release 7, Tulsa, OK, USA). Following verification of the normal distribution of data, the Student's *t*-test and analysis of variance (repeated measures or factorial ANOVA) were used and extended by Tukey's honest significant difference (HSD) test or Fisher's least square difference (LSD), as specified in the results. The level of significance was set at $p < 0.05$. The data are displayed as mean ± SEM (standard error of the mean).

3. Results

3.1. Tolerability of Treatments

The antibodies were injected subcutaneously into 68 rats (33 females and 35 males) and allocated equally to the fremanezumab and the isotype control antibody groups. The injection of either antibody did not cause any unusual behaviour during the following days. In 34 animals, GTN was i.p. injected on the day of the experiment. The injections were not followed by any unusual behaviour or licking of the injection sites. No signs of irritation or inflammation at the injection sites were observed. After GTN injection, the rats were not observed to show any unusual behaviour or decrease in their motor activities. None of the animals died during the waiting time following the injection of antibodies.

3.2. Body Weight

The rats in all the groups gained body weight during the experiment; the males with the higher body weight (on average 325.6 g) gained 13.8 ± 5.0 g after 3 days and 53.3 ± 5.6 g after 10 days; the females (269.4 g) gained 7.5 ± 1.6 g after 3 days and 17.1 ± 3.5 g after 10 days. Using factorial ANOVA, there was a significant difference between the sexes ($F_{1,52} = 11.72$, $p < 0.005$) and between the waiting days after antibody treatment ($F_{2,52} = 30.14$, $p < 0.0001$), as expected due to general body growth, but not between the fremanezumab- and the isotype control antibody-treated groups ($F_{1,52} = 1.27$, $p = 0.266$).

3.3. CGRP Release from the Dura Mater

3.3.1. Exclusion of Antibody-Assay Interactions

To test whether the antibodies interacted directly with the EIA kit, 30 µL of fremanezumab (10 mg/mL in saline), or the control antibody at the same dose, was diluted with 70 µL SIF and 25 µL EIA buffer and processed. The apparent CGRP concentration of the fremanezumab solution was 6.2 pg/mL, that of control antibody was 8.4 pg/mL. The apparent CGRP concentration of SIF alone ($n = 4$) was 8.3 ± 0.71 pg/mL, i.e., values below 8 pg/mL reflect a virtually zero CGRP concentration according to the manufacturer. Thus, neither the fremanezumab nor the isotype control antibody interacted directly with the assay to influence the virtual CGRP concentration in the solution.

3.3.2. Exclusion of Side Difference

Equal numbers of males and females (total $n = 48$) were used for the release experiments. First, we tested whether there was a systematic difference between the two skull halves of the animals, which were both used for examining the CGRP release from the dura mater. Therefore, we applied ANOVA with repeated measurements to compare the sequential release values of all the experiments and set the side of the head (left-right) as the independent factor (Figure 1A). As expected, the sequential CGRP values varied significantly ($F_{3,282} = 682.61$, $p < 0.0001$), with a significant increase following the capsaicin application (Tukey post hoc test, $p < 0.0001$; Figure 1A, ***), but there was no difference between the two basal CGRP levels measured at 5 and 10 min (Tukey, $p = 1.00$). There was also no difference between the two skull halves of the same animals ($F_{1,94} = 0.19$, $p = 0.597$). Therefore, the factor side was eliminated from further statistical evaluations, and the measurements of both sides were independently used for the further analysis of the other factors, i.e., the numbers used for statistics are the numbers of the skull halves.

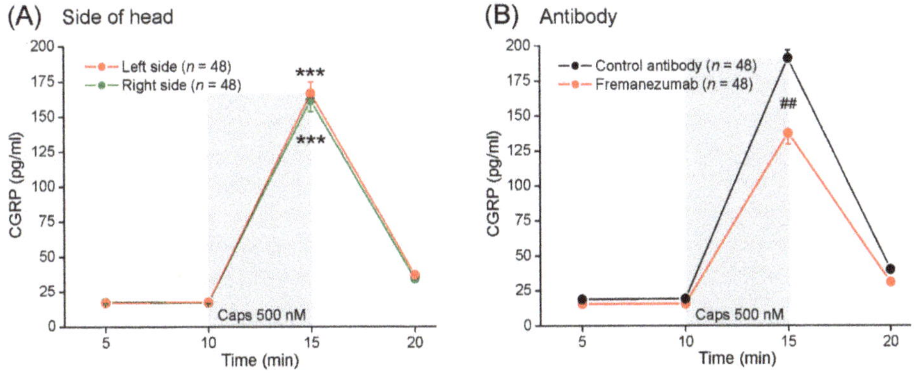

Figure 1. Sequential CGRP release from the dura mater 10 days after injection of antibodies. Data were grouped according to the head side of animals (**A**) and control antibody or fremanezumab treatment, respectively (**B**). Basal release was measured 5 and 10 min after application of physiological solution (SIF) and stimulated release after capsaicin application (Caps 500 nM). Repeated measures ANOVA with Tukey post hoc test, ***: $p < 0.0001$ to basal value (**A**) and ##: $p < 0.001$ between antibodies (**B**).

3.3.3. Impact of Antibodies on CGRP Release

We analysed whether the type of the antibody pre-treatment had an impact on the CGRP release in the course of the experiment, irrespective of the other factor treatments, the sex, and the waiting time (Figure 1B). The basal CGRP release (mean of the measurements at 5 and 10 min) was 19.2 ± 0.9 pg/mL in the animals injected with the control antibody (n = 24) and 15.6 ± 0.7 pg/mL in the animals injected with fremanezumab (n = 24). The application of capsaicin (500 nM) was followed by a 10-fold increase, approximately, in CGRP release in all groups. The mean (± SEM) of all the stimulated release values was 190.9 ± 8.2 pg/mL in the animals treated with the control antibody and 137.5 ± 5.8 pg/mL in the animals treated with fremanezumab; the difference is clearly significant (repeated measures ANOVA and Tukey test, $F_{1,94} = 32.14$, $p < 0.001$; Figure 1B, ##). After removing the capsaicin solution, the CGRP release values fell to 40.2 ± 1.1 pg/mL in the animals treated with the control antibody and 31.0 ± 1.1 pg/mL in the animals treated with fremanezumab.

3.3.4. Impact of Treatment, Sex, and Waiting Time on Basal CGRP Release

In addition to the two antibody types, each of the other factors (GNT/vehicle injection, sex, and day after antibody injection) seemed to influence the basal CGRP release (Figure 2). The factorial ANOVA and the Tukey post hoc tests were used to test the differences in basal CGRP release (mean of values at 5 and 10 min) specifically caused by these factors interacting with the factor antibody. In the animals pre-treated with the isotype control antibody (n = 24), the basal CGRP release was lower after injection with GTN compared to the vehicle injection ($F_{1,72} = 8.07$, $p < 0.01$; Figure 2A, *), which was aggravated in the animals (n = 24) pre-treated with fremanezumab ($p < 0.001$; Figure 2A, **). In the control animals, a lower basal CGRP release was also found in the males compared to the females ($F_{1,72} = 6.68$, $p < 0.001$; Figure 2B, **). In addition, a lowered basal CGRP release was found after waiting 3 and 10 days compared to 1 day ($F_{2,72} = 0.35$, $p < 0.001$; Figure 2C, **). Following the fremanezumab pre-treatment, the basal CGRP release appeared to be lower in most groups compared to the rats treated with the isotype control antibody ($F_{1,72} = 31.55$, $p < 0.001$; Figure 2A–C). The difference was statistically significant after GTN injection ($p < 0.001$; Figure 2A, ##), in females ($p < 0.001$; Figure 2B, ##), and after waiting times of 3 and 10 days, $p < 0.05$; Figure 2C, #).

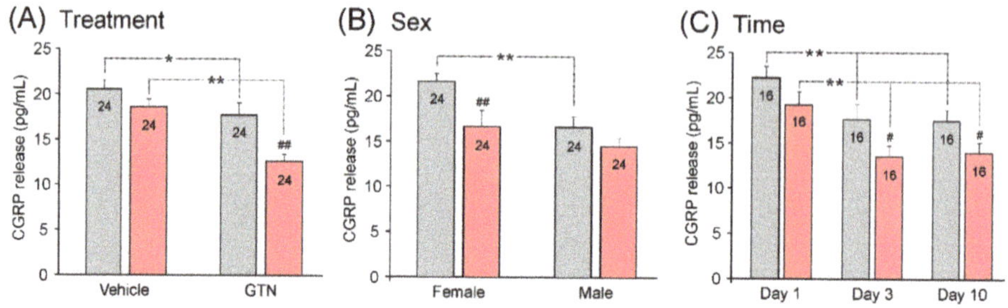

Figure 2. Basal CGRP release from the dura mater of animals treated with control antibody (grey) or fremanezumab (red) depending on injection of vehicle or GTN (**A**), animal's sex (**B**), and waiting time (day) after antibody treatment (**C**). Data are means ± SEM of averaged release values measured at 5 and 10 min; numbers of experiments shown in the columns; and significant differences between groups (factorial ANOVA and Tukey post hoc test, *: $p < 0.01$, **: $p < 0.001$) as well as between control antibody and fremanezumab (# $p < 0.05$, ## $p < 0.001$).

3.3.5. Impact of Treatment, Sex, and Waiting Time on Stimulated CGRP Release

As for the basal release, each of the other factors (GTN injection, sex, and days) may have had an influence on the stimulated CGRP release (Figure 3). Factorial ANOVA extended by the Tukey post hoc test was used to test differences in the capsaicin-evoked CGRP release (at 15 min) specifically caused by these factors interacting with the factor antibody. After the GTN injection, the capsaicin-induced increase in CGRP release was higher compared to the vehicle-injected samples only in the group of animals pre-treated with the isotype control antibody ($F_{1,72} = 7.70$, $p < 0.01$; Figure 3A, *). Likewise, the capsaicin-induced CGRP release was higher in the female rats, but only in these control antibody-treated animals ($F_{1,72} = 9.58$, $p < 0.001$; Figure 3B, **). The stimulated CGRP release was lower in all the groups of animals that had received fremanezumab compared with the control antibody (Figure 3A–C). The lowering effect of fremanezumab was more significant after the GTN injection ($p < 0.001$) than after the vehicle ($p < 0.05$) (Figure 3A, ## and #, respectively) and in the females ($p < 0.001$) than in the males ($p < 0.05$) (Figure 2B, ## and #, respectively). Regarding the waiting time, the CGRP release fell from day 1 to day 3 only in the fremanezumab-treated group ($p < 0.05$; Figure 1C, *) but was stable in the isotype control antibody-treated group. Furthermore, the lowering CGRP release following fremanezumab compared to the control antibody was already significantly reduced at day 1 ($p < 0.05$), with higher significance at days 3 and 10 ($p < 0.001$) (Figure 3C, # and ##, respectively).

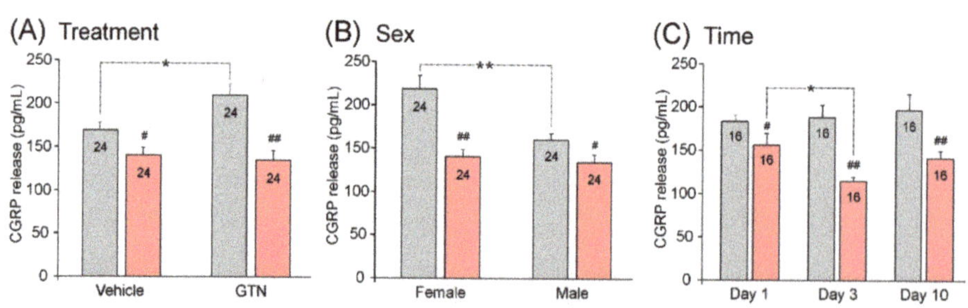

Figure 3. Stimulated CGRP release (mean ± SEM) from the dura mater of animals treated with control antibody (grey) or fremanezumab (red) depending on injection of vehicle or GTN (**A**), animal's sex (**B**), and waiting time after antibody treatment (**C**). Numbers within columns denote counts of experiments; significant difference between groups (*: $p < 0.05$, **: $p < 0.001$) and between control antibody and fremanezumab (#: $p < 0.05$, ##: $p < 0.001$).

Thus, fremanezumab lowered the capsaicin-evoked CGRP release, particularly after the GTN treatment and in female animals after a waiting time of 3 days.

3.3.6. Additional Experiments with Longer Waiting Time

In addition to the main release experiments, four female and six male animals were used to examine the CGRP release from the skull halves 28–30 days after the antibody injection and 4 h after the GTN treatment. The capsaicin-evoked CGRP release in females was 177.1 ± 17.2 pg/mL after pretreatment with the isotype control antibody and 154.9 ± 5.0 pg/mL after fremanezumab; in males, the respective values were 318.1 ± 30.8 pg/mL and 185.9 ± 16.5 pg/mL. Factorial ANOVA extended by the Tukey post hoc test, using the factors of antibody and sex, showed significant differences both for the antibody (ANOVA, $F_{1,16} = 11.28$, $p < 0.01$) and the sex ($F_{1,16} = 13.99$; $p < 0.01$). Thus, the fremanezumab injection reduced the stimulated CGRP release for at least 4 weeks.

3.4. Meningeal Blood Flow

3.4.1. Basal Blood Flow

The basal blood flow depends largely on the sites where the probes are positioned, i.e., it is generally higher when larger arteries are measured. Because the maximal flow value is the 1000 arbitrary units (AU) that can be displayed by the flowmeter, we aimed to set the probes onto sites with a flow value that warranted a wide scope of flow changes, which is usually the case at one of the main branches of the middle meningeal artery. Thus, most of the values were between 250 and 500 AU. For the flow measurements, 10 animals (5 females and 5 males) were used. In each animal, the blood flow was measured with two flow probes, which were positioned on different arteries, yielding two independent measurements; the number of measurements related to the application of the substances is seen in Figures 4 and 6. The basal blood flow was compared with factorial ANOVA, including the factors of sex, antibody, and stimulation (acrolein 100 µM, capsaicin 100 nM and 10 µM) extended by the Tukey HSD post hoc test. Regarding the basal flow, irrespective of the type of antibody treatment, there was a significant difference between the females (mean ± SEM: 301.8 ± 33.4 AU) and the males (403.2 ± 22.5 AU) ($F_{1,58} = 8.87$, $p < 0.01$). There was also a difference between the animals treated with the control antibody (mean ± SEM: 403.4 ± 26.5 AU) and fremanezumab (318.5 ± 28.5 AU) ($F_{1,58} = 5.55$, $p < 0.05$). The interaction between the sexes and the antibodies was also significant ($F_{1,58} = 26.06$, $p < 0.001$), and the Tukey post hoc test showed that the difference between the antibodies was solely based on a difference within the females ($p < 0.001$) but not the males ($p = 0.337$) (Figure 4A). Finally, there was no difference in the baseline flow values in the same animal before the serial application of acrolein and capsaicin at the two concentrations ($F_{1,58} = 1.28$, $p = 0.28$) (Figure 4B). Thus, the basal flow was lower in the female animals after treatment with fremanezumab, probably due to the lower unstimulated basal CGRP release.

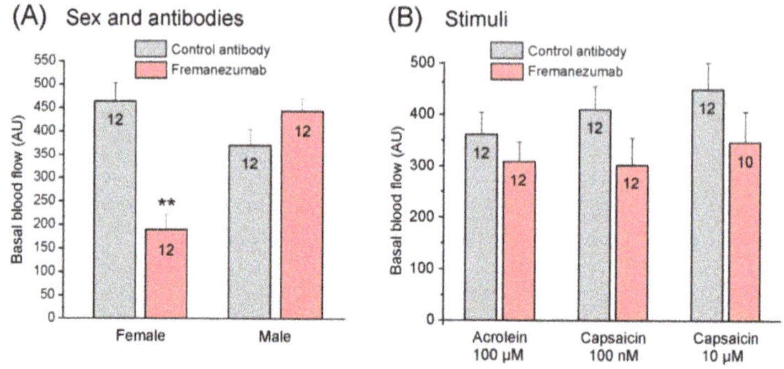

Figure 4. Basal blood flow of the dura mater (means ± SEM) of female and male rats 13–20 days after injection of either control antibody (grey) or fremanezumab (red) (**A**) and before stimulation with acrolein and capsaicin at two doses (**B**). Numbers within columns denote counts of experiments; **: $p < 0.001$ comparing the effect of control antibody and fremanezumab treatment on basal blood flow in females.

3.4.2. Stimulated Blood Flow

Stimulation with Acrolein

After application of the TRPA1 receptor agonist acrolein (100 μM), the meningeal blood flow increased slightly in the rats treated with the control antibody but did not significantly change in the animals treated with fremanezumab (Figures 5A and 6A left). Two-way repeated measures ANOVA of single-minute values (factor time) and the factor antibody showed a significant change over time ($F_{5,110} = 4.38$, $p < 0.005$) and a significant difference between the baseline and the values of minutes 3–5 in the control antibody experiments (LSD post hoc test, $p < 0.05$). In the fremanezumab-treated animals, the meningeal blood flow decreased transiently within the first minute ($p < 0.05$). The mean flow during the 5 min acrolein stimulation was 102.3% of the baseline in the control antibody experiments and 98.3% in the fremanezumab experiments (Figure 6A right). Thus, acrolein caused a moderate increase in the meningeal blood flow with a delay of 2–3 min in the rats treated with the antibody isotype but not in the animals treated with fremanezumab.

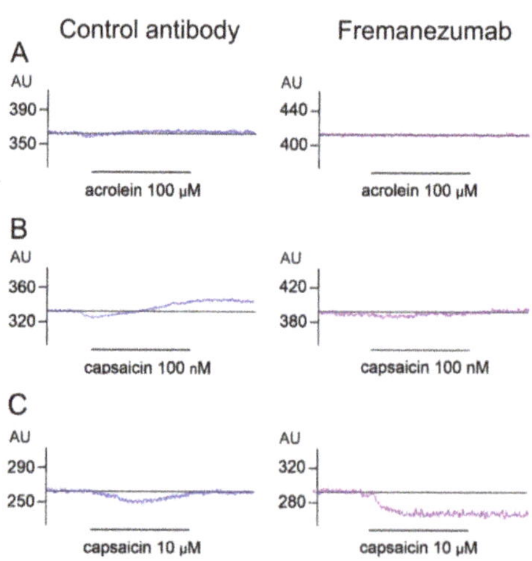

Figure 5. Examples of original blood flow recordings comparing responses to 5 min application of acrolein (**A**), capsaicin at low dose (**B**), and at high dose (**C**) after treatment with control antibody (left) and fremanezumab (right). AU, arbitrary perfusion units.

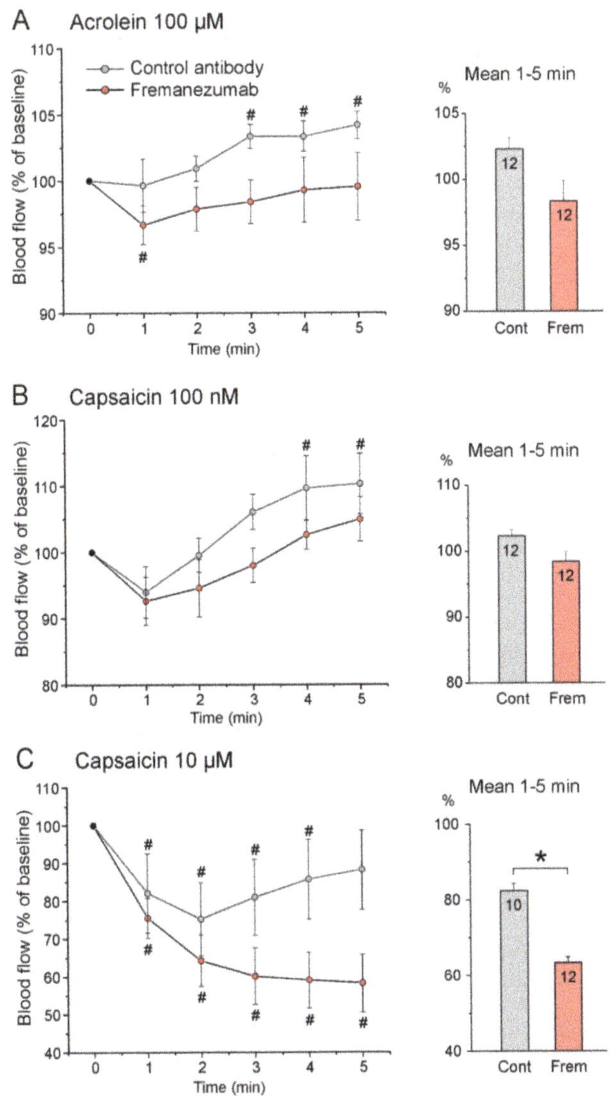

Figure 6. Stimulated blood flow of the dura mater (means ± SEM) of rats of both sexes 13–20 days after injection of either control antibody (grey) or fremanezumab (red) before stimulation with acrolein (**A**) and capsaicin at two doses (**B**,**C**). Blood flow was normalized to the baseline flow prior to stimulation; # significant difference to baseline; * significant difference between control antibody and fremanezumab experiments; numbers within columns denote counts of experiments.

Stimulation with Low-Dose Capsaicin

After application of the TRPV1 agonist capsaicin (100 nM), the blood flow tended to decrease transiently and then increased in the rats treated with the control antibody but showed no significant change in the rats treated with fremanezumab (Figures 5B and 6B left). Two-way repeated measures ANOVA showed a significant change over time ($F_{5,110} = 8.75$, $p < 0.0001$) and a significant difference between the baseline and the values of minutes 4 and 5 in the isotype control antibody experiments (LSD post hoc test, $p < 0.05$); there was no difference in any minute of the fremanezumab experiments. The mean blood flow during the 5 min capsaicin stimulation was 103.9% of the baseline in the control antibody experiments and 98.6% in the fremanezumab experiments (Figure 6B right). Thus, capsaicin at the low dose of 100 nM caused an increase in meningeal blood flow after 3–4 min of application in the rats treated with the isotype control antibody but not in the animals treated with fremanezumab.

Stimulation with High-Dose Capsaicin

After the application of capsaicin at 10 μM, which, in addition to the CGRP-releasing effect, vigorously stimulates the TRPV1 receptors of the smooth muscle cells of the meningeal blood vessels, the flow decreased transiently in the rats treated with the control antibody and permanently in the animals treated with fremanezumab (Figures 5C and 6C left). Two-way repeated measures ANOVA showed a significant change over time ($F_{5,110} = 17.07$, $p < 0.0001$) and a significant difference in the interaction of the factors of time and antibody ($F_{5,110} = 5.59$, $p < 0.0005$). The post hoc LSD test indicated that in the control antibody experiments, the values of minutes 1–4 were different to those of the baseline ($p < 0.05$), while in the fremanezumab experiments all the values were significantly lower than the baseline ($p < 0.0001$). The mean flow during the 5 min of stimulation was 82.4% of the baseline in the control antibody experiments and 63.4% in the fremanezumab experiments, which was significantly different (Student's t-test, df = 20, $p < 0.05$; Figure 6C right). Thus, capsaicin at the high dose of 10 μM caused a transient decrease in flow for 4 min in the rats treated with the isotype control antibody but a robust and sustained decrease in flow in the rats treated with fremanezumab.

4. Discussion

The present study was initiated in an attempt to further examine changes in the nociceptor function in the trigeminovascular system induced by the systemic administration of the monoclonal anti-CGRP antibody fremanezumab. We subcutaneously administered the anti-CGRP antibody fremanezumab to rats, in a manner similar to the human clinical use of this antibody for the prevention of chronic and frequent migraine. Rats treated with fremanezumab or an isotype control antibody were studied by applying well-established ex vivo and in vivo experimental models of meningeal nociception relevant to the pathophysiology of migraine headache. Fremanezumab did not cause any changes in behaviour, nor did it interfere with the growth of the animals in comparison with isotype control antibody-treated animals. The isotype control antibody does not bind to CGRP and was used as a negative control to assess the effect of specifically targeting CGRP with fremanezumab.

4.1. *Sex Difference in CGRP Release*

First, we examined the amount of CGRP released from the dura mater in our established ex vivo hemisected rat head preparation. The CGRP release was reduced as early as three days after the fremanezumab treatment, which applied to both the spontaneous (basal) release and the stimulated release following the application of the TRPV1 agonist capsaicin (500 nM), and this effect lasted up to four weeks. Being aware of the possible sex differences in CGRP signalling, we used male and female rats in equal parts. Although we did not test the oestrus cycle of the female animals, the females showed significantly higher basal CGRP release from the dura mater compared to the males; hence, the reduction observed with the fremanezumab treatment on the basal and stimulated CGRP release was more robust in the females than in the males. To our knowledge, a sex difference in CGRP release has not been published so far; however, it has been reported that female compared to male rodents are more sensitive to CGRP [71]. In the latter study [71], only female animals showed facial mechanical hypersensitivity and pain-like grimace behaviour when CGRP at low doses was directly applied onto the dura mater, as well as mechanical hypersensitivity of the hind paws after intra-plantar injection of CGRP, suggesting a generally higher susceptibility to CGRP in females. Earlier, it was reported that systemically or locally administered CGRP did not excite or sensitize meningeal afferents, but in these experiments only male rats were used [23]. In another study with both male and female mice, intraperitoneal CGRP injection caused grimace behaviour indicating pain, without a significant sex difference but with a trend of higher responsiveness in females [72]. In this case, sumatriptan reduced the response to CGRP only in male animals, while the anti-CGRP antibody ALD405 was effective in both sexes. These discrepancies may be partly due to the different modes of application of CGRP and the doses, as discussed [27], and demonstrate that differentiating sexes is important regarding the examination and therapeutic targeting of the CGRP release and signalling system.

4.2. Where Do Anti-CGRP Antibodies Act?

The migraine-preventing effect of CGRP-targeting antibodies is considered mainly as a peripheral effect in the trigeminovascular system as antibody penetration of the blood brain barrier (BBB) into the central nervous system is limited. CGRP is also released from the central terminals of the activated trigeminal afferents within the spinal trigeminal nucleus [36,37] and contributes to synaptic transmission [14,73]. Immunohistochemical labelling of a fusion protein of the CGRP receptor components RAMP1 and CLR suggests CGRP binding in the monkey spinal trigeminal nucleus, in addition to the known peripheral sites of CGRP receptor expression in the trigeminovascular system [74]. However, due to the limited access of IgG antibodies to central sites within the BBB, an effective central effect of CGRP-targeting antibodies appears unlikely. Moreover, despite some controversial clinical observations, no convincing evidence supports the assumption that migraine attacks enhance the permeability of the BBB, allowing the passage of chemical substances from the blood into the brain tissue [75]. Therefore, trigeminal ganglion neurons and satellite cells and the dura mater innervated by the peripheral axons of the trigeminal afferents—structures not protected by the BBB—are the most likely targets of the antibody treatment. While it is presumed that the anti-CGRP antibodies act in the periphery, the precise target of peripheral CGRP signalling to afferent structures is not known. Based on the immunohistochemical findings, it has been hypothesized that CGRP released from C-fibres binds to CGRP receptors located on the sensory axons of the Aδ-fibres, namely within the nodes of Ranvier, thereby sensitizing the membrane channels of the Aδ-fibres [76], but functional evidence for this assumption is lacking. The main problem with this hypothesis is that voltage-dependent conduction channels could possibly be sensitized to increased excitatory currents, as seen in experimental neuropathic conditions [77,78], but it does not explain the CGRP release from the adjacent C-fibres. The hypothesis is at least in line with the finding that fremanezumab inhibits mainly the Aδ fibres in a rat model of the cortical spreading depression-induced activation of second-order neurons in the rat spinal trigeminal nucleus [63]. However, the mechanism of CGRP involvement in cortical spreading depression is not yet clear [79]. Fremanezumab does not inhibit the arterial dilatation induced by cortical spreading depression in rat [80].

4.3. Meningeal Blood Flow Induced by the Stimulation of TRP Receptors

Our release experiments were complemented by in vivo meningeal blood flow recordings as functional measurements reflecting the blood flow-increasing effect of CGRP released from nociceptive afferents. Consistently with the CGRP release measurements, both the basal blood flow and the flow induced by the TRPA1 agonist acrolein and the low concentration of the TRPV1 agonist capsaicin (100 nM), which are both known to induce CGRP release, were reduced in the animals treated with fremanezumab compared to the isotype control antibody. Capsaicin at the high dose of 10 μM, which directly activates the TRPV1 receptors of the vascular smooth muscle cells, induced a transient decrease in flow for 4 min in the rats treated with the antibody isotype but a robust and sustained decrease in flow in the rats treated with fremanezumab. Topical application of capsaicin at this high dose has been shown to decrease meningeal blood flow [62], most likely by direct activation of the TRPV1 receptors of the vascular smooth muscle cells that induce calcium inflow and smooth muscle constriction [43]. The CGRP-releasing effect of capsaicin from the meningeal afferents counteracts this effect, which is partly effective in rats treated with the isotype control antibody but not in rats treated with fremanezumab.

4.4. Possible Effects of Anti-CGRP Antibodies beyond CGRP Neutralization

Anti-CGRP antibodies are thought to neutralize part of the CGRP molecules released by activated peptidergic afferents, thus lowering the capacity of the CGRP signalling. They most likely exert their effect within the tissues such as the dura mater and the trigeminal ganglion, probably directly at the sites of CGRP release. The mechanisms by which CGRP interacts with other nerve fibres is largely unknown, as discussed above; however, the effects of the gene expression within the trigeminal ganglion may contribute to sensitization. The CGRP receptors are expressed by both neurons and satellite glial cells in the trigeminal ganglion [81,82], and signalling between CGRP-releasing and CGRP-sensitive trigeminal ganglion cells seems possible [83]. CGRP may induce gene expression by ERK signalling [84], which could result, for example, in an increase in neuronal NO synthase and hence the production of NO in the ganglion [85]. The NO may back-signal to the CGRP-producing cells, which then increase their expression of CGRP [86] and the CGRP receptor components in other neurons [51]. In short, these mechanisms could induce a vicious circle, including the production of other sensitizing mediators such as brain-derived neurotrophic factor, which then are delivered by axonal transport to the peripheral nerve fibres and central terminals [87]. Anti-CGRP antibodies, which may act in the trigeminal ganglion, could interrupt this cross-signalling at an early state and prevent the peripheral sensitization dependent on the gene expression.

Alexa Fluor 594-conjugated fremanezumab has recently been found not only in the sensory but also in the autonomic ganglia of rats [21]. Early investigations have already revealed CGRP immunoreactivity in small numbers of neuronal cell bodies but in numerous stained axons in several parasympathetic ganglia [88]. Varicose CGRP-immunoreactive nerve fibres have been described as forming synaptic-like contacts with the somata of the sphenopalatine and ciliary ganglia [89,90], and CGRP receptor components have been localized in rat sphenopalatine ganglion, especially in the satellite glial cells [91]. These morphological findings may form the structural basis for the so-called trigemino-parasympathetic reflex [92], which is postulated to crucially contribute not only to the generation of the cluster headache [93] but also to migraine pain [94], but this has not been sufficiently examined so far. The assumed signalling between the CGRP-secreting nerve fibres and the glial cells in autonomic ganglia is certainly not an acute synaptic process but indicates, rather, an indirect control of neuronal processing, which is most likely based on gene regulation in the glia and/or neuronal cells.

5. Conclusions

Taken together, our results suggest that fremanezumab is able to neutralize part of the released CGRP in rat dura mater and possibly also in the trigeminal ganglion. In addition, fremanezumab significantly decreases the CGRP release evoked by noxious stimulation, thereby lowering meningeal blood flow. Fremanezumab's effect on CGRP release and meningeal blood flow is more pronounced in female than in male animals. Our data may provide a mechanism explaining the functional changes in the trigeminovascular system, leading to reduced pain susceptibility in migraine patients treated with CGRP-targeting monoclonal antibodies.

Author Contributions: Conceptualization, M.D. and K.M.; methodology, M.D., B.V. and K.M.; investigation, M.D., B.V. and A.K.; resources, M.D., K.D.M., J.S. and K.M.; writing—original draft preparation, M.D. and K.M.; writing—review and editing, M.D., K.D.M., J.S. and K.M.; visualization, M.D. and K.M.; supervision, M.D. and K.M.; project administration, K.M.; funding acquisition, M.D. and K.M. All authors have read and agreed to the published version of the manuscript.

Funding: These studies were part of a sponsored research collaboration with Teva, who provided research support and the anti-CGRP antibody, fremanezumab. The project was also partly supported by the Johannes and Frieda Marohn Foundation, the research grant K119597 project of the Hungarian National Research, Development and Innovation Office, and the Alexander von Humboldt Foundation (Research Group Linkage Program between the Institutes of Physiology of the FAU Erlangen-Nürnberg and the University of Szeged).

Institutional Review Board Statement: These studies were conducted in accordance with the Declaration of Helsinki. The animal study protocol was approved by the District Government of Mittelfranken (54-2532.1-21/12) after review by the Ethics Committee of the District Government of Unterfranken.

Data Availability Statement: Not applicable.

Conflicts of Interest: The authors declare no conflict of interest. K.D.M. and J.S. are employees of Teva Pharmaceuticals.

References

1. Edvinsson, L.; Goadsby, P.J. Neuropeptides in Migraine and Cluster Headache. *Cephalalgia Int. J. Headache* **1994**, *14*, 320–327. [CrossRef] [PubMed]
2. Vollesen, A.L.; Benemei, S.; Cortese, F.; Labastida-Ramírez, A.; Marchese, F.; Pellesi, L.; Romoli, M.; Ashina, M.; Lampl, C.; School of Advanced Studies of the European Headache Federation (EHF-SAS). Migraine and Cluster Headache—The Common Link. *J. Headache Pain* **2018**, *19*, 89. [CrossRef] [PubMed]
3. Carmine Belin, A.; Ran, C.; Edvinsson, L. Calcitonin Gene-Related Peptide (CGRP) and Cluster Headache. *Brain Sci.* **2020**, *10*, 30. [CrossRef] [PubMed]
4. Goadsby, P.J.; Edvinsson, L.; Ekman, R. Release of Vasoactive Peptides in the Extracerebral Circulation of Humans and the Cat during Activation of the Trigeminovascular System. *Ann. Neurol.* **1988**, *23*, 193–196. [CrossRef]
5. Ashina, M.; Bendtsen, L.; Jensen, R.; Schifter, S.; Olesen, J. Evidence for Increased Plasma Levels of Calcitonin Gene-Related Peptide in Migraine Outside of Attacks. *Pain* **2000**, *86*, 133–138. [CrossRef]
6. Sarchielli, P.; Alberti, A.; Codini, M.; Floridi, A.; Gallai, V. Nitric Oxide Metabolites, Prostaglandins and Trigeminal Vasoactive Peptides in Internal Jugular Vein Blood during Spontaneous Migraine Attacks. *Cephalalgia Int. J. Headache* **2000**, *20*, 907–918. [CrossRef]
7. Cernuda-Morollón, E.; Larrosa, D.; Ramón, C.; Vega, J.; Martínez-Camblor, P.; Pascual, J. Interictal Increase of CGRP Levels in Peripheral Blood as a Biomarker for Chronic Migraine. *Neurology* **2013**, *81*, 1191–1196. [CrossRef]
8. Lassen, L.H.; Jacobsen, V.B.; Haderslev, P.A.; Sperling, B.; Iversen, H.K.; Olesen, J.; Tfelt-Hansen, P. Involvement of Calcitonin Gene-Related Peptide in Migraine: Regional Cerebral Blood Flow and Blood Flow Velocity in Migraine Patients. *J. Headache Pain* **2008**, *9*, 151–157. [CrossRef]
9. Vollesen, A.L.H.; Snoer, A.; Beske, R.P.; Guo, S.; Hoffmann, J.; Jensen, R.H.; Ashina, M. Effect of Infusion of Calcitonin Gene-Related Peptide on Cluster Headache Attacks: A Randomized Clinical Trial. *JAMA Neurol.* **2018**, *75*, 1187–1197. [CrossRef]
10. Buzzi, M.G.; Moskowitz, M.A. Evidence for 5-HT1B/1D Receptors Mediating the Antimigraine Effect of Sumatriptan and Dihydroergotamine. *Cephalalgia Int. J. Headache* **1991**, *11*, 165–168. [CrossRef]
11. Hargreaves, R.; Olesen, J. Calcitonin Gene-Related Peptide Modulators—The History and Renaissance of a New Migraine Drug Class. *Headache* **2019**, *59*, 951–970. [CrossRef]
12. Dodick, D.W.; Lipton, R.B.; Ailani, J.; Lu, K.; Finnegan, M.; Trugman, J.M.; Szegedi, A. Ubrogepant for the Treatment of Migraine. *N. Engl. J. Med.* **2019**, *381*, 2230–2241. [CrossRef]
13. Croop, R.; Lipton, R.B.; Kudrow, D.; Stock, D.A.; Kamen, L.; Conway, C.M.; Stock, E.G.; Coric, V.; Goadsby, P.J. Oral Rimegepant for Preventive Treatment of Migraine: A Phase 2/3, Randomised, Double-Blind, Placebo-Controlled Trial. *Lancet* **2021**, *397*, 51–60. [CrossRef]
14. Storer, R.J.; Akerman, S.; Goadsby, P.J. Calcitonin Gene-Related Peptide (CGRP) Modulates Nociceptive Trigeminovascular Transmission in the Cat. *Br. J. Pharmacol.* **2004**, *142*, 1171–1181. [CrossRef]
15. Fischer, M.J.M.; Koulchitsky, S.; Messlinger, K. The Nonpeptide Calcitonin Gene-Related Peptide Receptor Antagonist BIBN4096BS Lowers the Activity of Neurons with Meningeal Input in the Rat Spinal Trigeminal Nucleus. *J. Neurosci.* **2005**, *25*, 5877–5883. [CrossRef]
16. Fischer, M.J.M.; Schmidt, J.; Koulchitsky, S.; Klussmann, S.; Vater, A.; Messlinger, K. Effect of a Calcitonin Gene-Related Peptide-Binding L-RNA Aptamer on Neuronal Activity in the Rat Spinal Trigeminal Nucleus. *J. Headache Pain* **2018**, *19*, 3. [CrossRef]
17. Koulchitsky, S.; Fischer, M.J.M.; Messlinger, K. Calcitonin Gene-Related Peptide Receptor Inhibition Reduces Neuronal Activity Induced by Prolonged Increase in Nitric Oxide in the Rat Spinal Trigeminal Nucleus. *Cephalalgia Int. J. Headache* **2009**, *29*, 408–417. [CrossRef]

18. Schuster, N.M.; Rapoport, A.M. New Strategies for the Treatment and Prevention of Primary Headache Disorders. *Nat. Rev. Neurol.* **2016**, *12*, 635–650. [CrossRef]
19. Raffaelli, B.; Neeb, L.; Reuter, U. Monoclonal Antibodies for the Prevention of Migraine. *Expert Opin. Biol. Ther.* **2019**, *19*, 1307–1317. [CrossRef]
20. Schoenen, J.; Manise, M.; Nonis, R.; Gérard, P.; Timmermans, G. Monoclonal Antibodies Blocking CGRP Transmission: An Update on Their Added Value in Migraine Prevention. *Rev. Neurol.* **2020**, *176*, 788–803. [CrossRef]
21. Noseda, R.; Schain, A.J.; Melo-Carrillo, A.; Tien, J.; Stratton, J.; Mai, F.; Strassman, A.M.; Burstein, R. Fluorescently-Labeled Fremanezumab Is Distributed to Sensory and Autonomic Ganglia and the Dura but Not to the Brain of Rats with Uncompromised Blood Brain Barrier. *Cephalalgia Int. J. Headache* **2020**, *40*, 229–240. [CrossRef] [PubMed]
22. Christensen, S.L.; Ernstsen, C.; Olesen, J.; Kristensen, D.M. No Central Action of CGRP Antagonising Drugs in the GTN Mouse Model of Migraine. *Cephalalgia Int. J. Headache* **2020**, *40*, 924–934. [CrossRef] [PubMed]
23. Levy, D.; Burstein, R.; Strassman, A.M. Calcitonin Gene-Related Peptide Does Not Excite or Sensitize Meningeal Nociceptors: Implications for the Pathophysiology of Migraine. *Ann. Neurol.* **2005**, *58*, 698–705. [CrossRef] [PubMed]
24. Covasala, O.; Stirn, S.L.; Albrecht, S.; De Col, R.; Messlinger, K. Calcitonin Gene-Related Peptide Receptors in Rat Trigeminal Ganglion Do Not Control Spinal Trigeminal Activity. *J. Neurophysiol.* **2012**, *108*, 431–440. [CrossRef] [PubMed]
25. Bhatt, D.K.; Ramachandran, R.; Christensen, S.L.T.; Gupta, S.; Jansen-Olesen, I.; Olesen, J. CGRP Infusion in Unanesthetized Rats Increases Expression of C-Fos in the Nucleus Tractus Solitarius and Caudal Ventrolateral Medulla, but Not in the Trigeminal Nucleus Caudalis. *Cephalalgia Int. J. Headache* **2015**, *35*, 220–233. [CrossRef]
26. Wattiez, A.-S.; Wang, M.; Russo, A.F. CGRP in Animal Models of Migraine. In *Calcitonin Gene-Related Peptide (CGRP) Mechanisms (Handbook of Experimental Pharmacology)*; Brain, S.D., Geppetti, P., Eds.; Springer International Publishing: Cham, Switzerland, 2018; Volume 255, pp. 85–107. ISBN 978-3-030-21453-1.
27. Moehring, F.; Sadler, K.E. Female-Specific Effects of CGRP Suggest Limited Efficacy of New Migraine Treatments in Males. *J. Neurosci.* **2019**, *39*, 9062–9064. [CrossRef]
28. Lassen, L.H.; Haderslev, P.A.; Jacobsen, V.B.; Iversen, H.K.; Sperling, B.; Olesen, J. CGRP May Play a Causative Role in Migraine. *Cephalalgia Int. J. Headache* **2002**, *22*, 54–61. [CrossRef]
29. Hansen, J.M.; Hauge, A.W.; Olesen, J.; Ashina, M. Calcitonin Gene-Related Peptide Triggers Migraine-like Attacks in Patients with Migraine with Aura. *Cephalalgia Int. J. Headache* **2010**, *30*, 1179–1186. [CrossRef]
30. Christensen, C.E.; Younis, S.; Deen, M.; Khan, S.; Ghanizada, H.; Ashina, M. Migraine Induction with Calcitonin Gene-Related Peptide in Patients from Erenumab Trials. *J. Headache Pain* **2018**, *19*, 105. [CrossRef]
31. Ashina, H.; Schytz, H.W.; Ashina, M. CGRP in Human Models of Migraine. *Handb. Exp. Pharmacol.* **2019**, *255*, 109–120. [CrossRef]
32. O'Connor, T.P.; van der Kooy, D. Enrichment of a Vasoactive Neuropeptide (Calcitonin Gene Related Peptide) in the Trigeminal Sensory Projection to the Intracranial Arteries. *J. Neurosci.* **1988**, *8*, 2468–2476. [CrossRef]
33. Hoheisel, U.; Mense, S.; Scherotzke, R. Calcitonin Gene-Related Peptide-Immunoreactivity in Functionally Identified Primary Afferent Neurones in the Rat. *Anat. Embryol.* **1994**, *189*, 41–49. [CrossRef]
34. Bennett, D.L.; Dmietrieva, N.; Priestley, J.V.; Clary, D.; McMahon, S.B. TrkA, CGRP and IB4 Expression in Retrogradely Labelled Cutaneous and Visceral Primary Sensory Neurones in the Rat. *Neurosci. Lett.* **1996**, *206*, 33–36. [CrossRef]
35. Iyengar, S.; Johnson, K.W.; Ossipov, M.H.; Aurora, S.K. CGRP and the Trigeminal System in Migraine. *Headache* **2019**, *59*, 659–681. [CrossRef]
36. Jenkins, D.W.; Langmead, C.J.; Parsons, A.A.; Strijbos, P.J. Regulation of Calcitonin Gene-Related Peptide Release from Rat Trigeminal Nucleus Caudalis Slices In Vitro. *Neurosci. Lett.* **2004**, *366*, 241–244. [CrossRef]
37. Kageneck, C.; Nixdorf-Bergweiler, B.E.; Messlinger, K.; Fischer, M.J. Release of CGRP from Mouse Brainstem Slices Indicates Central Inhibitory Effect of Triptans and Kynurenate. *J. Headache Pain* **2014**, *15*, 7. [CrossRef]
38. Messlinger, K.; Ebersberger, A.; Schaible, H.G. Release of Immunoreactive Substance P in the Brain Stem upon Stimulation of the Cranial Dura Mater with Low PH—Inhibition by the Serotonin (5-HT1) Receptor Agonist CP 93,129. *Br. J. Pharmacol.* **1998**, *125*, 1726–1732. [CrossRef]
39. Ebersberger, A.; Averbeck, B.; Messlinger, K.; Reeh, P.W. Release of Substance P, Calcitonin Gene-Related Peptide and Prostaglandin E2 from Rat Dura Mater Encephali Following Electrical and Chemical Stimulation In Vitro. *Neuroscience* **1999**, *89*, 901–907. [CrossRef]
40. Eberhardt, M.; Neeb, L.; Vogel, E.-M.; Tiegs, G.; Reuter, U.; Messlinger, K.; Fischer, M.J.M. Glyceroltrinitrate Facilitates Stimulated CGRP Release but Not Gene Expression of CGRP or Its Receptor Components in Rat Trigeminal Ganglia. *Neuropeptides* **2009**, *43*, 483–489. [CrossRef]
41. Hoffmann, J.; Wecker, S.; Neeb, L.; Dirnagl, U.; Reuter, U. Primary Trigeminal Afferents Are the Main Source for Stimulus-Induced CGRP Release into Jugular Vein Blood and CSF. *Cephalalgia Int. J. Headache* **2012**, *32*, 659–667. [CrossRef]
42. Durham, P.L.; Masterson, C.G. Two Mechanisms Involved in Trigeminal CGRP Release: Implications for Migraine Treatment. *Headache* **2013**, *53*, 67–80. [CrossRef]
43. Marics, B.; Peitl, B.; Varga, A.; Pázmándi, K.; Bácsi, A.; Németh, J.; Szilvássy, Z.; Jancsó, G.; Dux, M. Diet-Induced Obesity Alters Dural CGRP Release and Potentiates TRPA1-Mediated Trigeminovascular Responses. *Cephalalgia Int. J. Headache* **2016**, *37*, 581–591. [CrossRef]

44. Dux, M.; Will, C.; Eberhardt, M.; Fischer, M.J.M.; Messlinger, K. Stimulation of Rat Cranial Dura Mater with Potassium Chloride Causes CGRP Release into the Cerebrospinal Fluid and Increases Medullary Blood Flow. *Neuropeptides* **2017**, *64*, 61–68. [CrossRef]
45. Christiansen, I.; Thomsen, L.L.; Daugaard, D.; Ulrich, V.; Olesen, J. Glyceryl Trinitrate Induces Attacks of Migraine without Aura in Sufferers of Migraine with Aura. *Cephalalgia Int. J. Headache* **1999**, *19*, 660–667; discussion 626. [CrossRef]
46. Juhasz, G.; Zsombok, T.; Modos, E.A.; Olajos, S.; Jakab, B.; Nemeth, J.; Szolcsanyi, J.; Vitrai, J.; Bagdy, G. NO-Induced Migraine Attack: Strong Increase in Plasma Calcitonin Gene-Related Peptide (CGRP) Concentration and Negative Correlation with Platelet Serotonin Release. *Pain* **2003**, *106*, 461–470. [CrossRef]
47. Sances, G.; Tassorelli, C.; Pucci, E.; Ghiotto, N.; Sandrini, G.; Nappi, G. Reliability of the Nitroglycerin Provocative Test in the Diagnosis of Neurovascular Headaches. *Cephalalgia Int. J. Headache* **2004**, *24*, 110–119. [CrossRef]
48. Tassorelli, C.; Greco, R.; Wang, D.; Sandrini, M.; Sandrini, G.; Nappi, G. Nitroglycerin Induces Hyperalgesia in Rats—A Time-Course Study. *Eur. J. Pharmacol.* **2003**, *464*, 159–162. [CrossRef]
49. Ramachandran, R.; Bhatt, D.K.; Ploug, K.B.; Hay-Schmidt, A.; Jansen-Olesen, I.; Gupta, S.; Olesen, J. Nitric Oxide Synthase, Calcitonin Gene-Related Peptide and NK-1 Receptor Mechanisms Are Involved in GTN-Induced Neuronal Activation. *Cephalalgia Int. J. Headache* **2014**, *34*, 136–147. [CrossRef]
50. Capuano, A.; Greco, M.C.; Navarra, P.; Tringali, G. Correlation between Algogenic Effects of Calcitonin-Gene-Related Peptide (CGRP) and Activation of Trigeminal Vascular System, in an In Vivo Experimental Model of Nitroglycerin-Induced Sensitization. *Eur. J. Pharmacol.* **2014**, *740*, 97–102. [CrossRef]
51. Seiler, K.; Nusser, J.I.; Lennerz, J.K.; Neuhuber, W.L.; Messlinger, K. Changes in Calcitonin Gene-Related Peptide (CGRP) Receptor Component and Nitric Oxide Receptor (SGC) Immunoreactivity in Rat Trigeminal Ganglion Following Glyceroltrinitrate Pretreatment. *J. Headache Pain* **2013**, *14*, 74. [CrossRef]
52. Kurosawa, M.; Messlinger, K.; Pawlak, M.; Schmidt, R.F. Increase of Meningeal Blood Flow after Electrical Stimulation of Rat Dura Mater Encephali: Mediation by Calcitonin Gene-Related Peptide. *Br. J. Pharmacol.* **1995**, *114*, 1397–1402. [CrossRef] [PubMed]
53. Williamson, D.J.; Hargreaves, R.J.; Hill, R.G.; Shepheard, S.L. Intravital Microscope Studies on the Effects of Neurokinin Agonists and Calcitonin Gene-Related Peptide on Dural Vessel Diameter in the Anaesthetized Rat. *Cephalalgia Int. J. Headache* **1997**, *17*, 518–524. [CrossRef] [PubMed]
54. Petersen, K.A.; Birk, S.; Doods, H.; Edvinsson, L.; Olesen, J. Inhibitory Effect of BIBN4096BS on Cephalic Vasodilatation Induced by CGRP or Transcranial Electrical Stimulation in the Rat. *Br. J. Pharmacol.* **2004**, *143*, 697–704. [CrossRef] [PubMed]
55. Edvinsson, L.; Jansen, I.; Cunha e Sa, M.; Gulbenkian, S. Demonstration of Neuropeptide Containing Nerves and Vasomotor Responses to Perivascular Peptides in Human Cerebral Arteries. *Cephalalgia Int. J. Headache* **1994**, *14*, 88–96. [CrossRef]
56. Kunkler, P.E.; Ballard, C.J.; Oxford, G.S.; Hurley, J.H. TRPA1 Receptors Mediate Environmental Irritant-Induced Meningeal Vasodilatation. *Pain* **2011**, *152*, 38–44. [CrossRef]
57. Dux, M.; Deák, É.; Tassi, N.; Sántha, P.; Jancsó, G. Endovanilloids Are Potential Activators of the Trigeminovascular Nocisensor Complex. *J. Headache Pain* **2016**, *17*, 53. [CrossRef]
58. Dux, M.; Rosta, J.; Messlinger, K. TRP Channels in the Focus of Trigeminal Nociceptor Sensitization Contributing to Primary Headaches. *Int. J. Mol. Sci.* **2020**, *21*, 342. [CrossRef]
59. Marics, B.; Peitl, B.; Pázmándi, K.; Bácsi, A.; Németh, J.; Oszlács, O.; Jancsó, G.; Dux, M. Diet-Induced Obesity Enhances TRPV1-Mediated Neurovascular Reactions in the Dura Mater. *Headache* **2017**, *57*, 441–454. [CrossRef]
60. Denner, A.C.; Vogler, B.; Messlinger, K.; De Col, R. Role of Transient Receptor Potential Ankyrin 1 Receptors in Rodent Models of Meningeal Nociception—Experiments In Vitro. *Eur. J. Pain Lond. Engl.* **2016**, *21*, 843–854. [CrossRef]
61. Deák, É.; Rosta, J.; Boros, K.; Kis, G.; Sántha, P.; Messlinger, K.; Jancsó, G.; Dux, M. Chronic Adriamycin Treatment Impairs CGRP-Mediated Functions of Meningeal Sensory Nerves. *Neuropeptides* **2018**, *69*, 46–52. [CrossRef]
62. Dux, M.; Sántha, P.; Jancsó, G. Capsaicin-Sensitive Neurogenic Sensory Vasodilatation in the Dura Mater of the Rat. *J. Physiol.* **2003**, *552*, 859–867. [CrossRef]
63. Melo-Carrillo, A.; Strassman, A.M.; Nir, R.-R.; Schain, A.J.; Noseda, R.; Stratton, J.; Burstein, R. Fremanezumab-A Humanized Monoclonal Anti-CGRP Antibody-Inhibits Thinly Myelinated (Aδ) but Not Unmyelinated (C) Meningeal Nociceptors. *J. Neurosci.* **2017**, *37*, 10587–10596. [CrossRef]
64. Bigal, M.E.; Rapoport, A.M.; Silberstein, S.D.; Walter, S.; Hargreaves, R.J.; Aycardi, E. From LBR-101 to Fremanezumab for Migraine. *CNS Drugs* **2018**, *32*, 1025–1037. [CrossRef]
65. Bhakta, M.; Vuong, T.; Taura, T.; Wilson, D.S.; Stratton, J.R.; Mackenzie, K.D. Migraine Therapeutics Differentially Modulate the CGRP Pathway. *Cephalalgia Int. J. Headache* **2021**, *41*, 499–514. [CrossRef]
66. Melo-Carrillo, A.; Noseda, R.; Nir, R.-R.; Schain, A.J.; Stratton, J.; Strassman, A.M.; Burstein, R. Selective Inhibition of Trigeminovascular Neurons by Fremanezumab: A Humanized Monoclonal Anti-CGRP Antibody. *J. Neurosci.* **2017**, *37*, 7149–7163. [CrossRef]
67. González-Hernández, A.; Marichal-Cancino, B.A.; García-Boll, E.; Villalón, C.M. The Locus of Action of CGRPergic Monoclonal Antibodies against Migraine: Peripheral over Central Mechanisms. *CNS Neurol. Disord. Drug Targets* **2020**, *19*, 344–359. [CrossRef]
68. Zeller, J.; Poulsen, K.T.; Sutton, J.E.; Abdiche, Y.N.; Collier, S.; Chopra, R.; Garcia, C.A.; Pons, J.; Rosenthal, A.; Shelton, D.L. CGRP Function-Blocking Antibodies Inhibit Neurogenic Vasodilatation without Affecting Heart Rate or Arterial Blood Pressure in the Rat. *Br. J. Pharmacol.* **2008**, *155*, 1093–1103. [CrossRef]

69. Lionetto, L.; Curto, M.; Cisale, G.Y.; Capi, M.; Cipolla, F.; Guglielmetti, M.; Martelletti, P. Fremanezumab for the Preventive Treatment of Migraine in Adults. *Expert Rev. Clin. Pharmacol.* **2019**, *12*, 741–748. [CrossRef]
70. Gupta, S.; Amrutkar, D.V.; Mataji, A.; Salmasi, H.; Hay-Schmidt, A.; Sheykhzade, M.; Messlinger, K.; Olesen, J.; Jansen-Olesen, I. Evidence for CGRP Re-Uptake in Rat Dura Mater Encephali. *Br. J. Pharmacol.* **2010**, *161*, 1885–1898. [CrossRef]
71. Avona, A.; Burgos-Vega, C.; Burton, M.D.; Akopian, A.N.; Price, T.J.; Dussor, G. Dural Calcitonin Gene-Related Peptide Produces Female-Specific Responses in Rodent Migraine Models. *J. Neurosci.* **2019**, *39*, 4323–4331. [CrossRef]
72. Rea, B.J.; Wattiez, A.-S.; Waite, J.S.; Castonguay, W.C.; Schmidt, C.M.; Fairbanks, A.M.; Robertson, B.R.; Brown, C.J.; Mason, B.N.; Moldovan-Loomis, M.-C.; et al. Peripherally Administered Calcitonin Gene-Related Peptide Induces Spontaneous Pain in Mice: Implications for Migraine. *Pain* **2018**, *159*, 2306–2317. [CrossRef]
73. Zheng, F.; Nixdorf-Bergweiler, B.E.; van Brederode, J.; Alzheimer, C.; Messlinger, K. Excitatory Effects of Calcitonin Gene-Related Peptide (CGRP) on Superficial Sp5C Neurons in Mouse Medullary Slices. *Int. J. Mol. Sci.* **2021**, *22*, 3794. [CrossRef]
74. Miller, S.; Liu, H.; Warfvinge, K.; Shi, L.; Dovlatyan, M.; Xu, C.; Edvinsson, L. Immunohistochemical localization of the calcitonin gene-related peptide binding site in the primate trigeminovascular system using functional antagonist antibodies. *Neuroscience* **2016**, *328*, 165–183. [CrossRef]
75. Yamanaka, G.; Suzuki, S.; Morishita, N.; Takeshita, M.; Kanou, K.; Takamatsu, T.; Suzuki, S.; Morichi, S.; Watanabe, Y.; Ishida, Y.; et al. Role of Neuroinflammation and Blood-Brain Barrier Permutability on Migraine. *Int. J. Mol. Sci.* **2021**, *22*, 8929. [CrossRef]
76. Edvinsson, J.C.A.; Warfvinge, K.; Krause, D.N.; Blixt, F.W.; Sheykhzade, M.; Edvinsson, L.; Haanes, K.A. C-Fibers May Modulate Adjacent Aδ-Fibers through Axon-Axon CGRP Signaling at Nodes of Ranvier in the Trigeminal System. *J. Headache Pain* **2019**, *20*, 105. [CrossRef]
77. Hong, S.; Wiley, J.W. Altered Expression and Function of Sodium Channels in Large DRG Neurons and Myelinated A-Fibers in Early Diabetic Neuropathy in the Rat. *Biochem. Biophys. Res. Commun.* **2006**, *339*, 652–660. [CrossRef]
78. Huang, Z.-J.; Song, X.-J. Differing Alterations of Sodium Currents in Small Dorsal Root Ganglion Neurons after Ganglion Compression and Peripheral Nerve Injury. *Mol. Pain* **2008**, *4*, 20. [CrossRef]
79. Close, L.N.; Eftekhari, S.; Wang, M.; Charles, A.C.; Russo, A.F. Cortical Spreading Depression as a Site of Origin for Migraine: Role of CGRP. *Cephalalgia Int. J. Headache* **2019**, *39*, 428–434. [CrossRef]
80. Schain, A.J.; Melo-Carrillo, A.; Stratton, J.; Strassman, A.M.; Burstein, R. CSD-Induced Arterial Dilatation and Plasma Protein Extravasation Are Unaffected by Fremanezumab: Implications for CGRP's Role in Migraine with Aura. *J. Neurosci.* **2019**, *39*, 6001–6011. [CrossRef]
81. Lennerz, J.K.; Rühle, V.; Ceppa, E.P.; Neuhuber, W.L.; Bunnett, N.W.; Grady, E.F.; Messlinger, K. Calcitonin Receptor-like Receptor (CLR), Receptor Activity-Modifying Protein 1 (RAMP1), and Calcitonin Gene-Related Peptide (CGRP) Immunoreactivity in the Rat Trigeminovascular System: Differences between Peripheral and Central CGRP Receptor Distribution. *J. Comp. Neurol.* **2008**, *507*, 1277–1299. [CrossRef]
82. Eftekhari, S.; Salvatore, C.A.; Calamari, A.; Kane, S.A.; Tajti, J.; Edvinsson, L. Differential Distribution of Calcitonin Gene-Related Peptide and Its Receptor Components in the Human Trigeminal Ganglion. *Neuroscience* **2010**, *169*, 683–696. [CrossRef] [PubMed]
83. Messlinger, K.; Balcziak, L.K.; Russo, A.F. Cross-Talk Signaling in the Trigeminal Ganglion: Role of Neuropeptides and Other Mediators. *J. Neural Transm.* **2020**, *127*, 431–444. [CrossRef] [PubMed]
84. Benarroch, E.E. CGRP: Sensory Neuropeptide with Multiple Neurologic Implications. *Neurology* **2011**, *77*, 281–287. [CrossRef] [PubMed]
85. Vause, C.V.; Durham, P.L. Calcitonin Gene-Related Peptide Differentially Regulates Gene and Protein Expression in Trigeminal Glia Cells: Findings from Array Analysis. *Neurosci. Lett.* **2010**, *473*, 163–167. [CrossRef]
86. Dieterle, A.; Fischer, M.J.M.; Link, A.S.; Neuhuber, W.L.; Messlinger, K. Increase in CGRP- and NNOS-Immunoreactive Neurons in the Rat Trigeminal Ganglion after Infusion of an NO Donor. *Cephalalgia Int. J. Headache* **2011**, *31*, 31–42. [CrossRef]
87. Buldyrev, I.; Tanner, N.M.; Hsieh, H.; Dodd, E.G.; Nguyen, L.T.; Balkowiec, A. Calcitonin Gene-Related Peptide Enhances Release of Native Brain-Derived Neurotrophic Factor from Trigeminal Ganglion Neurons. *J. Neurochem.* **2006**, *99*, 1338–1350. [CrossRef]
88. Beckers, H.J.; Klooster, J.; Vrensen, G.F.; Lamers, W.P. Ultrastructural Identification of Trigeminal Nerve Terminals in the Pterygopalatine Ganglion of Rats: An Anterograde Tracing and Immunohistochemical Study. *Brain Res.* **1991**, *557*, 22–30. [CrossRef]
89. Motosugi, H.; Chiba, T.; Konno, A.; Kaneko, T. Distribution of Neuropeptides in Rat Pterygopalatine Ganglion: Light and Electron Microscopic Immunohistochemical Studies. *Arch. Histol. Cytol.* **1992**, *55*, 513–524. [CrossRef]
90. Kirch, W.; Neuhuber, W.; Tamm, E.R. Immunohistochemical Localization of Neuropeptides in the Human Ciliary Ganglion. *Brain Res.* **1995**, *681*, 229–234. [CrossRef]
91. Csati, A.; Tajti, J.; Tuka, B.; Edvinsson, L.; Warfvinge, K. Calcitonin Gene-Related Peptide and Its Receptor Components in the Human Sphenopalatine Ganglion—Interaction with the Sensory System. *Brain Res.* **2012**, *1435*, 29–39. [CrossRef]
92. Gottselig, R.; Messlinger, K. Noxious Chemical Stimulation of Rat Facial Mucosa Increases Intracranial Blood Flow through a Trigemino-Parasympathetic Reflex—An Experimental Model for Vascular Dysfunctions in Cluster Headache. *Cephalalgia Int. J. Headache* **2004**, *24*, 206–214. [CrossRef]

93. Leone, M.; Ferraro, S.; Proietti Cecchini, A. The Neurobiology of Cluster Headache. *Handb. Clin. Neurol.* **2021**, *182*, 401–414. [CrossRef]
94. Khan, S.; Schoenen, J.; Ashina, M. Sphenopalatine Ganglion Neuromodulation in Migraine: What Is the Rationale? *Cephalalgia Int. J. Headache* **2014**, *34*, 382–391. [CrossRef]

cells

MDPI

Article

Citalopram Neuroendocrine Challenge Shows Altered Tryptophan and Kynurenine Metabolism in Migraine

Kinga Gecse [1,2], Andrea Edit Édes [1,2], Tamás Nagy [1,2,3], Adrienn Katalin Demeter [1,2], Dávid Virág [4], Márton Király [4], Borbála Dalmadi Kiss [4], Krisztina Ludányi [4], Zsuzsanna Környei [5], Adam Denes [5], Gyorgy Bagdy [1,6,7,†] and Gabriella Juhasz [1,2,*,†]

1 Department of Pharmacodynamics, Faculty of Pharmacy, Semmelweis University, Üllői Street 26, H-1085 Budapest, Hungary
2 SE-NAP2 Genetic Brain Imaging Migraine Research Group, Hungarian Brain Research Program, Semmelweis University, Üllői Street 26, H-1085 Budapest, Hungary
3 Department of Measurement and Information Systems, Faculty of Electrical Engineering and Informatics, Budapest University of Technology and Economics, Magyar Tudósok krt.2, H-1521 Budapest, Hungary
4 Department of Pharmaceutics, Faculty of Pharmacy, Semmelweis University, Üllői Street 26, H-1085 Budapest, Hungary
5 Momentum Laboratory of Neuroimmunology, Institute of Experimental Medicine, Szigony Street 43, H-1083 Budapest, Hungary
6 NAP-2-SE New Antidepressant Target Research Group, Hungarian Brain Research Program, Semmelweis University, Üllői Street 26, H-1085 Budapest, Hungary
7 MTA-SE Neuropsychopharmacology and Neurochemistry Research Group, Hungarian Brain Research Program, Semmelweis University, Üllői Street 26, H-1085 Budapest, Hungary
* Correspondence: juhasz.gabriella@pharma.semmelweis-univ.hu; Tel.: +36-1-4591500
† These authors contributed equally to the work.

Citation: Gecse, K.; Édes, A.E.; Nagy, T.; Demeter, A.K.; Virág, D.; Király, M.; Dalmadi Kiss, B.; Ludányi, K.; Környei, Z.; Denes, A.; et al. Citalopram Neuroendocrine Challenge Shows Altered Tryptophan and Kynurenine Metabolism in Migraine. *Cells* **2022**, *11*, 2258. https://doi.org/10.3390/cells11142258

Academic Editors: László Vécsei, Bernadett Tuka and Masaru Tanaka

Received: 18 May 2022
Accepted: 18 July 2022
Published: 21 July 2022

Abstract: Altered tryptophan (TRP) metabolism may have an important role in migraine susceptibility through its main metabolites, serotonin and kynurenine (KYN). Both affect pain processing and stress response by interfering with neural and brain hypersensitivity and by interacting with chemokines and cytokines that control vascular and inflammatory processes. The involvement of these pathways in migraine has been widely studied, but acute citalopram neuroendocrine challenge on TRP metabolism and cytokine profile has not been investigated yet. In our study, females with episodic migraine without aura and healthy controls were studied before and after acute citalopram or placebo in a double-blind setting. At baseline, increased TRP/large neutral amino acid (LNAA) ratio and decreased RANTES chemokine concentration were detected in migraine patients compared to controls. The challenge induced a significant increase in TRP, KYN, and TRP/LNAA in healthy controls, but not in migraine patients. Furthermore, migraine attack frequency negatively correlated with KYN/TRP ratio and positively correlated with the neuroendocrine-challenge-induced KYN concentration increase. Our results support a decreased breakdown of TRP via KYN pathway and a failure to modulate TRP–KYN pathway during citalopram-induced acute stress together with an increased vascular sensitivity in migraine. These mechanisms may provide useful drug targets for future drug development.

Keywords: headache; stress; biomarker; RANTES; neuroendocrine challenge; cytokine

1. Introduction

Migraine is considered the primary cause of disability in young women, despite the advances in diagnosis and treatment in recent decades [1]. Different theories of migraine have been developed over the years, but nowadays migraine is described as a neurovascular disorder accompanied by altered brain sensory processing [2]. The involvement of serotonin in migraine pathophysiology was among the first to be discovered [3,4]. The observation of increased sensitivity to alterations in serotonergic neurotransmission in migraine patients has led to the development of the first migraine-specific drugs, called triptans [5], which act not only as vasoconstrictors as initially thought, but also through neural mechanisms by inhibiting trigeminal afferents and central trigeminal neurons [6]. Despite their effectiveness [7], we still do not know who will respond to triptan treatment, but there are promising studies to find markers of responsiveness [8]. In spite of the fact that potential cardiac side effects related to triptans directed drug discovery to other targets, serotonin remains in sight of migraine research [9].

Since serotonin does not cross the blood–brain barrier, dietary intake and plasma concentration of L-tryptophan (TRP), its essential precursor, affect brain serotonin synthesis [10] and potentially migraine susceptibility. TRP uptake to the brain is dependent on plasma concentration of other large neutral amino acids (LNAAs), as they compete for a common transporter [11]. The ratio of TRP/LNAA is widely used to describe the portion of TRP uptake to the brain; however, plasma TRP/LNAA ratio in migraine has not been investigated yet. Studies concerning the blood TRP concentration of migraine patients are not consistent; both higher [12–15] and lower [16,17] TRP concentrations were reported in migraineurs compared to controls. Furthermore, migraine patients are suggested to be sensitive to TRP concentration changes; thus, TRP depletion induces intense headache, nausea, and photophobia in migraineurs [18]. In addition, consuming relatively less TRP increases the risk for developing migraine in susceptible persons [19]. However, TRP supplementation alone has not proved to be an effective anti-migraine therapy [20,21].

One potential explanation of the uncertain therapeutic effect of TRP in migraine could be related to its complex metabolism. Although a small fraction of TRP serves serotonin synthesis, the main metabolic pathway of TRP is the kynurenine (KYN) pathway, which is also known to be involved in migraine [14,17,22,23]. Ninety-five percent of TRP is metabolized into N-formyl-kynurenine via two enzymes: tryptophan-2,3-dioxygenase (TDO) in the liver or indolamine-2,3-dioxyganese (IDO) in peripheral tissues and the nervous system [10]. After this rate-limiting step, L-kynurenine (KYN) is formed by formamidase enzyme. KYN is the main component of this pathway, as several active metabolites are synthesized from KYN, such as the neuroprotective kynurenic acid (KYNA) and the excitotoxic quinolinic acid (QUINA) [10].

The KYN–KYNA pathway may exert its effect on migraine via influencing glutamatergic neurotransmission. In migraine, glutamate acting on N-methyl-D-aspartate (NMDA) receptors induces the dilation of intra- and extracranial vessels [24] and also has an important modulatory effect on the descending pain control system in the brainstem, which influences trigeminal activation and nociception [2]. As KYNA is an endogenous NMDA receptor antagonist, it exerts an antinociceptive effect in the trigeminovascular system and may be a possible target of anti-migraine therapy [25,26]. In previous studies, decreased plasma KYN concentration was reported in both episodic [17] and chronic [14] migraine patients in interictal period compared to controls, suggesting a metabolic imbalance in TRP pathways of migraineurs [22]. KYN concentration is directly proportional to TRP concentration, and the KYN/TRP ratio is widely used to represent the shift of TRP metabolism from serotonin synthesis to KYN pathway [27]; previously, it was used as an indicator of IDO activity [28]. Even though the involvement of KYN pathway in migraine is well-known, there is still no anti-migraine therapeutic agent in the market acting on this pathway.

Serotonin and KYN may also have a notable role in migraine pathophysiology due to their regulatory influence on inflammatory processes [22,29]. KYN contributes to chemotactic activation of peripheral monocytes and also has a potential neuroimmunoregulatory role [28,30], while accumulating evidence suggests that alteration of the serotonin level in the body by serotonin reuptake inhibitors (SSRIs) modulates immune cell function and cytokine production [31]. However, the exact role of inflammatory processes in migraine pathophysiology is intensely debated [32]. The sterile neurogenic inflammation is hypothesized to contribute to the development of migraine attacks [33]. During a migraine attack, the activation of trigeminal neurons induces neuropeptide release; among others, calcitonin gene-related peptide (CGRP) [34,35] is a successful drug target for novel antimigraine therapies [36], leading to increased vascular permeability, vasodilation, and sensitization of the trigeminal neurons [37]. Furthermore, recent in vitro studies suggest that CGRP may increase the production of pro-inflammatory mediators, among them IL-1β, IL-6, and TNF-α, in the trigeminal ganglia [38]. Cytokines could act as pain mediators in neurovascular inflammation and play a role in pain generation in migraine [39]. The increased level of chemokines reinforces the stimulation of trigeminal neurons and thus the release of vasoactive peptides [40]. In addition, pro-inflammatory cytokines shift the metabolism of tryptophan toward the KYN pathway by activating the IDO enzyme [41]. Nevertheless, the investigation of cytokines in human migraine provided conflicting results, suggesting a possible increase in serum IL-1β, IL-6, and TNF-α levels in migraineurs compared to controls [42], depending on the ictal or interictal phase [32].

To further investigate the role of TRP metabolism and cytokine production in migraine and to identify potential migraine biomarkers, we investigated the baseline plasma concentration of tryptophan pathway and cytokine profile in interictal migraine patients without aura compared to healthy controls at two time-points at least two weeks apart in order to determine trait-like alterations. In addition, we examined serotonergic responsivity using a low-dose acute citalopram challenge, which is a safe neuroendocrine probe [43,44]. Citalopram is the most selective serotonin-reuptake inhibitor (SSRI) and increases extracellular brain serotonin availability without binding to any receptor type, and it also acts peripherally on serotonin-transporter-expressing cells. Intravenous administration of the drug is favored to avoid pharmacokinetic issues [43]. Acute citalopram challenge induces an increased arousal-like state [45] and a neuroendocrine stress response by increasing prolactin, ACTH, and cortisol release by stimulating the hypothalamic–pituitary–adrenal (HPA) axis [46,47]. Therefore, it is used as a neuroendocrine challenge to examine acute stress response [43,46,47]. Our previous study demonstrated that migraine patients are more sensitive to citalopram challenge showing increased brain activation in the anterior cingulate cortex compared to controls, for which observation suggests that increased stress-sensitivity in migraineurs is partially mediated by the serotonergic system [48]. However, to the best of our knowledge, the neuroendocrine response of migraine patients in acute citalopram challenge has not been investigated yet, nor has this challenge been used to determine the alterations of tryptophan metabolites and cytokines during acute serotonin level increase. Our additional aim was to investigate the correlation between clinical parameters of migraine and plasma concentration of tryptophan pathway and cytokine profile in baseline condition and after acute citalopram neuroendocrine challenge.

2. Materials and Methods

2.1. Study Design

Participants received 7.5 mg citalopram infusion in a randomized, double-blind crossover design. Normal saline or citalopram was infused over 7.5 min. The experimental days were separated by at least two weeks to evaluate them as independent sessions. In previous studies, 7.5 mg dose intravenous citalopram was proven to be an effective neuroendocrine challenge. Blood samples were collected 10 min before the infusion as well as 20 and 60 min after the infusion. The first blood sample was collected around 50 min after the participants' arrival to let them acclimatize and to reduce the effect of experiment-induced stress (Figure 1).

Every five minutes, participants were asked about their subjective experiences, and they answered with yes/no for the following statements: anxious, nauseous, drowsy, lightheaded, restless, uncomfortable.

The experiments were conducted between 4:00 pm and 8:00 pm in order to avoid the impact of circadian variations in cortisol and DHEA-S concentration [49,50].

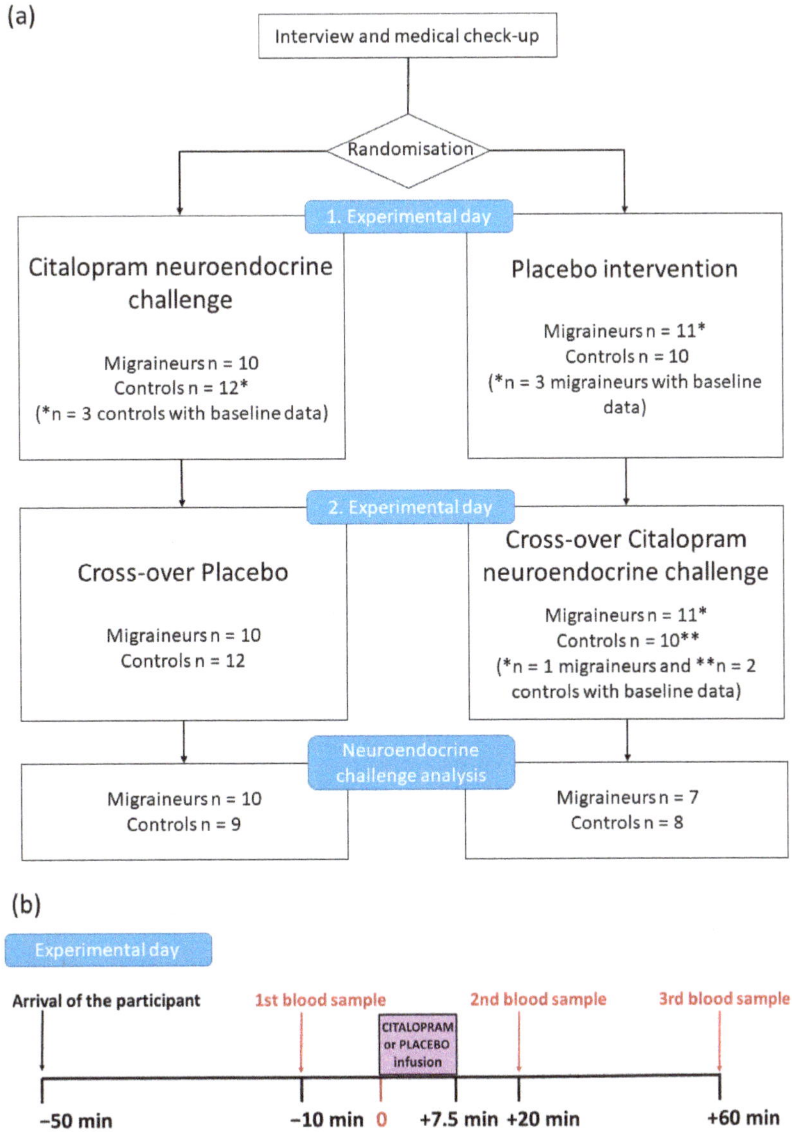

Figure 1. The study design. (**a**) Overall design of the randomized, cross-over study. Participants with only baseline data had missing variables; therefore, we could not include them in neuroendocrine challenge analysis. They were only included in baseline data analysis. (**b**) The design of each experimental day.

2.2. Participants

Forty-three female participants were included in the study: 21 patients with episodic migraine without aura and 22 healthy controls, between 18–50 years of age. The recruitment was based on university advertisements and newspaper articles, and some migraineurs volunteered from Headache Clinics. The participants underwent a diagnostic interview and medical screening, and their mental health status was checked using the Mini-International Neuropsychiatric Interview [51]. Exclusion criteria were the following: having any past or current serious medical, neurological (except migraine without aura), or psychiatric disorders; use of any daily medication (except contraceptives). All participants refrained from alcohol for at least 24 h and caffeine for at least 4 h before taking the first blood sample. In the case of four migraineurs and five controls, there were technical problems in the blood sample collection during the citalopram neuroendocrine challenge. Their data were included in the baseline concentration analysis but excluded from the neuroendocrine challenge analyses. Among the included participants, 17 migraineurs and 17 controls successfully completed the study without missing data and were included in the neuroendocrine challenge analyses.

2.3. Clinical Variables of Migraine

The diagnosis of episodic migraine without aura was given by expert neurologists based on the International Classification of Headache Disorders-III criteria [52]. Migraine with aura patients were excluded from the study. Participants were pain- and medication-free 48 h before and 24 h after the examination days and they were not taking any preventive migraine medications. They were asked to provide information about the clinical indicators of migraine such as age at disease onset and frequency of attacks per month.

2.4. Ethics

The study was conducted in accordance with the Declaration of Helsinki. Each participant was informed regarding the aim of the study, the substance used, and possible aversive side effects, and they provided written informed consent. The study protocol was approved by the Scientific and Research Ethics Committee of the Medical Research Council, Hungary (number: 23609-1/2011-EKU, 23421-1/2015-EKU).

2.5. Biological Samples

The blood samples were collected into K3EDTA tubes. After immediate centrifugation (10 min at 2000 RPM), plasma samples were kept frozen at −80 °C until the assay. For the quantitative determination of tryptophan, kynurenine, and other large amino acid (valine, leucine, isoleucine, phenylalanine, and tyrosine) concentration, LC–MS/MS method was used. The detailed methodology was published by Virág et al. [53]. Plasma cortisol and dehydroepiandrosterone sulfate (DHEA-S) levels were measured by competitive ELISA kits from NovaTec Immunodiagnostica GmbH (Dietzenbach, Germany) according to the manufacturer's instructions [54]. The coefficient of variation expressing inter- and intra-assay variance were as follows: cortisol, 11.0% and 5.1%; DHEA-S, 10.4% and 7.9%, respectively. The cortisol antibody showed 46.2% cross-reaction with prednisolone, while the DHEA-S antibody showed cross-reaction with DHEA (100%) and androstenedione (59%).

The concentrations of cytokines and the main chemokines induced by them [55,56] (granulocyte colony-stimulating factor (GCSF), interleukin 1 alpha (IL1a), interleukin 1 beta (IL1b), interleukin 6 (IL6), interleukin 8 (IL8), interleukin 10 (IL10), tumor necrosis factor alpha (TNFa), monocyte chemoattractant protein-1 (MCP-1), and regulated upon activation, normal T-cell-expressed, and presumably secreted CCL5 (RANTES)) were determined by BD Cytometric Bead Array (CBA) using BD CBA Flex Sets according to the manufacturer's instructions. Samples were acquired using a BD FACSVerse flow cytometer (BD, Becton, Dickinson and Company, Franklin Lakes, NJ, USA), and the results were analyzed by FCAP Array software (BD, Becton, Dickinson and Company, Franklin Lakes, NJ, USA) [57]. Average blanks for the cytokines and chemokines were the following: GCSF: 2.57 pg/mL, IL1a: 2.38 pg/mL, IL1b: 0 pg/mL, IL6: 0 pg/mL, IL8: 0 pg/mL, IL10: 2.89 pg/mL, MCP-1: 0.58 pg/mL, RANTES: 7.46 pg/mL, and TNFa: 1.40 pg/mL.

The concentration of plasma citalopram was measured with liquid-chromatography-tandem mass spectrometry (Agilent 1260 Infinity LC system, Agilent Technologies, Santa Clara, CA, USA).

2.6. Statistical Methods

Statistical analysis of the data was conducted in SPSS (SPSS Statistics for Windows, Version 27.0, IBM Corp., Armonk, NY, USA) Mann–Whitney tests were used to determine any differences between migraine and control groups regarding age and baseline plasma concentrations. The sum of LNAA affecting tryptophan blood–brain barrier crossing (namely tyrosine, phenylalanine, leucine, isoleucine, and valine) was calculated after Fernstrom et al. [11]. The ratio of tryptophan and LNAA (TRP/LNAA), the ratio of kynurenine and tryptophan (KYN/TRP), and the ratio of cortisol (CORT) and DHEA-S (CORT/DHEA-S) were calculated. The ratio of CORT/DHEA-S was used as an index of the neurobiological stress response [58]. Considering the non-normally distributed data, we used the Friedman test to determine the effect of the citalopram neuroendocrine challenge on tryptophan pathway and inflammatory markers in all participants and in separately analyzing the control and migraine group. For detected significant differences in plasma concentration, post-hoc analysis with Wilcoxon signed-rank tests was applied with a Bonferroni correction. Kruskal–Wallis test was applied to investigate the difference in citalopram plasma concentrations between migraine and control groups. The age at migraine onset and migraine frequency were correlated with baseline plasma concentrations and the neuroendocrine-challenge-induced concentration changes (calculated as the difference of the 20-min, 60-min, and baseline concentrations) using Spearman correlations. The significance threshold was set at $p < 0.05$. Although our sample size was relatively low, previous studies suggested that we had sufficient power to detect the effect of the citalopram neuroendocrine challenge [43,44,46,47,59–61].

For data visualization, GraphPad Prism version 8.0.1 for Windows software (GraphPad Software, San Diego, CA, USA) was used.

A partial least-squares linear discriminant analysis (PLS-LDA) model [62] was established to pattern discrimination between migraine patients and healthy controls [17]. PLS is a potent dimension-reduction method that is applicable in scenarios with multicollinearity where the measure of relevance for individual variables is also important. Two PLS models were established, each producing three latent variables as output: (1) the baseline model, containing biological markers from the blood samples before the citalopram infusion, and (2) the neuroendocrine challenge model, containing markers from blood samples taken 20 min after citalopram administration. Il1b was not included in the PLS-LDA model, because it was 0 pg/mL for every participant. The latent variables resulting from the PLS step were used in an LDA classifier to test the diagnostic capabilities of the models. Because of the small amount of data, we used leave-one-out cross-validation to test the predictive power of the full model without overfitting. The contribution of each variable to the latent variables of the PLS model was described with the variable importance projection (VIP) score, where values above 1.0 are indicative of relevance. The PLSRegression and LinearDiscriminantAnalysis modules of the sklearn package (version 1.0.2) were used in Python 3.8.5 (Python Software Foundation, Wilmington, DE, USA).

3. Results

3.1. Descriptive Characteristics of the Participants

There was no significant difference in age between the two groups (migraine: 25.00 (23.00 to 28.00) (median years (95% CI)), control: 24.50 (23.00 to 27.00) (median years (95% CI)); U = 236.5, $p = 0.893$). The mean age at migraine onset was 15.48 ± 6.23 (mean years ± SD). The mean of migraine attack frequency was 3.57 ± 3.29 (mean attacks per month ± SD).

3.2. Baseline Plasma Concentrations

Both citalopram and placebo infusion were preceded by blood sampling to determine the baseline concentration of tryptophan (TRP), kynurenine (KYN), cortisol (CORT), and the ratio of TRP/LNAA, KYN/TRP, CORT/DHEA-S, and the concentration of inflammatory cytokines and chemokines (GCSF, RANTES, MCP-1, IL1a, Il1b, IL6, IL8, IL10, and TNFa) before any intervention (Table 1).

Table 1. Baseline concentrations of the measured biomarkers before citalopram neuroendocrine challenge and placebo infusions.

	Before Citalopram Neuroendocrine Challenge		*p*-Value	Before Placebo Infusion		*p*-Value
	Migraine	Control		Migraine	Control	
Tryptophan, kynurenine(μg/mL), cortisol (ng/mL) and cortisol/DHEA-S (μg/mL)						
TRP	8.92 (6.75–10.18)	7.12 (5.55–9.28)	0.075	8.20 (6.85–9.47)	7.15 (5.59–8.11)	0.080
TRP/LNAA	0.09 (0.06–0.11)	0.06 (0.05–0.08)	**0.012**	0.08 (0.07–0.11)	0.06 (0.05–0.09)	**0.033**
KYN	0.57 (0.40–0.67)	0.46 (0.39–0.54)	0.308	0.54 (0.46–0.62)	0.47 (0.39–0.64)	0.274
KYN/TRP	0.07 (0.05–0.07))	0.07 (0.06–0.07)	0.264	0.06 (0.06 to 0.08)	0.07 (0.06–0.08)	0.481
CORT	58.74 (40.23–118.0)	56.72 (38.15–80.86)	0.512	64.21 (51.13–111.0)	40.86 (31.23–66.54)	0.382
CORT/DHEA-S	46.49 (37.62–85.32)	42.10 (33.41–58.25)	0.662	34.48 (35.86–82.30)	28.88 (23.90–53.65)	0.423
Cytokines and chemokines (pg/mL)						
GCSF	4.19 (3.64–4.83)	3.95 (3.54–4.91)	0.894	4.18 (3.94–4.94)	4.18 (3.39–5.13)	0.423
RANTES	10641 (6336–11,588)	13456 (10,192–16,846)	**0.017**	10593 (7279–12,588)	12670 (10,091–19,710)	**0.049**
MCP-1	15.01 (10.70–29.94)	13.96 (8.70–19.99)	0.538	16.94 (9.24–32.08)	15.74 (8.32–22.04)	0.388
IL1a	2.69 (2.20–3.62)	2.42 (1.92–3.00)	0.224	2.96 (2.22–3.11)	2.85 (1.36–3.47)	0.696
IL1b	0 (0.00–0.00)	0 (0.00–0.00)	1.000	0 (0.00–0.00)	0 (0.00–0.00)	1.000
IL6	0 (0.00–0.00)	0 (0.00–0.00)	0.163	0 (0.00–0.00)	0 (0.00–0.00)	0.143
IL8	0 (0.00–7.85)	0 (0.00–6.81)	0.748	7.13 (0.00–8.67)	0 (0.00–0.00)	**0.014**
IL10	3.20 (3.14–3.39)	3.26 (3.14–3.46)	0.473	3.46 (3.32–3.61)	3.27 (3.11–3.60)	0.193
TNFa	1.96 (1.50–2.20)	1.41 (0.89–2.40)	0.233	1.96 (0.79–2.22)	1.26 (0.66–2.22)	0.752

Note: Values demonstrate median ±95% confidence intervals; Mann–Whitney U test was used to compare the plasma concentrations between migraine and control groups with a $p < 0.05$ significance threshold (bold). TRP: tryptophan, LNAA: large neutral amino acids, KYN: kynurenine, CORT: cortisol, DHEA-S: dehydroepiandrosterone sulfate, GCSF: granulocyte colony-stimulating factor, RANTES: regulated upon activation, normal T-cell-expressed, and presumably secreted CCL5, MCP-1: monocyte chemoattractant protein-1, IL1a: interleukin 1 alpha, IL1b: interleukin 1 beta, IL6: interleukin 6, IL8: interleukin 8, IL10: interleukin 10, TNFa: tumor necrosis factor alpha.

The TRP/LNAA ratio was significantly higher in the migraine group compared to controls at both blood samplings. The TRP concentration was not significantly different between migraine patients and controls, but it tended to be higher in migraineurs at both blood-sampling occasions. The RANTES concentration was lower in the migraine group compared to controls on both experiment days. The IL8 concentration was higher in the migraine group compared to the control before placebo infusion; however, there was no difference between the two groups before citalopram intervention. There were no further differences between migraine patients and controls at baseline.

3.3. *Citalopram Neuroendocrine Challenge*

3.3.1. Main Effect of Citalopram Neuroendocrine Challenge

The citalopram neuroendocrine challenge significantly increased the plasma citalopram level in the whole study population (20 min: median [ng/mL] (95% CI) = 18.74 (13.81 to 21.98); 60 min: median [ng/mL] (95% CI) = 10.62 (8.71 to 13.18)) (Figure 2a), and the plasma concentrations corresponded well with previous studies [43,63]. Post hoc analysis showed that there was no significant difference in plasma citalopram concentration between migraine (20 min: median [ng/mL] (95% CI) = 20.32 (15.76 to 26.32); 60 min: median [ng/mL] (95% CI) = 10.79 (8.42 to 13.18)) and control (20 min: median [ng/mL] (95% CI) = 14.82 (10.21 to 27.00); 60 min: median [ng/mL] (95% CI) = 10.51 (8.61 to 19.20)) groups (20 min: H(1) = 2.75, $p = 0.097$; 60 min: H(1) = 0.15, $p = 0.904$) (Figure 2b).

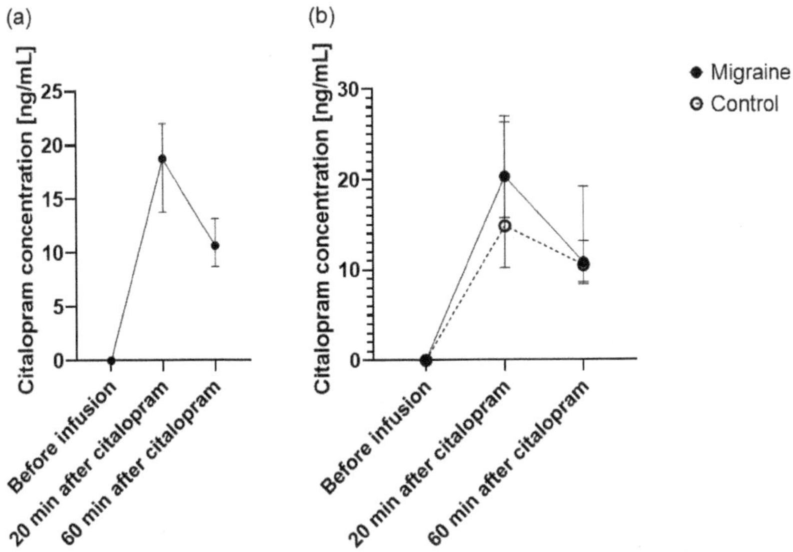

Figure 2. Effect of citalopram neuroendocrine challenge on plasma citalopram concentration. The median plasma concentration of citalopram (±95% CI) 20 min and 60 min after infusion (**a**) in the whole population and (**b**) separately in the migraine and control groups.

Administration of IV citalopram significantly increased the concentration of plasma TRP ($\chi2(2) = 6.35$, $p = 0.042$), KYN ($\chi2(2) = 11.53$, $p = 0.003$), RANTES ($\chi2(2) = 11.12$, $p = 0.004$) and TRP/LNAA ratio ($\chi2(2) = 6.59$, $p = 0.037$) in the population. During the placebo condition, CORT concentration ($\chi2(2) = 31.83$, $p < 0.001$) and CORT/DHEA-S ratio ($\chi2(2) = 26.65$, $p < 0.001$) showed significant time-dependent decrease that cannot be seen after the citalopram neuroendocrine challenge.

3.3.2. Differences between the Effects of Citalopram Neuroendocrine Challenge in Migraine Patients and Controls

There were no differences between the migraine and the control groups regarding the subjective experiences during the citalopram neuroendocrine challenge. Of the migraine patients, 12.5% (2/16) (one migraine patient did not answer) and 5.89% (1/17) of controls reported anxiety ($p = 0.601$), 18.75% of migraine patients and 23.53% of controls reported nausea ($p = 1.000$), 68.75 % of migraine patients and 58.82% of controls reported drowsiness ($p = 0.721$), 25% of migraine patients and 23.53% of controls reported dizziness ($p = 1.000$), 31.25% of migraine patients and 29.41% of controls reported restlessness ($p = 1.000$), and 56.25% of migraine patients and 41.18% of controls reported discomfort ($p = 0.494$) as a side-effect during the citalopram neuroendocrine challenge.

The citalopram neuroendocrine challenge showed a diagnosis-dependent effect on TRP, KYN concentration and on TRP/LNAA, KYN/TRP ratios (Figure 3). Separately analyzing the control group, we found that the citalopram neuroendocrine challenge induced significant increase in the TRP ($\chi2(2) = 14.94$, $p < 0.001$), KYN ($\chi2(2) = 8.94$, $p = 0.011$) concentration and TRP/LNAA ratio ($\chi2(2) = 13.18$, $p < 0.001$). There was a significant time-dependent increase in the TRP ($\chi2(2) = 7.18$, $p = 0.028$) concentration of control subjects after placebo, although the magnitude of change was lower compared to the citalopram neuroendocrine challenge. In the migraine group, the KYN/TRP ratio ($\chi2(2) = 7.41$, $p = 0.025$) decreased significantly after the citalopram neuroendocrine challenge, and no other significant change could be detected regarding the TRP pathway. The CORT concentration (migraine: $\chi2(2) = 23.48$, $p < 0.001$; control: $\chi2(2) = 9.88$, $p = 0.007$) and CORT/DHEA-S ratio (migraine: $\chi2(2) = 20.24$, $p < 0.001$; control: $\chi2(2) = 7.88$, $p = 0.019$) showed the expected circadian decrease in both groups during placebo condition, but this time-dependent change was not observed in the neuroendocrine challenge condition. As to the cytokines and chemokines, the changes of RANTES concentration after the neuroendocrine challenge showed no difference between the migraine ($\chi2(2) = 5.77$, $p = 0.056$) and the control groups ($\chi2(2) = 5.77$, $p = 0.056$) (Figure 3).

In the control group, there were no significant differences after post-hoc analysis between baseline and 60-min values of TRP concentration ($Z = 0.06$, $p = 1.000$) after citalopram. However, there was a statistically significant elevation of TRP concentration 20 min after citalopram infusion compared to baseline ($Z = -1.12$, $p = 0.003$) and a significant reduction in TRP concentration 60 min later compared to the blood sample taken after 20 min ($Z = 1.18$, $p = 0.002$).

Regarding the KYN concentration changes in the control group, there were no significant differences between the baseline and the citalopram concentration of KYN ($Z = 0.12$, $p = 1.000$) after 60 min. However, there was a statistically significant elevation of KYN concentration 20 min after citalopram infusion compared to baseline ($Z = -0.82$, $p = 0.049$) and a significant reduction in 60 min KYN concentration compared to the blood sample taken after 20 min ($Z = 0.94$, $p = 0.018$).

In the control group, there were no significant differences between the baseline and the citalopram ratio of TRP/LNAA ($Z = -0.24$, $p = 1.000$) after 60 min. However, there was a statistically significant elevation of TRP/LNAA ratio 20 min after citalopram infusion compared to baseline ($Z = -1.18$, $p = 0.002$) and a significant reduction of TRP/LNAA ratio after 60 min compared to the blood sample taken after 20 min ($Z = 0.94$, $p = 0.018$).

Post-hoc test of the KYN/TRP ratio changes in the migraine group showed no significant difference between the baseline and the 20-min KYN/TRP ($Z = -0.71$, $p = 0.119$) and between the baseline and the 60-min KYN/TRP ratio ($Z = 0.18$, $p = 1.000$). However, a significant reduction was observed between the 20-min KYN/TRP ratio and the 0 min KYN/TRP ratio ($Z = 0.88$, $p = 0.003$).

3.4. The Relationship of Migraine Parameters with the Measured Biomarkers

The frequency of migraine attacks was negatively correlated with KYN/TRP ratio before citalopram neuroendocrine challenge ($r_s = -0.451$ $p = 0.046$) and placebo ($r_s = -0.525$ $p = 0.018$) infusions (Figure 4).

The citalopram neuroendocrine-challenge-induced concentration changes of the plasma KYN 20 min after the start of the infusion showed positive correlation with the frequency of migraine attacks ($r_s = 0.766$ $p < 0.001$) (Figure 5).

There were no significant associations between other biomarkers and migraine frequency or age at migraine onset.

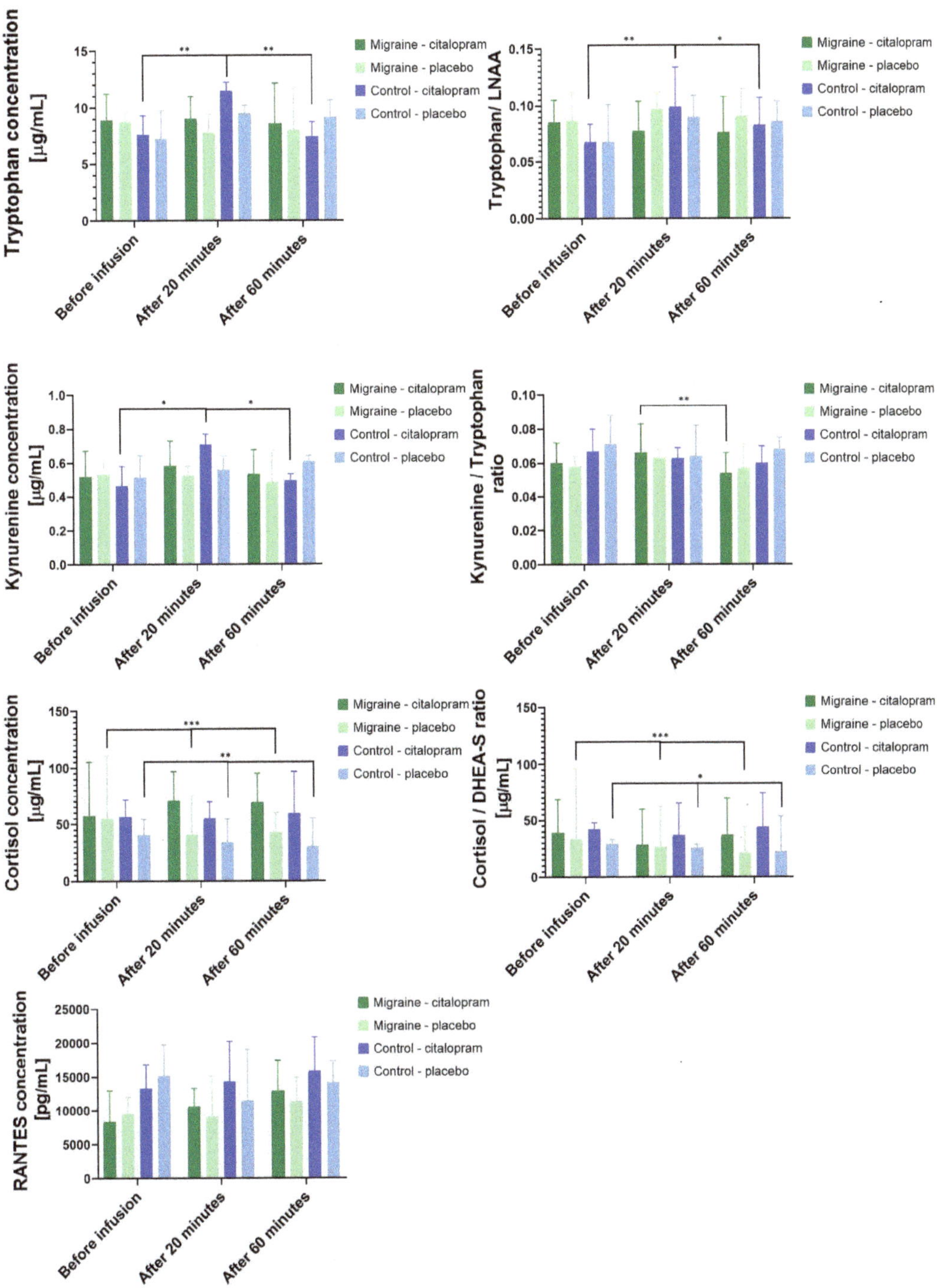

Figure 3. Significant effect of citalopram neuroendocrine challenge on the measured biomarkers. Medians and 95% confidence interval for tryptophan, kynurenine, cortisol, and RANTES plasma concentrations and tryptophan/LNAA, kynurenine/tryptophan, and cortisol/DHEA-S ratios. LNAA: large neutral amino acids, DHEA-S: dehydroepiandrosterone sulfate * $p < 0.05$, ** $p < 0.01$, *** $p < 0.001$.

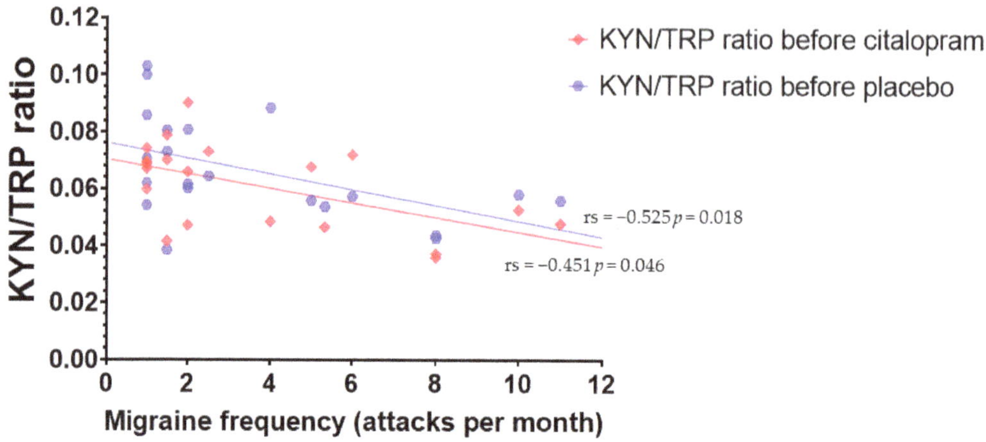

Figure 4. Correlation between number of migraine attacks per month and the ratio of kynurenine (KYN) and tryptophan (TRP) before citalopram neuroendocrine challenge ($r_s = -0.451$ $p = 0.046$) and placebo ($r_s = -0.525$ $p = 0.018$) infusions.

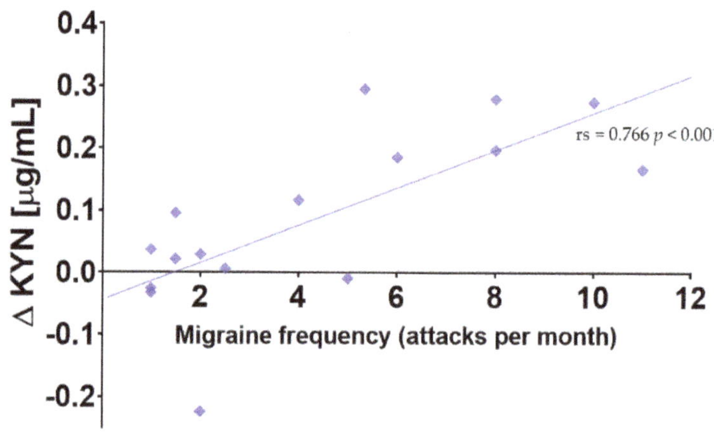

Figure 5. Correlation between number of migraine attacks per month and citalopram neuroendocrine-challenge-induced kynurenine (KYN) concentration changes ($r_s = 0.766$ $p < 0.001$). ДKYN = The difference between the concentrations measured at 20 min after citalopram infusion started and baseline.

3.5. Discriminating Migraine Patients from Controls with PLS-LDA

The multivariate classification method was used to identify the variables that play a role in the differentiation between migraine and control group. The baseline model contained the biological data before any intervention (Figure 6A). RANTES, TRP/LNAA, KYN/TRP, IL6, TNFa, and MCP-1 showed greater VIP score than 1, indicating relevance in migraine diagnosis. The citalopram neuroendocrine challenge model used the biological data 20 min after citalopram infusion (Figure 6B). In that case, TRP, TRP/LNAA ratio, KYN, IL6, KYN/TRP ratio, GCSF, and RANTES showed greater VIP score than 1. The resulting diagnostic accuracies of the full model were 79% for the baseline model and 71% for the citalopram model.

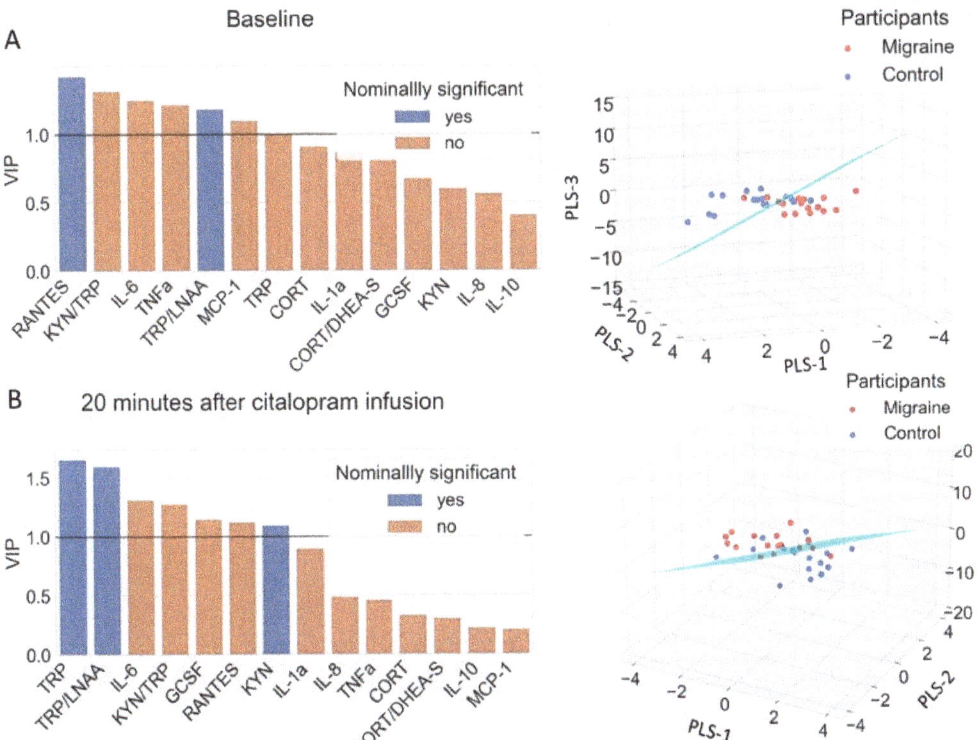

Figure 6. The bar plots of partial least-squares linear discriminant analysis (PLS-LDA) and 3D projection plots. (**A**) Baseline model representing the biological data before any intervention. The variable importance projection scores (VIP) above 1 indicates relevance in migraine. (**B**) Citalopram neuroendocrine challenge model representing the biological data 20 min after citalopram infusion. The classification hyperplane distinguishes the migraine group compared to controls. TRP: tryptophan, LNAA: large neutral amino acids, KYN: kynurenine, CORT: cortisol, DHEA-S: dehydroepiandrosterone sulfate, GCSF: granulocyte colony-stimulating factor, RANTES: regulated upon activation, normal T-cell-expressed and presumably secreted CCL5, MCP-1: monocyte chemoattractant protein-1, IL1a: interleukin 1 alpha, IL1b: interleukin 1 beta, IL6: interleukin 6, IL8: interleukin 8, IL10: interleukin 10, TNFa: tumor necrosis factor alpha.

4. Discussion

According to our knowledge, this is the first study about TRP pathway and cytokine profiles of episodic migraine without aura patients before and after the citalopram neuroendocrine challenge. Trait-like alterations were identified in migraine patients, namely increased TRP/LNAA ratio and decreased RANTES concentration, compared to controls in two independent baseline plasma samples. The acute neuroendocrine citalopram challenge that can also reflect stress-responsivity showed significant differences between migraine patients compared to controls. More precisely, the citalopram neuroendocrine challenge induced the increase of TRP and KYN concentration and TRP/LNAA ratio in healthy controls, but not in migraine patients. Furthermore, the frequency of migraine attacks showed negative correlation with baseline KYN/TRP ratio, another trait-like marker in two independent blood samples, but it positively correlated with the citalopram neuroendocrine-challenge-induced KYN concentration increase. Thus, our study supports a conclusion that the downregulation of the TRP–KYN pathway, probably partially through a maladaptive stress response, may contribute to migraine pathophysiology.

4.1. Alterations of the TRP Metabolism in Migraine

Increased plasma TRP/LNAA ratio was detected in interictal episodic migraine without aura patients compared to healthy controls at two independent blood samplings. To the best of our knowledge, our study is the first investigation of plasma TRP/LNAA ratio in migraine patients, as previous studies only reported results related to plasma TRP concentration [12–17,64]. In this study, the TRP concentration was not significantly higher in the migraine group, but it tended to be higher compared to controls at both blood samplings. Our results are consistent with previous studies reporting elevated TRP concentration in episodic [12,15] and chronic migraine patients [13,14]. However, there are some studies reporting the opposite results [16,17]. The opposing results may be due to different study populations and different time points for blood sampling. In our study, only patients with episodic migraine without aura who do not take prophylactic medication were included, and samples were collected in the afternoon, in contrast to other studies where samples were collected from headache clinic patients in the morning [17], and they were allowed to use prophylactic drugs [14,17]. Plasma TRP concentration shows changes with circadian rhythm, so the time of blood sampling could have an effect on the results [65]. Moreover, TRP uptake into the brain is affected by plasma LNAA concentration [11]; thus, plasma TRP concentration might not be an adequate biomarker in migraine. However, the TRP/LNAA ratio is a reliable marker of TRP brain uptake [11].

One possible explanation of the increased plasma TRP/LNAA ratio could be the decreased breakdown of TRP due to the impairment of TDO and IDO enzyme activity in migraine [22]. TDO and IDO enzymes are the rate-limiting step of KYN formation from TRP [10]. TDO is expressed in the liver, and astrocytes in the brain, while IDO is expressed in peripheral tissues, immune system cells, and microglia [66]. In a nitroglycerin administration rodent model, Nagy-Grócz et al. [67] demonstrated that migraine may be related to decreased expression of KYN pathway enzymes (kynurenine amino transferase-II (KAT-II), IDO, TDO, KYNU, KMO). Thus, our results could be a further support to the notion of downregulation of KYN enzymes in migraine.

The elevated TRP/LNAA ratio suggests an increased TRP intake by interictal migraine brain [12,14,64]. The increased TRP availability might be a protective factor between migraine attacks providing adequate TRP level for brain serotonin synthesis. TRP depletion studies demonstrated that diminished brain TRP level through decreased serotonin synthesis provokes intense headache, nausea, and photophobia in migraine patients [18]. Furthermore, relatively less TRP consumption increases the risk for developing migraine in susceptible people [19]. However, in our previous study, we did not find any correlation between migraine indicators (attack frequency and age at onset) and plasma TRP concentration, but an association between TRP concentration with depressive symptoms and trait-anxiety was apparent in migraine patients [15], which in turn may affect the recurrence of attacks and the processing of pain stimuli. Moreover, the increased plasma TRP concentration (corrected for plasma LNAA concentration) was associated with decreased functional connectivity of periaqueductal gray matter with regions implicated in fear-cascade and pain processing but increased functional connectivity with frontal emotion and pain regulating areas, thereby supporting TRP having an important role in optimizing stress coping [15]. Our previous results might suggest that serotonergic pathways synthetized from TRP could be more responsible for emotional symptoms and stress-processing in migraine than for the attack generation itself. Indeed, in the present study, the KYN pathway showed association with migraine frequency, as the decreased KYN/TRP ratio negatively correlated with the increasing number of migraine attacks per month at two independent blood samplings. Previously, the ratio of KYN/TRP was used to measure IDO activity; however, the current view is that the KYN/TRP ratio is more suitable to demonstrate the TRP pathway shift from serotonin to KYN [27]. Therefore, our results further suggest a less active kynurenine pathway in the background of migraine.

4.2. Migraine Patients Failed to Activate TRP Pathway during Citalopram Neuroendocrine Challenge

This is the first study reporting plasma TRP pathway and cytokine profile alterations during citalopram neuroendocrine challenge in humans. Acute citalopram challenge is a neuroendocrine probe to test serotonergic responsivity, and it induces acute stress [43]. We detected an immediate elevation of plasma TRP, KYN and TRP/LNAA ratio in healthy controls during the citalopram neuroendocrine challenge, but not in migraine patients, supporting the role of maladaptive stress-responsivity in migraine [68,69].

Acute stress increases the plasma concentration of TRP and KYN via different processes [70]. Previous studies showed that activation of the HPA axis increases plasma amino acid levels [71]; however, in our study, plasma TRP elevation preceded the significant increase in cortisol level. Thus, a more plausible explanation for our results is that the activation of the sympathetic nervous system may contribute to TRP increase [72]. The increase in the TRP/LNAA ratio suggests that the brain influx of TRP is also increased during the short-term effects of citalopram. Only free, non-protein-bound TRP can enter the brain [11]. The majority of plasma TRP is bound to circulating albumin and competes for its binding site with non-esterified fatty acids (NEFA) or drugs [71]. Elevation of NEFA level could happen under several circumstances, such as upon activation of sympathetic nervous system during physical exercise [73] or during neuroendocrine challenge, and this could displace TRP from protein binding [10,71], eventually leading to increased TRP brain uptake, which promotes serotonin synthesis and stress-coping [74].

The increase in TRP and KYN, without a change in the KYN/TRP ratio, could also represent an acute increase of the KYN pathway activity, in which induction of IDO and TDO enzymes by citalopram neuroendocrine challenge may play a role [75,76]. Indeed, acute physiological and psychological stress, such as neuroendocrine stress, could also induce KYN pathway enzymes [77]. TDO is mainly induced by cortisol, IDO induced by cytokines, and these processes enhance brain input of KYN for further metabolic transformations [66]. Based on animal experiments and our results, the intricate balance between TRP and KYN metabolites in healthy controls may contribute to stress resilience [78]. However, our results also suggest that these changes are temporary in healthy controls because the short-term increase in TRP and KYN concentration was followed by a rapid decrease 60 min after citalopram infusion, presumably due to the activation of the TRP and KYN degradation enzymes [78].

In migraine patients, TRP and TRP/LNAA plasma concentration did not increase from the citalopram neuroendocrine challenge. A likely explanation for this observation is that the already-elevated TRP concentration in migraine patients, which might be necessary to sustain the interictal phase or could be a sign of increased stress sensitivity [15], could not be elevated further by the acute activation of the sympathetic nervous system. Moreover, the induction of IDO and TDO enzymes may also be less effective, since enzymes of the KYN pathway are expressed in a reduced manner in migraine, resulting only in a modest increase in KYN plasma level [67]. Therefore, the dampened response of the TRYP–KYN pathway in migraineurs for the citalopram neuroendocrine stress challenge further emphasizes that the downregulated TRYP–KYN pathway may have a pathophysiologic role in migraine.

Interestingly, KYN concentration did not increase significantly in migraine patients during the citalopram neuroendocrine challenge, but the magnitude of the increase was associated with migraine frequency. Although we did not measure it in our study, previous research suggested that KAT expression is lower in migraine patients; thus, KYN conversion to KYNA is diminished [79]. KYNA has an important antinociceptive effect both in the periphery and in the brain [22]; by blocking first-order neurons and CGRP release [80], it could also reduce the activation of serotonergic neurons in the raphe nuclei [81] and have analgesic effect when injected directly into the PAG [82]. The decreased KYNA production could cause increased NMDA receptor hyperactivity in migraine that leads to brain hyperexcitability [66]. Therefore, despite increased plasma TRP level interictally, the insufficiently increased plasma KYN concentration during acute stress and the downregulated KYN metabolic pathway may lead to higher migraine attack frequency.

4.3. Inflammatory Biomarkers in Migraine

Although the neurogenic inflammation theory of migraine has been debated intensely [2], cytokines and chemokines may contribute to the development of migraine attacks [33], and they may have a role in migraine chronification [32]. Our classical statistical analysis showed that our subjects have no systemic inflammation based on their low IL-6, Il-1, and TNFa cytokine concentrations; furthermore, no significant difference appeared in plasma cytokines (namely GCSF, IL1a, IL1b, IL6, IL10, and TNFa) between migraineurs and healthy controls in the interictal phase. The plasma concentration of three chemokines was also determined in this study, namely MCP-1, RANTES, and IL8. Higher IL8 concentration was found in migraine patients compared to controls, but this was not replicated at the second blood sampling. However, decreased RANTES plasma concentration in migraine patients was replicated at both blood samplings. Furthermore, the multivariate analysis of our data suggested that RANTES, IL6, TNFa, MCP-1, and GCSF, together with the TRP–KYN pathway, might also be important in shaping the vulnerability to migraine by modulating inflammatory and vascular processes [83].

In recent decades, RANTES has gained more attention in headache research since Fidan et al. [84] demonstrated that RANTES levels rise during migraine attacks, but they found no difference between attacks in migraine patients compared to controls. Similarly, another study showed that the expression of RANTES gene (*CCL5*) is upregulated during migraine attacks, contributing to platelet activation [85]. Regarding other cytokines and chemokines, our results are in concordance with Fidan et al.'s [84] study, because they also reported no difference in IL10 and MCP-1 concentration between interictal migraineurs and healthy controls. Sarchielli et al. [86] examined the chemokine levels in jugular venous blood of migraine patients during attacks, but they did not find any difference in MCP-1 and RANTES concentration between the ictal and interictal state. They involved only eight migraine patients in the study, so the low number of participants may explain the contradictory findings. Domingues et. al. [87] reported an increased RANTES concentration in migraine patients compared to tension-type headache patients and suggested that increased anxiety and depressive symptoms of migraine patients could explain the difference. In our study, both migraine and control participants were free from any mental disorders, including anxiety or depression diagnosis. This is one possible explanation for our results differing from previous studies where this was not an exclusion criterion; the other is that we investigated the concentration of cytokines and chemokines in attack-free periods of migraineurs.

On the basis of the neurovascular theory of migraine pathophysiology, the hyperexcitability of trigeminovascular system is the key point of migraine attacks, although the nature of the primary triggers is still debated [2]. The trigeminal neurons innervate the dural blood vessels and contribute to the release of neuroinflammatory mediators parallel with the activation of the trigeminal system, which could be responsible for migraine pain and allodynia, as the release of the neuroinflammatory mediators is accompanied by dilation of the dural blood vessels [2]. Although vasodilation alone is not sufficient to initiate migraine attacks [88], accumulating evidence suggests that different vascular mechanisms are instrumental in migraine pathophysiology. For example, in previous studies, lower RANTES concentrations were associated with higher flow-mediated dilation (FMD) in high cardiovascular risk people [89], and higher FMD was observed in migraineurs with aura compared to migraineurs without aura and controls [90]. However, another study found no difference in FMD between migraine and non-migraine individuals [91]. In our study, episodic migraine without aura patients had decreased plasma RANTES concentration compared to healthy controls at two independent blood samplings. Contrary to previous studies, our results suggest that low RANTES concentration, presumably through an increased vascular sensitivity to blood-flow-induced dilation [89], may also contribute to episodic migraine without aura, which is in line with the neurovascular hypothesis of migraine. The low RANTES level is also consistent with GWAS studies of migraine, in which genes related to the vascular system are over-represented in migraine patients, but vascular genetic risk factors that showed an association with migraine did not increase the risk of hypertension or cardiovascular diseases [92]. Thus, our results, together with the previous findings, suggest that decreased RANTES concentration may contribute to interictal vascular hypersensitivity in migraine patients, predisposing them for migraine attacks, and the ictally increased RANTES concentration might be associated with the platelet activation that was observed during attacks.

In addition, we detected an increase in RANTES concentration during the citalopram neuroendocrine challenge both in migraine and control participants. Acute citalopram administration by inducing a neuroendocrine response leads to an acute stress reaction [43,47]. Our results suggest that during a neuroendocrine stress condition, RANTES concentration rises independently from migraine diagnosis. Increased RANTES is suggested to play a role in dysfunctional vascular regulation by the modulation of perivascular inflammation [89,93], so the increasing RANTES concentration might be a part of the vascular response in neuroendocrine stress. However, in our study, RANTES concentration remained lower in migraine patients (median: 10,580 pg/mL (13,292–5882) 95% CI) compared to controls (median: 14,340 pg/mL (20,233–7260) 95% CI), even during the neuroendocrine stress challenge. However, the difference was not significant, suggesting a permanent, trait-like RANTES-related increased vascular sensitivity in migraine patients, which supports the vascular component of neurovascular migraine theory [2]. The exact role of RANTES in migraine generation is still unknown; however, our results suggest that it might act through pro-inflammatory, vascular, and nociceptive effects.

4.4. Limitations

Our study has some important limitations. The first is the relatively low number of participants involved in the analysis, which is partially due to the complexity of the neuroendocrine citalopram challenge paradigm. However, the randomized, double-blind crossover design (Figure 1) and the availability of two independent baseline samples per participant strengthen our results. The second limitation of our study is the fact that we measured plasma total TRP concentration and did not investigate the free and bound TRP ratio. However, we determined plasma LNAA concentrations because the TRP/LNAA ratio is a much better marker for the TRP influx into the brain. Finally, not measuring the expression and the activity of the KYN pathway enzymes and the plasma concentration of other KYN metabolites is another limitation of our study. Instead, we focused on chemokines and cytokines beside TRP, KYN, and LNAA in order to capture the effect of the citalopram neuroendocrine challenge in migraineurs not only on the KYN pathway but also on neuroinflammation. Indeed, our multivariate analysis supports the role of these biological systems in migraine vulnerability.

5. Conclusions

We have provided further evidence for KYN pathway downregulation in migraine through trait-like decrease in TRP and KYN metabolism. There is increasing evidence that a decreased amount of KYN pathway metabolites plays an important role in migraine pathogenesis affecting glutamate signaling. Our results also demonstrated an increased susceptibility to vascular changes, indicated by decreased RANTES concentration in migraine patients at trait level in interictal period. Furthermore, an altered response of KYN pathway was observed in migraineurs during the citalopram neuroendocrine challenge, which distinguishes migraine patients from controls. Further studies are needed to investigate whether migraine prevention is possible by influencing the TRP–KYN pathway. In ongoing clinical trials, glutamate receptor (mGLUR5) modulators and AMPA/kainite receptor antagonists are promising treatments for acute migraine, as are NMDA and kainate receptor antagonists for migraine prophylaxis. These targets are under the influence of KYN pathway metabolites.

Author Contributions: The study was designed and conceived by G.J. A.E.É. took part in subject recruitment and data collection. K.G. contributed to the biological data collection, K.G. and T.N. to the data analysis. Biological sample measurement was performed by M.K., D.V., K.L., B.D.K., Z.K. and A.D. K.G. and G.J. drafted the manuscript, K.G., A.K.D., A.D., G.J. and G.B. contributed to the interpretation of the results. All authors have read and agreed to the published version of the manuscript.

Funding: This study was supported by the Hungarian Academy of Sciences (MTA-SE Neuropsychopharmacology and Neurochemistry Research Group); the Hungarian Brain Research Program (Grants: 2017-1.2.1-NKP-2017-00002; KTIA_13_NAPA-II/14); the National Development Agency (Grant: KTIA_NAP_13-1-2013- 0001); by the Hungarian Academy of Sciences, Hungarian National Development Agency, Semmelweis University and the Hungarian Brain Research Program (Grant: KTIA_NAP_13-2- 2015-0001; MTA-SE-NAP B Genetic Brain Imaging Migraine Research Group); by the Ministry of Innovation and Technology of Hungary from the National Research, Development and Innovation Fund, financed under the TKP2021-EGA funding scheme (TKP2021-EGA-25; TKP2021-EGA-24); and by the National Research, Development and Innovation Office, Hungary (2019-2.1.7-ERA-NET-2020-00005), under the frame of ERA PerMed (ERAPERMED2019-108). TN was supported by the ÚNKP-21-3-I-SE-32 New National Excellence Program of the Ministry for Innovation and Technology from the source of the National Research, Development and Innovation Fund.

Institutional Review Board Statement: The study was conducted in accordance with the Declaration of Helsinki and approved by the Scientific and Research Ethics Committee of the Medical Research Council, Hungary (number: 23609-1/2011-EKU, 23421-1/2015-EKU).

Informed Consent Statement: A written informed consent was obtained from all subjects involved in the study.

Data Availability Statement: The datasets generated and analyzed during the current study are not publicly available due to ongoing analysis for future publication, but are available from the corresponding author upon reasonable request.

Acknowledgments: The authors would like to thank to Terézia Zsombók, Dorottya Pap, Ádám György Békésy-Szabó, Lajos Rudolf Kozák, Krisztina Kovács, Dóra Török and Gyöngyi Kökönyei for their contribution. The authors thank the Cell Biology Center at the Institute of Experimental Medicine for providing support for cytokine measurement.

Conflicts of Interest: György Bagdy is a member of the Board of directors at Gedeon Richter Plc., and Andrea Edit Édes is an employee of Gedeon Richter Plc. Medical Division, but the company did not provide any funding or had any further role in the preparation of the article. The other authors declare no competing interest.

References

1. GBD DALYs; HALE Collaborators. Global, regional, and national disability-adjusted life-years (DALYs) for 359 diseases and injuries and healthy life expectancy (HALE) for 195 countries and territories, 1990–2017: A systematic analysis for the Global Burden of Disease Study 2017. *Lancet* **2018**, *392*, 1859–1922. [CrossRef]
2. Goadsby, P.J.; Holland, P.R.; Martins-Oliveira, M.; Hoffmann, J.; Schankin, C.; Akerman, S. Pathophysiology of Migraine: A Disorder of Sensory Processing. *Physiol. Rev.* **2017**, *97*, 553–622. [CrossRef] [PubMed]
3. Sicuteri, F.; Testi, A.; Anselmi, B. Biochemical Investigations in Headache: Increase in the Hydroxyindoleacetic Acid Excretion During Migraine Attacks. *Int. Arch. Allergy Immunol.* **1961**, *19*, 55–58. [CrossRef]
4. Ferrari, M.D.; Odink, J.; Tapparelli, C.; Van Kempen, G.M.; Pennings, E.J.; Bruyn, G.W. Serotonin metabolism in migraine. *Neurology* **1989**, *39*, 1239–1242. [CrossRef]
5. Panconesi, A. Serotonin and migraine: A reconsideration of the central theory. *J. Headache Pain* **2008**, *9*, 267–276. [CrossRef]
6. Hoskin, K.L.; Kaube, H.; Goadsby, P.J. Sumatriptan can inhibit trigeminal afferents by an exclusively neural mechanism. *Brain* **1996**, *119 Pt 5*, 1419–1428. [CrossRef]
7. Cameron, C.; Kelly, S.; Hsieh, S.C.; Murphy, M.; Chen, L.; Kotb, A.; Peterson, J.; Coyle, D.; Skidmore, B.; Gomes, T.; et al. Triptans in the Acute Treatment of Migraine: A Systematic Review and Network Meta-Analysis. *Headache* **2015**, *55* (Suppl. 4), 221–235. [CrossRef]
8. Wu, J.W.; Lai, P.Y.; Chen, Y.L.; Wang, Y.F.; Lirng, J.F.; Chen, S.T.; Lai, K.L.; Chen, W.T.; Wu, Y.T.; Wang, S.J. The Use of Neuroimaging for Predicting Sumatriptan Treatment Response in Patients With Migraine. *Front. Neurol.* **2022**, *13*, 798695. [CrossRef]
9. de Vries, T.; Villalon, C.M.; MaassenVanDenBrink, A. Pharmacological treatment of migraine: CGRP and 5-HT beyond the triptans. *Pharmacol. Ther.* **2020**, *211*, 107528. [CrossRef]

10. Palego, L.; Betti, L.; Rossi, A.; Giannaccini, G. Tryptophan Biochemistry: Structural, Nutritional, Metabolic, and Medical Aspects in Humans. *J. Amino Acids* **2016**, *2016*, 8952520. [CrossRef]
11. Fernstrom, J.D. Large neutral amino acids: Dietary effects on brain neurochemistry and function. *Amino Acids* **2013**, *45*, 419–430. [CrossRef] [PubMed]
12. Alam, Z.; Coombes, N.; Waring, R.H.; Williams, A.C.; Steventon, G.B. Plasma levels of neuroexcitatory amino acids in patients with migraine or tension headache. *J. Neurol. Sci.* **1998**, *156*, 102–106. [CrossRef]
13. Salmon, S.; Fanciullacci, M.; Bonciani, M.; Sicuteri, F. Plasma tryptophan in migraine. *Headache* **1978**, *17*, 238–241. [CrossRef] [PubMed]
14. Curto, M.; Lionetto, L.; Negro, A.; Capi, M.; Fazio, F.; Giamberardino, M.A.; Simmaco, M.; Nicoletti, F.; Martelletti, P. Altered kynurenine pathway metabolites in serum of chronic migraine patients. *J. Headache Pain* **2015**, *17*, 47. [CrossRef]
15. Gecse, K.; Dobos, D.; Aranyi, C.S.; Galambos, A.; Baksa, D.; Kocsel, N.; Szabo, E.; Pap, D.; Virag, D.; Ludanyi, K.; et al. Association of plasma tryptophan concentration with periaqueductal gray matter functional connectivity in migraine patients. *Sci. Rep.* **2022**, *12*, 739. [CrossRef]
16. Ren, C.; Liu, J.; Zhou, J.; Liang, H.; Wang, Y.; Sun, Y.; Ma, B.; Yin, Y. Low levels of serum serotonin and amino acids identified in migraine patients. *Biochem. Biophys. Res. Commun.* **2018**, *496*, 267–273. [CrossRef]
17. Tuka, B.; Nyari, A.; Cseh, E.K.; Kortesi, T.; Vereb, D.; Tomosi, F.; Kecskemeti, G.; Janaky, T.; Tajti, J.; Vecsei, L. Clinical relevance of depressed kynurenine pathway in episodic migraine patients: Potential prognostic markers in the peripheral plasma during the interictal period. *J. Headache Pain* **2021**, *22*, 60. [CrossRef]
18. Drummond, P.D. Tryptophan depletion increases nausea, headache and photophobia in migraine sufferers. *Cephalalgia* **2006**, *26*, 1225–1233. [CrossRef]
19. Razeghi Jahromi, S.; Togha, M.; Ghorbani, Z.; Hekmatdoost, A.; Khorsha, F.; Rafiee, P.; Shirani, P.; Nourmohammadi, M.; Ansari, H. The association between dietary tryptophan intake and migraine. *Neurol. Sci.* **2019**, *40*, 2349–2355. [CrossRef]
20. Kangasniemi, P.; Falck, B.; Långvik, V.A.; Hyyppä, M.T. Levotryptophan treatment in migraine. *Headache* **1978**, *18*, 161–165. [CrossRef]
21. Hasselmark, L.; Malmgren, R.; Hannerz, J. Effect of a carbohydrate-rich diet, low in protein-tryptophan, in classic and common migraine. *Cephalalgia* **1987**, *7*, 87–92. [CrossRef] [PubMed]
22. Fejes, A.; Párdutz, A.; Toldi, J.; Vécsei, L. Kynurenine metabolites and migraine: Experimental studies and therapeutic perspectives. *Curr. Neuropharmacol.* **2011**, *9*, 376–387. [CrossRef] [PubMed]
23. Vecsei, L.; Szalardy, L.; Fulop, F.; Toldi, J. Kynurenines in the CNS: Recent advances and new questions. *Nat. Rev. Drug Discov.* **2013**, *12*, 64–82. [CrossRef] [PubMed]
24. Vikelis, M.; Mitsikostas, D.D. The role of glutamate and its receptors in migraine. *CNS Neurol. Disord. Drug Targets* **2007**, *6*, 251–257. [CrossRef] [PubMed]
25. Vamos, E.; Pardutz, A.; Varga, H.; Bohar, Z.; Tajti, J.; Fulop, F.; Toldi, J.; Vecsei, L. l-kynurenine combined with probenecid and the novel synthetic kynurenic acid derivative attenuate nitroglycerin-induced nNOS in the rat caudal trigeminal nucleus. *Neuropharmacology* **2009**, *57*, 425–429. [CrossRef]
26. Greco, R.; Demartini, C.; Zanaboni, A.M.; Redavide, E.; Pampalone, S.; Toldi, J.; Fulop, F.; Blandini, F.; Nappi, G.; Sandrini, G.; et al. Effects of kynurenic acid analogue 1 (KYNA-A1) in nitroglycerin-induced hyperalgesia: Targets and anti-migraine mechanisms. *Cephalalgia* **2017**, *37*, 1272–1284. [CrossRef]
27. Badawy, A.A.; Guillemin, G. The Plasma [Kynurenine]/[Tryptophan] Ratio and Indoleamine 2,3-Dioxygenase: Time for Appraisal. *Int. J. Tryptophan Res.* **2019**, *12*, 1178646919868978. [CrossRef]
28. Mithaiwala, M.N.; Santana-Coelho, D.; Porter, G.A.; O'Connor, J.C. Neuroinflammation and the Kynurenine Pathway in CNS Disease: Molecular Mechanisms and Therapeutic Implications. *Cells* **2021**, *10*, 1548. [CrossRef]
29. Wu, H.; Denna, T.H.; Storkersen, J.N.; Gerriets, V.A. Beyond a neurotransmitter: The role of serotonin in inflammation and immunity. *Pharmacol. Res.* **2019**, *140*, 100–114. [CrossRef]
30. Zang, X.; Zheng, X.; Hou, Y.; Hu, M.; Wang, H.; Bao, X.; Zhou, F.; Wang, G.; Hao, H. Regulation of proinflammatory monocyte activation by the kynurenine-AhR axis underlies immunometabolic control of depressive behavior in mice. *FASEB J.* **2018**, *32*, 1944–1956. [CrossRef]
31. Walker, F.R. A critical review of the mechanism of action for the selective serotonin reuptake inhibitors: Do these drugs possess anti-inflammatory properties and how relevant is this in the treatment of depression? *Neuropharmacology* **2013**, *67*, 304–317. [CrossRef] [PubMed]
32. Edvinsson, L.; Haanes, K.A.; Warfvinge, K. Does inflammation have a role in migraine? *Nat. Rev. Neurol.* **2019**, *15*, 483–490. [CrossRef] [PubMed]
33. Spekker, E.; Tanaka, M.; Szabo, A.; Vecsei, L. Neurogenic Inflammation: The Participant in Migraine and Recent Advancements in Translational Research. *Biomedicines* **2021**, *10*, 76. [CrossRef] [PubMed]
34. Juhasz, G.; Zsombok, T.; Jakab, B.; Nemeth, J.; Szolcsanyi, J.; Bagdy, G. Sumatriptan causes parallel decrease in plasma calcitonin gene-related peptide (CGRP) concentration and migraine headache during nitroglycerin induced migraine attack. *Cephalalgia* **2005**, *25*, 179–183. [CrossRef] [PubMed]

35. Juhasz, G.; Zsombok, T.; Modos, E.A.; Olajos, S.; Jakab, B.; Nemeth, J.; Szolcsanyi, J.; Vitrai, J.; Bagdy, G. NO-induced migraine attack: Strong increase in plasma calcitonin gene-related peptide (CGRP) concentration and negative correlation with platelet serotonin release. *Pain* **2003**, *106*, 461–470. [CrossRef]
36. Edvinsson, L.; Haanes, K.A.; Warfvinge, K.; Krause, D.N. CGRP as the target of new migraine therapies-successful translation from bench to clinic. *Nat. Rev. Neurol.* **2018**, *14*, 338–350. [CrossRef]
37. Moskowitz, M.A. Defining a pathway to discovery from bench to bedside: The trigeminovascular system and sensitization. *Headache* **2008**, *48*, 688–690. [CrossRef]
38. Afroz, S.; Arakaki, R.; Iwasa, T.; Oshima, M.; Hosoki, M.; Inoue, M.; Baba, O.; Okayama, Y.; Matsuka, Y. CGRP Induces Differential Regulation of Cytokines from Satellite Glial Cells in Trigeminal Ganglia and Orofacial Nociception. *Int. J. Mol. Sci.* **2019**, *20*, 711. [CrossRef]
39. Empl, M.; Sostak, P.; Riedel, M.; Schwarz, M.; Muller, N.; Forderreuther, S.; Straube, A. Decreased sTNF-RI in migraine patients? *Cephalalgia* **2003**, *23*, 55–58. [CrossRef]
40. Christopherson, K., II; Hromas, R. Chemokine regulation of normal and pathologic immune responses. *Stem Cells* **2001**, *19*, 388–396. [CrossRef]
41. Suneson, K.; Lindahl, J.; Chamli Harsmar, S.; Soderberg, G.; Lindqvist, D. Inflammatory Depression-Mechanisms and Non-Pharmacological Interventions. *Int. J. Mol. Sci.* **2021**, *22*, 1640. [CrossRef] [PubMed]
42. Geng, C.; Yang, Z.; Xu, P.; Zhang, H. Aberrations in peripheral inflammatory cytokine levels in migraine: A systematic review and meta-analysis. *J. Clin. Neurosci.* **2022**, *98*, 213–218. [CrossRef] [PubMed]
43. Lotrich, F.E.; Bies, R.; Muldoon, M.F.; Manuck, S.B.; Smith, G.S.; Pollock, B.G. Neuroendocrine response to intravenous citalopram in healthy control subjects: Pharmacokinetic influences. *Psychopharmacology* **2005**, *178*, 268–275. [CrossRef] [PubMed]
44. Seifritz, E.; Baumann, P.; Muller, M.J.; Annen, O.; Amey, M.; Hemmeter, U.; Hatzinger, M.; Chardon, F.; Holsboer-Trachsler, E. Neuroendocrine effects of a 20-mg citalopram infusion in healthy males. A placebo-controlled evaluation of citalopram as 5-HT function probe. *Neuropsychopharmacology* **1996**, *14*, 253–263. [CrossRef]
45. Edes, A.E.; McKie, S.; Szabo, E.; Kokonyei, G.; Pap, D.; Zsombok, T.; Hullam, G.; Gonda, X.; Kozak, L.R.; McFarquhar, M.; et al. Spatiotemporal brain activation pattern following acute citalopram challenge is dose dependent and associated with neuroticism: A human phMRI study. *Neuropharmacology* **2020**, *170*, 107807. [CrossRef]
46. Bhagwagar, Z.; Hafizi, S.; Cowen, P.J. Acute citalopram administration produces correlated increases in plasma and salivary cortisol. *Psychopharmacology* **2002**, *163*, 118–120. [CrossRef]
47. Attenburrow, M.J.; Mitter, P.R.; Whale, R.; Terao, T.; Cowen, P.J. Low-dose citalopram as a 5-HT neuroendocrine probe. *Psychopharmacology* **2001**, *155*, 323–326. [CrossRef]
48. Edes, A.E.; McKie, S.; Szabo, E.; Kokonyei, G.; Pap, D.; Zsombok, T.; Magyar, M.; Csepany, E.; Hullam, G.; Szabo, A.G.; et al. Increased activation of the pregenual anterior cingulate cortex to citalopram challenge in migraine: An fMRI study. *BMC Neurol.* **2019**, *19*, 237. [CrossRef]
49. Laudenslager, M.L.; Calderone, J.; Philips, S.; Natvig, C.; Carlson, N.E. Diurnal patterns of salivary cortisol and DHEA using a novel collection device: Electronic monitoring confirms accurate recording of collection time using this device. *Psychoneuroendocrinology* **2013**, *38*, 1596–1606. [CrossRef]
50. Weitzman, E.D.; Fukushima, D.; Nogeire, C.; Roffwarg, H.; Gallagher, T.F.; Hellman, L. Twenty-four hour pattern of the episodic secretion of cortisol in normal subjects. *J. Clin. Endocrinol. Metab.* **1971**, *33*, 14–22. [CrossRef]
51. Sheehan, D.V.; Lecrubier, Y.; Sheehan, K.H.; Amorim, P.; Janavs, J.; Weiller, E.; Hergueta, T.; Baker, R.; Dunbar, G.C. The Mini-International Neuropsychiatric Interview (M.I.N.I.): The development and validation of a structured diagnostic psychiatric interview for DSM-IV and ICD-10. *J. Clin. Psychiatry* **1998**, *59* (Suppl. 20), 22–33. [PubMed]
52. Arnold, M. Headache Classification Committee of the International Headache Society (IHS) The International Classification of Headache Disorders. *Cephalalgia* **2018**, *38*, 1–211. [CrossRef] [PubMed]
53. Virág, D.; Király, M.; Drahos, L.; Édes, A.E.; Gecse, K.; Bagdy, G.; Juhász, G.; Antal, I.; Klebovich, I.; Dalmadi Kiss, B.; et al. Development, validation and application of LC-MS/MS method for quantification of amino acids, kynurenine and serotonin in human plasma. *J. Pharm. Biomed. Anal.* **2020**, *180*, 113018. [CrossRef] [PubMed]
54. Kokonyei, G.; Galambos, A.; Kocsel, N.; Szabo, E.; Edes, A.E.; Gecse, K.; Baksa, D.; Pap, D.; Kozak, L.R.; Bagdy, G.; et al. Inter-individual differences in pain anticipation and pain perception in migraine: Neural correlates of migraine frequency and cortisol-to-dehydroepiandrosterone sulfate (DHEA-S) ratio. *PLoS ONE* **2021**, *16*, e0261570. [CrossRef] [PubMed]
55. Pyrillou, K.; Burzynski, L.C.; Clarke, M.C.H. Alternative Pathways of IL-1 Activation, and Its Role in Health and Disease. *Front. Immunol.* **2020**, *11*, 613170. [CrossRef]
56. Panopoulos, A.D.; Watowich, S.S. Granulocyte colony-stimulating factor: Molecular mechanisms of action during steady state and 'emergency' hematopoiesis. *Cytokine* **2008**, *42*, 277–288. [CrossRef]
57. Tóth, K.; Lénárt, N.; Berki, P.; Fekete, R.; Szabadits, E.; Pósfai, B.; Cserép, C.; Alatshan, A.; Benkő, S.; Kiss, D.; et al. The NKCC1 ion transporter modulates microglial phenotype and inflammatory response to brain injury in a cell-autonomous manner. *PLoS Biol.* **2022**, *20*, e3001526. [CrossRef]
58. Kamin, H.S.; Kertes, D.A. Cortisol and DHEA in development and psychopathology. *Horm. Behav.* **2017**, *89*, 69–85. [CrossRef]

59. Anderson, I.M.; Juhasz, G.; Thomas, E.; Downey, D.; McKie, S.; Deakin, J.F.; Elliott, R. The effect of acute citalopram on face emotion processing in remitted depression: A pharmacoMRI study. *Eur. Neuropsychopharmacol. J. Eur. Coll. Neuropsychopharmacol.* **2011**, *21*, 140–148. [CrossRef]
60. Del-Ben, C.M.; Deakin, J.F.W.; McKie, S.; Delvai, N.A.; Williams, S.R.; Elliott, R.; Dolan, M.; Anderson, I.M. The Effect of Citalopram Pretreatment on Neuronal Responses to Neuropsychological Tasks in Normal Volunteers: An fMRI Study. *Neuropsychopharmacology* **2005**, *30*, 1724–1734. [CrossRef]
61. Klomp, A.; van Wingen, G.A.; de Ruiter, M.B.; Caan, M.W.A.; Denys, D.; Reneman, L. Test–retest reliability of task-related pharmacological MRI with a single-dose oral citalopram challenge. *NeuroImage* **2013**, *75*, 108–116. [CrossRef] [PubMed]
62. Yi, L.Z.; He, J.; Liang, Y.Z.; Yuan, D.L.; Chau, F.T. Plasma fatty acid metabolic profiling and biomarkers of type 2 diabetes mellitus based on GC/MS and PLS-LDA. *FEBS Lett.* **2006**, *580*, 6837–6845. [CrossRef] [PubMed]
63. Corregiari, F.M.; Gattaz, W.F.; Bernik, M. Acute hormonal changes after IV citalopram and treatment response in OCD. *Psychopharmacology* **2007**, *193*, 487–494. [CrossRef] [PubMed]
64. Hyypp, M.T.; Kangasniemi, P. Variation of plasma free tryptophan and CSF 5-HIAA during migraine. *Headache* **1977**, *17*, 25–27. [CrossRef] [PubMed]
65. Wurtman, R.J.; Rose, C.M.; Chou, C.; Larin, F.F. Daily rhythms in the concentrations of various amino acids in human plasma. *N. Engl. J. Med.* **1968**, *279*, 171–175. [CrossRef] [PubMed]
66. Guo, S.; Vecsei, L.; Ashina, M. The L-kynurenine signalling pathway in trigeminal pain processing: A potential therapeutic target in migraine? *Cephalalgia* **2011**, *31*, 1029–1038. [CrossRef]
67. Nagy-Grocz, G.; Laborc, K.F.; Veres, G.; Bajtai, A.; Bohar, Z.; Zadori, D.; Fejes-Szabo, A.; Spekker, E.; Vecsei, L.; Pardutz, A. The Effect of Systemic Nitroglycerin Administration on the Kynurenine Pathway in the Rat. *Front. Neurol.* **2017**, *8*, 278. [CrossRef]
68. Borsook, D.; Maleki, N.; Becerra, L.; McEwen, B. Understanding migraine through the lens of maladaptive stress responses: A model disease of allostatic load. *Neuron* **2012**, *73*, 219–234. [CrossRef]
69. Maleki, N.; Becerra, L.; Borsook, D. Migraine: Maladaptive brain responses to stress. *Headache* **2012**, *52* (Suppl. 2), 102–106. [CrossRef] [PubMed]
70. Miura, H.; Ozaki, N.; Sawada, M.; Isobe, K.; Ohta, T.; Nagatsu, T. A link between stress and depression: Shifts in the balance between the kynurenine and serotonin pathways of tryptophan metabolism and the etiology and pathophysiology of depression. *Stress* **2008**, *11*, 198–209. [CrossRef]
71. Ruddick, J.P.; Evans, A.K.; Nutt, D.J.; Lightman, S.L.; Rook, G.A.; Lowry, C.A. Tryptophan metabolism in the central nervous system: Medical implications. *Expert Rev. Mol. Med.* **2006**, *8*, 1–27. [CrossRef] [PubMed]
72. Dunn, A.J.; Welch, J. Stress- and endotoxin-induced increases in brain tryptophan and serotonin metabolism depend on sympathetic nervous system activity. *J. Neurochem.* **1991**, *57*, 1615–1622. [CrossRef]
73. Fernstrom, J.D.; Fernstrom, M.H. Exercise, serum free tryptophan, and central fatigue. *J. Nutr.* **2006**, *136*, 553S–559S. [CrossRef] [PubMed]
74. Hoglund, E.; Overli, O.; Winberg, S. Tryptophan Metabolic Pathways and Brain Serotonergic Activity: A Comparative Review. *Front. Endocrinol.* **2019**, *10*, 158. [CrossRef] [PubMed]
75. Ara, I.; Bano, S. Citalopram decreases tryptophan 2,3-dioxygenase activity and brain 5-HT turnover in swim stressed rats. *Pharmacol. Rep.* **2012**, *64*, 558–566. [CrossRef]
76. Bano, S.; Gitay, M.; Ara, I.; Badawy, A. Acute effects of serotonergic antidepressants on tryptophan metabolism and corticosterone levels in rats. *Pak. J. Pharm. Sci.* **2010**, *23*, 266–272.
77. Herhaus, B.; Joisten, N.; Wessels, I.; Zimmer, P.; Petrowski, K. How acute physical and psychological stress differentially influence the kynurenine pathway: A randomized cross-over trial. *Psychoneuroendocrinology* **2021**, *134*, 105433. [CrossRef]
78. O'Farrell, K.; Harkin, A. Stress-related regulation of the kynurenine pathway: Relevance to neuropsychiatric and degenerative disorders. *Neuropharmacology* **2017**, *112*, 307–323. [CrossRef]
79. Nagy-Grocz, G.; Tar, L.; Bohar, Z.; Fejes-Szabo, A.; Laborc, K.F.; Spekker, E.; Vecsei, L.; Pardutz, A. The modulatory effect of anandamide on nitroglycerin-induced sensitization in the trigeminal system of the rat. *Cephalalgia* **2016**, *36*, 849–861. [CrossRef]
80. Vamos, E.; Fejes, A.; Koch, J.; Tajti, J.; Fulop, F.; Toldi, J.; Pardutz, A.; Vecsei, L. Kynurenate derivative attenuates the nitroglycerin-induced CamKIIalpha and CGRP expression changes. *Headache* **2010**, *50*, 834–843. [CrossRef]
81. Levine, E.S.; Jacobs, B.L. Neurochemical afferents controlling the activity of serotonergic neurons in the dorsal raphe nucleus: Microiontophoretic studies in the awake cat. *J. Neurosci.* **1992**, *12*, 4037–4044. [CrossRef]
82. Jiang, M.; Behbehani, M.M. Physiological characteristics of the projection pathway from the medial preoptic to the nucleus raphe magnus of the rat and its modulation by the periaqueductal gray. *Pain* **2001**, *94*, 139–147. [CrossRef]
83. Conti, P.; D'Ovidio, C.; Conti, C.; Gallenga, C.E.; Lauritano, D.; Caraffa, A.; Kritas, S.K.; Ronconi, G. Progression in migraine: Role of mast cells and pro-inflammatory and anti-inflammatory cytokines. *Eur. J. Pharmacol.* **2019**, *844*, 87–94. [CrossRef] [PubMed]
84. Fidan, I.; Yuksel, S.; Ymir, T.; Irkec, C.; Aksakal, F.N. The importance of cytokines, chemokines and nitric oxide in pathophysiology of migraine. *J. Neuroimmunol.* **2006**, *171*, 184–188. [CrossRef] [PubMed]
85. Kogelman, L.J.A.; Falkenberg, K.; Buil, A.; Erola, P.; Courraud, J.; Laursen, S.S.; Michoel, T.; Olesen, J.; Hansen, T.F. Changes in the gene expression profile during spontaneous migraine attacks. *Sci. Rep.* **2021**, *11*, 8294. [CrossRef]
86. Sarchielli, P.; Alberti, A.; Vaianella, L.; Pierguidi, L.; Floridi, A.; Mazzotta, G.; Floridi, A.; Gallai, V. Chemokine levels in the jugular venous blood of migraine without aura patients during attacks. *Headache* **2004**, *44*, 961–968. [CrossRef]

87. Domingues, R.B.; Duarte, H.; Senne, C.; Bruniera, G.; Brunale, F.; Rocha, N.P.; Teixeira, A.L. Serum levels of adiponectin, CCL3/MIP-1alpha, and CCL5/RANTES discriminate migraine from tension-type headache patients. *Arq. Neuropsiquiatr.* **2016**, *74*, 626–631. [CrossRef]
88. Humphrey, P.P. The discovery of a new drug class for the acute treatment of migraine. *Headache* **2007**, *47* (Suppl. 1), S10–S19. [CrossRef]
89. Mikolajczyk, T.P.; Nosalski, R.; Szczepaniak, P.; Budzyn, K.; Osmenda, G.; Skiba, D.; Sagan, A.; Wu, J.; Vinh, A.; Marvar, P.J.; et al. Role of chemokine RANTES in the regulation of perivascular inflammation, T-cell accumulation, and vascular dysfunction in hypertension. *FASEB J.* **2016**, *30*, 1987–1999. [CrossRef]
90. Vernieri, F.; Moro, L.; Altamura, C.; Palazzo, P.; Antonelli Incalzi, R.; Rossini, P.M.; Pedone, C. Patients with migraine with aura have increased flow mediated dilation. *BMC Neurol.* **2010**, *10*, 18. [CrossRef]
91. Larsen, J.S.; Skaug, E.A.; Wisloff, U.; Ellingsen, O.; Stovner, L.J.; Linde, M.; Hagen, K. Migraine and endothelial function: The HUNT3 Study. *Cephalalgia* **2016**, *36*, 1341–1349. [CrossRef] [PubMed]
92. Hautakangas, H.; Winsvold, B.S.; Ruotsalainen, S.E.; Bjornsdottir, G.; Harder, A.V.E.; Kogelman, L.J.A.; Thomas, L.F.; Noordam, R.; Benner, C.; Gormley, P.; et al. Genome-wide analysis of 102,084 migraine cases identifies 123 risk loci and subtype-specific risk alleles. *Nat. Genet.* **2022**, *54*, 152–160. [CrossRef] [PubMed]
93. Singh, S.; Bruder-Nascimento, A.; Belin de Chantemele, E.J.; Bruder-Nascimento, T. CCR5 antagonist treatment inhibits vascular injury by regulating NADPH oxidase 1. *Biochem. Pharmacol.* **2022**, *195*, 114859. [CrossRef] [PubMed]

cells MDPI

Review

ATP-Sensitive Potassium Channels in Migraine: Translational Findings and Therapeutic Potential

Amalie Clement [1], Song Guo [1,2], Inger Jansen-Olesen [1] and Sarah Louise Christensen [1,*]

1 Glostrup Research Institute, Department of Neurology, Danish Headache Center, Copenhagen University Hospital—Rigshospitalet, Nordstjernevej 42, Glostrup, 2600 Copenhagen, Denmark
2 Department of Odontology, Panum Institute, Faculty of Health, University of Copenhagen, 2300 Copenhagen, Denmark
* Correspondence: schr0227@regionh.dk

Abstract: Globally, migraine is a leading cause of disability with a huge impact on both the work and private life of affected persons. To overcome the societal migraine burden, better treatment options are needed. Increasing evidence suggests that ATP-sensitive potassium (K_{ATP}) channels are involved in migraine pathophysiology. These channels are essential both in blood glucose regulation and cardiovascular homeostasis. Experimental infusion of the K_{ATP} channel opener levcromakalim to healthy volunteers and migraine patients induced headache and migraine attacks in 82–100% of participants. Thus, this is the most potent trigger of headache and migraine identified to date. Levcromakalim likely induces migraine via dilation of cranial arteries. However, other neuronal mechanisms are also proposed. Here, basic K_{ATP} channel distribution, physiology, and pharmacology are reviewed followed by thorough review of clinical and preclinical research on K_{ATP} channel involvement in migraine. K_{ATP} channel opening and blocking have been studied in a range of preclinical migraine models and, within recent years, strong evidence on the importance of their opening in migraine has been provided from human studies. Despite major advances, translational difficulties exist regarding the possible anti-migraine efficacy of K_{ATP} channel blockage. These are due to significant species differences in the potency and specificity of pharmacological tools targeting the various K_{ATP} channel subtypes.

Keywords: K_{ATP} channels; provoked migraine; SUR; Kir6.x; levcromakalim; glibenclamide; human migraine model; in vivo models; migraine

Citation: Clement, A.; Guo, S.; Jansen-Olesen, I.; Christensen, S.L. ATP-Sensitive Potassium Channels in Migraine: Translational Findings and Therapeutic Potential. *Cells* **2022**, *11*, 2406. https://doi.org/10.3390/cells11152406

Academic Editor: Thierry Coppola

Received: 12 July 2022
Accepted: 30 July 2022
Published: 4 August 2022

1. Introduction

According to the World Health Organization (WHO), more than a billion people are living with migraine, and among the 15–49 year-old population, headache disorders is the most burdensome of all disorders [1,2]. Migraine attacks are characterized by pulsating head pain of moderate to severe intensity, photo- and/or phonophobia, nausea, vomiting, and aggravation by routine physical activity [3]. Migraine has a tremendous impact on quality of life for sufferers and may affect sleep [4], cognitive function [5], and private and professional life [6]. Despite huge individual suffering and socioeconomic impact, the pathophysiological mechanisms of migraine remain incompletely understood and highly debated [7]. The brain is generally thought of as non-nociceptive, but plexuses of nociceptive nerve fibers from the trigeminal ganglion innervate the blood vessels of the meninges (dura, pia, and arachnoid mater), linking pain perception and the brain vascular system in what is described as the trigeminovascular system [8]. Nowadays, many think of migraine as a neurovascular [7] sensory threshold disease [9]. The identification of calcitonin gene-related peptide (CGRP) involvement in migraine is a translational success story culminating in the marketing of monoclonal antibodies targeting CGRP or its receptor as well as small molecule receptor antagonists [10]. However, these CGRP-targeting migraine preventatives are only effective in approximately 60% of patients [11–15], stressing the importance of continued research and drug development.

A range of migraine-provoking substances have been identified in human experiments. Common for all, is the dilation of cephalic arteries [16–21] via downstream opening of vascular smooth muscle ATP-sensitive potassium (K_{ATP}) channels [7]. The finding that the K_{ATP} channel opener levcromakalim is the most potent trigger of experimental migraine tested to date [22,23] has fueled interest in K_{ATP} channel involvement in migraine pain generation and, within recent years, a significant number of studies have addressed this topic.

The aim of this review is to collectively present the evidence on K_{ATP} channel involvement in migraine pain and review the underlying hypotheses of where and how the K_{ATP} channels are involved in migraine pathophysiology. Logically, the possibility of targeting antimigraine therapeutics against these channels will also be discussed.

2. Molecular Basis and Physiological Function of K_{ATP} Channels

2.1. Molecular Structure and Regulation of Channel Activity

The K_{ATP} channels were first identified in cardiac muscle cells by A. Noma in the early 1980s [24]. These channels were later shown in other tissues, such as pancreas, smooth muscle cells, and the nervous system [25–27]. They belong to the family of transmembrane potassium inward-rectifying (Kir) channels, which are predominantly found on the plasma membranes but are also present on the mitochondrial inner membrane [28]. Seven subfamilies within the Kir family have been identified with different molecular and physiological functions (Kir1.x through to Kir7.x), where ATP-sensitive K^+ channels belong to the Kir6.x subfamily and are strongly associated with cellular metabolism and membrane electrophysiology [27]. Kir6.x have two subtypes, namely Kir6.1 and Kir6.2, which are expressed in various tissues [29].

Kir channels have two transmembrane spanning regions (TM1 and TM2) with an extracellular pore-forming region (H5) and both the amino and carboxyl terminal are cytosolic (Figure 1A) [29,30]. However, to obtain a functional channel, four Kir subunits are necessary, and the activity of the channel is regulated by four sulfonylurea receptors (SUR), thus creating a hetero-octameric structure (Figure 1B) [29,31]. These SUR receptors are ATP-binding cassettes (ABCs) or transport ATPases and have 17 transmembrane regions arranged into three domains (TMD0, TMD1 and TMD2) together with two intracellular nucleotide binding domains (NBD1 and NBD2); see Figure 1A. SUR subunits SUR2A and SUR2B only differ at the carboxyl terminal 42 amino acids (C42), while the SUR1 subunit is more unique [27,29,30].

The inward-rectifying function is a result of an intracellular blockage of the pore by Mg^{2+} or polyamines, which blocks the efflux of K^+. During channel activation the blockage is removed and K^+ efflux can occur [27]. High concentrations of ATP will inhibit the channel, while reduced ATP levels will activate and open the channel [32]. The activity of the channel is controlled by the SUR subunits due to their NBDs, where MgATP binds to NBD2 and MgADP binds to NBD1 [29,33]. Phosphatidylinositol 4,5-bisphosphate (PIP_2) is suggested to play a role in channel activity regulation, as PIP_2 activates the channel and reduces its sensitivity to ATP, thus counteracting the inhibitory effect at ATP [34–36]. Lastly, Kir6.x channels can be activated via phosphorylation by protein kinase A (PKA) or protein kinase G (PKG) [37–40]. These kinases are downstream targets of cAMP and cGMP, respectively, and have been suggested as molecular pathways in migraine pathophysiology (Figure 2) [41,42].

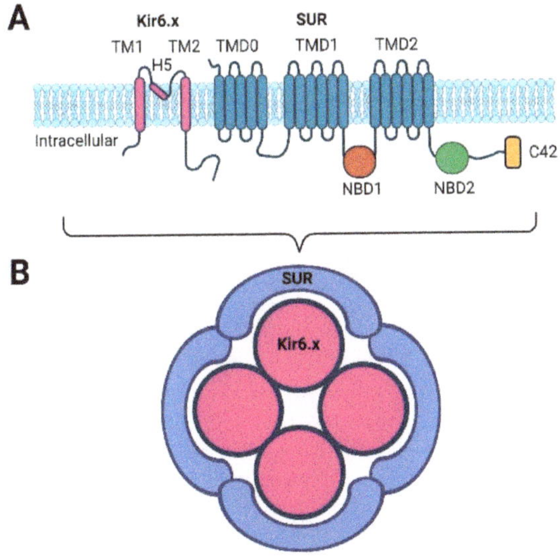

Figure 1. Simple structure of the K_{ATP} channel. (**A**) The Kir6.x subunit is composed of a two transmembrane region (TM1 and TM2) connected by a pore-forming region (H5). The SURx subunit is composed of three domains of either five transmembrane regions (TMD0) or six transmembrane regions (TMD1 and TMD2). The nucleotide binding domains are found intracellularly (NBD1 and NBD2). SUR2A and SUR2B only differ in their C-terminal end (C42). (**B**) The functional K_{ATP} channel is formed by four Kir6.x subunits and four SURx subunits (created using BioRender.com).

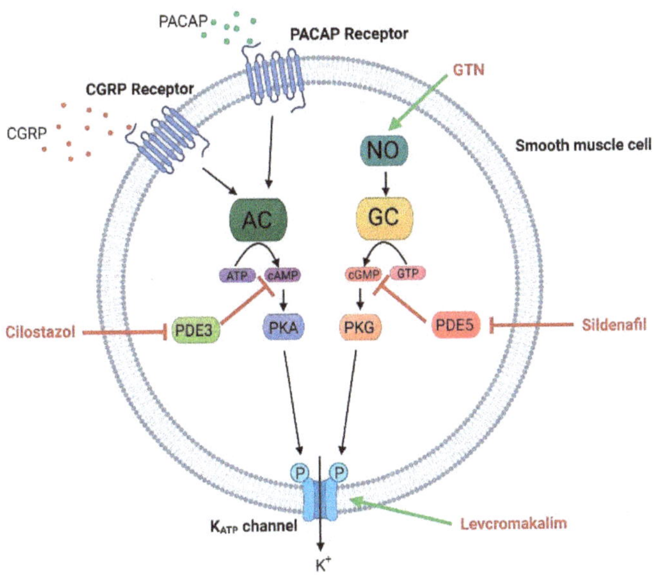

Figure 2. Molecular pathways and pharmacological agents leading to the opening of the K_{ATP} channel in vascular smooth muscle. The neuropeptides PACAP and CGRP activate K_{ATP} channels via the adenylyl cyclase pathway, while the NO donor GTN (glyceryl trinitrate) activates the channel via the guanylyl cyclase pathway. Cilostazol and Sildenafil are blockers of the phosphodiesterase 3 and 5 (PDE3 and PDE5), respectively, causing accumulation of cAMP and cGMP, which promote the opening of K_{ATP} channels. Levcromakalim causes vasodilation by direct action on the K_{ATP} channels (created using BioRender.com).

2.2. Tissue Distribution

The K_{ATP} channels are expressed throughout the body but the combination of the different subunits of Kir6.x and SURx vary in different tissues, such as the vascular system, neuronal system, and pancreas (see Table 1). The pancreatic β-cells express Kir6.2/SUR1, which control the glucose-stimulated insulin secretion (and represents the most studied channel), while the Kir6.2/SUR2A channels are the predominant form found in myocardia [25,29,33]. The vascular smooth muscle cells express Kir6.1/SUR2B and these have distinct structural features from the pancreatic Kir6.2/SUR1 isoforms, as the Kir6.1 cytoplasmic regions is placed too far from the membrane to interfere with the membrane-bound PIP_2, which is known to activate or open the Kir6.2/SUR1 channels in pancreatic β-cells [33]. Furthermore, Kir6.1 channels do not show spontaneous channel activity, while pancreatic and myocardial Kir6.2 channels open spontaneously when ATP levels are low or absent [33,38,43]. In most tissue, channels are composed of two homogenous Kir subunits and four homogenous SUR subunits; however, examples of more heterogenous compositions have been reported [44]. The different compositions of K_{ATP} channel subunits in different tissues potentially allow for more specific therapeutic targets in the development of novel drug candidates for specific pathologies.

Kir6.1/SUR2B is found in the smooth muscle cells of the vascular system, and are the dominant form in brain arteries and dura mater [45], where they are involved in vasodilation and constriction. For this reason, Kir6.1/SUR2B have been suggested as a target for migraine pain intervention [46,47].

Table 1. Subunits composition and tissue expression of K_{ATP} channels. For a more detailed overview of subunit composition, tissue distribution and physiological function, please see [48].

Channel Subunit Composition	Tissue	References
Kir6.1/SUR1	Retina	[49]
	Nervous system	[46,48]
Kir6.1/SUR2B	Vascular smooth muscle	[43,45,50–52]
	Non-vascular smooth muscle	[48,53]
	Conduction system of the heart	[48,54]
Kir6.2/SUR1	Pancreatic β-cells	[52,55]
	Arterial cardiac myocytes	[52,56]
	Nervous system	[48,52,57,58]
	Skeletal muscle	[48,59]
Kir6.2/SUR2A	Ventricular myocytes	[54,60]
	Skeletal muscle	[48,59]
Kir6.2/SUR2B	Non-vascular smooth muscle	[53]
	Nervous system	[48,57,61]
	Conduction system of the heart	[54,62]
	Skeletal muscle	[59]

2.3. Physiological Functions of K_{ATP} Channels

In a physiological resting state, K_{ATP} channels are blocked but allow a small inward current of K^+, while the active or open state of the channel results in the efflux of K^+, resulting in hyperpolarization of the membrane. Below, the physiological consequence of this in different cell and tissue types is presented.

2.3.1. Vascular System

The tone of the vascular system is controlled by a sophisticated relationship of molecular functions causing vasoconstriction or vasodilation. K_{ATP} channels have long been known as the target of vasodilatory drugs like diazoxide and pinacidil [63] and their role in vasodilation have likewise been studied for decades [33,38,63–65].

In vascular smooth muscle cells, hyperpolarization caused by K^+ efflux upon K_{ATP} channel opening will cause the inhibition of voltage-operated Ca^{2+} channels (VOCC), reducing Ca^{2+} influx and consequently causing smooth muscle relaxation and vasodilation (Figure 2) [33,66–68]. Many vasodilating substances target receptors or second messengers upstream from K_{ATP} channels. Nitric oxide (NO) binds to guanylyl cyclase (GC), which in turn converts GTP to cGMP, and cGMP can subsequently phosphorylate and open the K_{ATP} channels [68–70]. Additionally, K_{ATP} channels, located in the endothelium, mediate vasodilation to some extent [71]. The potent vasodilators CGRP and pituitary adenylate cyclase-activating peptide (PACAP) bind to their respective G-protein coupled receptors on vascular smooth muscle to activate the adenylyl cyclase (AC) enzyme, causing cAMP to be converted from ATP [40,72,73]. In addition, inhibitors of phosphodiesterase (PDE) type 3 and 5 (cilostazol and sildenafil) are marketed for their vasodilating effect for different indications [74,75]. PDEs degrade cAMP and cGMP; thus, the inhibition of PDEs causes the accumulation of cAMP and/or cGMP and downstream phosphorylation of the K_{ATP} channel [7]. Interestingly, all these drugs or substances have headache as a primary side effect [16,17,74–76]. Thus, the above mentioned mechanisms have been speculated to be important in migraine pathophysiology.

Genetic manipulations of specific subunits of the functional K_{ATP} channel may result in vascular issues like knockout of the *Kcnj8*, the gene that encodes Kir6.1, which was shown to cause sudden early death associated with an atrioventricular blockage that could not be rescued by the K_{ATP} channel opener pinacidil but was worsened by the vasoconstrictive agent methylergometrine [77]. A later in vivo study showed that the selective deletion of Kir6.1 in vascular smooth muscle cells resulted in hypertension and a loss of response to pinacidil but did not cause sudden death [64]. Thus, sudden death was likely related to the global deletion of Kir6.1. It is important to keep in mind that global knockout of specific genes might cause unexpected phenotypic traits due to a lack of expression of the gene during embryonic development and consequences thereof, which might not be expected if the gene had been expressed in the embryonic stages and silenced later in life [78–80].

2.3.2. Neuronal Function

K_{ATP} channels (Kir6.2/SUR1, Kir6.2/SUR2A, and Kir6.2/SUR2B) are also expressed in the brain, especially in neurons where their activation causes hyperpolarization and reduced excitability [27,29,30], which is often related to a reduction in neurotransmitter release [81,82]. Hyperpolarization may lead to the activation of hyperpolarization-activated cyclic nucleotide-gated (HCN) channels of which the consequence is dependent on the given situation [83]. The excitatory neurotransmitter glutamate will cause an influx of Ca^{2+}, which can result in an elevation in mitochondrial Ca^{2+} levels, depolarization of the mitochondria, loss of oxidative phosphorylation, ATP depletion, swelling and rupture of mitochondrial membrane and, subsequently, release of pro-apoptotic species, ultimately leading to neuronal cell death [81,84]. This cascade can be blocked or regulated by activation of the mitochondrial K_{ATP} channels due to their ability to cause hyperpolarization and reduced glutamate release, thus playing a role in neuronal protection from excitatory toxicity [81]. Moreover, neuronal mitochondrial K_{ATP} channels regulate intracellular Ca^{2+} concentrations during hypoxia, thus protecting the neuron from hypoxia [84]. The K_{ATP} channel opener, iptakalim, was used to rescue stress-induced mitochondrial damage and alleviate a depressive-like phenotype in the rodent chronic mild stress model of depression [85]. Altogether, activating K_{ATP} channels in neurons most likely serve a neuroprotective function during stressful stimuli like ischemia and oxidative stress [48,57,86].

2.3.3. Analgesia, Antinociception, and Opioid Signaling

K_{ATP} channels are, furthermore, involved in opioid signaling, which appears to be selective to these types of channels as no other K^+ channels have been illustrated to bear similar properties [87]. Nevertheless, other K^+ channels are implicated in pain perception but are beyond the scope of this review. For an extensive review on K^+ channels in the pathophysiology of pain, please see [88].

The K_{ATP} channels have been shown to be downstream targets of the opioid receptor via regulation of nitric oxide synthase and NO generation and subsequent efflux of K^+ and hyperpolarization [89]. This NO/cGMP/K_{ATP} channel pathway has further been suggested to be an important factor in the inflammatory nociceptive response [70,90,91]. Additionally, loss of K_{ATP} channels or their function in peripheral neurons has been implicated in pain perception in multiple studies [57,58,92,93]. For instance, expression of Kir6.1, SUR1, and SUR2 in rat spinal cord were downregulated after nerve injury [94] and knockout of SUR1 subtype resulted in the loss of the antinociceptive effect of morphine in mice [95,96]. Furthermore, the expression of Kir6.1/SUR2B were shown to be regulated by the inflammatory toll-like receptor 4 and NF-κB-dependent signaling, which was suggested to be a factor in the poor vasoconstriction during sepsis [97]. Furthermore, the K_{ATP} channel opener, cromakalim administered centrally, reduced astrocyte activation and lowered expression of IL-1β and TNF-α, thereby reducing neuroinflammation [98], and relieved injury-induced neuropathic pain both acutely (lasting hours) and chronically (lasting days) [94]. Khanna et al., 2011 showed the analgesic effect of the systemic administration of cromakalim in the formalin test to the same degree as morphine but only high doses of cromakalim (1 mg/kg and 5 mg/kg) induced an analgesic effect in the tail flick test [87]. Combining morphine and cromakalim only showed additive analgesic effects in the formalin test (inflammatory pain) and not in the tail flick test (heat sensitivity) [87]. Likewise, the K_{ATP} channel blocker, glibenclamide, increased the pain response to the formalin test but not in the tail flick test [87]. Interestingly, central administration of cromakalim did not induce an analgesic effect in the tail flick test [98]. One could speculate that the reason for this is the lack of inflammatory agents in the tail flick test, which would suggest that the antinociceptive effect of K_{ATP} channel openers is related to anti-inflammatory properties, resulting in analgesia in inflammatory pain rather than neuropathic pain.

Overall, the involvement of K_{ATP} channels in analgesic or antinociceptive mechanisms appear to be closely related to their ability to hyperpolarize the membrane and reduce hyperexcitability induced by inflammatory events. However, even though K_{ATP} channels in pain research have been studied for decades, no therapeutic agent has reached the market, and we speculate that the reason for this lack of drug development may be due to the high complexity of these mechanisms, the abundant distribution of K_{ATP} channels throughout the body and lack of subtype specific pharmacological tools.

Interestingly, when levcromakalim (K_{ATP} channel opener) was systemically administered, humans developed headache or migraine [22,23] and mice became hypersensitive [45,47,99]. When levcromakalim was centrally administered in mice, analgesic effects became evident [45,100]; see Section 4. (K_{ATP} Channels and Headache).

2.3.4. Insulin Secretion and Glucose Metabolism

The pancreatic islet β-cells are responsible for the secretion of insulin, and Kir6.2/SUR1 K_{ATP} channels play an important role in this mechanism. When blood glucose levels rise, cell metabolism increases, and the higher ATP levels will trigger the closing of K_{ATP} channels, resulting in membrane depolarization and the activation of voltage-gated Ca^{2+} channels and Ca^{2+} influx, which ultimately leads to insulin secretion [101,102]. It is generally accepted that the loss of function of K_{ATP} channels in islet β-cells leads to poor or no response of K_{ATP} channels to changes in ATP/ADP ratio, leaving the channels closed and the membrane at a depolarized state. This leads to a rise in cytosolic Ca^{2+} and the secretion of insulin, causing the phenotype of hyperinsulinism [103,104]. In contrast, gain-of-function mutations can lead to reduced or absent secretion of insulin, hyperglycemia, and diabetes due to a high degree of K^+ efflux and membrane hyperpolarization [102]. Congenital hyperinsulinism is linked to the loss of function of the Kir6.2/SUR1 K_{ATP} channel expression in the β-cells [101,104].

Type 2 diabetes mellitus is linked to chronic low-grade systemic inflammation and oxidative stress in β-cell, leading to low or no insulin secretion due to a loss of β-cells [105,106]. Several studies have reported that reactive oxygen species (ROS) and reactive nitrogen species (RNS) modulate K_{ATP} channel activity by inhibition of mitochondrial ATP production [107–111]. One study illustrated that oxidative stress-induced loss of β-cells caused high blood glucose levels in wild-type (WT) mice, while in $SUR1^{-/-}$ mice, glucose levels only rose slightly over control levels and these mice had a significantly better survival rate compared to WT mice [107]. As the SUR subunit of the K_{ATP} channel holds the nucleotide binding sites, this illustrates that the mitochondrial ATP production is involved in the negative effects of oxidative stress on the insulin secretion pathway. One could speculate that oxidative stress in other tissues also disturbs the K_{ATP} channel function, for instance, in the migraine-relevant trigeminovascular system [112,113].

3. Pharmacological Tools Targeting K_{ATP} Channels

Pharmacological investigation using the different and more or less selective openers and inhibitors of the K_{ATP} channel is crucial for the understanding of channel function and interpretation of results. Both categories include several chemical classes. In summary, there is some selectivity of the K_{ATP} channel openers and blockers, but one should keep in mind that selectivity is commonly not an all or nothing phenomena but a function of dose.

3.1. K_{ATP} Channel Openers

Levcromakalim or cromakalim belongs to the benzopyran class and primarily opens smooth muscle and cardiomyocyte K_{ATP} channels via affinity to TMD2 on SUR2 [48]. PK_i values for levcromakalim is 6.37 ± 0.04 on SUR2A and 6.95 ± 0.03 on SUR2B [114]. In vasomotor experiments, levcromakalim has a reported pEC_{50} of 6.36 ± 0.09 on human pial arteries and 6.32 ± 0.3 on omental arteries [115]. In rats, similar studies found values of 6.32 ± 0.09 and 5.46 ± 0.17 for levcromakalim on basilar and middle cerebral arteries, respectively [116], and 7.14 ± 0.11 on middle meningeal arteries [117].

The cyanoguanidines (pinacidil, P-1075) also display selectivity to SUR2 and have potent hypotensive effects [118,119], but like levcromakalim, pinacidil did not reverse glibenclamide-induced SUR1-mediated hyperglycemia [120].

In contrast, diazoxide (benzothiadiazine) is more specific for SUR1, but also activates SUR2B [121]. Accordingly, it displayed both hypotensive and hyperglycemic effects [120,122,123].

3.2. K_{ATP} Channel Blockers

The sulfonylureas (glibenclamide, glicazide, and gliquidone) are compounds that target the SUR subunits to inhibit K_{ATP} channels via NBD2 (Figure 1A) [124]. Sulfonylureas are used clinically to promote insulin secretion in type 2 diabetes [125]. Glibenclamide has a 50-fold higher affinity towards SUR1 over SUR2A and SUR2B [126], and the inhibition of Kir6.2/SUR1 is poorly reversed, whereas the blocking of Kir6.2/SUR2A is rapidly reversed [127]. In addition, gliquidone and glicazide display lower IC_{50} values in pancreatic cells than in cardiomyocytes and vascular smooth muscle, suggesting a higher affinity for SUR1 [128]. Species differences are reported on the activity of glibenclamide on the human and mouse Kir6.1/SUR1 channels. pIC_{50} is 8.37 on the human and 5.7 on the mouse channel [126,129] suggest a much larger potency of glibenclamide on human compared to mouse pancreatic K_{ATP} channels.

The thiazolidinediones (rosiglitazone, pioglitazone, etc.) are another class of drugs used clinically for the treatment of type 2 diabetes via their effect on peroxisome proliferator-activated receptors (PPARs) [125]. However, patch clamp experiments on HEK cells expressing various subtypes of K_{ATP} channels have shown that these compounds also block vascular K_{ATP} channels to various degrees [130]. Rosiglitazone was found to inhibit all isoforms of K_{ATP} channels. The IC_{50} was 10 μM for the Kir6.1/SUR2B channel and ~45 μM for KIR6.2/SURx channels. The inhibition was also present without the SUR subunit. Additionally, rosiglitazone had no effect on Kir1.1, Kir2.1, and Kir4.1 channels, suggesting that the channel inhibitory effect is selective for Kir6.x channels [131]. In a subsequent study, the same group compared a large number of PPAR agonists and found some to potently inhibit Kir6.1/SUR2B. The most potent agonists were AS-252424, englitazone, A6355, rosiglitazone, and cay10415 with IC_{50} values of 4 μM, 7 μM, 8 μM, 12 μM, and 15 μM, respectively. Lastly, the morpholinoguanidine PNU-37883A needs mentioning. Both in vitro and in vivo evidence suggest that this compound is selective for vascular K_{ATP} channels over pancreatic ones and that it acts on the Kir6.1 rather than SUR components [132].

4. K_{ATP} Channels and Headache

Clinical trials with K_{ATP} channel openers for the treatment of hypertension and asthma had headache as a primary side effect [63,133,134]. A range of migraine- and headache-triggering substances (NO, CGRP, PACAP, cilostazol, sildenafil, and more) activate K_{ATP} channels downstream from target binding (Figure 2). This agrees with the theory that the arteries of the trigeminovascular system, are involved in the generation of migraine pain [7].

4.1. *Levcromakalim Is a Potent Trigger of Experimental Headache and Migraine*

In an experimental setting, recognized as the human model of migraine [135], intravenous infusion of levcromakalim (1 mg over 20 min) was studied in healthy volunteers [136], including migraine without aura (MO) patients [22] and migraine with aura (MA) patients [23].

After levcromakalim infusion, 12 of 14 healthy volunteers reported a headache with a median time to onset of 30 min (range 10–60 min) compared to 1 of 6 participants after placebo [136]. In MO patients, migraine attacks without aura were induced in 16 out of 16 patients contra 1 out of 16 participants after placebo [22]. The median time to migraine onset was 3 h (range 1-9 h) after levcromakalim infusion. In MA patients, attacks were induced in 14 out of 17 participants, and 1 participant after placebo. Four attacks were MO whereas ten were MA. Median time of onset for MO was 2.8 h (range 1–4 h) and 44 min (range 20–120 min) for MA [23].

Apart from headache characteristics, hemodynamic parameters and circumference or blood flow velocity of selected arteries were reported in the above-mentioned clinical studies. In healthy participants, the middle meningeal artery (MMA) had a 7–22% larger circumference throughout the 5 h test period measured with 3.0 Tesla (magnetic resonance angiography (MRA)). The superficial temporal artery (STA) also dilated, but less robustly throughout the test period. The middle cerebral artery (MCA) dilated but it was not significantly different from placebo. Heart rate (HR) AUC_{0-290} was significantly increased but mean arterial blood pressure (MAP) AUC_{0-290} was not significantly lowered [136]. However, in a larger study, all arteries (STA, MMA, and MCA) dilated, and HR and MAP also changed significantly in response to levcromakalim [137]. In MO and MA patients both HR AUC_{0-120} and MAP AUC_{0-120} were also significantly altered. In the MO patient study [22], STA and radial artery diameters were measured by ultrasonography and blood flow velocity of the MCA was measured by transcranial doppler as a proxy for arterial circumference. Thus, this method is inferior to MRA. Only the STA was found to dilate in response to levcromakalim. Effects on arterial dilation was not published in the MA study [23].

4.2. *K_{ATP} Channel Opening in Preclinical Migraine Models*

Preclinical models in which the effect of K_{ATP} channel opening and inhibition have been studied in the context of migraine include models of vasoactivity, CGRP release, mast cells degranulation and behavior in rodents.

4.2.1. Dilatory Effects on Cranial Arteries

The effect of K_{ATP} channel opening on cranial arteries have been studied both ex vivo using the wire myograph technique and in vivo using intravital microscopy through a closed cranial window in anesthetized rats. The latter allows simultaneous imaging of dural and pial arteries in an intact animal. It was found that levcromakalim infusion (0.1 mg/kg i.v.) increased dural artery diameter by 130 ± 24%, pial artery diameter by 18 ± 3%, and lowered MAP by 31% [138]. For pinacidil (0.38 mg/kg i.v.), the figures were 126 ± 8% and 17 ± 3%, respectively. The response was significantly lower in pial than in dural arteries for both K_{ATP} channel openers [138]. A lower dose of levcromakalim (0.025 mg/kg i.v.) sub-maximally dilated the MMA and decreased MAP by 29% [117]. In ex vivo artery preparations, the picture was similar. Levcromakalim (3 μM) induced relaxation of 74 ± 9% in rat dural arteries and 38 ± 8% in middle cerebral arteries, and pinacidil (3 μM) induced relaxation that amounted to 55 ± 11% and 26 ± 4%. Again, the dilatory responses were significantly different between the dural and cerebral arteries [138]. This is also reflected in the pEC_{50} values; see Section 3.1. (K_{ATP} Channel Openers), revealing an approximate 10-fold higher potency of levcromakalim on meningeal over cerebral arteries [117,139]. The in vivo findings could be explained by poor blood-brain barrier passage of levcromakalim and pinacidil, but this cannot explain the ex vivo difference as the blood-brain barrier is bypassed in the wire myograph technique. Therefore, it was suggested that K_{ATP} channels are heterogeneously distributed between cranial arteries [138].

4.2.2. Stimulation of CGRP Release

The CGRP release assay is an ex vivo technique in which CGRP release from isolated tissue can be investigated. In migraine research, CGRP release from relevant structures as the trigeminal ganglia, trigeminal nucleus caudalis and dura mater are commonly investigated [140]. Levcromakalim (1 μM) and diazoxide (10 μM) have been tested for their ability to stimulate CGRP release from all three tissues in rats [141]. Levcromakalim (0.1–100 μM) was tested in mouse trigeminal ganglia and brain stem [99]. In neither species nor tissue preparation did levcromakalim induce release of CGRP, supporting the hypothesis that K_{ATP} channel opening is a downstream event upon CGRP receptor activation within single cells [142]. For review of the effect of K_{ATP} channel blockage on CGRP release, see Section 4.3.1. (Effect of K_{ATP} Channel Blockers in Preclinical Models). From the CGRP release model, it is evident that the headache-inducing effect of levcromakalim is not caused by direct stimulation of neuronal CGRP release.

4.2.3. Mast Cell Degranulation

Dural mast cells may be involved in migraine attacks [143] and therefore mast cell degranulation assays and markers have been applied to study this possible aspect of headache mechanisms. Levcromakalim and diazoxide (10 μM) failed to degranulate rat dural mast cells in situ and, likewise, both drugs (0.01 μM–10 μM) failed to degranulate rat peritoneal mast cells in vitro [141]. Thus, degranulation of mast cells is not the primary mechanism in headache caused by levcromakalim.

4.2.4. In Vivo Mouse Model

Many migraine triggering substances defined in the human migraine model have also been used in mice where they induce a state of hypersensitivity to cutaneous stimulation with von Frey filaments [99,144–146]. This model is considered to be the mouse parallel to the human model of provoked migraine. Repeated (every 48 h) injections of levcromakalim (1 mg/kg i.p.) induce both cephalic and hind paw hypersensitivity to von Frey stimulation, peaking 2 h after the 3rd injection [45,47,99], whereas levcromakalim administered locally in the hind paw did not induce hypersensitivity, and intracerebroventricular administration provided analgesia on the hotplate [45]. The observed hypersensitivity is at odds with the study by Khanna et al., 2011 showing an antinociceptive effect of cromakalim and diazoxide delivered i.p. [87].

4.3. K_{ATP} Channel Blockage as Therapeutic Target in Migraine

The opening of K_{ATP} channels by systemic levcromakalim induces headache and migraine attacks with and without aura. Accordingly, blocking K_{ATP} channels may abort migraine attacks. A convincing amount of preclinical evidence suggests that K_{ATP} channel blockage is a promising drug target for migraine. However, translation to patients is pending better pharmacological tools. Table 2 summarizes the studies reviewed in the following sections and provides an overview of the doses of applied test substances expressed as µmol/kg and the ratio between blocker (glibenclamide) and headache trigger substance.

Table 2. Details of human and rodent studies on KATP channel blockage in different migraine models. Rows in same color are compared. The ratio of blocker/migraine trigger are used for rough assessment of effectiveness across models. Effective Y/N/P: Y = yes, N = no, P = partially. Percentwise changes of arterial circumference and diameter are the same. Thus, 20% change in diameter = 20% change in circumference. Dose mol/kg = (dose g/kg)/(MW g/mol), dose umol/kg = (dose mol/kg) × 10^6. Glibenclamide 494 g/mol, levcromakalim 286 g/mol, PACAP 4534 g/mol, CGRP 3798 g/mol, PNU 382 g/mol. * Glibenclamide given after PACAP, # CGRP is accumulated dose in man/bolus in rat. Ratio will increase if the 1 min dose of CGRP is applied. $ Possible first pass metabolism of levcromakalim i.p will increase the mouse ratio, due to a smaller denominator. & PACAP s.c. may result in lower plasma concentrations than i.v. which will increase the mouse ratio.

Species	Endpoint	Headache Trigger mg/kg	Headache Trigger, umol/kg	Blocker mg/kg	Blocker, umol/kg	Ratio (Blocker/Trigger)	Effective Y/N/P
Rat	MMA diameter	Levcromakalim 0.025 mg/kg iv over 10 min	0.087	PNU-37883A 0.5 mg/kg i.v. over 10 min	1.3	15	P
Rat	MMA diameter	Levcromakalim 0.1 mg/kg iv over 20 min	0.35	Glibenclamide 20 mg/kg iv over 20 min	40.5	116	P
Rat	MMA diameter	Levcromakalim 0.1 mg/kg iv over 20 min	0.35	Glibenclamide 30 mg/kg iv over 20 min	60.7	174	Y
Human	MMA, STA, MCA circumference	Levcromakalim 0.014 mg/kg iv over 20 min	0.049	Glibenclamide 0.14 mg/kg p.o.	0.3	5.8	N
Rat	MMA diameter	CGRP 0.3 ug/kg iv bolus	0.000079	Glibenclamide 7 mg/kg iv over 20 min	14.2	178,968	P
Rat	MMA diameter	CGRP 0.3 ug/kg iv bolus	0.000079	Glibenclamide 30 mg/kg iv over 20 min	60.7	767,004	Y
Human	STA and RA diameter	CGRP 0.43 ug/kg iv over 20 min	0.000011	Glibenclamide 0.14 mg/kg p.o.	0.3	24,972 #	N
Human	STA and RA diameter	CGRP 0.02 ug/kg/min i.v.	0.0000053	Glibenclamide 0.14 mg/kg p.o.	0.3	53,690	N
Human	MMA circumference	PACAP 200 picomol/kg over 20 min	0.2	Glibenclamide 0.14 mg/kg p.o. *	0.3	1.4	N
Human	Headache	Levcromakalim 0.014 mg/kg iv over 20 min	0.049	Glibenclamide 0.14 mg/kg p.o.	0.3	5.8	N/P
Mouse	Tactile hyper-sensitivity	Levcromakalim 1 mg/kg i.p $	3.5	Glibenclamide 1 mg/kg i.p.	2	0.6	Y
Human	Headache	PACAP 200 picomol/kg over 20 min	0.2	Glibenclamide 0.14 mg/kg p.o.	0.3	1.4	N
Mouse	Tactile hyper-sensitivity	PACAP 0.2 ug/kg s.c. &	0.000044	Glibenclamide 1 mg/kg i.p.	2	45,891	P
Human	Headache	CGRP 0.43 ug/kg iv over 20 min	0.000011	Glibenclamide 0.14 mg/kg p.o.	0.3	24,972	N/P

4.3.1. Effect of K_{ATP} Channel Blockers in Preclinical Models

The preclinical evidence suggesting the relevance of K_{ATP} channel blockage in migraine is based on evidence from studies of (a) cranial arteries, (b) CGRP release, (c) behavioral models, and (d) a genetically modified model:

(a) High dose glibenclamide (30 mg/kg i.v.) effectively blocked vasodilation in rats induced by levcromakalim and pinacidil in both dural and pial arteries in vivo [31,138]. Glibenclamide also inhibited dilation caused by migraine-triggering peptides CGRP [65] and PACAP [147] that support K_{ATP} channel activation by phosphorylation via cAMP and PKA [148]. PNU-37883A effectively inhibited dilatory responses to stimulation with K_{ATP} channel openers in various arteries of different species including the MMA in vitro and in vivo [117,132]. Interestingly, glibenclamide [65] and PNU-37883A [149] failed to inhibit arterial dilation caused by NO-donors in some reports whereas others did find a relationship between NO (cGMP) and K_{ATP} channel-mediated arterial dilation [45,150].

(b) Glibenclamide (3 μM) inhibited ex vivo capsaicin-induced CGRP release from trigeminal ganglia and dura mater from spontaneous trigeminal allodynic (STA) [151,152] rats via an unknown mechanism [47].

(c) Also, in the STA rat model of migraine, glibenclamide (1–10 mg/kg i.p.) and gliquidone (10–100 mg/kg i.p.) reversed spontaneous trigeminal allodynia [47]. In the mouse models of provoked migraine, glibenclamide (1 mg/kg i.p.) was highly effective against levcromakalim, cilostazol, and glyceryl-trinitrate (GTN)-induced tactile hypersensitivity [47,99], whereas it only partially blocked the effect of PACAP-38 [146].

(d) Mice lacking the Kir6.1 subunit in smooth muscle cells were less sensitive to CGRP [64], levcromakalim and GTN [45] induced vasodilation and hypersensitivity.

4.3.2. Clinical Effect of K_{ATP} Channel Inhibition in Human Migraine Models

Glibenclamide (10 mg p.o.) has been tested against levcromakalim [137,153], CGRP [154], and PACAP-38 [155] induced migraine or headache in healthy volunteers. Glibenclamide was given 2 h prior to levcromakalim and CGRP infusions, but after the PACAP infusion. Headache data and hemodynamic measures were obtained. In all studies, subjects continuously received glucose to counteract the pronounced drop in blood glucose caused by glibenclamide. Overall, glibenclamide was found ineffective against both hemodynamic changes and headache induction after infusion of all three migraine triggering compounds.

The three above-mentioned studies were all cross-over studies but with variations in experimental design. In the levcromakalim study, NCT03886922 [137,153,156], three study arms were included: placebo-placebo, glibenclamide-placebo, and glibenclamide-levcromakalim. The study did not have a placebo-levcromakalim group, making the conclusions a bit distorted. In total, 12/15 participants (80%) reported headache after glibenclamide-levcromakalim, 5/15 (33%) after glibenclamide-placebo, and 1/15 (7%) following placebo-placebo. Thus, glibenclamide itself did not induce headache at a rate significantly different from placebo. To test if glibenclamide protected against levcromakalim-induced headache, comparison was made to a previous study showing headache induction in 12/14 of participants (86%) after levcromakalim versus 1/6 (17%) after placebo [136]. Hence, glibenclamide pretreatment did not inhibit headache development but, noteworthily, the median time to headache onset was 30 min (range 10–60) after levcromakalim infusion without pretreatment and 180 min (range 20–600) with glibenclamide pretreatment (p = 0.007) [153]. Glibenclamide did not influence HR, MAP, nor the circumference of neither STA,, MMA, nor MCA, which, in this study, all significantly changed in response to levcromakalim [137].

The study on CGRP and glibenclamide included two experimental groups in a randomized cross-over design: placebo-CGRP and glibenclamide-CGRP. The incidence of headache on the placebo-CGRP day was 19/20 (95%) vs. 14/20 (70%) on the glibenclamide-CGRP day (p = 0.06). Biologically, this was a 25% reduction in headache inductions, but power was set to detect 50% reduction in the study; thus, we cannot with certainly say if this finding happened by chance. Glibenclamide clearly did not influence CGRP-mediated changes on the hemodynamic parameters arterial diameter, HR, MAP, and facial skin blood flow [154]. Similar findings were obtained with glibenclamide as posttreatment when headache was induced by PACAP-38 [155]. Here, 19/20 participants (95%) reported headache compared to 18/20 (90%) on the placebo-PACAP day ($p = 0.698$).

5. Discussion

Direct comparison between human and animal experiments is not straight forward [157]. Apart from the rule of thumb conversion factor for doses based on body surface area to account for a generally faster metabolism in smaller animals [158], several other factors may be of importance. For the studies reviewed here, the difficulties concern: (1) different routes of administration and lack of pharmacokinetic data to safely interpret their impact, (2) different measuring endpoints, (3) a lack of evidence on the exact K_{ATP} channel distribution and expression levels in different tissues and species, and (4) the potency of test compounds on different channel subtypes and receptors across species. In Table 2, these are mentioned with the possible effect it may have on the conclusions.

5.1. K_{ATP} Channel Opening Has Similar Effect in Preclinical and Clinical Studies

Across species, K_{ATP} channel openers dose-dependently dilate arteries and decrease blood pressure [48,137,138]. The headache- or migraine-inducing effect of levcromakalim infusion in humans was modeled in the mouse model of migraine, but with major differences that need mentioning. Humans received a single dose of 0.014 mg/kg i.v. to induce headache or migraine [22,23,136], whereas mice received one, two, or sometimes three i.p. injections of 1 mg/kg before hypersensitivity to tactile stimuli was evident [45,47,99]. Thus, translation is complicated by different routes of administration and different measuring endpoints, and no common readout to which the other measures can be related to. Some degree of first pass metabolism following i.p. administration of levcromakalim is likely due to portal absorption [159], thus somewhat lowering the actual mouse dose. The net conclusion is that, in addition to hemodynamic effects, migraine-relevant nociceptive pathways are also replicated in the mouse model but at a higher dosing regimen.

5.2. Discrepant Results on K_{ATP} Channel Inhibition in Preclinical and Clinical Studies

Evidently, there has been a poor translation between preclinical and clinical studies looking at K_{ATP} channel blockage with glibenclamide in a migraine context, both on the hemodynamic parameters and headache readouts (hypersensitivity to tactile stimulation in rodents). These discrepancies may be explained by differences in dosing regimens, pharmacodynamic action of glibenclamide, subunit distributions, and trigger potency on various receptors.

5.2.1. Discrepant Effect of Glibenclamide on Cranial Arteries

In animal models using intravital microscopy of dural arteries, high doses of glibenclamide (7–30 mg/kg i.v.) were needed to prevent arterial dilation and a decrease in blood pressure followed by levcromakalim (0.1 mg/kg i.v.), pinacidil (0.38 mg/kg i.v.) [137], and CGRP (0.3 μg/kg i.v.) [65]. Looking at the glibenclamide/levcromakalim dose ratio (Table 2), it was 174 for full blockage and 116 for partial blockage of dural artery dilation in rats [138]. In the human equivalent study, the ratio was 5.8 (glibenclamide given to non-fasting participants has a high oral bioavailability, therefore this ratio is not adjusted). For CGRP, the ratio was 178,968 for the partially (but non-significant) effective dose of glibenclamide, and 767,004 for the fully effective dose in rats [65]. In the human experiment, the ratio was 24,972–53,690 depending on whether the total or 1 min CGRP dose was used. In animals, glibenclamide has only been tested against PACAP-induced arterial dilation in vitro. Here, the dose ratio calculation would not be meaningful to compare to the clinical data. In summary, the relative dose of glibenclamide contra trigger (levcromakalim, CGRP) was much higher in the rat studies. Given the SUR1 preference of glibenclamide, the relative low dose of glibenclamide given in humans may explain the lack of an effect of glibenclamide on human cranial arteries expressing SUR2B, while the higher dose applied to rats was sufficient to also inhibit SUR2B.

5.2.2. Discrepant Effect of Glibenclamide on Headache Measures

In terms of headache measures, the effectiveness of glibenclamide in mouse and rat models of migraine has not been seen in the human model of provoked migraine. In both rats and mice, glibenclamide (1 mg/kg i.p.) was sufficient to inhibit cutaneous tactile hypersensitivity, but in the human studies, 10 mg p.o. (0.14 mg/kg) was not convincingly effective. The rodent dose was 7 times higher than the human dose (assuming equal bioavailability), which is the typical conversion factor between rats and humans based on body surface area [158]. Looking at the glibenclamide/trigger ratio in the mouse model and the human model, we found that in mice, the ratio was 0.58 for levcromakalim and 45,891 for PACAP, the latter only partially preventing hypersensitivity. In the human studies, the ratios were 5.79 for levcromakalim and 1.42 for PACAP. Despite different routes of administration, we get a clear indication that the glibenclamide/PACAP ratio was smaller in the human study compared to the mouse study, which may explain the lack of translation between results. However, for levcromakalim, the ratio was 10-fold higher in the human experiment, which leaves no simple explanation for the lack of efficacy on the primary readout. Recall, however, that in this study [153], a comparison was made to a previous study [136]. A highly significant effect was found on a secondary output, median time to headache onset, which was 30 min after levcromakalim and 180 min after glibenclamide-levcromakalim, suggesting that glibenclamide was effective when plasma levels were high [153]. Glibenclamide has not been directly tested against CGRP in the mouse model of provoked migraine.

5.2.3. Target Engagement

The 10 mg dose of glibenclamide applied in the human studies was clearly insufficient in terms of blocking vascular K_{ATP} channels (Kir6.1/SUR2B), but the effect on blood glucose was shown to indicate the efficient blocking of pancreatic (Kir6.2/SUR1) channels in line with SUR1 selectivity of glibenclamide described in Section 3.2. (K_{ATP} channel blockers). In rats, the applied (high) dose was able to block the vascular K_{ATP} channels. In humans, larger doses cannot be applied due to the adverse effect on blood glucose [153]. The effect of blood glucose was not reported in the rat studies looking at hemodynamics after glibenclamide 7–30 mg/kg, i.v. [65,138]. In another rat study, both glibenclamide 1 mg/kg and 10 mg/kg i.p. decreased blood glucose from 7 mmol/L (vehicle treatment) to 3 mmol/L 2 h post-treatment. In mice, acute injection of glibenclamide (10 and 30 µg/mouse) caused a rapid, dose-dependent drop in blood glucose levels from approximately 170 to 120 mg/dL, peaking at 60 min. The two highest concentrations of glibenclamide caused a similar marked reduction of fed blood glucose after an extended period, consistent with a saturated effect of the drug in vivo [160]. Glibenclamide 5 mg/kg/day (delivered by a subcutaneous minipump) did not affect blood glucose in vivo [161]. The pronounced effect on human (but not mouse) glucose level is likely due to the reported > 100-fold higher potency of glibenclamide on human SUR1 over mouse SUR1 [126,162].

The higher selectivity of PNU-37883A on the vascular channels compared to glibenclamide is evident when the inhibition of levcromakalim is related across two studies from the same laboratory using the same in vivo model to study dilation of the MMA [117,138]. Here, 1.3 µmol/kg of PNU-37883A partly inhibited 0.087 µmol/kg of levcromakalim-induced dilation (ratio 15) and 40.5 µmol/kg of glibenclamide partly inhibited 0.35 µmol/kg of levcromakalim (ratio 116).

In both rats and mice, the lower dose of glibenclamide (1 mg/kg i.p. = 2 µmol/kg, ratio 0.58) was effective in different behavioral models of migraine [47,99]. The human studies on headache prevention by glibenclamide following provocation with levcromakalim,

CGRP, and PACAP were negative on the primary outcome. Nevertheless, partial efficacy may have been present in the former two experiments; see Section 4.3.2. (Clinical effect of K_{ATP} channel inhibition in human migraine models). Partial inhibition of CGRP and PACAP induced alterations may be expected as the downstream effect from both neuropeptides likely also involve the opening of other ion channels [163].

5.3. Possible Mechanism of Headache Induction and Prevention

Different theories about where and how the opening of K_{ATP} channels causes headache exist. These are further fueled by the non-clarified effect of channel inhibition on headache readouts. Collectively, the interpretation regarding subunit contribution to headache is difficult as rodent and, to some extent, human data suggest that glibenclamide may inhibit headache (hypersensitivity in rodents) to some degree, without the relevant effect on the vascular Kir6.1/SUR2B channel, which is the proposed mediator of levcromakalim-induced migraine [22]. An effect on neuronal Kir6.2/SUR2 channels is also a possibility [164], albeit smooth muscle Kir6.1 subunits were identified as important [45].

5.3.1. Dilation of Meningeal Arteries

Dilation of intracranial arteries within the trigeminovascular system is the leading hypothesis of levcromakalim-induced headache and migraine [7,45,165]. Two proposed mechanisms are currently at play: (a) mechanical activation of trigeminal nociceptors by arterial dilation or (b) chemical activation of trigeminal nociceptors by high $[K^+]$ in the microenvironment between arteries and nerve endings [23]. These hypotheses need testing using a selective blocker of Kir6.1/SUR2B suitable for use in humans.

5.3.2. Effect on CGRP Signaling

A few preclinical studies suggest that K_{ATP} channels may affect CGRP signaling in different manners. In ex vivo organ preparations, glibenclamide inhibited the release of CGRP from trigeminal ganglia and dura mater [47]. In contrast, K_{ATP} channel openers did not directly stimulate the release of CGRP [99,141]. However, in vivo, the hypersensitivity induced by levcromakalim was abolished both by treatment with a CGRP-neutralizing antibody, and in genetically modified mice, by not expressing the CGRP receptor component RAMP1 [99], suggesting that CGRP is released by inter-tissue communication (not found ex vivo) following levcromakalim treatment and that this drives hypersensitivity. As an alternative to the vascular theory, this specific release of CGRP may in fact be what is inhibited by glibenclamide in vivo via its affinity to SUR1. This may also explain the speculative effect of glibenclamide on headache, in spite of the clear lack of a vascular effect on hemodynamics in human experiments.

5.3.3. Hyperpolarization-Activated Cyclic Nucleotide-Gated (HCN) Channels

Another alternative to the vascular theory of migraine induction by K_{ATP} channel opening is the involvement of HCN channels in the trigeminal nervous system [164]. Sustained hyperpolarization of neurons may engage HCN channels, and blockage of these have been suggested as therapeutic targets in diabetic neuropathy [166] and neuropathic [167] and inflammatory pain [168]. The proposed mechanism is that K_{ATP} channel openers lead to the long-lasting hyperpolarization of trigeminal nerves, in turn activating HCN channels that leads to augmented neuronal excitability and firing of the neurons [169,170]. In contrast, this hypothesis consists of the fact that, in CNS, the opening of K_{ATP} channels is involved in analgesia. Moreover, HCN channels are expressed in trigeminal and dorsal root ganglia, and systemic exposure to levcromakalim induces cephalic but not peripheral pain [22,23,136] and local administration of levcromakalim was unable to induce pain [171]. The HCN theory is currently being evaluated in a clinical trial (NCT04853797), testing HCN channel blocker ivabradine against levcromakalim-induced headache [156].

5.4. Clinical Therapeutic Perspectives

Ion channels are regarded as an important class of drug targets for modulating pain and is localized in primary sensory neurons and other key structures in pain processing [172]. K_{ATP} channels are probably the most diverse ion channel type, and each subtype has a specific physiological role. Developing drugs targeting all K_{ATP} channels may therefore be impossible since they are widespread and undesirable severe side effects would be expected. Thus, subtype selectivity is key and may be a very attractive target for the development of novel therapeutics for the acute and preventive treatment of migraine.

Accordingly, Kir6.1/SUR2B subunits are dominantly expressed in the vascular smooth muscle. In contrast, Kir6.2/SUR1 are expressed in the CNS and pancreas, and Kir6.2/SUR2A are expressed in cardiac and skeletal muscle [50,117]. However, the lack of detailed structural and functional insight of these channels poses a challenge for the development of selective drug candidates. The Kir6.1 selective K_{ATP} channel blocker, PNU-37883A, was developed as an orally effective non-kaliuretic diuretic in rats [149,173]. Because of its cardiac depressant activity, possibly related to its blockade of coronary artery Kir6.1 channels in animal experiments, the drug never advanced to human studies [174]. Thus, K_{ATP} channel blockers selective for Kir6.1 alone should be carefully considered. In addition, SUR2 null mice exhibited elevated resting blood pressure and sudden death from ST segment elevation and coronary artery vasospasm [175]. In SUR2, in null mice with a transgenic restoration of SUR2B, the above-mentioned side effects persisted [176]. Thus, these side effects seem to be caused by SUR2A knockout and are likely not related to knockout of the SUR2B subunit. A K_{ATP} channel blocker for the treatment of migraine should therefore preferably have an exclusive selectivity for Kir6.1/SUR2B K_{ATP} channels.

To date, most ion channel drug development has focused on identifying and developing small molecule and peptide modulators, mainly through serendipitous discovery [177]. Despite vastly improved screening tools for small molecule or compound libraries, only two novel ion channel drugs have been approved by the FDA since the 1990s [178]. The well-known disadvantage of small molecules is that they can bind to off-molecular targets, leading to more side effects and toxicity. Alternative modalities for targeting ion channels have recently included monoclonal antibodies (mAbs), which offer many additional advantages to selectivity and bioavailability. Yet, despite considerable interest, there are currently no marketed mAbs therapies that target an ion channel. This lack of success is mainly attributable to two important technical challenges. First, the ion channels have short extracellular loops displaying small epitope target areas over the plasma membrane, causing them to be challenging binding targets for large protein antibodies. Additionally, these extracellular loops tend to be highly conserved at the primary amino acid sequence level, and thus lack sufficient immunogenicity to generate robust antibody responses in mammalian hosts [178].

A major challenge and concern in developing K_{ATP} channel blockers is cardiac side effects. K_{ATP} channels are abundant in the myocardium and K_{ATP} channel openers have proven useful in ischemic heart disease through direct actions on the myocardium [179] and may prevent arrhythmias [180]. To overcome this problem, it is important to test with several heart assays, such as the ex vivo Langendorff heart model (perfused isolated heart model), to evaluate the direct effects of compounds on cardiac function and to ensure cardiovascular safety of new drug candidates [181].

In conclusion, K_{ATP} channels are recognized as promising therapeutic targets for migraine treatment but remain a major challenge for drug discovery. To move forward, we need further studies on the specific subtypes of the K_{ATP} channel to enable a deeper understanding of their structures, functions and distribution for more selective and successful drug development. Furthermore, knowledge on the consequences of activation or blockage of the K_{ATP} channels on a molecular- or pathway-specific level in the pathophysiology of migraine, is necessary to fully comprehend and predict the potential of this novel therapeutic target.

Author Contributions: Conceptualization, S.L.C. and A.C.; writing—original draft preparation, S.L.C., A.C. and S.G.; writing—review and editing, S.L.C., A.C., S.G. and I.J.-O.; supervision, I.J.-O. All authors have read and agreed to the published version of the manuscript.

Funding: This research was funded by the Candys Foundation. Song Guo is supported by the BRIDGE – Translational Excellence Programme (bridge.ku.dk) at the Faculty of Health and Medical Sciences, University of Copenhagen, funded by the Novo Nordisk Foundation. Grant agreement no. NNF20SA0064340 (2021 fellows).

Institutional Review Board Statement: Not applicable.

Informed Consent Statement: Not applicable.

Data Availability Statement: Not applicable.

Acknowledgments: Thanks to Jes Olesen for proofreading the manuscript.

Conflicts of Interest: The authors declare no conflict of interest. The funders had no role in the drafting, proofreading, and finalizing of the manuscript.

References

1. Steiner, T.J.; Stovner, L.J.; Jensen, R.; Uluduz, D.; Katsarava, Z. Migraine Remains Second among the World's Causes of Disability, and First among Young Women: Findings from GBD2019. *J. Headache Pain* **2020**, *21*, 4–7. [CrossRef] [PubMed]
2. James, S.L.; Abate, D.; Abate, K.H.; Abay, S.M.; Abbafati, C.; Abbasi, N.; Abbastabar, H.; Abd-Allah, F.; Abdela, J.; Abdelalim, A.; et al. Global, Regional, and National Incidence, Prevalence, and Years Lived with Disability for 354 Diseases and Injuries for 195 Countries and Territories, 1990–2017: A Systematic Analysis for the Global Burden of Disease Study 2017. *Lancet* **2018**, *392*, 1789–1858. [CrossRef]
3. Olesen, J. Headache Classification Committee of the International Headache Society (IHS) The International Classification of Headache Disorders, 3rd Edition. *Cephalalgia* **2018**, *38*, 1–211. [CrossRef]
4. Vgontzas, A.; Pavlović, J.M. Sleep Disorders and Migraine: Review of Literature and Potential Pathophysiology Mechanisms. *Headache* **2018**, *58*, 1030–1039. [CrossRef]
5. Vuralli, D.; Ayata, C.; Bolay, H. Cognitive Dysfunction and Migraine 17 Psychology and Cognitive Sciences 1701 Psychology 11 Medical and Health Sciences 1103 Clinical Sciences 11 Medical and Health Sciences 1109 Neurosciences. *J. Headache Pain* **2018**, *19*, 109. [CrossRef] [PubMed]
6. Ashina, M.; Katsarava, Z.; Do, T.P.; Buse, D.C.; Pozo-Rosich, P.; Özge, A.; Krymchantowski, A.V.; Lebedeva, E.R.; Ravishankar, K.; Yu, S.; et al. Migraine: Epidemiology and Systems of Care. *Lancet* **2021**, *397*, 1485–1495. [CrossRef]
7. Ashina, M. Migraine. *N. Engl. J. Med.* **2020**, *383*, 1866–1876. [CrossRef]
8. Ashina, M.; Hansen, J.M.; Do, T.P.; Melo-Carrillo, A.; Burstein, R.; Moskowitz, M.A. Migraine and the Trigeminovascular System-40 Years and Counting. *Lancet. Neurol.* **2019**, *18*, 795–804. [CrossRef]
9. Peng, K.P.; May, A. Migraine Understood as a Sensory Threshold Disease. *Pain* **2019**, *160*, 1494–1501. [CrossRef]
10. Edvinsson, L.; Haanes, K.A.; Warfvinge, K.; Krause, D.N. CGRP as the Target of New Migraine Therapies—Successful Translation from Bench to Clinic. *Nat. Rev. Neurol.* **2018**, *14*, 338–350. [CrossRef]
11. Stauffer, V.L.; Dodick, D.W.; Zhang, Q.; Carter, J.N.; Ailani, J.; Conley, R.R. Evaluation of Galcanezumab for the Prevention of Episodic Migraine: The EVOLVE-1 Randomized Clinical Trial. *JAMA Neurol.* **2018**, *75*, 1080–1088. [CrossRef] [PubMed]
12. Förderreuther, S.; Zhang, Q.; Stauffer, V.L.; Aurora, S.K.; Láinez, M.J.A. Preventive Effects of Galcanezumab in Adult Patients with Episodic or Chronic Migraine Are Persistent: Data from the Phase 3, Randomized, Double-Blind, Placebo-Controlled EVOLVE-1, EVOLVE-2, and REGAIN Studies. *J. Headache Pain* **2018**, *19*, 121. [CrossRef] [PubMed]
13. Ashina, M.; Goadsby, P.J.; Reuter, U.; Silberstein, S.; Dodick, D.W.; Xue, F.; Zhang, F.; Paiva da Silva Lima, G.; Cheng, S.; Mikol, D.D. Long-Term Efficacy and Safety of Erenumab in Migraine Prevention: Results from a 5-Year, Open-Label Treatment Phase of a Randomized Clinical Trial. *Eur. J. Neurol.* **2021**, *28*, 1716–1725. [CrossRef] [PubMed]
14. Silberstein, S.D.; Dodick, D.W.; Bigal, M.E.; Yeung, P.P.; Goadsby, P.J.; Blankenbiller, T.; Grozinski-Wolff, M.; Yang, R.; Ma, Y.; Aycardi, E. Fremanezumab for the Preventive Treatment of Chronic Migraine. *N. Engl. J. Med.* **2017**, *377*, 2113–2122. [CrossRef] [PubMed]
15. Ashina, M.; Saper, J.; Cady, R.; Schaeffler, B.A.; Biondi, D.M.; Hirman, J.; Pederson, S.; Allan, B.; Smith, J. Eptinezumab in Episodic Migraine: A Randomized, Double-Blind, Placebo-Controlled Study (PROMISE-1). *Cephalalgia* **2020**, *40*, 241–254. [CrossRef]
16. Lassen, L.H.; Haderslev, P.A.; Jacobsen, V.B.; Iversen, H.K.; Sperling, B.; Olesen, J. CGRP May Play a Causative Role in Migraine. *Cephalalgia* **2002**, *22*, 54–61. [CrossRef]
17. Schytz, H.W.; Birk, S.; Wienecke, T.; Kruuse, C.; Olesen, J.; Ashina, M. PACAP38 Induces Migraine-like Attacks in Patients with Migraine without Aura. *Brain* **2009**, *132*, 16–25. [CrossRef]
18. Thomsen, L.L. Investigations into the Role of Nitric Oxide and the Large Intracranial Arteries in Migraine Headache. *Cephalalgia* **1997**, *17*, 873–895. [CrossRef]

19. Pellesi, L.; Al-Karagholi, M.A.M.; Chaudhry, B.A.; Lopez, C.L.; Snellman, J.; Hannibal, J.; Amin, F.M.; Ashina, M. Two-Hour Infusion of Vasoactive Intestinal Polypeptide Induces Delayed Headache and Extracranial Vasodilation in Healthy Volunteers. *Cephalalgia* **2020**, *40*, 1212–1223. [CrossRef]
20. Kruuse, C.; Thomsen, L.L.; Birk, S.; Olesen, J. Migraine Can Be Induced by Sildenafil without Changes in Middle Cerebral Artery Diameter. *Brain* **2003**, *126*, 241–247. [CrossRef]
21. Guo, S.; Olesen, J.; Ashina, M. Phosphodiesterase 3 Inhibitor Cilostazol Induces Migraine-like Attacks via Cyclic AMP Increase. *Brain* **2014**, *137*, 2951–2959. [CrossRef]
22. Al-Karagholi, M.A.M.; Hansen, J.M.; Guo, S.; Olesen, J.; Ashina, M. Opening of ATP-Sensitive Potassium Channels Causes Migraine Attacks: A New Target for the Treatment of Migraine. *Brain* **2019**, *142*, 2644–2654. [CrossRef]
23. Al-Karagholi, M.A.M.; Ghanizada, H.; Nielsen, C.A.W.; Hougaard, A.; Ashina, M. Opening of ATP Sensitive Potassium Channels Causes Migraine Attacks with Aura. *Brain* **2021**, *144*, 2322–2332. [CrossRef]
24. Noma, A. ATP-Regulated K+ Channels in Cardiac Muscle. *Nature* **1983**, *305*, 147–148. [CrossRef] [PubMed]
25. Miki, T.; Nagashima, K.; Seino, S. The Structure and Function of the ATP-Sensitive K+ Channel in Insulin-Secreting Pancreatic Beta-Cells. *J. Mol. Endocrinol.* **1999**, *22*, 113–123. [CrossRef]
26. Proks, P.; Ashcroft, F.M. Modeling KATP Channel Gating and Its Regulation. *Prog. Biophys. Mol. Biol.* **2009**, *99*, 7–19. [CrossRef]
27. Hibino, H.; Inanobe, A.; Furutani, K.; Murakami, S.; Findlay, I.; Kurachi, Y. Inwardly Rectifying Potassium Channels: Their Structure, Function, and Physiological Roles. *Physiol. Rev.* **2010**, *90*, 291–366. [CrossRef] [PubMed]
28. Choma, K.; Bednarczyk, P.; Koszela-Piotrowska, I.; Kulawiak, B.; Kudin, A.; Kunz, W.S.; Dołowy, K.; Szewczyk, A. Single Channel Studies of the ATP-Regulated Potassium Channel in Brain Mitochondria. *J. Bioenerg. Biomembr.* **2009**, *41*, 323–334. [CrossRef]
29. Babenko, A.P.; Aguilar-Bryan, L.; Bryan, J. A View of Sur/Kir6.X, KATP Channels. *Annu. Rev. Physiol* **1998**, *60*, 667–687. [CrossRef]
30. Seino, S.; Miki, T. Physiological and Pathophysiological Roles of ATP-Sensitive K+ Channels. *Prog. Biophys. Mol. Biol.* **2003**, *81*, 133–176. [CrossRef]
31. Syed, A.U.; Koide, M.; Brayden, J.E.; Wellman, G.C. Tonic Regulation of Middle Meningeal Artery Diameter by ATP-Sensitive Potassium Channels. *J. Cereb. Blood Flow Metab.* **2019**, *39*, 670–679. [CrossRef]
32. Standen, N.B.; Quayle, J.M.; Davies, N.W.; Brayden, J.E.; Huang, Y.; Nelson, M.T. Hyperpolarizing Vasodilators Activate ATP-Sensitive K+ Channels in Arterial Smooth Muscle. *Science* **1989**, *245*, 177–180. [CrossRef]
33. Sung, M.W.; Yang, Z.; Driggers, C.M.; Patton, B.L.; Mostofian, B.; Russo, J.D.; Zuckerman, D.M.; Shyng, S.L. Vascular KATP Channel Structural Dynamics Reveal Regulatory Mechanism by Mg-Nucleotides. *Proc. Natl. Acad. Sci. USA* **2021**, *118*. [CrossRef]
34. Ribalet, B.; John, S.A.; Xie, L.H.; Weiss, J.N. Regulation of the ATP-Sensitive K Channel Kir6.2 by ATP and PIP(2). *J. Mol. Cell. Cardiol.* **2005**, *39*, 71–77. [CrossRef]
35. Baukrowitz, T.; Schulte, U.; Oliver, D.; Herlitze, S.; Krauter, T.; Tucker, S.J.; Ruppersberg, J.P.; Fakler, B. PIP2 and PIP as Determinants for ATP Inhibition of KATP Channels. *Science* **1998**, *282*, 1141–1144. [CrossRef]
36. Larsson, O.; Barker, C.J.; Berggren, P.O. Phosphatidylinositol 4,5-Bisphosphate and ATP-Sensitive Potassium Channel Regulation: A Word of Caution. *Diabetes* **2000**, *49*, 1409–1412. [CrossRef]
37. Shi, Y.; Cui, N.; Shi, W.; Jiang, C. A Short Motif in Kir6.1 Consisting of Four Phosphorylation Repeats Underlies the Vascular KATP Channel Inhibition by Protein Kinase C. *J. Biol. Chem.* **2008**, *283*, 2488–2494. [CrossRef]
38. Shi, W.-W.; Yang, Y.; Shi, Y.; Chun, J. KATP Channel Action in Vascular Tone Regulation: From Genetics to Diseases. *Sheng Li Xue Bao* **2012**, *64*, 1–13. [PubMed]
39. Quinn, K.V.; Giblin, J.P.; Tinker, A. Multisite Phosphorylation Mechanism for Protein Kinase A Activation of the Smooth Muscle ATP-Sensitive K+ Channel. *Circ. Res.* **2004**, *94*, 1359–1366. [CrossRef]
40. Shi, Y.; Chen, X.; Wu, Z.; Shi, W.; Yang, Y.; Cui, N.; Jiang, C.; Harrison, R.W. CAMP-Dependent Protein Kinase Phosphorylation Produces Interdomain Movement in SUR2B Leading to Activation of the Vascular KATP Channel. *J. Biol. Chem.* **2008**, *283*, 7523–7530. [CrossRef]
41. Kokoti, L.; Al-Karagholi, M.A.M.; Ashina, M. Latest Insights into the Pathophysiology of Migraine: The ATP-Sensitive Potassium Channels. *Curr. Pain Headache Rep.* **2020**, *24*, 77. [CrossRef]
42. Schytz, H.W.; Schoonman, G.G.; Ashina, M. What Have We Learnt from Triggering Migraine? *Curr. Opin. Neurol.* **2010**, *23*, 259–265. [CrossRef] [PubMed]
43. Yamada, M.; Isomoto, S.; Matsumoto, S.; Kondo, C.; Shindo, T.; Horio, Y.; Kurachi, Y. Sulphonylurea Receptor 2B and Kir6.1 Form a Sulphonylurea-Sensitive but ATP-Insensitive K+ Channel. *J. Physiol.* **1997**, *499*, 715–720. [CrossRef]
44. Teramoto, N.; Zhu, H.-L.; Shibata, A.; Aishima, M.; Walsh, E.J.; Nagao, M.; Cole, W.C. ATP-Sensitive K+ Channels in Pig Urethral Smooth Muscle Cells Are Heteromultimers of Kir6.1 and Kir6.2. *Am. J. Physiol Ren. Physiol* **2009**, *296*, 107–117. [CrossRef]
45. Christensen, S.L.; Rasmussen, R.H.; Cour, S.L.; Ernstsen, C.; Hansen, T.F.; Kogelman, L.J.; Lauritzen, S.P.; Guzaite, G.; Styrishave, B.; Janfelt, C.; et al. Smooth Muscle ATP-Sensitive Potassium Channels Mediate Migraine-Relevant Hypersensitivity in Mouse Models. *Cephalalgia* **2022**, *42*, 93–107. [CrossRef] [PubMed]
46. Al-Karagholi, M.A.-M.; Hansen, J.M.; Severinsen, J.; Jansen-Olesen, I.; Ashina, M. The KATP Channel in Migraine Pathophysiology: A Novel Therapeutic Target for Migraine. *J. Headache Pain* **2017**, *18*, 90. [CrossRef]
47. Christensen, S.L.; Munro, G.; Petersen, S.; Shabir, A.; Jansen-Olesen, I.; Kristensen, D.M.; Olesen, J. ATP Sensitive Potassium (KATP) Channel Inhibition: A Promising New Drug Target for Migraine. *Cephalalgia* **2020**, *40*, 650–664. [CrossRef] [PubMed]

48. Li, Y.; Aziz, Q.; Tinker, A. The Pharmacology of ATP-Sensitive K + Channels (K ATP). *Handb. Exp. Pharmacol.* **2021**, *267*, 357–378. [CrossRef]
49. Skatchkov, S.N.; Rojas, L.; Eaton, M.J.; Orkand, R.K.; Biedermann, B.; Bringmann, A.; Pannicke, T.; Veh, R.W.; Reichenbach, A. Functional Expression of Kir 6.1/SUR1-KATP Channels in Frog Retinal Müller Glial Cells. *Glia* **2002**, *38*, 256–267. [CrossRef]
50. Ploug, K.B.; Sørensen, M.A.; Strøbech, L.; Klaerke, D.A.; Hay-Schmidt, A.; Sheykhzade, M.; Olesen, J.; Jansen-Olesen, I. KATP Channels in Pig and Human Intracranial Arteries. *Eur. J. Pharmacol.* **2008**, *601*, 43–49. [CrossRef]
51. Aziz, Q.; Li, Y.; Tinker, A. ATP-Sensitive Potassium Channels and Vascular Function. *Channels* **2015**, *9*, 3–4. [CrossRef] [PubMed]
52. Ploug, K.B.; Baun, M.; Hay-Schmidt, A.; Olesen, J.; Jansen-Olesen, I. Presence and Vascular Pharmacology of KATP Channel Subtypes in Rat Central and Peripheral Tissues. *Eur. J. Pharmacol.* **2010**, *637*, 109–117. [CrossRef] [PubMed]
53. Rodrigo, G.; Standen, N. ATP-Sensitive Potassium Channels. *Curr. Pharm. Des.* **2005**, *11*, 1915–1940. [CrossRef] [PubMed]
54. Bao, L.; Kefaloyianni, E.; Lader, J.; Hong, M.; Morley, G.; Fishman, G.I.; Sobie, E.A.; Coetzee, W.A. Unique Properties of the ATP-Sensitive K^+ Channel in the Mouse Ventricular Cardiac Conduction System. *Circ. Arrhythm. Electrophysiol.* **2011**, *4*, 926–935. [CrossRef] [PubMed]
55. Aguilar-Bryan, L.; Nichols, C.G.; Wechsler, S.W.; Clement IV, J.P.; Boyd, A.E.; González, G.; Herrera-Sosa, H.; Nguy, K.; Bryan, J.; Nelson, D.A. Cloning of the Beta Cell High-Affinity Sulfonylurea Receptor: A Regulator of Insulin Secretion. *Science* **1995**, *268*, 423–426. [CrossRef]
56. Flagg, T.P.; Kurata, H.T.; Masia, R.; Caputa, G.; Magnuson, M.A.; Lefer, D.J.; Coetzee, W.A.; Nichols, C.G. Differential Structure of Atrial and Ventricular KATP: Atrial KATP Channels Require SUR1. *Circ. Res.* **2008**, *103*, 1458–1465. [CrossRef] [PubMed]
57. Zoga, V.; Kawano, T.; Liang, M.Y.; Bienengraeber, M.; Weihrauch, D.; McCallum, B.; Gemes, G.; Hogan, Q.; Sarantopoulos, C. KATPchannel Subunits in Rat Dorsal Root Ganglia: Alterations by Painful Axotomy. *Mol. Pain* **2010**, *6*. [CrossRef]
58. Kawano, T.; Zoga, V.; McCallum, J.B.; Wu, H.E.; Gemes, G.; Liang, M.Y.; Abram, S.; Kwok, W.M.; Hogan, Q.H.; Sarantopoulos, C.D. ATP-Sensitive Potassium Currents in Rat Primary Afferent Neurons: Biophysical, Pharmacological Properties, and Alterations by Painful Nerve Injury. *Neuroscience* **2009**, *162*, 431–443. [CrossRef] [PubMed]
59. Tricarico, D.; Mele, A.; Lundquist, A.L.; Desai, R.R.; George, A.L.; Conte Camerino, D. Hybrid Assemblies of ATP-Sensitive K+ Channels Determine Their Muscle-Type-Dependent Biophysical and Pharmacological Properties. *Proc. Natl. Acad. Sci. USA* **2006**, *103*, 1118–1123. [CrossRef]
60. Inagaki, N.; Gonoi, T.; Clement IV, J.P.; Wang, C.Z.; Aguilar-Bryan, L.; Bryan, J.; Seino, S. A Family of Sulfonylurea Receptors Determines the Pharmacological Properties of ATP-Sensitive K+ Channels. *Neuron* **1996**, *16*, 1011–1017. [CrossRef]
61. Sun, H.S.; Feng, Z.P.; Barber, P.A.; Buchan, A.M.; French, R.J. Kir6.2-Containing ATP-Sensitive Potassium Channels Protect Cortical Neurons from Ischemic/Anoxic Injury in Vitro and in Vivo. *Neuroscience* **2007**, *144*, 1509–1515. [CrossRef] [PubMed]
62. Jovanović, S.; Ballantyne, T.; Du, Q.; Blagojević, M.; Jovanović, A. Phenylephrine Preconditioning in Embryonic Heart H9c2 Cells Is Mediated by Up-Regulation of SUR2B/Kir6.2: A First Evidence for Functional Role of SUR2B in Sarcolemmal KATP Channels and Cardioprotection. *Int. J. Biochem. Cell Biol.* **2016**, *70*, 23–28. [CrossRef] [PubMed]
63. Jahangir, A.; Terzic, A. K(ATP) Channel Therapeutics at the Bedside. *J. Mol. Cell. Cardiol.* **2005**, *39*, 99–112. [CrossRef]
64. Aziz, Q.; Thomas, A.M.; Gomes, J.; Ang, R.; Sones, W.R.; Li, Y.; Ng, K.E.; Gee, L.; Tinker, A. The ATP-Sensitive Potassium Channel Subunit, Kir6.1, in Vascular Smooth Muscle Plays a Major Role in Blood Pressure Control. *Hypertension* **2014**, *64*, 523–529. [CrossRef]
65. Gozalov, A.; Jansen-Olesen, I.; Klaerke, D.; Olesen, J. Role of KATP Channels in Cephalic Vasodilatation Induced by Calcitonin Gene-Related Peptide, Nitric Oxide, and Transcranial Electrical Stimulation in the Rat. *Headache* **2008**, *48*, 1202–1213. [CrossRef]
66. Nelson, M.T.; Quayle, J.M. Physiological Roles and Properties of Potassium Channels in Arterial Smooth Muscle. *Am. J. Physiol. Cell Physiol.* **1995**, *268*, 799–822. [CrossRef]
67. Tinker, A.; Aziz, Q.; Thomas, A. The Role of ATP-Sensitive Potassium Channels in Cellular Function and Protection in the Cardiovascular System. *Br. J. Pharmacol.* **2014**, *171*, 12–23. [CrossRef] [PubMed]
68. Murphy, M.E.; Brayden, J.E. Nitric Oxide Hyperpolarizes Rabbit Mesenteric Arteries via ATP-Sensitive Potassium Channels. *J. Physiol.* **1995**, *486*, 58. [CrossRef] [PubMed]
69. Olesen, J. The Role of Nitric Oxide (NO) in Migraine, Tension-Type Headache and Cluster Headache. *Pharmacol. Ther.* **2008**, *120*, 157–171. [CrossRef]
70. Duarte, I.D.G.; dos Santos, I.R.; Lorenzetti, B.B.; Ferreira, S.H. Analgesia by Direct Antagonism of Nociceptor Sensitization Involves the Arginine-Nitric Oxide-CGMP Pathway. *Eur. J. Pharmacol.* **1992**, *217*, 225–227. [CrossRef]
71. Aziz, Q.; Li, Y.; Anderson, N.; Ojake, L.; Tsisanova, E.; Tinker, A. Molecular and Functional Characterization of the Endothelial ATP-Sensitive Potassium Channel. *J. Biol. Chem.* **2017**, *292*, 17587–17597. [CrossRef] [PubMed]
72. De Logu, F.; Landini, L.; Janal, M.N.; Li Puma, S.; De Cesaris, F.; Geppetti, P.; Nassini, R. Migraine-Provoking Substances Evoke Periorbital Allodynia in Mice. *J. Headache Pain* **2019**, *20*, 18. [CrossRef] [PubMed]
73. Schytz, H.W.; Olesen, J.; Ashina, M. The PACAP Receptor: A Novel Target for Migraine Treatment. *Neurotherapeutics* **2010**, *7*, 191–196. [CrossRef] [PubMed]
74. Brown, T.; Forster, R.B.; Cleanthis, M.; Mikhailidis, D.P.; Stansby, G.; Stewart, M. Cilostazol for Intermittent Claudication. *Cochrane Database Syst. Rev.* **2021**, *2021*. [CrossRef]
75. Moore, R.A.; Edwards, J.E.; McQuay, H.J. Sildenafil (Viagra) for Male Erectile Dysfunction: A Meta-Analysis of Clinical Trial Reports. *BMC Urol.* **2002**, *2*, 1–12. [CrossRef]

76. Trainor, D.C.; Jones, R.C. Headaches in Explosive Magazine Workers. *Arch. Environ. Health* **1966**, *12*, 231–234. [CrossRef]
77. Miki, T.; Suzuki, M.; Shibasaki, T.; Uemura, H.; Sato, T.; Yamaguchi, K.; Koseki, H.; Iwanaga, T.; Nakaya, H.; Seino, S. Mouse Model of Prinzmetal Angina by Disruption of the Inward Rectifier Kir6.1. *Nat. Med.* **2002**, *8*, 466–472. [CrossRef] [PubMed]
78. Porter, A. Controlling Your Losses: Conditional Gene Silencing in Mammals. *Trends Genet.* **1998**, *14*, 73–79. [CrossRef]
79. Sauer, B. Inducible Gene Targeting in Mice Using the Cre/Lox System. *Methods* **1998**, *14*, 381–392. [CrossRef]
80. Utomo, A.R.H.; Nikitin, A.Y.; Lee, W.H. Temporal, Spatial, and Cell Type-Specified Control of Cre-Mediated DNA Recombination in Transgenic Mice. *Nat. Biotechnol.* **1999**, *17*, 1091–1096. [CrossRef]
81. Soundarapandian, M.M.; Zhong, X.; Peng, L.; Wu, D.; Lu, Y. Role of K ATP Channels in Protection against Neuronal Excitatory Insults. *J. Neurochem* **2007**, *103*, 1721–1729. [CrossRef] [PubMed]
82. Yildirim, C.; Özkaya, B.; Bal, R. KATP and TRPM2-like Channels Couple Metabolic Status to Resting Membrane Potential of Octopus Neurons in the Mouse Ventral Cochlear Nucleus. *Brain Res. Bull.* **2021**, *170*, 115–128. [CrossRef] [PubMed]
83. Kase, D.; Imoto, K. The Role of HCN Channels on Membrane Excitability in the Nervous System. *J. Signal. Transduct.* **2012**, *2012*, 1–11. [CrossRef] [PubMed]
84. Country, M.W.; Jonz, M.G. Mitochondrial KATP Channels Stabilize Intracellular Ca2+ during Hypoxia in Retinal Horizontal Cells of Goldfish (Carassius Auratus). *J. Exp. Biol.* **2021**, *224*. [CrossRef]
85. Guo, W.; Tang, Z.Y.; Cai, Z.Y.; Zhao, W.E.; Yang, J.; Wang, X.P.; Ji, J.; Huang, X.X.; Sun, X.L. Iptakalim Alleviates Synaptic Damages via Targeting Mitochondrial ATP-Sensitive Potassium Channel in Depression. *FASEB J.* **2021**, *35*. [CrossRef]
86. Toulorge, D.; Guerreiro, S.; Hirsch, E.C.; Michel, P.P. KATP Channel Blockade Protects Midbrain Dopamine Neurons by Repressing a Glia-to-Neuron Signaling Cascade That Ultimately Disrupts Mitochondrial Calcium Homeostasis. *J. Neurochem.* **2010**, *114*, 553–564. [CrossRef]
87. Khanna, N.; Malhotra, R.S.; Mehta, A.K.; Garg, G.R.; Halder, S.; Sharma, K.K. Interaction of Morphine and Potassium Channel Openers on Experimental Models of Pain in Mice. *Fundam. Clin. Pharmacol.* **2011**, *25*, 479–484. [CrossRef]
88. Tsantoulas, C.; McMahon, S.B. Opening Paths to Novel Analgesics: The Role of Potassium Channels in Chronic Pain. *Trends Neurosci.* **2014**, *37*, 146–158. [CrossRef]
89. Cunha, T.M.; Roman-Campos, D.; Lotufo, C.M.; Duarte, H.L.; Souza, G.R.; Verri, W.A.; Funez, M.I.; Dias, Q.M.; Schivo, I.R.; Domingues, A.C.; et al. Morphine Peripheral Analgesia Depends on Activation of the PI3Kγ/AKT/NNOS/NO/KATP Signaling Pathway. *Proc. Natl. Acad. Sci. USA* **2010**, *107*, 4442–4447. [CrossRef]
90. Kusuda, R.; Carreira, E.U.; Ulloa, L.; Cunha, F.Q.; Kanashiro, A.; Cunha, T.M. Choline Attenuates Inflammatory Hyperalgesia Activating Nitric Oxide/CGMP/ATP-Sensitive Potassium Channels Pathway. *Brain Res.* **2020**, *1727*, 146567. [CrossRef]
91. Fakhri, S.; Ahmadpour, Y.; Rezaei, H.; Kooshki, L.; Moradi, S.Z.; Iranpanah, A.; Gravandi, M.M.; Abbaszadeh, F.; Ghanbarveisi, F. The Antinociceptive Mechanisms of Melatonin: Role of L-Arginine/Nitric Oxide/Cyclic GMP/KATP Channel Signaling Pathway. *Behav. Pharmacol.* **2020**, *31*, 728–737. [CrossRef] [PubMed]
92. Gong, L.; Gao, F.; Li, J.; Li, J.; Yu, X.; Ma, X.; Zheng, W.; Cui, S.; Liu, K.; Zhang, M.; et al. Oxytocin-Induced Membrane Hyperpolarization in Pain-Sensitive Dorsal Root Ganglia Neurons Mediated by Ca(2+)/NNOS/NO/KATP Pathway. *Neuroscience* **2015**, *289*, 417–428. [CrossRef] [PubMed]
93. Sarantopoulos, C.; McCallum, B.; Sapunar, D.; Kwok, W.M.; Hogan, Q. ATP-Sensitive Potassium Channels in Rat Primary Afferent Neurons: The Effect of Neuropathic Injury and Gabapentin. *Neurosci. Lett.* **2003**, *343*, 185–189. [CrossRef]
94. Wu, X.F.; Liu, W.T.; Liu, Y.P.; Huang, Z.J.; Zhang, Y.K.; Song, X.J. Reopening of ATP-Sensitive Potassium Channels Reduces Neuropathic Pain and Regulates Astroglial Gap Junctions in the Rat Spinal Cord. *Pain* **2011**, *152*, 2605–2615. [CrossRef]
95. Sakamaki, G.; Johnson, K.; Mensinger, M.; Hmu, E.; Klein, A.H. Loss of SUR1 Subtype KATP Channels Alters Antinociception and Locomotor Activity after Opioid Administration. *Behav. Brain Res.* **2021**, *414*. [CrossRef]
96. Fisher, C.; Johnson, K.; Okerman, T.; Jurgenson, T.; Nickell, A.; Salo, E.; Moore, M.; Doucette, A.; Bjork, J.; Klein, A.H. Morphine Efficacy, Tolerance, and Hypersensitivity Are Altered After Modulation of SUR1 Subtype KATP Channel Activity in Mice. *Front. Neurosci.* **2019**, *13*. [CrossRef]
97. Shi, W.; Cui, N.; Wu, Z.; Yang, Y.; Zhang, S.; Gai, H.; Zhu, D.; Jiang, C. Lipopolysaccharides Up-Regulate Kir6.1/SUR2B Channel Expression and Enhance Vascular KATP Channel Activity via NF-KappaB-Dependent Signaling. *J. Biol. Chem.* **2010**, *285*, 3021–3029. [CrossRef]
98. Cao, Z.; Dai, W.; Zhang, R.; Chen, L.; Yang, X.; Hu, L.; Chiang, L.Y.; Liu, W. Opening of the Adenosine Triphosphate-Sensitive Potassium Channel Attenuates Morphine Tolerance by Inhibiting JNK and Astrocyte Activation in the Spinal Cord. *Clin. J. Pain* **2016**, *32*, 617–623. [CrossRef]
99. Christensen, S.L.; Rasmussen, R.H.; Ernstsen, C.; La Cour, S.; David, A.; Chaker, J.; Haanes, K.A.; Christensen, S.T.; Olesen, J.; Kristensen, D.M. CGRP-Dependent Signalling Pathways Involved in Mouse Models of GTN- Cilostazol- and Levcromakalim-Induced Migraine. *Cephalalgia* **2021**, *41*, 1413–1426. [CrossRef]
100. Narita, M.; Suzuki, T.; Misawa, M.; Nagase, H.; Nabeshima, A.; Ashizawa, T.; Ozawa, H.; Saito, T.; Takahata, N. Role of Central ATP-Sensitive Potassium Channels in the Analgesic Effect and Spinal Noradrenaline Turnover-Enhancing Effect of Intracerebroventricularly Injected Morphine in Mice. *Brain Res.* **1992**, *596*, 209–214. [CrossRef]
101. Saint-Martin, C.; Arnoux, J.B.; de Lonlay, P.; Bellanné-Chantelot, C. KATP Channel Mutations in Congenital Hyperinsulinism. *Semin. Pediatr. Surg.* **2011**, *20*, 18–22. [CrossRef] [PubMed]

102. De Franco, E.; Saint-Martin, C.; Brusgaard, K.; Knight Johnson, A.E.; Aguilar-Bryan, L.; Bowman, P.; Arnoux, J.B.; Larsen, A.R.; May, S.; Greeley, S.A.W.; et al. Update of Variants Identified in the Pancreatic β-Cell KATP Channel Genes KCNJ11 and ABCC8 in Individuals with Congenital Hyperinsulinism and Diabetes. *Hum. Mutat.* **2020**, *41*, 884–905. [CrossRef] [PubMed]
103. Kane, C.; Shepherd, R.M.; Squires, P.E.; Johnson, P.R.V.; James, R.F.L.; Milla, P.J.; Aynsley-Green, A.; Lindley, K.J.; Dunne, M.J. Loss of Functional KATP Channels in Pancreatic Beta-Cells Causes Persistent Hyperinsulinemic Hypoglycemia of Infancy. *Nat. Med.* **1996**, *2*, 1344–1347. [CrossRef] [PubMed]
104. Li, C.; Ackermann, A.M.; Boodhansingh, K.E.; Bhatti, T.R.; Liu, C.; Schug, J.; Doliba, N.; Han, B.; Cosgrove, K.E.; Banerjee, I.; et al. Functional and Metabolomic Consequences of K ATP Channel Inactivation in Human Islets. *Diabetes* **2017**, *66*, 1901–1913. [CrossRef]
105. Lontchi-Yimagou, E.; Sobngwi, E.; Matsha, T.E.; Kengne, A.P. Diabetes Mellitus and Inflammation. *Curr. Diab. Rep.* **2013**, *13*, 435–444. [CrossRef]
106. Duncan, B.B.; Schmidt, M.I. The Epidemiology of Low-Grade Chronic Systemic Inflammation and Type 2 Diabetes. *Diabetes Technol. Ther.* **2006**, *8*, 7–17. [CrossRef]
107. Gier, B.; Krippeit-Drews, P.; Sheiko, T.; Aguilar-Bryan, L.; Bryan, J.; Düfer, M.; Drews, G. Suppression of KATP Channel Activity Protects Murine Pancreatic Beta Cells against Oxidative Stress. *J. Clin. Investig.* **2009**, *119*, 3246–3256. [CrossRef]
108. Krippeit-Drews, P.; Lang, F.; Häussinger, D.; Drews, G. H2O2 Induced Hyperpolarization of Pancreatic B-Cells. *Pflug. Arch.* **1994**, *426*, 552–554. [CrossRef]
109. Maechler, P.; Jornot, L.; Wollheim, C.B. Hydrogen Peroxide Alters Mitochondrial Activation and Insulin Secretion in Pancreatic Beta Cells. *J. Biol. Chem.* **1999**, *274*, 27905–27913. [CrossRef]
110. Nakazaki, M.; Kakei, M.; Koriyama, N.; Tanaka, H. Involvement of ATP-Sensitive K+ Channels in Free Radical-Mediated Inhibition of Insulin Secretion in Rat Pancreatic Beta-Cells. *Diabetes* **1995**, *44*, 878–883. [CrossRef]
111. Edalat, A.; Schulte-Mecklenbeck, P.; Bauer, C.; Undank, S.; Krippeit-Drews, P.; Drews, G.; Düfer, M. Mitochondrial Succinate Dehydrogenase Is Involved in Stimulus-Secretion Coupling and Endogenous ROS Formation in Murine Beta Cells. *Diabetologia* **2015**, *58*, 1532–1541. [CrossRef] [PubMed]
112. Benemei, S.; Fusi, C.; Trevisan, G.; Geppetti, P. The TRPA1 Channel in Migraine Mechanism and Treatment. *Br. J. Pharmacol.* **2014**, *171*, 2552–2567. [CrossRef] [PubMed]
113. Akerman, S.; Salvemini, D.; Romero-Reyes, M. Targeting Reactive Nitroxidative Species in Preclinical Models of Migraine. *Cephalalgia* **2021**, *41*, 1187–1200. [CrossRef] [PubMed]
114. Hambrock, A.; Löffler-Walz, C.; Kloor, D.; Delabar, U.; Horio, Y.; Kurachi, Y.; Quast, U. ATP-Sensitive K+ Channel Modulator Binding to Sulfonylurea Receptors SUR2A and SUR2B: Opposite Effects of MgADP. *Mol. Pharmacol.* **1999**, *55*, 832–840. [PubMed]
115. Petersson, J.; Ryman, T.; Högestätt, E.D. Vasodilator Effects of KRN2391, Levcromakalim and 3-Morpholino- Sydnonimin in Human Pial and Omental Arteries. *Naunyn. Schmiedebergs. Arch. Pharmacol.* **2000**, *362*, 68–73. [CrossRef]
116. Jansen-Olesen, I.; Mortensen, C.H.; El-Bariaki, N.; Ploug, K.B. Characterization of KATP-Channels in Rat Basilar and Middle Cerebral Arteries: Studies of Vasomotor Responses and MRNA Expression. *Eur. J. Pharmacol.* **2005**, *523*, 109–118. [CrossRef]
117. Ploug, K.B.; Boni, L.J.; Baun, M.; Hay-Schmidt, A.; Olesen, J.; Jansen-Olesen, I. KATP Channel Expression and Pharmacological in Vivo and in Vitro Studies of the KATP Channel Blocker PNU-37883A in Rat Middle Meningeal Arteries. *Br. J. Pharmacol.* **2008**, *154*, 72–81. [CrossRef]
118. Carlsen, J.E.; Kardel, T.; Jensen, H.; Tangø, M.; Trap-Jensen, J. Pinacidil, a New Vasodilator: Pharmacokinetics and Pharmacodynamics of a New Retarded Release Tablet in Essential Hypertension. *Eur. J. Clin. Pharmacol.* **1983**, *25*, 557–561. [CrossRef] [PubMed]
119. Ward, J.W.; McBurney, A.; Farrow, P.R.; Sharp, P. Pharmacokinetics and Hypotensive Effect in Healthy Volunteers of Pinacidil, a New Potent Vasodilator. *Eur. J. Clin. Pharmacol.* **1984**, *26*, 603–608. [CrossRef]
120. Clapham, J.C.; Trail, B.K.; Hamilton, T.C. K+ Channel Activators, Acute Glucose Tolerance and Glibenclamide-Induced Hypoglycaemia in the Hypertensive Rat. *Eur. J. Pharmacol.* **1994**, *257*, 79–85. [CrossRef]
121. Mannhold, R. KATP Channel Openers: Structure-Activity Relationships and Therapeutic Potential. *Med. Res. Rev.* **2004**, *24*, 213–266. [CrossRef] [PubMed]
122. Wolff, F. Diazoxide Hyperglycaemia and Its Continued Relief by Tolbutamide. *Lancet* **1964**, *1*, 309–310. [CrossRef]
123. Rubin, A.A.; Roth, F.E.; Taylor, R.M.; Rosenkilde, H. Pharmacology of Diazoxide, an Antihypertensive, Nondiuretic Benzothiadiazine. *J. Pharmacol. Exp. Ther.* **1962**, *136*, 344–352. [PubMed]
124. Lodwick, D.; Rainbow, R.D.; Rubaiy, H.N.; Al Johi, M.; Vuister, G.W.; Norman, R.I. Sulfonylurea Receptors Regulate the Channel Pore in ATP-Sensitive Potassium Channels via an Intersubunit Salt Bridge. *Biochem. J.* **2014**, *464*, 343–354. [CrossRef] [PubMed]
125. Maruthur, N.M.; Tseng, E.; Hutfless, S.; Wilson, L.M.; Suarez-Cuervo, C.; Berger, Z.; Chu, Y.; Iyoha, E.; Segal, J.B.; Bolen, S. Diabetes Medications as Monotherapy or Metformin-Based Combination Therapy for Type 2 Diabetes: A Systematic Review and Meta-Analysis. *Ann. Intern. Med.* **2016**, *164*, 740–751. [CrossRef]
126. Coghlan, M.J.; Carroll, W.A.; Gopalakrishnan, M. Recent Developments in the Biology and Medicinal Chemistry of Potassium Channel Modulators: Update from a Decade of Progress. *J. Med. Chem.* **2001**, *44*, 1627–1653. [CrossRef]
127. Gribble, F.M.; Tucker, S.J.; Seino, S.; Ashcroft, F.M. Tissue Specificity of Sulfonylureas Studies on Cloned Cardiac and β- Cell K(ATP) Channels. *Diabetes* **1998**, *47*, 1412–1418. [CrossRef]

128. Liu, S.Y.; Tian, H.M.; Liao, D.Q.; Chen, Y.F.; Gou, Z.P.; Xie, X.Y.; Li, X.J. The Effect of Gliquidone on KATP Channels in Pancreatic β-Cells, Cardiomyocytes, and Vascular Smooth Muscle Cells. *Diabetes Res. Clin. Pract.* **2015**, *109*, 334–339. [CrossRef]
129. Glibenclamide | Ligand Activity Charts | IUPHAR/BPS Guide to PHARMACOLOGY. Available online: https://www.guidetopharmacology.org/GRAC/LigandActivityRangeVisForward?ligandId=2414 (accessed on 26 June 2022).
130. Wang, Y.; Yu, L.; Cui, N.; Jin, X.; Zhu, D.; Jiang, C. Differential Sensitivities of the Vascular K(ATP) Channel to Various PPAR Activators. *Biochem. Pharmacol.* **2013**, *85*, 1495–1503. [CrossRef]
131. Yu, L.; Jin, X.; Cui, N.; Wu, Y.; Shi, Z.; Zhu, D.; Jiang, C. Rosiglitazone Selectively Inhibits KATP Channels by Acting on the KIR6 Subunit. *Br. J. Pharmacol.* **2012**, *167*, 26. [CrossRef]
132. Humphrey, S.J. Pharmacology of the K-ATP Channel Blocking Morpholinoguanidine PNU- 37883A. *Cardiovasc. Drug Rev.* **1999**, *17*, 295–328. [CrossRef]
133. D'Arcy, V.; Laher, M.; McCoy, D.; Sullivan, P.; Walsh, C.H.; Hickey, M.P. Pinacidil, a New Vasodilator, in the Treatment of Mild to Moderate Essential Hypertension. *Eur. J. Clin. Pharmacol.* **1985**, *28*, 347–349. [CrossRef] [PubMed]
134. Kidney, J.C.; Fuller, R.W.; Worsdell, Y.M.; Lavender, E.A.; Chung, K.F.; Barnes, P.J. Effect of an Oral Potassium Channel Activator, BRL 38227, on Airway Function and Responsiveness in Asthmatic Patients: Comparison with Oral Salbutamol. *Thorax* **1993**, *48*, 130–133. [CrossRef] [PubMed]
135. Ashina, M.; Hansen, J.M.; Olesen, J. Pearls and Pitfalls in Human Pharmacological Models of Migraine: 30 Years' Experience. *Cephalalgia* **2013**, *33*, 540–553. [CrossRef]
136. Al-Karagholi, M.A.-M.; Ghanizada, H.; Hansen, J.M.; Skovgaard, L.T.; Olesen, J.; Larsson, H.B.W.; Amin, F.M.; Ashina, M. Levcromakalim, an Adenosine Triphosphate-Sensitive Potassium Channel Opener, Dilates Extracerebral but Not Cerebral Arteries. *Headache* **2019**, *59*, 1468–1480. [CrossRef] [PubMed]
137. Al-Karagholi, M.A.M.; Ghanizada, H.; Nielsen, C.A.W.; Ansari, A.; Gram, C.; Younis, S.; Vestergaard, M.B.; Larsson, H.B.W.; Skovgaard, L.T.; Amin, F.M.; et al. Cerebrovascular Effects of Glibenclamide Investigated Using High-Resolution Magnetic Resonance Imaging in Healthy Volunteers. *J. Cereb. Blood Flow Metab.* **2021**, *41*, 1328–1337. [CrossRef]
138. Gozalov, A.; Petersen, K.A.; Mortensen, C.; Jansen-Olesen, I.; Klaerke, D.; Olesen, J. Role of KATP Channels in the Regulation of Rat Dura and Pia Artery Diameter. *Cephalalgia* **2005**, *25*, 249–260. [CrossRef]
139. Ploug, K.B.; Edvinsson, L.; Olesen, J.; Jansen-Olesen, I. Pharmacological and Molecular Comparison of KATP Channels in Rat Basilar and Middle Cerebral Arteries. *Eur. J. Pharmacol.* **2006**, *553*, 254–262. [CrossRef]
140. Rasmussen, R.H.; Jansen-Olesen, I.; Kristensen, D.M.; Christensen, S.L. Ex Vivo Release of Calcitonin Gene-Related Peptide from the Trigeminovascular System in Rodents | Protocol. *J. Vis. Exp.* **2022**, *183*. [CrossRef]
141. Ploug, K.B.; Amrutkar, D.V.; Baun, M.; Ramachandran, R.; Iversen, A.; Lund, T.M.; Gupta, S.; Hay-Schmidt, A.; Olesen, J.; Jansen-Olesen, I. K ATP Channel Openers in the Trigeminovascular System. *Cephalalgia* **2012**, *32*, 55–65. [CrossRef]
142. Russell, F.A.; King, R.; Smillie, S.-J.; Kodji, X.; Brain, S.D. Calcitonin Gene-Related Peptide: Physiology and Pathophysiology. *Physiol. Rev.* **2014**, *94*, 1099–1142. [CrossRef] [PubMed]
143. Sicuteri, F. Mast cells and their active substances: Their role in the pathogenesis of migraine. *Headache* **1963**, *3*, 86–92. [CrossRef] [PubMed]
144. Bates, E.A.; Nikai, T.; Brennan, K.C.; Fu, Y.-H.; Charles, A.C.; Basbaum, A.I.; Ptácek, L.J.; Ahn, A.H. Sumatriptan Alleviates Nitroglycerin-Induced Mechanical and Thermal Allodynia in Mice. *Cephalalgia* **2010**, *30*, 170–178. [CrossRef] [PubMed]
145. Pradhan, A.A.; Smith, M.L.; McGuire, B.; Tarash, I.; Evans, C.J.; Charles, A. Characterization of a Novel Model of Chronic Migraine. *PAIN®* **2014**, *155*, 269–274. [CrossRef] [PubMed]
146. Ernstsen, C.; Christensen, S.L.; Rasmussen, R.H.; Nielsen, B.S.; Jansen-Olesen, I.; Olesen, J.; Kristensen, D.M. The PACAP Pathway Is Independent of CGRP in Mouse Models of Migraine: Possible New Drug Target? *Brain* **2022**, *145*, 2450–2460. [CrossRef] [PubMed]
147. Bruch, L.; Bychkov, R.; Kästner, A.; Bülow, T.; Ried, C.; Gollasch, M.; Baumann, G.; Luft, F.C.; Haller, H. Pituitary Adenylate-Cyclase-Activating Peptides Relax Human Coronary Arteries by Activating K(ATP) and K(Ca) Channels in Smooth Muscle Cells. *J. Vasc. Res.* **1997**, *34*, 11–18. [CrossRef] [PubMed]
148. Hayabuchi, Y.; Davies, N.W.; Standen, N.B. Angiotensin II Inhibits Rat Arterial KATP Channels by Inhibiting Steady-State Protein Kinase A Activity and Activating Protein Kinase Cε. *J. Physiol.* **2001**, *530*, 193. [CrossRef]
149. Meisheri, K.D.; Humphrey, S.J.; Khan, S.A.; Cipkus-Dubray, L.A.; Smith, M.P.; Jones, A.W. 4-Morpholinecarboximidine-N-1-Adamantyl-N'-Cyclohexylhydrochloride (U- 37883A): Pharmacological Characterization of a Novel Antagonist of Vascular ATP-Sensitive K+ Channel Openers. *J. Pharmacol. Exp. Ther.* **1993**, *266*, 655–665.
150. Armstead, W.M. Role of Nitric Oxide, Cyclic Nucleotides, and the Activation of ATP-Sensitive K+ Channels in the Contribution of Adenosine to Hypoxia-Induced Pial Artery Dilation. *J. Cereb. Blood Flow Metab.* **1997**, *17*, 100–108. [CrossRef]
151. Oshinsky, M.L.; Sanghvi, M.M.; Maxwell, C.R.; Gonzalez, D.; Spangenberg, R.J.; Cooper, M.; Silberstein, S.D. Spontaneous Trigeminal Allodynia in Rats: A Model of Primary Headache. *Headache* **2012**, *52*, 1336–1349. [CrossRef]
152. Munro, G.; Petersen, S.; Jansen-Olesen, I.; Olesen, J. A Unique Inbred Rat Strain with Sustained Cephalic Hypersensitivity as a Model of Chronic Migraine-like Pain. *Sci. Rep.* **2018**, *8*, 1836. [CrossRef] [PubMed]
153. Al-Karagholi, M.A.M.; Ghanizada, H.; Kokoti, L.; Paulsen, J.S.; Hansen, J.M.; Ashina, M. Effect of KATP Channel Blocker Glibenclamide on Levcromakalim-Induced Headache. *Cephalalgia* **2020**, *40*, 1045–1054. [CrossRef] [PubMed]

154. Coskun, H.; Elbahi, F.A.; Al-Karagholi, M.A.M.; Ghanizada, H.; Sheykhzade, M.; Ashina, M. The Effect of KATP Channel Blocker Glibenclamide on CGRP-Induced Headache and Hemodynamic in Healthy Volunteers. *Front. Physiol.* **2021**, *12*, 652136. [CrossRef] [PubMed]
155. Kokoti, L.; Al-Mahdi Al-Karagholi, M.; Elbahi, F.A.; Coskun, H.; Ghanizada, H.; Amin, F.M.; Ashina, M. Effect of KATP Channel Blocker Glibenclamide on PACAP38-Induced Headache and Hemodynamic. *Cephalalgia* **2022**, *42*, 846–858. [CrossRef]
156. Home—ClinicalTrials.Gov. Available online: https://www.clinicaltrials.gov/ (accessed on 27 June 2022).
157. Nair, A.; Jacob, S. A Simple Practice Guide for Dose Conversion between Animals and Human. *J. Basic Clin. Pharm.* **2016**, *7*, 27. [CrossRef] [PubMed]
158. Reagan-Shaw, S.; Nihal, M.; Ahmad, N. Dose Translation from Animal to Human Studies Revisited. *FASEB J.* **2008**, *22*, 659–661. [CrossRef]
159. Lukas, G.; Brindle, S.D.; Greengard, P. The Route of Absorption of Intraperitoneally Administered Compounds. *J. Pharmacol. Exp. Ther.* **1971**, *178*, 562–564. [PubMed]
160. Remedi, M.S.; Nichols, C.G. Chronic Antidiabetic Sulfonylureas In Vivo: Reversible Effects on Mouse Pancreatic β-Cells. *PLoS Med.* **2008**, *5*, e206. [CrossRef]
161. Lahmann, C.; Kramer, H.B.; Ashcroft, F.M. Systemic Administration of Glibenclamide Fails to Achieve Therapeutic Levels in the Brain and Cerebrospinal Fluid of Rodents. *PLoS ONE* **2015**, *10*, 1–18. [CrossRef]
162. Ashcroft, F.M.; Gribble, F.M. ATP-Sensitive K+ Channels and Insulin Secretion: Their Role in Health and Disease. *Diabetologia* **1999**, *42*, 903–919. [CrossRef]
163. Sheykhzade, M.; Berg Nyborg, N.C. Mechanism of CGRP-Induced Relaxation in Rat Intramural Coronary Arteries. *Br. J. Pharmacol.* **2001**, *132*, 1235–1246. [CrossRef] [PubMed]
164. Haanes, K.A.; Edvinsson, L. Hyperpolarization through ATP-Sensitive Potassium Channels; Relevance to Migraine Pathology. *Brain* **2020**, *143*, e13. [CrossRef]
165. Al-Mahdi Al-Karagholi, M.; Olesen, J.; Ashina, M. Reply: Hyperpolarization through ATP-Sensitive Potassium Channels; Relevance to Migraine Pathology. *Brain* **2020**, *143*, E14. [CrossRef] [PubMed]
166. Tsantoulas, C.; Lainez, S.; Wong, S.; Mehta, I.; Vilar, B.; McNaughton, P.A. Hyperpolarization-Activated Cyclic Nucleotide-Gated 2 (HCN2) Ion Channels Drive Pain in Mouse Models of Diabetic Neuropathy. *Sci. Transl. Med.* **2017**, *9*. [CrossRef] [PubMed]
167. Bernard Healey, S.A.; Scholtes, I.; Abrahams, M.; McNaughton, P.A.; Menon, D.K.; Lee, M.C. Role of Hyperpolarization-Activated Cyclic Nucleotide-Gated Ion Channels in Neuropathic Pain: A Proof-of-Concept Study of Ivabradine in Patients with Chronic Peripheral Neuropathic Pain. *Pain Rep.* **2021**, *6*, e967. [CrossRef] [PubMed]
168. Young, G.T.; Emery, E.C.; Mooney, E.R.; Tsantoulas, C.; McNaughton, P.A. Inflammatory and Neuropathic Pain Are Rapidly Suppressed by Peripheral Block of Hyperpolarisation-Activated Cyclic Nucleotide-Gated Ion Channels. *Pain* **2014**, *155*, 1708–1719. [CrossRef]
169. Tu, H.; Deng, L.; Sun, Q.; Yao, L.; Han, J.S.; Wan, Y. Hyperpolarization-Activated, Cyclic Nucleotide-Gated Cation Channels: Roles in the Differential Electrophysiological Properties of Rat Primary Afferent Neurons. *J. Neurosci. Res.* **2004**, *76*, 713–722. [CrossRef]
170. Momin, A.; Cadiou, H.; Mason, A.; Mcnaughton, P.A. Role of the Hyperpolarization-Activated Current Ih in Somatosensory Neurons. *J. Physiol.* **2008**, *586*, 5911–5929. [CrossRef] [PubMed]
171. Al-Karagholi, M.A.M.; Ghanizada, H.; Hansen, J.M.; Aghazadeh, S.; Skovgaard, L.T.; Olesen, J.; Ashina, M. Extracranial Activation of ATP-Sensitive Potassium Channels Induces Vasodilation without Nociceptive Effects. *Cephalalgia* **2019**, *39*, 1789–1797. [CrossRef] [PubMed]
172. Zhang, J.; Yao, J.; Rong, M. Editorial: Role of Ion Channels in Pain. *Front. Pharmacol.* **2022**, *13*. [CrossRef] [PubMed]
173. Perricone, S.C.; Humphrey, S.J.; Skaletzky, L.L.; Graham, B.E.; Zandt, R.A.; Zins, G.R. Synthesis and Diuretic Activity of Alkyl- and Arylguanidine Analogs of N,N′-Dicyclohexyl-4-Morpholinecarboxamidine in Rats and Dogs. *J. Med. Chem.* **1994**, *37*, 3693–3700. [CrossRef] [PubMed]
174. Humphrey, S.J.; Smith, M.P.; Cimini, M.G.; Buchanan, L.V.; Gibson, J.K.; Khan, S.A.; Meisheri, K.D. Cardiovascular Effects of the K-ATP Channel Blocker U-37883A and Structurally Related Morpholinoguanidines. *Methods Find. Exp. Clin. Pharmacol.* **1996**, *18*, 247–260. [PubMed]
175. Chutkow, W.A.; Pu, J.; Wheeler, M.T.; Wada, T.; Makielski, J.C.; Burant, C.F.; McNally, E.M. Episodic Coronary Artery Vasospasm and Hypertension Develop in the Absence of Sur2 K(ATP) Channels. *J. Clin. Investig.* **2002**, *110*, 203–208. [CrossRef]
176. Kakkar, R.; Ye, B.; Stoller, D.A.; Smelley, M.; Shi, N.Q.; Galles, K.; Hadhazy, M.; Makielski, J.C.; McNally, E.M. Spontaneous Coronary Vasospasm in KATP Mutant Mice Arises from a Smooth Muscle-Extrinsic Process. *Circ. Res.* **2006**, *98*, 682–689. [CrossRef]
177. Kalia, J.; Milescu, M.; Salvatierra, J.; Wagner, J.; Klint, J.K.; King, G.F.; Olivera, B.M.; Bosmans, F. From Foe to Friend: Using Animal Toxins to Investigate Ion Channel Function. *J. Mol. Biol.* **2015**, *427*, 158–175. [CrossRef] [PubMed]
178. Hutchings, C.J.; Colussi, P.; Clark, T.G. Ion Channels as Therapeutic Antibody Targets. *MAbs* **2019**, *11*, 265–296. [CrossRef]
179. Gross, G.J.; Peart, J.N. KATP Channels and Myocardial Preconditioning: An Update. *Am. J. Physiol. Hear. Circ. Physiol.* **2003**, *285*, 921–930. [CrossRef]

180. Antzelevitch, C.; Di Diego, J.M. Role of K+ Channel Activators in Cardiac Electrophysiology and Arrhythmias. *Circulation* **1992**, *85*, 1627–1629. [CrossRef]
181. King, D.R.; Hardin, K.M.; Hoeker, G.S.; Poelzing, S. Re-Evaluating Methods Reporting Practices to Improve Reproducibility: An Analysis of Methodological Rigor for the Langendorff Whole-Heart Technique. *Am. J. Physiol. Circ. Physiol.* **2022**. [CrossRef]

cells

MDPI

Review

Could Experimental Inflammation Provide Better Understanding of Migraines?

Philip Victor Reducha [1,2], Lars Edvinsson [1,3] and Kristian Agmund Haanes [1,2,*]

1 Department of Clinical Experimental Research, Glostrup Research Institute, Copenhagen University Hospital, Rigshospitalet Glostrup, 2600 Glostrup, Denmark
2 Department of Biology, Section of Cell Biology and Physiology, University of Copenhagen, 1017 Copenhagen, Denmark
3 Division of Experimental Vascular Research, Department of Clinical Sciences, Lund University Hospital, 221 00 Lund, Sweden
* Correspondence: kristian.agmund.haanes@regionh.dk

Abstract: Migraines constitute a common neurological and headache disorder affecting around 15% of the world's population. In addition to other mechanisms, neurogenic neuroinflammation has been proposed to play a part in migraine chronification, which includes peripheral and central sensitization. There is therefore considerable evidence suggesting that inflammation in the intracranial meninges could be a key element in addition to calcitonin gene-related peptide (CGRP), leading to sensitization of trigeminal meningeal nociceptors in migraines. There are several studies that have utilized this approach, with a strong focus on using inflammatory animal models. Data from these studies show that the inflammatory process involves sensitization of trigeminovascular afferent nerve terminals. Further, by applying a wide range of different pharmacological interventions, insight has been gained on the pathways involved. Importantly, we discuss how animal models should be used with care and that it is important to evaluate outcomes in the light of migraine pathology.

Keywords: inflammation; migraine; CGRP; CFA; inflammatory soup

Citation: Reducha, P.V.; Edvinsson, L.; Haanes, K.A. Could Experimental Inflammation Provide Better Understanding of Migraines? *Cells* **2022**, *11*, 2444. https://doi.org/10.3390/cells11152444

Academic Editors: Juan Pablo de Rivero Vaccari and Wolfgang Jost

Received: 17 June 2022
Accepted: 4 August 2022
Published: 6 August 2022

1. Pathophysiology of Migraine

Migraine is a common neurological and headache disorder affecting around 15% of the world's population (≈1 billion people, reported migraine in the previous year) and are three times more prevalent in women (around 18%) than in men (around 6%) [1,2]. It is the second cause of disability worldwide (behind only lower back pain) and the leading cause of neurological disability. Migraines share some similarities with other primary headache disorders in multiple aspects which can lead to misdiagnosis, but migraine also have several unique features. Various probable migraine triggers have been identified including genetic factors such as channelopathies that are familial or sporadic: maternal inheritance of mitochondrial encephalopathy, lactic acidosis, and stroke-like episodes; angiopathy inheritance, to name a few [3].

For a long time, the vascular theory of migraine was dominant. The theory was coined by Harold Wolff in the 1940s. He hypothesized that vasodilation of cranial vessels was the cause of migraines [4]. However, the vascular theory of migraine has since been challenged and by many replaced by neuronal theories which include the central and/or the peripheral nervous systems in migraine pathophysiology. Nevertheless, the vasculature has remained an important factor in migraine pathophysiology as the neurovascular theory of migraine is currently dominating. The neurovascular theory suggests involvement of the trigeminovascular system (TGVS) in the ictal phase of a migraine attack [5]. Neurogenic inflammation (NI) has also been proposed to play a part in migraine chronification. NI would consist of vasodilation, mast cell degranulation, plasma protein extravasation, platelet aggregation and the involvement of vasoactive peptides and inflammatory mediators such as calcitonin gene-related peptide (CGRP), substance P, and neurokinin A, as well as of prostaglandins [6]. However, besides CGRP, the involvement in migraine pathophysiology of the other neuropeptides, as well as mast cell degranulation and plasma extravasation is unclear and needs further research [7]. Peripheral and central sensitization have been proposed as likely causes of pain and involved in the process of chronification of migraine [8].

Sensitization is the process where the response to a stimulus is facilitated and the response itself increased in magnitude. The activation of endogenous inflammatory mediators (which are yet to be fully recognized) is believed to sensitize peripheral trigeminovascular neurons (or first-order neurons) through constant innervation of the dural meninges. Peripheral sensitization likely causes the throbbing nature of the headache, and CGRP has been observed to be involved in initiating and sustaining peripheral sensitization. The constant stimulation of the peripheral TGVS causes sensitization of second-order neurons (trigeminocervical complex, trigeminal nucleus caudalis (TNC)), third-order neurons (trigeminothalamic), as well as neurons in the brainstem and thalamus, causing central sensitization. Central sensitization may cause muscle tenderness, aversion to touch, as well as cephalic and extracephalic allodynia and could possibly be the reason that migraines turn chronic [9,10]. The neurovascular theory builds on the view that vasodilatation of cranial arteries excites the sensory perivascular nerves which cause either immediate or delayed headache pain [11]. The view suggests that vasodilatation activates cyclic nucleotides and potassium channels, resulting in a local rise in potassium ions which causes activation of the sensory nerve terminals in the dura mater [12].

The initiation of migraines is also unclear but likely occurs centrally, as fMRI scans have shown increased activation of the hypothalamus during the prodrome phase [13]. Cortical spreading depression (CSD) has been proposed to play a significant role in the initiation of a migraine attack. However, only 25% of migraineurs experience aura symptoms [14]. In migraines without aura (MoA), if a minor reduction in brain blood flow occurs it is not observed with current methods. Therefore, arguments have been made that CSD only causes aura in patients that have a specific genetic predisposition, which would explain the lack of experience of aura symptoms in MoA patients. Therefore, the role of CSD as a general migraine trigger has been questioned [15]. Regarding CSD and inflammation, there is no evidence of the depression itself being caused by inflammation. However, experimental CSD can trigger neurogenic meningeal inflammation and subsequently activate the trigeminovascular system [16,17]. However, this has not been confirmed in humans.

Although the origin and cause of migraine is not fully known, CGRP is an established key component in migraine attacks. For example, the plasma levels of CGRP are greatly increased during migraine attacks [18], and inhibition of CGRP release decreases both plasma levels of CGRP and the severity of migraine symptoms [18,19]. The role of CGRP in migraines is associated with CGRP being a strong vasodilator and involved in pain transmission from various parts of the head to the central nervous system. Therapies targeting CGRP release or its canonical receptor in the TGVS have been proven to ease symptoms and alleviate pain [5,20,21]. Binding of CGRP to its receptor (between RAMP1 and CLR) leads to conformational changes of the CGRP receptor, and activation of the $G\alpha_s$ subunit. The activated $G\alpha_s$ subunit binds to adenylate cyclase (transmembrane protein), which converts adenosine triphosphate (ATP) into the second messenger known as cyclic adenosine 3′,5′-cyclic monophosphate (cAMP). Intracellular cAMP levels are increased and causes an intracellular downstream signalling cascade. The increase in intracellular cAMP activates cAMP-dependent protein kinase A (PKA) signalling, which results in the activation of cAMP response element-binding protein (CREB), which finally affects gene transcription [22]. In the context of migraines, the increase in intracellular cAMP levels has been suggested to cause opening of hyperpolarization-activated cyclic nucleotide gated channels (HCN) or ATP-sensitive potassium (K_{ATP}) channels that could lead to hyperexcitability of the membrane and possibly be the reason for migraine pain [23].

CGRP has been observed in regions of the central nervous system (cortex, thalamus, hypothalamus, TNC etc.) [24,25]. Additionally, CGRP release has been linked to peripheral areas relevant to pain sensation during the ictal phase of migraine, such as the dura mater and trigeminal ganglion (TG) [18,26]. CGRP and the CGRP receptor are localized in proximity and in parallel to each other, as CGRP is found in great numbers in unmyelinated C-fibers, while CGRP receptors are localized in the Nodes of Ranvier of myelinated Aδ-fibers, both in the TG and dura mater [27]. The TGVS, where CGRP is found in high concentrations, consist of the trigeminal nerve (or fifth cranial nerve, CN V) which has three distinctive branches: the ophthalmic (V1), the maxillary (V2), and the mandibular (V3) branch. As reviewed by Walker and colleagues [28], V1 innervates and supplies strictly sensory information to intracranial dura matter and brain vessels as well as extracranially to the upper facial areas forehead and the scalp, and to orbit and the eyes, upper nasal cavities and skin above the eyes and upper parts of the nose. V2 innervates and conveys strictly sensory information to mid-facial regions, including inferior nasal cavities and sinuses, as well as the upper jaw. V3 innervates sensory information and has motor functions by innervating the muscles of lower facial areas, and the mandibles. In the context of migraines, it is believed that nociceptive information caused by the excitation of C-fibers and in turn of Aδ-fibers, passes along the trigeminal nerves (V1, and to lesser extend V2-V3) in a series of action potentials and merges in the TG. From the TG, the sensory information is sent further to second-order neurons in the trigeminocervical complex (TCC which contains: TNC, dorsal horn and upper cervical spinal horn C1-C2). The nociceptive information is passed on deeper into the central areas and processed by the nuclei of the brain stem, the thalamus and the hypothalamus. Visual, auditory, olfactory, somatosensory, and motor cortices receive the nociceptive signals and are believed to cause the characteristic migraine symptoms such as phonophobia, photophobia, headache pain or cognitive dysfunction [9].

CGRP is believed to play a big role during this communication of pain information [5,29]. It is postulated that during the ictal phase of migraines, neurons in the dura mater and other peripheral neurons receive increased stimulation causing the release of CGRP from C-fibers. CGRP binds to the CGRP receptor located in the Aδ-fibers, which further stimulates neurons, causing extended depolarization. Depolarization and CGRP release are continued along the trigeminovascular pathway, creating nociceptive signalling. The initial release of CGRP in the periphery is likely caused by the excitation of neural membranes from action potentials or through antidromic signalling, causing an opening of ion channels such as voltage gated sodium (Na^+) ion channels. The opening of voltage-gated ion channels leads to an influx of calcium ions (Ca^{2+}) into the cytosol provoking exocytosis of CGRP from vesicles and into the synaptic cleft; that CGRP can later be detected in blood samples from migraine patients [18].

2. Inflammation in Migraine

During the years there have been much work, mainly in rodents, suggesting a significant role of NI in migraines. Overall, the current view is that in single migraine attacks there is limited evidence of inflammation. However, many migraine cases evolve over time to be more severe and even enter a chronic phase. It is therefore possible that inflammation might play a role in migraine chronification, which is the current working hypothesis (Figure 1). To understand such a notion would assist in providing better therapy.

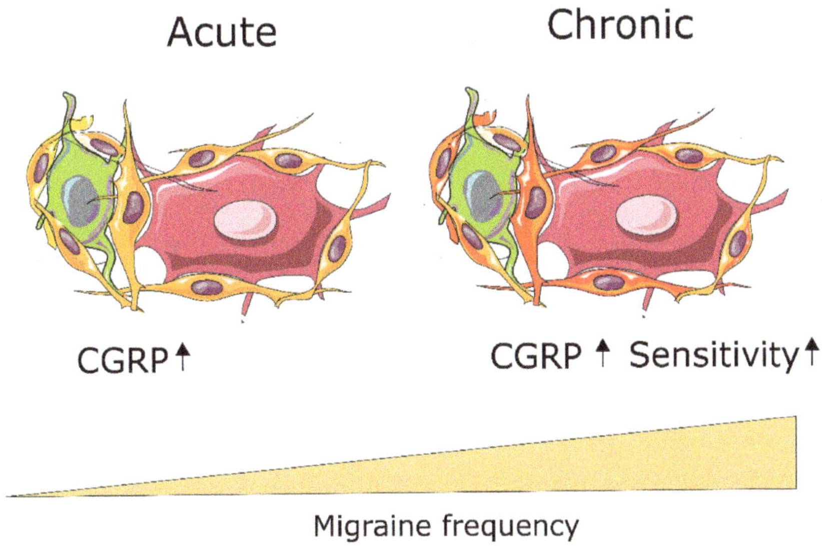

Figure 1. Chronification of migraine. Small CGRP-containing neurons (green), and larger CGRP receptor-containing neurons (pink) are surrounded by satellite glial cells. We postulate that activation of the satellite glial cells (indicated by a change in color from orange to red) sensitizes the glia/neuron subunit, thereby decreasing the migraine threshold and leading to increased migraine frequency.

In the 1980s, the theory "sterile neurogenic inflammation" of the dural meninges was presented [30]. The involvement of inflammation in migraines was further supported by preclinical studies that showed significant inhibition of plasma protein extravasation with classic anti-migraine drugs, ergots and sumatriptan [31–34]. Considerable evidence suggested that NI in the intracranial meninges could be a key element resulting in the sensitization of trigeminal meningeal nociceptors in migraine [35–38].

Major cytokines have been implicated in the TGVS pathway resulting in NI, including tumor necrosis factor (TNF)-α, IL-1β and IL-6 [39,40]. There are studies that observed inflammatory molecules, in particular TNF-α, in plasma, serum or urine samples. These data suggested presence of inflammation in patients when comparing the levels during attacks and in attack-free intervals [41,42]. In animals and particularly in the knock-in animal model of familial hemiplegic migraine type-1, it has been shown that mRNA expression of TNF-α was increased in the TG [43].

Elevated TNF-α serum levels in humans were found even outside of attacks, supporting a role of TNF-α in migraine [44] the. However, data is variable as some have shown that TNF-α is only elevated immediately after the attack and is lower in-between attacks [42]. For other cytokines, such as IL-1β and IL-6, these were reported to be increased during a migraine attack [45]. Further, an increase in IL-6 and TNF-α has been detected in the jugular venous blood during a migraine attack [46]. Although changes in TNF-α and other cytokines can be confirmed, it is not clear as to their significance in migraine pathology. In migraine patients, there is often an increase in inflammatory markers over the course of the disease.

As CGRP might be released during the 72 h of the migraine attack [18], it could potentially evoke continuous activation of C- and Aδ-fibers. This in turn would lead to an increase in expression of inflammatory cytokines, not only in the dura mater, but possibly also in neuronal cell bodies, which are localized in the TG and the TNC. Here the term "neurogenic neuroinflammation", which is defined by inflammatory reactions in the nervous system in response to neuronal activity, can be applied [47]. Combined with central effects which play an important part in causing pericranial allodynia, hyperpathia and muscular pericranial hyperactivity, part of the pathology could also be linked to the peripheral effects of CGRP (or inflammatory sensitization). For example, migraine-proving agents generate periorbital allodynia, for example in animals [48,49]. This is likely due to the release of inflammatory mediators (bradykinin, prostaglandins, etc.) from nerve endings or cells of the immune system [50–52]. Worth noting is that peptidergic fibers were not observed along or in between the striated muscle bundles of the temporal or the occipital regions, only in the skin or accompanying the blood vessels [53]. Furthermore, a part of this sensitization could also occur at the nodes of Ranvier [27]. How the nodes of Ranvier could relate to inflammation remains to be explored. There are several studies that have utilized the case that migraines are linked to meningeal/dural activation of nociceptive fibers. Data from these models supports that sensitization of trigeminovascular afferent nerve terminals. This has been studied using several different methods and particularly electrophysiology has given insight into changes in meningeal excitability [54], and has provided insight in modulation of ion channel excitability [54,55] and responses to migraine therapeutics [55]. Data using an inflammatory model has given evidence that triptans can induce hyperpolarizing shifts in meningeal nociceptive fibers [55]. Further, some sex specific difference can also be observed using this model, even in electrophysiological properties [56], adding value of this model in relation to sex differences in migraine pathology in humans.

3. The Inflammatory Soup (IS) Model and Its Variants

The most common inflammatory migraine model is to surgically insert a cannula that reaches the dura of anesthetized animals and inject inflammatory mediators through the cannula. This model was first applied by Oshinsky and colleagues [57] and has been applied in various migraine research papers since. Certain variations of the model have been used, notably in the type of animals used, changes in surgical procedures or variations in injection volume, time points and frequency. Besides using cannulas, some groups use supradural catheters [58].

The application of chemical stimulants through this model and its variants has led to various physiological and behavioural changes in animals. These findings allow us to evaluate the relevance of this model as a tool for migraine research and act as reference points if one desires to use this model. Here we will present some of the main findings. First, we look at the effect of IS. IS applied to the dura causes periorbital and/or plantar allodynia [57,58], however most often the allodynia is observed after chronic administration. Additionally, other behavioural impairments have been observed. There are reports of increased face rubbing [59] and grooming [60] which are indicative of head pain and/or irritancy. Furthermore, increased rest and freezing behaviour, and decreased exploratory behaviour have been reported [61], which are behaviours also experienced by migraine patients. On the contrary, Liu and colleagues [62] observed increased distance travelled by IS rats in the Open Field Test (OFT), which was argued to be due to higher anxiety levels.

Some other notable findings are the changes at the cellular level, and we have summarized all the below findings in supplementary tables, notably the effects in TG (supplementary Table S1), brain/spinal cord (supplementary Table S2) and TNC (supplementary Table S3). The levels of c-Fos, a marker of neuronal activation and nociception, have been observed at higher expression levels in the TG and TNC [63–65]. Increases of other nociceptive and neuronal activation markers in the TGVS have been observed, such as increases in cAMP [64], PKA [64], $ERK_{1/2}$ [64], CREB [64], NGF [66], PKC [66], tyrosine receptor kinase B (TrkB) [67] and EAAT3 [67]. Increases of inflammatory markers have also been observed in the TGVS, notably an increase in IL-1β [65,68], IL-18 [69], TNF-α [65,70], p38 [67], BDNF [67], I$\kappa\beta$ [71], TRIF [71], TLR4 [71], Myd88 [71] and NF-$\kappa\beta$ [71], FKN [72], CX3CR1 [72] and PGE_2 [73]. Also, an increased presence of microglia [59,67,69–71,74,75] and astrocytes [69,74,75] has been reported. As mentioned previously, releases of neuropeptides such as CGRP, substance P and PACAP-38 are notable components of NI and these peptides are also linked to migraine pathophysiology, especially CGRP. Chronic application of IS on the dura causes upregulation of CGRP genes [62,63,66,67,73,75–83] and

Substance P [76,80–82] in the TGVS. PACAP-38 is increased in the plasma of migraineurs during attacks but reduced in chronic migraineurs [84,85]. Long-term IS infusion decreased PACAP-38 which the authors argue is due to PACAP being depleted after chronic administration [86].

Other notable pathological alterations caused by IS application to the dura include a decrease of 5-HT in the brain [65], an increase in the 5-HT_7 receptor expression [64], a decrease in α7nACh receptor expression in the hippocampus [75], an increase in nitric oxide (NO) [79], increases in the warm receptor TRPV1 [65] and cold receptor TRPM8 [87] expression, decreases in SIRT1 and PGC-1α [80], mitochondrial dysfunction [80,88], an altered functional connectivity between various brain regions [68,89–92], an increase in EphB2/EphrinB2 receptors [81], increased synaptic plasticity in TNC [76,79,81] but decreased plasticity in the hippocampus [93], an increase in glutamate in the TNC [94], an increase in mGluR5 [95], a decrease in GABA [94], decreases in GABABR1 and GABABR2 [94], increases in mTOR and autophagy [95], an increase in ASIC3 receptor expression [66,82], the increase of nNOS in the brain of rhesus monkeys [83], the impairment of descending inhibitory pathways [96], an increase in nerve growth factor (NGF) [66], increases of the $P2Y_{14}$ receptor [59] and the $P2X_4$ receptor expression [67], an increase in BBB permeability [74,97], the increase of VEGF [97], the sensitization of trigeminovascular neurons [96,98–101], an increase in white matter in different brain regions [102] and changes in the gut microbiota [65]. The application of other chemical stimulants that are inflammation-relevant have also been investigated. A chronic dural application of IL-6, a pro-inflammatory cytokine increased in the blood and serum of migraineurs [103], has been reported to cause periorbital and plantar allodynia [104–107], upregulate ERK1 [107] in the TG and sensitize dural afferents [107].

Treatments that can reduce the pathological changes of the chemical stimulants may give indications of what could be potential migraine treatments and could be used as positive controls in future studies that will use this model. Treatments that managed to reduce periorbital and/or plantar allodynia or other pathologies induced by IS application to the dura include sumatriptan [65,97,108], ketoprofen (NSAID) [73], nimesulide (NSAID) [73], etoricoxib (NSAID) [73], baclofen ($GABA_B$ receptor agonist) [95], propranolol (β-blocker) [101], amitriptyline (tricyclic antidepressant) [93], minocycline (tetracycline-derived antibiotic) [70], rapamycin (mTOR inhibitor) [95], anti-IL18 [69], anti-NGF [66], flunarizine (calcium channel blocker) [62], H89 (PKA inhibitor) [94], chelerythrin (PKC blocker) [66], APETx2 (ASIC3 inhibitor) [82], MPEP (mGluR5 antagonist) [95], PNU-282987 (α7nACh receptor agonist) [75], TNP-ATP (P2X inhibitor) [67], ibudilast (PDE inhibitor) [70], electroacupuncture [64,100,109], EphB1-Fc (EphB receptor inhibitor; only 0.5 μg) [81], TAK-242 (TLR4 inhibitor) [71], naltrexone (TLR4 antagonist) [70], PP2 (NR2B-pTyr inhibitor) [76], genistein (protein tyrosine kinase inhibitor) [76], ANA-12 (TrkB antagonist) [67], SIRT1720 (SIRT1 activator) [80], xiongmatang extract [62], and wuzhuyu decoction [65]. Meanwhile, treatments using APETx2 [104], ANA-12 [104], anisomycin (protein synthesis inhibitor) [110], 4EGI-1 (translation initiation inhibitor) [110] or U0126 (MEK inhibitor) [107] alleviated IL-6 induced periorbital/plantar allodynia.

4. Dural Activation by Complete Freund's Adjuvant (CFA)

We have, in addition to experiments with IS, used additional preclinical models involving inflammatory stimulation of the peripheral parts of the trigeminal system to simulate the chronic nature of migraines without receptor activation. Firstly, we studied inflammatory pathways in cultured rat TG cells [111,112]. Secondly, we administered CFA (Complete Freund's Adjuvant) into the temporomandibular joint (TMJ), which elicited activation of TG [113]. Thirdly, we developed a new animal model for trigeminal activation using chemical stimulation of the dura mater with CFA [114] to test whether application of CFA on a small part of the surface of the dura mater can cause long-term activation of the TG, and thus provide a model of migraine chronification [115]. Lastly, we evaluated whether activation of TNC and central sensitization occurs following CFA-induced activation of the dura mater [116].

Using cultures of isolated trigeminal neurons as a model for studies of neurons and glial cells, we found enhanced expression of CGRP in both neurons and satellite glia cells (SGCs) following inflammation. One of the most notable changes was in the mitogen-activated protein (MAP) kinase phosphorylation. The findings indicate that activation of a MAP kinase–dependent inflammatory signal pathway is involved in over-expression of CGRP in nociceptive neurons and could participate in generating pain hypersensitivity [117]. Looking further into in vivo inflammation, we showed that administration of CFA into the TMJ elicits activation of TG by increased expression of pERK1/2 (phosphorylated extracellular signal-regulated protein kinase), pp38 (phosphorylated p38 MAPK/ERK signaling pathway), CaMKII (Calmodulin-dependent protein kinase II), NF-κB and DREAM (Downstream regulatory element antagonist modulator) after 2 and 10 days. By applying CFA to induce local inflammation in the TMJ, this caused an upstream inflammation response in the TG where the TMJ sensory fibers have their cell bodies [113].

The inflammatory response does not only involve neurons, but importantly also SGCs, which together represent one anatomical and functional unit [113]. The idea of stimulating the nerves with inflammatory mediators was first shown by Takeda and colleagues, who injected CFA in the TMJ, and detected trigeminal activation. Importantly IL-1 receptor increased in the TG neurons, and this was combined with a potentiated excitability of Aδ-/C- fibers [118]. Furthermore, they later showed that the increased Aδ-fiber activity caused by inflammation could be blocked by an IL-1 receptor type 1 antagonist [119]. Others have shown that application of inflammatory substances onto the dura mater or chemical stimulation of the dural receptive fields causes hypersensitivity to mechanical and thermal stimulation together with direct activation of the TG [120].

The main application on the inflammatory investigation in rodents was the idea of applying the local inflammation of dura mater nociceptive fibers and the following activation in the TG. In the first papers we showed that the application of CFA [120,121] onto the dural surface indeed activated the TG, with the most remarkable changes seen in the expression of pERK1/2, IL-1β and CGRP positive nerve fibers in the TG [115,122]. This illustrates that the application of CFA onto the dura mater could be used as an animal model for long-term activation of the TGVS [115]. We further used this model to show that the application of CFA also induced activation (increased expression of c-Fos) of the central part of the TGVS: the TNC and C_1-C_2 regions of the spinal cord [116]. Interestingly, this inflammation could be blocked by the administration of a kynurenic acid analogue (SZR72), which is a precursor of an excitotoxin antagonist and anti-inflammatory substance [116,122]. Linking CFA to neurogenic neuroinflammation, masseteric injection of CFA caused spontaneous orofacial pain behaviors, neuronal activation in the TNC, and the release of interleukin-6 (IL-6). In this short-term study, pretreatment with a CGRP antagonist reduced pain behaviors. Nevertheless, IL-6 was unaffected by MK-8825 [123] but the study would probably have benefitted from a longer time frame than the 24 h used.

5. IS vs. CFA

Questions should also be asked about what type of inflammation model would best fit a migraine model. There are some notable differences between the IS and CFA. Usually, the IS consists of inflammatory mediators endogenous to what the body would release during an immune response like histamine, bradykinin, prostaglandins E2 and serotonin, and they are usually mixed into acidic PBS. Meanwhile, CFA is a water/oil emulsion consisting of dried and inactivated mycobacteria (most often *Mycobacterium tuberculosis*). IS induces a sterile inflammatory environment while CFA can go beyond and involve the adaptive immunity as well [124]. CFA is therefore generating local inflammatory markers, without direct receptor-induced activation which is opposite to IS.

One could argue that in the case of most migraines, it is more likely that a sterile inflammation is occurring caused by tissue damage and repair mechanisms initiated from prolonged neuronal activity and surplus of energy used, giving the upper hand to IS as being more relevant in migraines. However, CFA should not be dismissed, as CFA has been reliably used for multiple inflammation and pain models, and as mentioned earlier CFA creates similar cellular and sensitizing effects as IS when applied to the dura [115,116,125]. Further, one could question if some components of the IS could lead to indirect release of other signaling molecules such as ATP [126,127]. In addition to reactions to histamine and bradykinin, this has also been reported for other compounds known to induce itch [128,129] or pain [130]. Secondary effects from ATP could then affect mast cells [131], sensory neurons [132,133] or for example stimulate the release of CGRP [134], further complicating the interpretation of the data.

One could also consider whether the recipe in the IS should be revised, adding some inflammatory mediators that have been observed in the serum of migraine patients like IL-6, CGRP and TNF-α. Arguments could be made whether 5-HT should be removed from the IS mixture as 5-HT have potential anti-migraine effects, as triptans and ditans are agonists of 5-$HT_{1B/D/F}$. receptors. However, as mentioned earlier, applying IS reduces 5-HT in the TGVS in one paper [65] and it is still uncertain whether 5-HT levels fluctuates in migraineurs [135]. Another aspect to take under consideration when using an inflammation model, is the possibility of surgical procedures creating inflammation on their own, and so using SHAM animals should be highly considered. In addition, anaesthesia, antibiotics and painkillers used for recovery could potentially have a modulatory effect on the induced inflammation.

6. Evaluating Outcomes in Animal Inflammation Models

The final goal of the animal research into migraines and the inflammation aspects thereof is to gain mechanistic insight into the pathology, we therefore address some important considerations when evaluating outcomes in inflammation studies. It is important when looking for translational aspects and identifying novel therapeutic targets that the model and outcomes represent true relatable outcomes. The main concept is that inflammatory stimulation of the dura will affect the nociceptive input from the meninges, with a headache most likely occurring in the animals due to the activated nociceptors from the meninges (Figure 2).

The inflammatory approach usually includes a craniotomy, which is an invasive procedure, might have relevance in interpreting the data. As an example, the paper by Laborc and colleagues highlights that the mere attachment of a rat in a stereotaxic frame can activate the trigeminal system [136]. This highlights the additions made by using both fresh controls and also shams in the uncovering of the true role of inflammation, and its separation from the invasive measures [137]. As we have seen above, some of the endpoints are similar as seen in human migraine attacks, but these studies will always be limited by the lack of direct communication with animals. Below, we discuss some limitations linked to outcomes in animal models.

Although photophobia is a known symptom of migraines [138], the general mechanism of migraine-induced photophobia is still unclear. The convergence of optical signals from retinal photoreceptors and nociceptive signals from the TGVS to the CNS is the most likely cause of photophobic experiences in migraineurs [139]. Investigating photophobia is generally a hard task due to the lack of well-established investigation tools and the naturally light-aversive behaviour of laboratory rodents as they are anxious about being seen by predators [140]. Nonetheless, a Light/Dark (LD) box test has previously been used in migraine research [141].

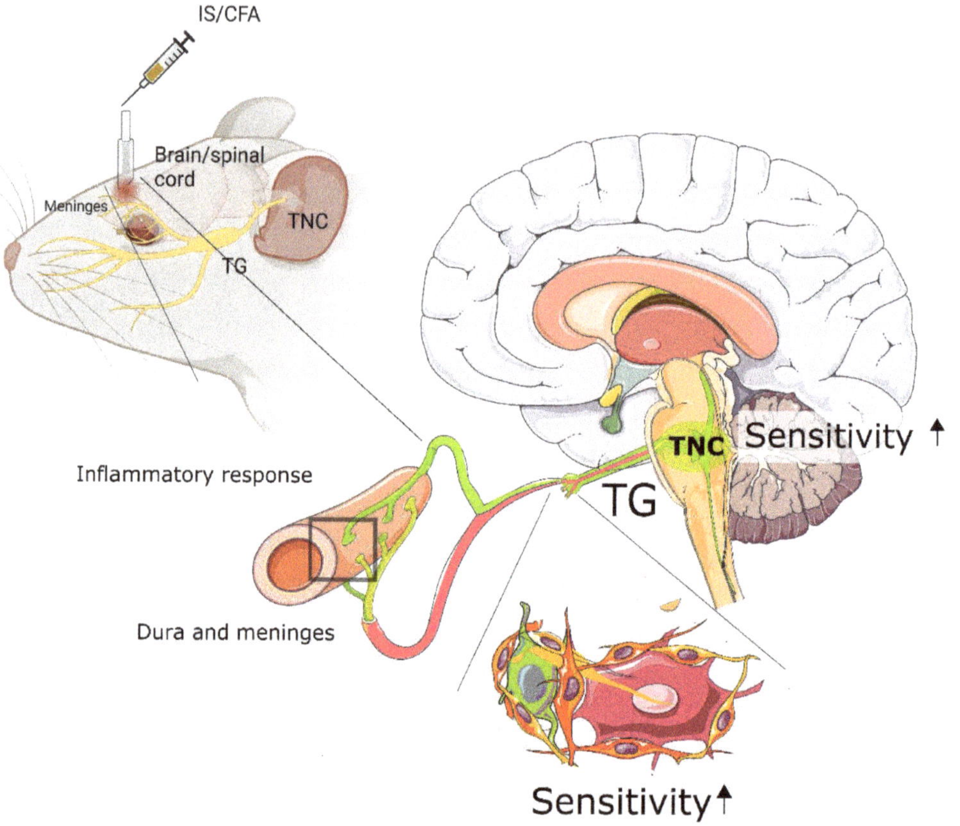

Figure 2. Sensitization in animal models. The chronification of migraines can to some extent be mimicked by the application of inflammatory substances such as Inflammatory Soup (IS) or Complete Freund's Adjuvant (CFA) to the dura and meninges of rodents. Experimental data suggest that this activates satellite glial cells (indicated by a change in colour from orange to red) sensitizes (black arrow) the trigeminal ganglion (TG), and the trigeminal nucleus caudalis (TNC).

The LD box test does not exclusively capture photophobic behaviour, as it also can capture anxiety-like behaviour and measure locomotor activity [142]. A parameter like "time spend in light zone" is the most indicative of photophobic behaviour. In most studies to date, no difference was observed in similar LD box tests, in rodents where CFA has been administered [143,144]. However, CFA administration in those studies were not used in the context of migraine research and were targeted at inducing inflammation in the plantar region. Studies using other inflammatory mediators, such as NTG, have reported more light-aversive behaviours in NTG treated rodents compared to untreated rodents [145].

Other known behavioural migraine symptoms include anxiety and aversion of physical activity because of experienced pain [146]. The OFT has been reported to capture exploratory and anxiety-like behaviour, as well been able to measure locomotor activity [147]. One of the strongest arguments for including this outcome is the concept that physical activity is negatively influenced by headaches, which could be detected in the lack of exploratory behaviour. This is exemplified by part of the diagnostic criterion for migraines, where the headache is supposedly worsened by physical activity, or that the patient will avoid activity [148].

Although like the LD box test in this regard, the OFT differs by having a larger area for exploration and the lack of a high-intensity light. Anxious or depressed rats are expected to spend less time in the centre zone, while worse locomotor and exploratory performances are more associated with experiencing pain. Care should be taken with sequential testing as the animal would lose the novelty and exploratory features of OFTs, admittedly causing inconsistency in performances and/or defeating the purpose of the test. Either way, different strategies should be approached before initial tests to make sure to produce the most consistent results possible.

Previous migraine studies reported both periorbital and plantar allodynia when rodents were administered with migraine factors to the dura, such as IS [78,149]. The main question is whether the Von Frey threshold is capturing headache and not a cutaneous hypersensitivity, as both are seen in migraine patients. One important issue is that allodynia is not headache, and therefore therapeutic approaches might lead to treatments for allodynia and not for headache.

7. Concluding on CGRP and Inflammation—A Perspective

It is no surprise that inflammation is believed to play a potential part in migraine pathophysiology when CGRP is both a key player of migraine pathophysiology and NI. However, it is still unclear whether CGRP is an inducer of inflammation or a by-product of it. Subsequently, the question arises whether inflammation follows the activated areas in migraines (for example is caused by CGRP) or if inflammation causes the activation (and CGRP release). Nevertheless, one would postulate that anatomical separation between the hemispheres would occur, to give rise to the one-sidedness of migraine pain. CGRP even exerts anti-inflammatory properties in some tissue, questioning its role as a pro-inflammatory mediator and its safety as a systemic drug target, as blocking the anti-inflammatory properties could result in inflammation-related consequences elsewhere [150].

Neuroinflammation is typically viewed as a protective response to tissue damage. The main question in migraine is therefore also what type of tissue damage could lead to this initial tissue damage. One possibility could be a neuronal energy deficit. Indeed, this has mainly been put forward by Borkum [151], in light of data on a suggested energy deficit in migraines, particularly by a lower phosphocreatine-to-creatine ratio and an increased concentration of ADP [152,153]. In addition, migraines are associated with mitochondrial disease [154] and with subjects undergoing a fasting state [155]. One potential explanation could be a link to increased membrane potential as the cells struggle to maintain the resting voltage, which could potentially lead to CSD [156] or increased sensitivity in peripheral nerve fibers [27]. Following a potential energy deficit, an increase in inflammatory markers would be expected to protect the brain [157]. This neuroinflammation would control the severity and possible progression, but could also have detrimental effects [158]. The following neuroinflammation has in other systems been showed both to induce and aggravate neurodegeneration, but at the same time aids in the recovery of neurons [159]. This balance is clearly important when addressing inflammation as a potential clinical target.

Nonetheless, using CGRP-targeted drugs on inflammation models could inform us both of the role of CGRP in inflammation and whether inflammation plays an important role (if at all) in migraine pathophysiology. From what we could gather, the effects of anti-CGRP on IS and CFA have yet been investigated. However, triptans appear to mitigate the effects of both IS [65,101,105] and CFA [160,161], but whether it is due to their CGRP targeting or separate anti-inflammatory effects is not clear. Similarly, α-CGRP$_{(8-37)}$, a CGRP receptor antagonist, was able to reduce allodynia from intraplantar CFA [162]. Mixed data exist on the effect of gepants on CFA. The gepant BIBN4096BS (olcegepant) applied topically, intravenously but not spinally [163], was able to alleviate inflammatory pain from subcutaneously plantar administered CFA. However, intrathecal administration of BIBN4096BS did not reduce the increased nNOS that intraplantar CFA caused [164].

Interestingly, when applying CGRP unto the dura mater, it only causes hypersensitivity in female rodents, at least at doses up to 3.8 µg [106], which could be caused by differences in oestrogen signalling [165]. Therefore, if one was to test the efficacy of CGRP-targeting drugs in the inflammation models, one should be wary to also include female rodents if no effects of the drugs is found in the males. An increase in CGRP release at baseline values or stimulated dura mater compared to fresh rats could indicate a sensitized dura mater. Additionally, a significant reduction in CGRP release could indicate a desensitized dura mater from repeated stimulation. To our knowledge, none of the inflammatory models have tested this outcome. An obesity model showcased an increase of basal CGRP in the dura in rats fed a high fat/high sucrose diet. In a previous study, an increase in CGRP expression was observed in the TG couple hours post dural in vivo CFA administration [115]. It could therefore be interesting to test whether basal levels of CGRP in an inflammation model would rise.

There are some studies showing that corticosteroids are useful in managing resistant, severe, recurrent or prolonged migraine attacks (such as status migraenosus) in the emergency department [166]. The questions in our opinion remain as to whether these migraine cases such as status migraenosus should be considered an acute migraine or as a short chronic migraine (at least when it comes to the biology/pathology). Likewise, one could argue that if NSAIDs are effective treatment in some migraine patients, inflammation must therefore play an important pathological role. However it should be noted that these drugs have modes of actions that goes beyond their anti-inflammatory properties, such as their analgesic and antipyretic effects [167], which should be addressed in future studies.

Supplementary Materials: The following supporting information can be downloaded at: https://www.mdpi.com/article/10.3390/cells11152444/s1, Table S1: Summary of pathological changes in the TG, Table S2: Summary of pathological changes in the Brain/Spinal cord, Table S3: Summary of pathological changes in the TNC.

Author Contributions: Conceptualization, P.V.R., L.E. and K.A.H.; methodology, P.V.R., L.E. and K.A.H.; validation, P.V.R., L.E. and K.A.H.; investigation, P.V.R., L.E. and K.A.H.; resources, L.E. and K.A.H.; writing—original draft preparation, P.V.R., L.E. and K.A.H.; writing—review and editing, P.V.R., L.E. and K.A.H.; visualization, P.V.R. and K.A.H.; supervision, L.E. and K.A.H.; project administration, K.A.H.; funding acquisition, L.E. and K.A.H. All authors have read and agreed to the published version of the manuscript.

Funding: P.V.R. and K.A.H. was supported by a Lundbeck foundation Fellowship (R345-2020-1977). The funders had no role in the design and content of the review.

Institutional Review Board Statement: Not applicable.

Informed Consent Statement: Not applicable.

Data Availability Statement: Not applicable.

Acknowledgments: Artwork was based on Servier Medical Art by Servier is licensed under a Creative Commons Attribution 3.0 Unported License (smart.servier.com, accessed on 16 June 2022) parts were created with BioRender.com (accessed on 16 June 2022).

Conflicts of Interest: The authors declare no conflict of interest.

References

1. Bigal, M.E.; Lipton, R.B. The epidemiology, burden, and comorbidities of migraine. *Neurol. Clin.* **2009**, *27*, 321–334. [CrossRef] [PubMed]
2. Ashina, M.; Katsarava, Z.; Do, T.P.; Buse, D.C.; Pozo-Rosich, P.; Özge, A.; Krymchantowski, A.V.; Lebedeva, E.R.; Ravishankar, K.; Yu, S.; et al. Migraine: Epidemiology and systems of care. *Lancet* **2021**, *397*, 1485–1495. [CrossRef]
3. Bron, C.; Sutherland, H.G.; Griffiths, L.R. Exploring the Hereditary Nature of Migraine. *Neuropsychiatr. Dis. Treat.* **2021**, *17*, 1183–1194. [CrossRef] [PubMed]
4. Evans, R.W. Wolff's "Headache and Other Head Pain": Happy 70th Anniversary! *Headache* **2019**, *59*, 443. [CrossRef]
5. Haanes, K.A.; Edvinsson, L. Pathophysiological Mechanisms in Migraine and the Identification of New Therapeutic Targets. *CNS Drugs* **2019**, *33*, 525–537. [CrossRef]
6. Matsuda, M.; Huh, Y.; Ji, R.R. Roles of inflammation, neurogenic inflammation, and neuroinflammation in pain. *J. Anesth.* **2019**, *33*, 131–139. [CrossRef]
7. Riesco, N.; Cernuda-Morollón, E.; Pascual, J. Neuropeptides as a Marker for Chronic Headache. *Curr. Pain Headache Rep.* **2017**, *21*, 18. [CrossRef]

8. Torres-Ferrus, M.; Ursitti, F.; Alpuente, A.; Brunello, F.; Chiappino, D.; Vries, T.D.; di Marco, S.; Ferlisi, S.; Guerritore, L.; Gonzalez-Garcia, N.; et al. School of Advanced Studies of European Headache, From transformation to chronification of migraine: Pathophysiological and clinical aspects. *J. Headache Pain* **2020**, *21*, 42. [CrossRef]
9. Dodick, D.W. A Phase-by-Phase Review of Migraine Pathophysiology. *Headache* **2018**, *58*, 4–16. [CrossRef]
10. Bernstein, C.; Burstein, R. Sensitization of the trigeminovascular pathway: Perspective and implications to migraine pathophysiology. *J. Clin. Neurol.* **2012**, *8*, 89–99. [CrossRef]
11. Ashina, M.; Hansen, J.M.; O, A.D.B.; Olesen, J. Human models of migraine—short-term pain for long-term gain. *Nat. Rev. Neurol.* **2017**, *13*, 713–724. [CrossRef]
12. Al-Karagholi, M.A.; Ghanizada, H.; Nielsen, C.A.W.; Hougaard, A.; Ashina, M. Opening of ATP sensitive potassium channels causes migraine attacks with aura. *Brain* **2021**, *144*, 2322–2332. [CrossRef] [PubMed]
13. Schulte, L.H.; May, A. The migraine generator revisited: Continuous scanning of the migraine cycle over 30 days and three spontaneous attacks. *Brain* **2016**, *139*, 1987–1993. [CrossRef] [PubMed]
14. Kikkeri, N.S.; Nagalli, S. *Migraine with Aura. Treasure Island (FL)*; StatPearls Publishing LLC: Treasure Island, FL, USA, 2022.
15. Borgdorff, P. Arguments against the role of cortical spreading depression in migraine. *Neurol. Res.* **2018**, *40*, 173–181. [CrossRef]
16. Bolay, H.; Reuter, U.; Dunn, A.K.; Huang, Z.; Boas, D.A.; Moskowitz, M.A. Intrinsic brain activity triggers trigeminal meningeal afferents in a migraine model. *Nat. Med.* **2002**, *8*, 136–142. [CrossRef] [PubMed]
17. Zhang, X.; Levy, D.; Kainz, V.; Noseda, R.; Jakubowski, M.; Burstein, R. Activation of central trigeminovascular neurons by cortical spreading depression. *Ann. Neurol.* **2011**, *69*, 855–865. [CrossRef]
18. Goadsby, P.J.; Edvinsson, L.; Ekman, R. Vasoactive peptide release in the extracerebral circulation of humans during migraine headache. *Ann. Neurol.* **1990**, *28*, 183–187. [CrossRef] [PubMed]
19. Edvinsson, L.; Haanes, K.A.; Warfvinge, K.; Krause, D.N. CGRP as the target of new migraine therapies—successful translation from bench to clinic. *Nat. Rev. Neurol.* **2018**, *14*, 338–350. [CrossRef]
20. Ramos, M.L.; Pascual, J. AMG 334 CGRP antibody for migraine: Time to celebrate? *Lancet Neurol.* **2016**, *15*, 347–349. [CrossRef]
21. Negro, A.; Martelletti, P. Gepants for the treatment of migraine. *Expert. Opin. Investig. Drugs* **2019**, *28*, 555–567. [CrossRef]
22. Russo, A.F. Calcitonin gene-related peptide (CGRP): A new target for migraine. *Annu. Rev. Pharmacol. Toxicol.* **2015**, *55*, 533–552. [CrossRef] [PubMed]
23. Haanes, K.A.; Edvinsson, L. Hyperpolarization through ATP-sensitive potassium channels; relevance to migraine pathology. *Brain* **2020**, *143*, e13. [CrossRef] [PubMed]
24. Mulderry, P.K.; Ghatei, M.A.; Bishop, A.E.; Allen, Y.S.; Polak, J.M.; Bloom, S.R. Distribution and chromatographic characterisation of CGRP-like immunoreactivity in the brain and gut of the rat. *Regul. Pept.* **1985**, *12*, 133–143. [CrossRef]
25. Warfvinge, K.; Edvinsson, L. Distribution of CGRP and CGRP receptor components in the rat brain. *Cephalalgia* **2019**, *39*, 342–353. [CrossRef] [PubMed]
26. Edvinsson, J.C.; Reducha, P.V.; Sheykhzade, M.; Warfvinge, K.; Haanes, K.A.; Edvinsson, L. Neurokinins and their receptors in the rat trigeminal system: Differential localization and release with implications for migraine pain. *Mol. Pain* **2021**, *17*, 17448069211059400. [CrossRef] [PubMed]
27. Edvinsson, J.C.A.; Warfvinge, K.; Krause, D.N.; Blixt, F.W.; Sheykhzade, M.; Edvinsson, L.; Haanes, K.A. C-fibers may modulate adjacent Adelta-fibers through axon-axon CGRP signaling at nodes of Ranvier in the trigeminal system. *J. Headache Pain* **2019**, *20*, 105. [CrossRef]
28. Walker, H.K. *Cranial Nerve V, Clinical Methods: The Trigeminal Nerve, the History, Physical, and Laboratory Examinations*; Walker, H.K., Hall, W.D., Hurst, J.W., Eds.; Butterworth Publishers: Boston, MA, USA, 1990.
29. Cottrell, G.S. CGRP Receptor Signalling Pathways. *Handb. Exp. Pharmacol.* **2019**, *255*, 37–64.
30. Moskowitz, M.A. The neurobiology of vascular head pain. *Ann. Neurol.* **1984**, *16*, 157–168. [CrossRef]
31. Markowitz, S.; Saito, K.; Moskowitz, M.A. Neurogenically mediated plasma extravasation in dura mater: Effect of ergot alkaloids. A possible mechanism of action in vascular headache. *Cephalalgia* **1988**, *8*, 83–91. [CrossRef]
32. Shepherd, S.L.; Williamson, D.J.; Beer, M.S.; Hill, R.G.; Hargreaves, R.J. Differential effects of 5-HT1B/1D receptor agonists on neurogenic dural plasma extravasation and vasodilation in anaesthetized rats. *Neuropharmacology* **1997**, *36*, 525–533. [CrossRef]
33. Buzzi, M.G.; Moskowitz, M.A. The antimigraine drug, sumatriptan (GR43175), selectively blocks neurogenic plasma extravasation from blood vessels in dura mater. *Br. J. Pharmacol.* **1990**, *99*, 202–206. [CrossRef] [PubMed]
34. Moskowitz, M.A. Neurogenic versus vascular mechanisms of sumatriptan and ergot alkaloids in migraine. *Trends Pharmacol. Sci.* **1992**, *13*, 307–311. [CrossRef]
35. Theoharides, T.C.; Spanos, C.; Pang, X.; Alferes, L.; Ligris, K.; Letourneau, R.; Rozniecki, J.J.; Webster, E.; Chrousos, G.P. Stress-induced intracranial mast cell degranulation: A corticotropin-releasing hormone-mediated effect. *Endocrinology* **1995**, *136*, 5745–5750. [CrossRef] [PubMed]
36. Theoharides, T.C.; Donelan, J.; Kandere-Grzybowska, K.; Konstantinidou, A. The role of mast cells in migraine pathophysiology. *Brain Res. Rev.* **2005**, *49*, 65–76. [CrossRef]
37. Dalkara, T.; Zervas, N.T.; Moskowitz, M.A. From spreading depression to the trigeminovascular system. *Neurol. Sci.* **2006**, *27*, S86–S90. [CrossRef]
38. Levy, D. Endogenous mechanisms underlying the activation and sensitization of meningeal nociceptors: The role of immuno-vascular interactions and cortical spreading depression. *Curr. Pain Headache Rep.* **2012**, *16*, 270–277. [CrossRef]

39. Johnson KW, B.H. *Neurogenic Inflammatory Mechanisms*; Lipincott Williams & Wilkins: Philadelphia, PA, USA, 2006.
40. Lassen, L.H.; Ashina, M.; Christiansen, I.; Ulrich, V.; Olesen, J. Nitric oxide synthase inhibition in migraine. *Lancet* **1997**, *349*, 401–402. [CrossRef]
41. Covelli, V.; Munno, I.; Pellegrino, N.M.; Altamura, M.; Decandia, P.; Marcuccio, C.; di Venere, A.; Jirillo, E. Are TNF-alpha and IL-1 beta relevant in the pathogenesis of migraine without aura? *Acta Neurol. Napoli* **1991**, *13*, 205–211.
42. Perini, F.; D'Andrea, G.; Galloni, E.; Pignatelli, F.; Billo, G.; Alba, S.; Bussone, G.; Toso, V. Plasma cytokine levels in migraineurs and controls. *Headache* **2005**, *45*, 926–931. [CrossRef]
43. Franceschini, A.; Vilotti, S.; Ferrari, M.D.; van den Maagdenberg, A.M.; Nistri, A.; Fabbretti, E. TNFalpha levels and macrophages expression reflect an inflammatory potential of trigeminal ganglia in a mouse model of familial hemiplegic migraine. *PLoS ONE* **2013**, *8*, e52394. [CrossRef]
44. Tanure, M.T.; Gomez, R.S.; Hurtado, R.C.; Teixeira, A.L.; Domingues, R.B. Increased serum levels of brain-derived neurotropic factor during migraine attacks: A pilot study. *J. Headache Pain* **2010**, *11*, 427–430. [CrossRef] [PubMed]
45. Yucel, M.; Kotan, D.; Ciftci, G.G.; Ciftci, I.H.; Cikriklar, H.I. Serum levels of endocan, claudin-5 and cytokines in migraine. *Eur. Rev. Med. Pharmacol. Sci.* **2016**, *20*, 930–936. [PubMed]
46. Sarchielli, P.; Alberti, A.; Baldi, A.; Coppola, F.; Rossi, C.; Pierguidi, L.; Floridi, A.; Calabresi, P. Proinflammatory cytokines, adhesion molecules, and lymphocyte integrin expression in the internal jugular blood of migraine patients without aura assessed ictally. *Headache* **2006**, *46*, 200–207. [CrossRef] [PubMed]
47. Edvinsson, L.; Haanes, K.A.; Warfvinge, K. Does inflammation have a role in migraine? *Nat. Rev. Neurol.* **2019**, *15*, 483–490. [CrossRef]
48. de Logu, F.; Landini, L.; Janal, M.N.; Puma, S.L.; de Cesaris, F.; Geppetti, P.; Nassini, R. Migraine-provoking substances evoke periorbital allodynia in mice. *J. Headache Pain* **2019**, *20*, 18. [CrossRef]
49. Nakamura-Craig, M.; Gill, B.K. Effect of neurokinin A, substance P and calcitonin gene related peptide in peripheral hyperalgesia in the rat paw. *Neurosci. Lett.* **1991**, *124*, 49–51. [CrossRef]
50. Birrell, G.J.; McQueen, D.S.; Iggo, A.; Coleman, R.A.; Grubb, B.D. PGI2-induced activation and sensitization of articular mechanonociceptors. *Neurosci. Lett.* **1991**, *124*, 5–8. [CrossRef]
51. Wang, H.; Ehnert, C.; Brenner, G.J.; Woolf, C.J. Bradykinin and peripheral sensitization. *Biol. Chem.* **2006**, *387*, 11–14. [CrossRef]
52. Schaible, H.G.; Schmidt, R.F. Excitation and sensitization of fine articular afferents from cat's knee joint by prostaglandin E2. *J. Physiol.* **1988**, *403*, 91–104. [CrossRef]
53. Uddman, R.; Edvinsson, L.; Jansen, I.; Stiernholm, P.; Jensen, K.; Olesen, J.; Sundler, F. Peptide-containing nerve fibres in human extracranial tissue: A morphological basis for neuropeptide involvement in extracranial pain? *Pain* **1986**, *27*, 391–399. [CrossRef]
54. Harriott, A.M.; Gold, M.S. Electrophysiological properties of dural afferents in the absence and presence of inflammatory mediators. *J. Neurophysiol.* **2009**, *101*, 3126–3134. [CrossRef] [PubMed]
55. Harriott, A.M.; Scheff, N.N.; Gold, M.S. The complex actions of sumatriptan on rat dural afferents. *Cephalalgia* **2012**, *32*, 738–749. [CrossRef] [PubMed]
56. Scheff, N.N.; Gold, M.S. Sex differences in the inflammatory mediator-induced sensitization of dural afferents. *J. Neurophysiol.* **2011**, *106*, 1662–1668. [CrossRef] [PubMed]
57. Oshinsky, M.L.; Gomonchareonsiri, S. Episodic dural stimulation in awake rats: A model for recurrent headache. *Headache* **2007**, *47*, 1026–1036. [CrossRef] [PubMed]
58. Edelmayer, R.M.; Vanderah, T.W.; Majuta, L.; Zhang, E.T.; Fioravanti, B.; de Felice, M.; Chichorro, J.G.; Ossipov, M.H.; King, T.; Lai, J.; et al. Medullary pain facilitating neurons mediate allodynia in headache-related pain. *Ann. Neurol.* **2009**, *65*, 184–193. [CrossRef]
59. Zhu, P.; Dong, X.; Xu, H.; Wan, Q.; Guo, Q.; Wang, J.; Xiao, H.; Yan, L. Microglial P2Y14 receptor contributes to central sensitization following repeated inflammatory dural stimulation. *Brain Res. Bull.* **2021**, *177*, 119–128. [CrossRef]
60. Hu, G.; Zhang, M.; Su, M.; Zhang, Q.; Wu, H.; Wang, X.; Dong, Z.; Yu, S. Wider range of allodynia in a rat model of repeated dural nociception compared with infraorbital nerve chronic constriction injury. *Neurosci. Lett.* **2018**, *666*, 120–126. [CrossRef]
61. Melo-Carrillo, A.; Lopez-Avila, A. A chronic animal model of migraine, induced by repeated meningeal nociception, characterized by a behavioral and pharmacological approach. *Cephalalgia* **2013**, *33*, 1096–1105. [CrossRef]
62. Liu, D.; Chang, L.; Wang, J.; Zhu, Q.; Jiang, N.; Azhar, M.; Ahmed, A.; Zeng, G. Effect of Xiongmatang Extract on Behavioral and TRPV1-CGRP/CGRP-R Pathway in Rats with Migraine. *Front. Pharmacol.* **2022**, *13*, 835187. [CrossRef]
63. Fan, S.; Xiao, Z.; Zhu, F.; He, X.; Lu, Z. A new comorbidity model and the common pathological mechanisms of migraine and epilepsy. *Am. J. Transl. Res.* **2017**, *9*, 2286–2295.
64. Liu, L.; Xu, X.B.; Qu, Z.Y.; Zhao, L.P.; Zhang, C.S.; Li, Z.J.; Lyu, T.L.; Wang, X.F.; Jing, X.H.; Li, B. Determining 5HT(7)R's Involvement in Modifying the Antihyperalgesic Effects of Electroacupuncture on Rats With Recurrent Migraine. *Front. Neurosci.* **2021**, *15*, 668616. [CrossRef] [PubMed]
65. Nan, N.; Gong, M.X.; Wang, Q.; Li, M.J.; Xu, R.; Ma, Z.; Wang, S.H.; Zhao, H.; Xu, Y.S. Wuzhuyu Decoction relieves hyperalgesia by regulating central and peripheral 5-HT in chronic migraine model rats. *Phytomedicine* **2022**, *96*, 153905. [CrossRef] [PubMed]
66. Zhou, H.; Wang, X.; Wang, S.; Liu, C.; Fu, Q.; Qin, G.; Zhou, J.; Chen, L. Inhibition of Nerve Growth Factor Signaling Alleviates Repeated Dural Stimulation-induced Hyperalgesia in Rats. *Neuroscience* **2019**, *398*, 252–262. [CrossRef]

67. Liu, C.; Zhang, Y.; Liu, Q.; Jiang, L.; Li, M.; Wang, S.; Long, T.; He, W.; Kong, X.; Qin, G.; et al. P2X4-receptor participates in EAAT3 regulation via BDNF-TrkB signaling in a model of trigeminal allodynia. *Mol. Pain* **2018**, *14*, 1744806918795930. [CrossRef] [PubMed]
68. Jia, Z.; Tang, W.; Zhao, D.; Yu, S. Disrupted functional connectivity between the periaqueductal gray and other brain regions in a rat model of recurrent headache. *Sci. Rep.* **2017**, *7*, 3960. [CrossRef]
69. Gong, Q.; Lin, Y.; Lu, Z.; Xiao, Z. Microglia-Astrocyte Cross Talk through IL-18/IL-18R Signaling Modulates Migraine-like Behavior in Experimental Models of Migraine. *Neuroscience* **2020**, *451*, 207–215. [CrossRef] [PubMed]
70. Wieseler, J.; Ellis, A.; McFadden, A.; Stone, K.; Brown, K.; Cady, S.; Bastos, L.F.; Sprunger, D.; Rezvani, N.; Johnson, K.; et al. Supradural inflammatory soup in awake and freely moving rats induces facial allodynia that is blocked by putative immune modulators. *Brain Res.* **2017**, *1664*, 87–94. [CrossRef] [PubMed]
71. Su, M.; Ran, Y.; He, Z.; Zhang, M.; Hu, G.; Tang, W.; Zhao, D.; Yu, S. Inhibition of toll-like receptor 4 alleviates hyperalgesia induced by acute dural inflammation in experimental migraine. *Mol. Pain* **2018**, *14*, 1744806918754612. [CrossRef]
72. Zhou, Y.; Zhang, L.; Hao, Y.; Yang, L.; Fan, S.; Xiao, Z. FKN/CX3CR1 axis facilitates migraine-Like behaviour by activating thalamic-cortical network microglia in status epilepticus model rats. *J. Headache Pain* **2022**, *23*, 42. [CrossRef]
73. Vellani, V.; Moschetti, G.; Franchi, S.; Giacomoni, C.; Sacerdote, P.; Amodeo, G. Effects of NSAIDs on the Release of Calcitonin Gene-Related Peptide and Prostaglandin E2 from Rat Trigeminal Ganglia. *Mediat. Inflamm.* **2017**, *2017*, 9547056. [CrossRef]
74. Fried, N.T.; Maxwell, C.R.; Elliott, M.B.; Oshinsky, M.L. Region-specific disruption of the blood-brain barrier following repeated inflammatory dural stimulation in a rat model of chronic trigeminal allodynia. *Cephalalgia* **2018**, *38*, 674–689. [CrossRef] [PubMed]
75. Liu, Q.; Liu, C.; Jiang, L.; Li, M.; Long, T.; He, W.; Qin, G.; Chen, L.; Zhou, J. α7 Nicotinic acetylcholine receptor-mediated anti-inflammatory effect in a chronic migraine rat model via the attenuation of glial cell activation. *J. Pain Res.* **2018**, *11*, 1129–1140. [CrossRef] [PubMed]
76. Wang, X.Y.; Zhou, H.R.; Wang, S.; Liu, C.Y.; Qin, G.C.; Fu, Q.Q.; Zhou, J.Y.; Chen, L.X. NR2B-Tyr phosphorylation regulates synaptic plasticity in central sensitization in a chronic migraine rat model. *J. Headache Pain* **2018**, *19*, 102. [CrossRef] [PubMed]
77. Wu, B.; Wang, S.; Qin, G.; Xie, J.; Tan, G.; Zhou, J.; Chen, L. Protein Kinase C γ Contributes to Central Sensitization in a Rat Model of Chronic Migraine. *J. Mol. Neurosci.* **2017**, *63*, 131–141. [CrossRef]
78. Stucky, N.L.; Gregory, E.; Winter, M.K.; He, Y.Y.; Hamilton, E.S.; McCarson, K.E.; Berman, N.E. Sex differences in behavior and expression of CGRP-related genes in a rodent model of chronic migraine. *Headache* **2011**, *51*, 674–692. [CrossRef]
79. Liang, X.; Wang, S.; Qin, G.; Xie, J.; Tan, G.; Zhou, J.; McBride, D.W.; Chen, L. Tyrosine Phosphorylation of NR2B Contributes to Chronic Migraines via Increased Expression of CGRP in Rats. *Biomed. Res. Int.* **2017**, *2017*, 7203458. [CrossRef]
80. Liang, J.; Zhou, X.; Wang, J.; Fei, Z.Y.; Qin, G.C.; Zhang, D.K.; Zhou, J.Y.; Chen, L.X. Upregulation of silent information regulator 1 alleviates mitochondrial dysfunction in the trigeminal nucleus caudalis in a rat model of chronic migraine. *Neuroreport* **2021**, *32*, 144–156. [CrossRef]
81. Wang, J.; Fei, Z.; Liang, J.; Zhou, X.; Qin, G.; Zhang, D.; Zhou, J.; Chen, L. EphrinB/EphB Signaling Contributes to the Synaptic Plasticity of Chronic Migraine Through NR2B Phosphorylation. *Neuroscience* **2020**, *428*, 178–191. [CrossRef]
82. Wang, S.; Wu, B.X.; Liu, C.Y.; Qin, G.C.; Yan, W.H.; Zhou, J.Y.; Chen, L.X. Expression of ASIC3 in the Trigeminal Nucleus Caudalis Plays a Role in a Rat Model of Recurrent Migraine. *J. Mol. Neurosci.* **2018**, *66*, 44–52. [CrossRef]
83. Chen, N.; Su, W.; Cui, S.H.; Guo, J.; Duan, J.C.; Li, H.X.; He, L. A novel large animal model of recurrent migraine established by repeated administration of inflammatory soup into the dura mater of the rhesus monkey. *Neural. Regen. Res.* **2019**, *14*, 100–106.
84. Tuka, B.; Helyes, Z.; Markovics, A.; Bagoly, T.; Szolcsányi, J.; Szabó, N.; Tóth, E.; Kincses, Z.T.; Vécsei, L.; Tajti, J. Alterations in PACAP-38-like immunoreactivity in the plasma during ictal and interictal periods of migraine patients. *Cephalalgia* **2013**, *33*, 1085–1095. [CrossRef] [PubMed]
85. Han, X.; Dong, Z.; Hou, L.; Wan, D.; Chen, M.; Tang, W.; Yu, S. Interictal plasma pituitary adenylate cyclase-activating polypeptide levels are decreased in migraineurs but remain unchanged in patients with tension-type headache. *Clin. Chim. Acta* **2015**, *450*, 151–154. [CrossRef] [PubMed]
86. Han, X.; Ran, Y.; Su, M.; Liu, Y.; Tang, W.; Dong, Z.; Yu, S. Chronic changes in pituitary adenylate cyclase-activating polypeptide and related receptors in response to repeated chemical dural stimulation in rats. *Mol. Pain* **2017**, *13*, 1744806917720361. [CrossRef] [PubMed]
87. Burgos-Vega, C.C.; Ahn, D.D.; Bischoff, C.; Wang, W.; Horne, D.; Wang, J.; Gavva, N.; Dussor, G. Meningeal transient receptor potential channel M8 activation causes cutaneous facial and hindpaw allodynia in a preclinical rodent model of headache. *Cephalalgia* **2016**, *36*, 185–193. [CrossRef] [PubMed]
88. Dong, X.; Guan, X.; Chen, K.; Jin, S.; Wang, C.; Yan, L.; Shi, Z.; Zhang, X.; Chen, L.; Wan, Q. Abnormal mitochondrial dynamics and impaired mitochondrial biogenesis in trigeminal ganglion neurons in a rat model of migraine. *Neurosci. Lett.* **2017**, *636*, 127–133. [CrossRef]
89. Jia, Z.; Yu, S.; Tang, W.; Zhao, D. Altered functional connectivity of the insula in a rat model of recurrent headache. *Mol. Pain* **2020**, *16*, 1744806920922115. [CrossRef]
90. Jia, Z.; Chen, X.; Tang, W.; Zhao, D.; Yu, S. Atypical functional connectivity between the anterior cingulate cortex and other brain regions in a rat model of recurrent headache. *Mol. Pain* **2019**, *15*, 1744806919842483. [CrossRef]
91. Becerra, L.; Bishop, J.; Barmettler, G.; Kainz, V.; Burstein, R.; Borsook, D. Brain network alterations in the inflammatory soup animal model of migraine. *Brain Res.* **2017**, *1660*, 36–46. [CrossRef]

92. Bishop, J.; Becerra, L.; Barmettler, G.; Chang, P.C.; Kainz, V.; Burstein, R.; Borsook, D. Modulation of brain networks by sumatriptan-naproxen in the inflammatory soup migraine model. *Pain* **2019**, *160*, 2161–2171. [CrossRef]
93. Zhang, M.; Liu, Y.; Hu, G.; Kang, L.; Ran, Y.; Su, M.; Yu, S. Cognitive impairment in a classical rat model of chronic migraine may be due to alterations in hippocampal synaptic plasticity and N-methyl-D-aspartate receptor subunits. *Mol. Pain* **2020**, *16*, 1744806920959582. [CrossRef]
94. Zeng, X.; Niu, Y.; Qin, G.; Zhang, D.; Zhou, J.; Chen, L. Deficiency in the function of inhibitory interneurons contributes to glutamate-associated central sensitization through GABABR2-SynCAM1 signaling in chronic migraine rats. *FASEB J.* **2020**, *34*, 14780–14798. [CrossRef] [PubMed]
95. Niu, Y.; Zeng, X.; Qin, G.; Zhang, D.; Zhou, J.; Chen, L. Downregulation of metabotropic glutamate receptor 5 alleviates central sensitization by activating autophagy via inhibiting mTOR pathway in a rat model of chronic migraine. *Neurosci. Lett.* **2021**, *743*, 135552. [CrossRef] [PubMed]
96. Boyer, N.; Dallel, R.; Artola, A.; Monconduit, L. General trigeminospinal central sensitization and impaired descending pain inhibitory controls contribute to migraine progression. *Pain* **2014**, *155*, 1196–1205. [CrossRef] [PubMed]
97. Mi, X.; Ran, L.; Chen, L.; Qin, G. Recurrent Headache Increases Blood-Brain Barrier Permeability and VEGF Expression in Rats. *Pain Physician.* **2018**, *21*, E633–E642. [PubMed]
98. Levy, D.; Zhang, X.C.; Jakubowski, M.; Burstein, R. Sensitization of meningeal nociceptors: Inhibition by naproxen. *Eur. J. Neurosci.* **2008**, *27*, 917–922. [CrossRef]
99. Lambert, G.A.; Davis, J.B.; Appleby, J.M.; Chizh, B.A.; Hoskin, K.L.; Zagami, A.S. The effects of the TRPV1 receptor antagonist SB-705498 on trigeminovascular sensitisation and neurotransmission. *Naunyn Schmiedeberg's Arch. Pharmcol.* **2009**, *380*, 311–325. [CrossRef]
100. Qu, Z.; Liu, L.; Yang, Y.; Zhao, L.; Xu, X.; Li, Z.; Zhu, Y.; Jing, X.; Wang, X.; Zhang, C.S.; et al. Electro-acupuncture inhibits C-fiber-evoked WDR neuronal activity of the trigeminocervical complex: Neurophysiological hypothesis of a complementary therapy for acute migraine modeled rats. *Brain Res.* **2020**, *1730*, 146670. [CrossRef]
101. Boyer, N.; Signoret-Genest, J.; Artola, A.; Dallel, R.; Monconduit, L. Propranolol treatment prevents chronic central sensitization induced by repeated dural stimulation. *Pain* **2017**, *158*, 2025–2034. [CrossRef]
102. Jia, Z.; Tang, W.; Zhao, D.; Hu, G.; Li, R.; Yu, S. Volumetric abnormalities of the brain in a rat model of recurrent headache. *Mol. Pain* **2018**, *14*, 1744806918756466. [CrossRef]
103. Togha, M.; Jahromi, S.R.; Ghorbani, Z.; Ghaemi, A.; Rafiee, P. Evaluation of Inflammatory State in Migraineurs: A Case-control Study. *Iran. J. Allergy Asthma Immunol.* **2020**, *19*, 83–90. [CrossRef]
104. Burgos-Vega, C.C.; Quigley, L.D.; Avona, A.; Price, T.; Dussor, G. Dural stimulation in rats causes brain-derived neurotrophic factor-dependent priming to subthreshold stimuli including a migraine trigger. *Pain* **2016**, *157*, 2722–2730. [CrossRef] [PubMed]
105. Avona, A.; Price, T.J.; Dussor, G. Interleukin-6 induces spatially dependent whole-body hypersensitivity in rats: Implications for extracephalic hypersensitivity in migraine. *J. Headache Pain* **2021**, *22*, 70. [CrossRef] [PubMed]
106. Avona, A.; Burgos-Vega, C.; Burton, M.D.; Akopian, A.N.; Price, T.J.; Dussor, G. Dural Calcitonin Gene-Related Peptide Produces Female-Specific Responses in Rodent Migraine Models. *J. Neurosci.* **2019**, *39*, 4323–4331. [CrossRef]
107. Yan, J.; Melemedjian, O.K.; Price, T.J.; Dussor, G. Sensitization of dural afferents underlies migraine-related behavior following meningeal application of interleukin-6 (IL-6). *Mol. Pain* **2012**, *8*, 6. [CrossRef]
108. Burgos-Vega, C.C.; Quigley, L.D.; Santos, G.T.D.; Yan, F.; Asiedu, M.; Jacobs, B.; Motina, M.; Safdar, N.; Yousuf, H.; Avona, A.; et al. Non-invasive dural stimulation in mice: A novel preclinical model of migraine. *Cephalalgia* **2019**, *39*, 123–134. [CrossRef]
109. Qu, Z.; Liu, L.; Zhao, L.; Xu, X.; Li, Z.; Zhu, Y.; Zhang, C.; Jing, X.; Wang, X.; Li, B.; et al. Prophylactic Electroacupuncture on the Upper Cervical Segments Decreases Neuronal Discharges of the Trigeminocervical Complex in Migraine-Affected Rats: An in vivo Extracellular Electrophysiological Experiment. *J. Pain Res.* **2020**, *13*, 25–37. [CrossRef]
110. Lackovic, J.; Price, T.J.; Dussor, G. De novo protein synthesis is necessary for priming in preclinical models of migraine. *Cephalalgia* **2021**, *41*, 237–246. [CrossRef] [PubMed]
111. Kristiansen, K.A.; Edvinsson, L. Neurogenic inflammation: A study of rat trigeminal ganglion. *J. Headache Pain* **2010**, *11*, 485–495. [CrossRef] [PubMed]
112. Kuris, A.; Xu, C.B.; Zhou, M.F.; Tajti, J.; Uddman, R.; Edvinsson, L. Enhanced expression of CGRP in rat trigeminal ganglion neurons during cell and organ culture. *Brain Res.* **2007**, *1173*, 6–13. [CrossRef] [PubMed]
113. Csati, A.; Edvinsson, L.; Vecsei, L.; Toldi, J.; Fulop, F.; Tajti, J.; Warfvinge, K. Kynurenic acid modulates experimentally induced inflammation in the trigeminal ganglion. *J. Headache Pain* **2015**, *16*, 99. [CrossRef]
114. Burstein, R.; Jakubowski, M. Analgesic triptan action in an animal model of intracranial pain: A race against the development of central sensitization. *Ann. Neurol.* **2004**, *55*, 27–36. [CrossRef] [PubMed]
115. Lukacs, M.; Haanes, K.A.; Majlath, Z.; Tajti, J.; Vecsei, L.; Warfvinge, K.; Edvinsson, L. Dural administration of inflammatory soup or Complete Freund's Adjuvant induces activation and inflammatory response in the rat trigeminal ganglion. *J. Headache Pain* **2015**, *16*, 564. [CrossRef] [PubMed]
116. Lukacs, M.; Warfvinge, K.; Tajti, J.; Fulop, F.; Toldi, J.; Vecsei, L.; Edvinsson, L. Topical dura mater application of CFA induces enhanced expression of c-fos and glutamate in rat trigeminal nucleus caudalis: Attenuated by KYNA derivate (SZR72). *J. Headache Pain* **2017**, *18*, 39. [CrossRef] [PubMed]

117. Tajti, J.; Kuris, A.; Vecsei, L.; Xu, C.B.; Edvinsson, L. Organ culture of the trigeminal ganglion induces enhanced expression of calcitonin gene-related peptide via activation of extracellular signal-regulated protein kinase 1/2. *Cephalalgia* **2011**, *31*, 95–105. [CrossRef] [PubMed]
118. Takeda, M.; Tanimoto, T.; Kadoi, J.; Nasu, M.; Takahashi, M.; Kitagawa, J.; Matsumoto, S. Enhanced excitability of nociceptive trigeminal ganglion neurons by satellite glial cytokine following peripheral inflammation. *Pain* **2007**, *129*, 155–166. [CrossRef] [PubMed]
119. Takeda, M.; Takahashi, M.; Matsumoto, S. Contribution of activated interleukin receptors in trigeminal ganglion neurons to hyperalgesia via satellite glial interleukin-1beta paracrine mechanism. *Brain Behav. Immun.* **2008**, *22*, 1016–1023. [CrossRef]
120. Burstein, R.; Yamamura, H.; Malick, A.; Strassman, A.M. Chemical stimulation of the intracranial dura induces enhanced responses to facial stimulation in brain stem trigeminal neurons. *J. Neurophysiol.* **1998**, *79*, 964–982. [CrossRef]
121. Strassman, A.M.; Raymond, S.A.; Burstein, R. Sensitization of meningeal sensory neurons and the origin of headaches. *Nature* **1996**, *384*, 560–564. [CrossRef]
122. Lukacs, M.; Warfvinge, K.; Kruse, L.S.; Tajti, J.; Fulop, F.; Toldi, J.; Vecsei, L.; Edvinsson, L. KYNA analogue SZR72 modifies CFA-induced dural inflammation- regarding expression of pERK1/2 and IL-1beta in the rat trigeminal ganglion. *J. Headache Pain* **2016**, *17*, 64. [CrossRef]
123. Romero-Reyes, M.; Pardi, V.; Akerman, S. A potent and selective calcitonin gene-related peptide (CGRP) receptor antagonist, MK-8825, inhibits responses to nociceptive trigeminal activation: Role of CGRP in orofacial pain. *Exp. Neurol.* **2015**, *271*, 95–103. [CrossRef]
124. Janeway, C.A., Jr. Approaching the asymptote? Evolution and revolution in immunology. *Cold Spring Harb. Symp. Quant. Biol.* **1989**, *54*, 1–13. [CrossRef] [PubMed]
125. Spekker, E.; Laborc, K.F.; Bohár, Z.; Nagy-Grócz, G.; Fejes-Szabó, A.; Szűcs, M.; Vécsei, L.; Párdutz, Á. Effect of dural inflammatory soup application on activation and sensitization markers in the caudal trigeminal nucleus of the rat and the modulatory effects of sumatriptan and kynurenic acid. *J. Headache Pain* **2021**, *22*, 17. [CrossRef] [PubMed]
126. Pinheiro, A.R.; Paramos-de-Carvalho, D.; Certal, M.; Costa, C.; Magalhães-Cardoso, M.T.; Ferreirinha, F.; Costa, M.A.; Correia-de-Sá, P. Bradykinin-induced Ca2+ signaling in human subcutaneous fibroblasts involves ATP release via hemichannels leading to P2Y12 receptors activation. *Cell Commun. Signal.* **2013**, *11*, 70. [CrossRef] [PubMed]
127. Pinheiro, A.R.; Paramos-de-Carvalho, D.; Certal, M.; Costa, M.A.; Costa, C.; Magalhaes-Cardoso, M.T.; Ferreirinha, F.; Sevigny, J.; Correia-de-Sa, P. Histamine induces ATP release from human subcutaneous fibroblasts, via pannexin-1 hemichannels, leading to Ca2+ mobilization and cell proliferation. *J. Biol. Chem.* **2013**, *288*, 27571–27583. [CrossRef] [PubMed]
128. Kowal, J.M.; Haanes, K.A.; Christensen, N.M.; Novak, I. Bile acid effects are mediated by ATP release and purinergic signalling in exocrine pancreatic cells. *Cell Commun. Signal* **2015**, *13*, 28. [CrossRef]
129. Beckel, J.M.; Gomez, N.M.; Lu, W.; Campagno, K.E.; Nabet, B.; Albalawi, F.; Lim, J.C.; Boesze-Battaglia, K.; Mitchell, C.H. Stimulation of TLR3 triggers release of lysosomal ATP in astrocytes and epithelial cells that requires TRPML1 channels. *Sci. Rep.* **2018**, *8*, 5726. [CrossRef]
130. Matsuka, Y.; Neubert, J.K.; Maidment, N.T.; Spigelman, I. Concurrent release of ATP and substance P within guinea pig trigeminal ganglia in vivo. *Brain Res.* **2001**, *915*, 248–255. [CrossRef]
131. Koroleva, K.; Gafurov, O.; Guselnikova, V.; Nurkhametova, D.; Giniatullina, R.; Sitdikova, G.; Mattila, O.S.; Lindsberg, P.J.; Malm, T.M.; Giniatullin, R. Meningeal Mast Cells Contribute to ATP-Induced Nociceptive Firing in Trigeminal Nerve Terminals: Direct and Indirect Purinergic Mechanisms Triggering Migraine Pain. *Front. Cell. Neurosci.* **2019**, *13*, 195. [CrossRef]
132. Haanes, K.A.; Edvinsson, L. Expression and characterization of purinergic receptors in rat middle meningeal artery-potential role in migraine. *PLoS ONE* **2014**, *9*, e108782. [CrossRef]
133. Yegutkin, G.G.; Guerrero-Toro, C.; Kilinc, E.; Koroleva, K.; Ishchenko, Y.; Abushik, P.; Giniatullina, R.; Fayuk, D.; Giniatullin, R. Nucleotide homeostasis and purinergic nociceptive signaling in rat meninges in migraine-like conditions. *Purinergic Signal* **2016**, *12*, 561–574. [CrossRef]
134. Haanes, K.A.; Labastida-Ramirez, A.; Blixt, F.W.; Rubio-Beltran, E.; Dirven, C.M.; Danser, A.H.; Edvinsson, L.; MaassenVanDen-Brink, A. Exploration of purinergic receptors as potential anti-migraine targets using established pre-clinical migraine models. *Cephalalgia* **2019**, *39*, 1421–1434. [CrossRef] [PubMed]
135. Dussor, G. Serotonin, 5HT1 agonists, and migraine: New data, but old questions still not answered. *Curr. Opin. Support Palliat Care* **2014**, *8*, 137–142. [CrossRef] [PubMed]
136. Laborc, K.F.; Spekker, E.; Bohar, Z.; Szucs, M.; Nagy-Grocz, G.; Fejes-Szabo, A.; Vecsei, L.; Pardutz, A. Trigeminal activation patterns evoked by chemical stimulation of the dura mater in rats. *J. Headache Pain* **2020**, *21*, 101. [CrossRef] [PubMed]
137. Harriott, A.M.; Strother, L.C.; Vila-Pueyo, M.; Holland, P.R. Animal models of migraine and experimental techniques used to examine trigeminal sensory processing. *J. Headache Pain* **2019**, *20*, 91. [CrossRef] [PubMed]
138. Vingen, J.V.; Pareja, J.A.; Støren, O.; White, L.R.; Stovner, L.J. Phonophobia in migraine. *Cephalalgia* **1998**, *18*, 243–249. [CrossRef]
139. Noseda, R.; Copenhagen, D.; Burstein, R. Current understanding of photophobia, visual networks and headaches. *Cephalalgia* **2019**, *39*, 1623–1634. [CrossRef]
140. Farajdokht, F.; Babri, S.; Karimi, P.; Mohaddes, G. Ghrelin attenuates hyperalgesia and light aversion-induced by nitroglycerin in male rats. *Neurosci. Lett.* **2016**, *630*, 30–37. [CrossRef]

141. Vuralli, D.; Wattiez, A.S.; Russo, A.F.; Bolay, H. Behavioral and cognitive animal models in headache research. *J. Headache Pain* **2019**, *20*, 11. [CrossRef]
142. Bourin, M.; Hascoët, M. The mouse light/dark box test. *Eur. J. Pharmacol.* **2003**, *463*, 55–65. [CrossRef]
143. Parent, A.J.; Beaudet, N.; Beaudry, H.; Bergeron, J.; Bérubé, P.; Drolet, G.; Sarret, P.; Gendron, L. Increased anxiety-like behaviors in rats experiencing chronic inflammatory pain. *Behav. Brain Res.* **2012**, *229*, 160–167. [CrossRef]
144. Hofmann, L.; Karl, F.; Sommer, C.; Üçeyler, N. Affective and cognitive behavior in the alpha-galactosidase A deficient mouse model of Fabry disease. *PLoS ONE* **2017**, *12*, e0180601. [CrossRef]
145. Sufka, K.J.; Staszko, S.M.; Johnson, A.P.; Davis, M.E.; Davis, R.E.; Smitherman, T.A. Clinically relevant behavioral endpoints in a recurrent nitroglycerin migraine model in rats. *J. Headache Pain* **2016**, *17*, 40. [CrossRef] [PubMed]
146. Lantéri-Minet, M.; Radat, F.; Chautard, M.H.; Lucas, C. Anxiety and depression associated with migraine: Influence on migraine subjects' disability and quality of life, and acute migraine management. *Pain* **2005**, *118*, 319–326. [CrossRef] [PubMed]
147. Kraeuter, A.K.; Guest, P.C.; Sarnyai, Z. The Open Field Test for Measuring Locomotor Activity and Anxiety-Like Behavior. *Methods Mol. Biol.* **2019**, *1916*, 99–103. [PubMed]
148. Ferrari, M.D. Migraine. *Lancet* **1998**, *351*, 1043–1051. [CrossRef]
149. Moye, L.S.; Pradhan, A.A.A. Animal Model of Chronic Migraine-Associated Pain. *Curr. Protoc. Neurosci.* **2017**, *80*, 9.60.1–9.60.9. [CrossRef]
150. Ray, J.C.; Allen, P.; Bacsi, A.; Bosco, J.J.; Chen, L.; Eller, M.; Kua, H.; Lim, L.L.; Matharu, M.S.; Monif, M.; et al. Inflammatory complications of CGRP monoclonal antibodies: A case series. *J. Headache Pain* **2021**, *22*, 121. [CrossRef]
151. Borkum, J.M. Brain Energy Deficit as a Source of Oxidative Stress in Migraine: A Molecular Basis for Migraine Susceptibility. *Neurochem. Res.* **2021**, *46*, 1913–1932. [CrossRef]
152. Reyngoudt, H.; Achten, E.; Paemeleire, K. Magnetic resonance spectroscopy in migraine: What have we learned so far? *Cephalalgia* **2012**, *32*, 845–859. [CrossRef]
153. Younis, S.; Hougaard, A.; Vestergaard, M.B.; Larsson, H.B.W.; Ashina, M. Migraine and magnetic resonance spectroscopy: A systematic review. *Curr. Opin. Neurol.* **2017**, *30*, 246–262. [CrossRef]
154. Tiehuis, L.H.; Koene, S.; Saris, C.G.J.; Janssen, M.C.H. Mitochondrial migraine; a prevalence, impact and treatment efficacy cohort study. *Mitochondrion* **2020**, *53*, 128–132. [CrossRef] [PubMed]
155. Dalkara, T.; Kilic, K. How does fasting trigger migraine? A hypothesis. *Curr. Pain Headache Rep.* **2013**, *17*, 368. [CrossRef]
156. Stuart, S.; Griffiths, L.R. A possible role for mitochondrial dysfunction in migraine. *Mol. Genet. Genom.* **2012**, *287*, 837–844. [CrossRef] [PubMed]
157. More, S.V.; Kumar, H.; Kim, I.S.; Song, S.Y.; Choi, D.K. Cellular and molecular mediators of neuroinflammation in the pathogenesis of Parkinson's disease. *Mediat. Inflamm* **2013**, *2013*, 952375. [CrossRef] [PubMed]
158. Kielian, T. Multifaceted roles of neuroinflammation: The need to consider both sides of the coin. *J. Neurochem.* **2016**, *136*, 5–9. [CrossRef]
159. Lucas, S.M.; Rothwell, N.J.; Gibson, R.M. The role of inflammation in CNS injury and disease. *Br. J. Pharmacol.* **2006**, *147*, S232–S240. [CrossRef]
160. Winters, B.L.; Jeong, H.J.; Vaughan, C.W. Inflammation induces developmentally regulated sumatriptan inhibition of spinal synaptic transmission. *Br. J. Pharmacol.* **2020**, *177*, 3730–3743. [CrossRef]
161. Cornelison, L.E.; Woodman, S.E.; Durham, P.L. Inhibition of Trigeminal Nociception by Non-invasive Vagus Nerve Stimulation: Investigating the Role of GABAergic and Serotonergic Pathways in a Model of Episodic Migraine. *Front Neurol.* **2020**, *11*, 146. [CrossRef]
162. Yarwood, R.E.; Imlach, W.L.; Lieu, T.; Veldhuis, N.A.; Jensen, D.D.; Herenbrink, C.K.; Aurelio, L.; Cai, Z.; Christie, M.J.; Poole, D.P.; et al. Endosomal signaling of the receptor for calcitonin gene-related peptide mediates pain transmission. *Proc. Natl. Acad. Sci. USA* **2017**, *114*, 12309–12314. [CrossRef]
163. Hirsch, S.; Corradini, L.; Just, S.; Arndt, K.; Doods, H. The CGRP receptor antagonist BIBN4096BS peripherally alleviates inflammatory pain in rats. *Pain* **2013**, *154*, 700–707. [CrossRef]
164. Wang, D.; Ruan, L.; Hong, Y.; Chabot, J.G.; Quirion, R. Involvement of PKA-dependent upregulation of nNOS-CGRP in adrenomedullin-initiated mechanistic pathway underlying CFA-induced response in rats. *Exp. Neurol.* **2013**, *239*, 111–119. [CrossRef] [PubMed]
165. Warfvinge, K.; Krause, D.N.; Maddahi, A.; Edvinsson, J.C.A.; Edvinsson, L.; Haanes, K.A. Estrogen receptors alpha, beta and GPER in the CNS and trigeminal system—molecular and functional aspects. *J. Headache Pain* **2020**, *21*, 131. [CrossRef] [PubMed]
166. Woldeamanuel, Y.W.; Rapoport, A.M.; Cowan, R.P. The place of corticosteroids in migraine attack management: A 65-year systematic review with pooled analysis and critical appraisal. *Cephalalgia* **2015**, *35*, 996–1024. [CrossRef] [PubMed]
167. Gunaydin, C.; Bilge, S.S. Effects of Nonsteroidal Anti-Inflammatory Drugs at the Molecular Level. *Eurasian J. Med.* **2018**, *50*, 116–121. [CrossRef]

cells

MDPI

Review

Pathophysiology and Therapy of Associated Features of Migraine

Maria Dolores Villar-Martinez [1,2] and Peter J. Goadsby [1,2,3,*]

1 Headache Group, Wolfson CARD, Institute of Psychiatry, Psychology and Neuroscience, King's College London, London WC2R 2LS, UK
2 NIHR King's Clinical Research Facility, SLaM Biomedical Research Centre, King's College Hospital, London SE5 9RS, UK
3 Department of Neurology, University of California, Los Angeles, CA 90095, USA
* Correspondence: peter.goadsby@kcl.ac.uk

Abstract: Migraine is a complex and debilitating disorder that is broadly recognised by its characteristic headache. However, given the wide array of clinical presentations in migraineurs, the headache might not represent the main troublesome symptom and it can even go unnoticed. Understanding migraines exclusively as a pain process is simplistic and certainly hinders management. We describe the mechanisms behind some of the most disabling associated symptoms of migraine, including the relationship between the central and peripheral processes that take part in nausea, osmophobia, phonophobia, vertigo and allodynia. The rationale for the efficacy of the current therapeutic arsenal is also depicted in this article. The associated symptoms to migraine, apart from the painful component, are frequent, under-recognised and can be more deleterious than the headache itself. The clinical anamnesis of a headache patient should enquire about the associated symptoms, and treatment should be considered and individualised. Acknowledging the associated symptoms as a fundamental part of migraine has permitted a deeper and more coherent comprehension of the pathophysiology of migraine.

Keywords: migraine pathophysiology; nausea; osmophobia; phonophobia; vertigo; allodynia

Citation: Villar-Martinez, M.D.; Goadsby, P.J. Pathophysiology and Therapy of Associated Features of Migraine. *Cells* **2022**, *11*, 2767. https://doi.org/10.3390/cells11172767

Academic Editors: László Vécsei, Bernadett Tuka and Masaru Tanaka

Received: 9 August 2022
Accepted: 31 August 2022
Published: 5 September 2022

1. Introduction

Migraine has been traditionally associated with the core symptom, headache [1]. Photophobia and vomiting, two of the canonical symptoms associated with migraine [2], are also widely accepted features of the typical migraine attack, as understood classically by patients and physicians [3]. However, reducing the understanding of migraine to a few symptoms would be as simplistic, perhaps, as reducing Parkinson's disease to tremors.

The way that migraineurs deal with their attacks provides valuable information about hypersensitivity to sensorial stimulation, including avoiding movement, light, sounds, touch or smells [4]. These are usually subjective, unpleasant experiences, unshared by family, friends or colleagues. Consequently, migraine patients presenting associated symptoms as prominent features can usually be labelled as sensitive. The Greek translation for sensitive, Ευαίσθητος "evahistos", can be separated into the following two parts: the prefix meaning good or well, and the rest meaning sense or perception. However, any positive connotation of the term has nowadays dissipated. Many of these "evahistic" manifestations can actually be the main symptom of the clinical picture in a patient with migraine, and imply a higher disability [5]. Migraine patients with sensory hypersensitivity may have more attention difficulties during daily activities [6], or more cranial autonomic symptoms associated to the headache [7], and the response to preventive treatments may vary [8]. Exogenous factors, such as stress, obesity, intestinal microbiota and even parental behaviour, have been speculated to play a role in the chronification and sensitization process [9–12].

In recent years, the study of non-headache symptoms has been useful in demonstrating their important role, yet independence from pain, in the pathophysiology of migraine. In this paper, we will focus on some frequently disabling associated symptoms, such as nausea, osmophobia, phonophobia, neuro-otological manifestations and cutaneous allodynia, and will spare comments on some premonitory-like symptoms, such as yawning. Photophobia has recently been reviewed elsewhere [13], and the usually omnipresent symptom in migraineurs, movement sensitivity, could be explained by some mechanisms that are commented on below.

2. Nausea and Vomiting

2.1. Nausea in Migraine and Conditions Related to Migraine

Nausea is one of the symptoms associated with migraine that is considered canonical, according to the International Classification of Headache Disorders, 3rd Edition (ICHD-3) [2]. Ictal and interictal nausea has a high impact on quality of life and economic cost [14,15], and is the second most bothersome migraine symptom, reported in 28% of patients, exceeded only by photophobia [16].

Up to half of the people with episodic migraine suffer from nausea in more than half of their headache episodes, and the attacks were accompanied by more headache symptoms and a higher impact, compared to patients with less frequency of nausea. The majority of those reporting high-frequency nausea were women [17] and had an increased risk of developing chronic migraine in 2 years [18].

Having migrainous biology could result in patients having more disability when presenting with other disorders that are generally associated with nausea and vomiting.

2.1.1. Cyclic Vomiting Syndrome

It is well known that there is a strong link between migraine and cyclic vomiting syndrome [19], with similar associated symptoms during the attacks, and triggers, as reported by the patients [20,21]. Both nausea and cyclic vomiting syndrome patients have a decreased connectivity between the sensorimotor network and the insula, which manages viscero-sensory processing [22] and may be regulated by the endocannabinoid system [23]. Cannabis can act as a pro-emetic or antiemetic and can cause cannabis hyperemesis syndrome, which shares similar features to cyclic vomiting syndrome [24], and whose recommended treatment is cannabis cessation [25]. Remarkably, the management of cyclic vomiting syndrome consists predominantly of treatments also used for migraine [26].

2.1.2. Motion Sickness

Motion sickness and migraine may share a similar pathophysiology, as patients with motion sickness have a robust migrainous biology [27] and around half of migraineurs present with motion sickness, in comparison with 20% of those with non-migrainous headaches [28]. Patients with "migrainous vertigo" had an improvement in severe motion sickness following rizatriptan [29]. Nociceptive stimulation in the trigeminal area is capable of increasing nausea during motion sickness caused by optokinetic stimulation [30], whereas nausea did not increase following extra-trigeminal nociceptive inputs [31]. Having a history of migraine has also been associated with developing post-operative nausea [32,33], and having motion sickness and being a female are independent risk factors for post-operative vomiting [32,34].

2.1.3. Pregnancy

Pregnancy is potentially a particularly disabling period for women with migraine. During pregnancy, one third of migrainous women require hospitalization due to hyperemesis gravidarum, and almost forty percent of women with hyperemesis reported migraine headaches [35].

Finally, a migrainous background may determine the quality of life related to nausea in palliative care, and migraine preventive treatments serve as efficacious relief in treating incoercible nausea in terminal patients with a history of migraine [36].

2.2. Neuroanatomy and Neuropharmacology

There is a matrix of neuro-anatomical structures involved in the onset and control of nausea, as well as several neurotransmitters that have been the main targets of antiemetic and acute treatment schemes.

Dopamine has been the main compound implicated in the pathophysiology of nausea associated with migraine, at least since the 1970s [37]. Patients with migraine are sensitive to dopaminergic pharmacological agents [38–40] and develop nausea and other classically considered dopaminergic symptoms, such as yawning, not necessarily accompanied by headache [38,40]. This propensity may entail a genetic predisposition, and a particular allelic distribution was found to be significantly different for the D_2 dopamine receptor in a subpopulation of migraineurs with prominent dopaminergic symptoms [41]. Among the dopaminergic symptoms, nausea, unlike yawning, is considered post-synaptic, and is triggered by apomorphine and inhibited by domperidone, which targets D_2 receptors [40]. Dopamine may also regulate headache pain, as dopaminergic neurons play a role in nociceptive control by modulating trigemino-vascular neurons [42].

Serotonin also has a major role in nausea, with the receptor 5-hydroxytryptamine-5-HT_3 as the main target not only of modern antiemetic pharmacological compounds, but also of natural antiemetics used for centuries, such as the gingerol compounds contained in ginger [43].

Hyporexia during headaches may be explained by the loss of appetite that can be observed during noxious dural stimulation, which activates the nucleus parabrachial and the ventromedial of the hypothalamus, and may be mediated by cholecystokinin [44]. However, nausea can also appear before the headache, during the premonitory phase, in almost a quarter of spontaneous attacks [45]. This percentage was doubled when headache attacks were triggered in a controlled environment [46].

Another intriguing component in migrainous nausea is substance P. Neurokinin 1 (NK-1) receptor antagonists can inhibit vomit produced by central or peripheral stimuli [47], and its central action may be mediated by inhibiting the substance P emetic effect [48], which may take place predominantly in the locus coeruleus [49].

Early pre-clinical experiments are good examples of the extent of anatomical structures that could be involved in the process of vomiting. Monkeys presented vomiting following the electrical stimulation of the olfactory tubercle, amygdala, septum, fornix and the thalamic ventral anterior nucleus [50]. In cats, lesions in the medulla abolished the characteristic pattern of respiratory motor nerve discharge, observed in vomiting [51], induced by emetic drugs and electrical vagal stimulation of abdominal afferents. This study suggested that the regions that control vomiting were localised between the obex and the retrofacial nucleus [52], both localized in the medulla.

In human neuroimaging studies, some brainstem areas showed significant activation with a $H_2{}^{15}O$ positron emission tomography (PET) scan in the premonitory phase of migraine participants with nausea, including the periaqueductal grey, dorsal motor nucleus of the vagus, nucleus ambiguous and nucleus tractus solitarius [53], as shown in the following paragraphs. Following a rostral-caudal approach, among them, the mesencephalic periaqueductal grey (PAG) deserves a special mention [53].

PAG has an important role in the descending modulation of the trigeminovascular processes (Figure 1) [54]. PAG has been related to other autonomic sympathetic activity [55,56], emotional perception of pain and aversive behaviours [57,58] cough [59] and breathing control [60]. It is involved in modulating the descending pain pathways [61–63]. This modulation has recently been shown to be activated by mu opioids by means of presynaptic disinhibition and reducing GABAergic postsynaptic currents [64]. It is yet unknown whether this area is related to the chronification observed in migraineurs with frequent use of opioids, as commented on below.

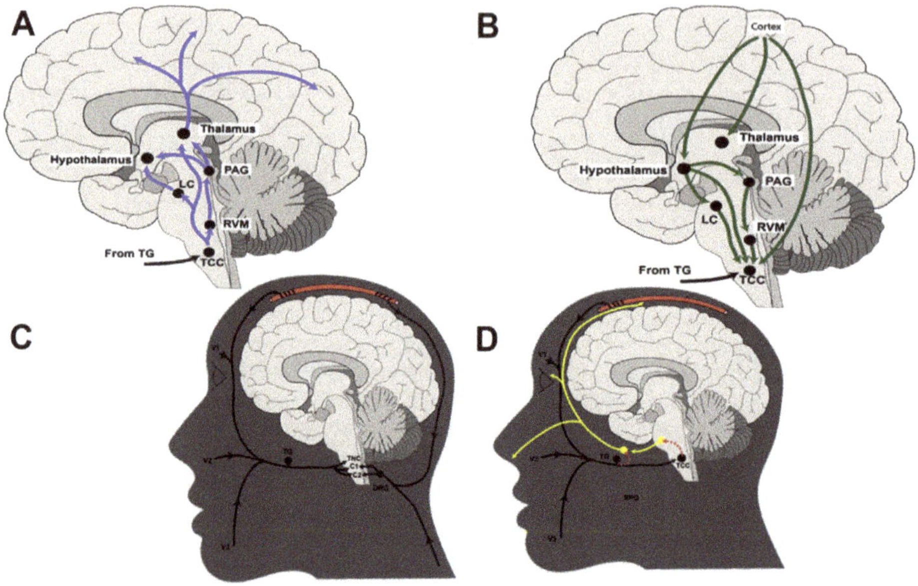

Figure 1. Schematic representation of ascending and descending mechanisms involved in the pathophysiology of migraine, interaction between peripheral and central nervous systems and the trigeminal autonomic reflex. (**A**) Ascending mechanisms; (**B**) Descending mechanisms; (**C**) Connection of dural, cervical and trigeminal inputs in the trigeminocervical complex; (**D**) Potential interfaces between trigeminal and parasympathetic arms of the trigeminal autonomic reflex. Cervical dermatomes (C1, C2); dorsal root ganglia (DRG); locus coeruleus (LC); periaqueductal gray (PAG); sphenopalatine ganglion (SPG); trigeminal ganglion (TG); trigeminocervical complex (TCC) rostral ventromedial medulla (RVM); ophthalmic, maxillary, and mandibular dermatomes of the trigeminal nerve (V1, V2, V3, respectively). Reproduced from Goadsby and Holland 2019 with permission.

More caudal areas in the rostral dorsal medulla were involved, including the dorsal motor nucleus of the vagus [53], which may relax the lower esophageal sphincter [65].

The nucleus tractus solitarius has connections with hypothalamic areas that play a role in autonomic control [66]. Both the nucleus tractus solitarius and dorsal motor nucleus of the vagus conform, along with the area postrema, the dorsal vagal complex, which is one of the main termination sites of the afferent fibres of the vagal nerve [67] and has a high distribution of dopamine $D_{2–4}$ receptors [68]. The area postrema is one of the sensory circumventricular organs with a possible chemoreceptive function, situated outside the blood–brain barrier and connected to the hypothalamus, which is thought to be essential in controlling neuroendocrine functions [69], is rich in type D_2 dopamine receptors [70] and is the brain area with the higher estimates of substance P [71].

2.3. *Treatment of Nausea*

The treatment of nausea during migraine attacks must be considered in every patient presenting with that symptom. When nausea does not respond to analgesic treatment, specific antiemetic treatment should focus on the pathways of the neurotransmitters described above (dopamine, serotonin, substance P) as main targets for treatment. Nevertheless, acute treatment can be essential in the management of nausea associated with migraine. NSAIDs could be effective in alleviating nausea in patients who have not taken any triptans [72] and there is a recent meta-analysis that supports gepants as an effective treatment for nausea in patients with episodic migraine [73]. Special attention must be paid to patients consuming opioids. Nausea is a recognised side effect following opioid use [74]. Patients with episodic migraines who are exposed to opioids have a twofold risk of migraine chronification [75], a likely reduction in the efficacy of triptans for acute treatment [76] and the issue of developing gastro-intestinal adverse events after long-term consumption [77]. For the treatment of nausea, we have focused on the three main neurotransmitters involved, serotonin, substance P and dopamine.

2.3.1. Serotonin

Triptans are serotonin 5-$HT_{1B/1D}$ receptor agonists, and can help in alleviating nausea, as exemplified by rizatriptan [78,79]. However, having a sensation of nausea pre-treatment predicts a low efficacy response [80], perhaps due to the delay in treatment intake, as discussed in the allodynia section. Ondansetron is a highly-specific 5-HT_3 receptor antagonist, although there are no randomized-controlled trials on migraine. Granisetron, however, was significantly more effective than placebo for nausea at 30 min [81], and was more effective than metoclopramide as an adjuvant treatment for acute migraine [82].

Ginger could be a reasonable "over the counter" serotonergic therapeutic strategy for patients trying to avoid chemical treatments. It might be effective in lowering nausea according to a meta-analysis of three studies [83], and headache relief similar to that of sumatriptan has been reported in a double-blind, randomized controlled study [84].

2.3.2. Dopamine

Among the several antiemetics available, metoclopramide is an antagonist of dopamine D_2 receptors and has also an antagonist effect on serotonin 5-HT_3 receptors [85]. Metoclopramide presents the highest passage of the blood–brain barrier, compared to domperidone or chlorpromazine [86]. Metoclopramide helps with the impairment of gastric motility during migraine attacks, improving the absorption rate of NSAIDs [87], and may also exert its effect as a pain relief agent [88], probably due to its action in the trigemino–cervical complex [89]. However, recent literature found conflicting results as a single therapeutic approach, with either an efficacy similar to that of NSAIDs [90], or no difference of intravenous metoclopramide compared to saline [91]. Prochlorperazine is a phenothiazine antipsychotic with antagonizing effect of dopamine D_2 receptors, similar to chlorpromazine [87,92] and might be the most effective intravenous antiemetic, which also has a higher risk of extrapyramidal adverse events [93]. Chlorpromazine is also an effective option to consider for the treatment of nausea in emergency settings [94].

2.3.3. Substance P

By inhibiting the substance P pathway, NK-1 receptor antagonists, such as aprepitant, have been used in the treatment of nausea generated by intravenous dihydroergotamine in patients with migraine [95]. NK1 receptor antagonists are potent antiemetics that have been approved for the treatment of severe nausea associated with chemotherapy [96], and are also recommended for cyclic vomiting syndrome, along with ondansetron or triptans [26].

3. Osmophobia

The perception of odour is certainly an extremely subjective experience, or we would all be wearing the same perfume. Being perhaps the less studied of the senses, the mechanisms behind the way a fragrance is perceived is not yet fully understood. A brief mention here is appropriate for two interesting theories that were proposed in the twentieth century, involving a lock-and-key system and vibrational wavelengths [97], which have not yet been fully developed.

There are several substances whose consumption or inhalation has been popularly related to headaches [98–101]. Remarkably, *Umbellularia californica* is a type of tree, commonly known as "the headache tree" [102], which contains umbellulone, a ketone that was reported of being capable of triggering cluster headache-like attacks in a gardener with a history of cluster headaches [103]. It was later discovered that this mechanism was mediated by the activation of the transient receptor potential (TRP) ankyrin 1 (TRPA1) [104,105], followed by the release of calcitonin gene-related peptide (CGRP) [104]. CGRP is also released through the activation of vanilloid receptors, following stimulation with nitric oxide [106] or ethanol [107,108], one of the most relevant cluster headache triggers. TRPA1 has also been involved in the responses to some inhaled chemicals, including the smoke of cigarettes [109], chloride [110,111] hydrogen peroxide-containing substances [111] or formalin, the noxious compound largely used in pain models [112].

It has been reported that up to 70% of migraineurs can develop a headache after the stimulation with some odorants, which happened around 25 minutes following the exposure [113], and there is a case report of migraine improvement following the imposition of mandatory masks in the workplace during the COVID-19 pandemic [114]. Increased sensitivity to smells can be part of the premonitory-like symptoms experienced by migraineurs; therefore, certain smells may be misinterpreted as the trigger for a migraine attack, which might not be a necessary factor for its occurrence [115,116]. As a consequence, the results of studies that assess migraine triggers have debatable interpretations.

Nevertheless, the presence of osmophobia may be related to more florid migraine phenotypes and greater disability, and a scale has been developed recently for the quantification of quality of life related to osmophobia [117]. Migraineurs that present with ictal osmophobia may have more painful headaches [118,119]. Ictal and interictal osmophobia have been associated with a longer history of migraines or high frequency of the attacks, as well as other associated symptoms, such as cranial allodynia [120–122], suggesting a central sensitization process [123]. Vomiting can also be more common in the presence of osmophobia [119,121]. Osmophobic migraineurs may also have a higher prevalence of psychiatric comorbidities than those without it [118,124–126].

Osmophobia has been proposed as a specific marker, helpful for the diagnosis of migraine [119,124,127–132]; however, it is not very sensitive [122]. Around half of the patients with migraines reported an increased sense of smell or reduced tolerability to smells [129,133]. Remarkable examples of patients reporting hyperosmia include the smell of a rose from more than 5 meters of distance, or soap from a different room, and the main scents triggers for osmophobia arose from food, specifically fried food and onions, cigarettes or self-care products, and perfume or paint specifically were reported as triggers [133]. More recently, forty percent of patients with chronic migraine reported osmophobia [134], and a similar number suggested odours or perfumes as potential triggers of a migraine attack [101].

Paradoxically, despite their hypersensitivity to smells, migraineurs have a lower capability for the threshold, identification and discrimination of smells [135,136]. Patients with episodic migraine were found to have a similar olfactory acuity to controls, and furthermore, around one fifth of them developed hyposmia during the attack [137]. Taste abnormalities in migraineurs [133] are a matter of debate [138].

Patients with migraine and osmophobia have neuroanatomical alterations. A significantly reduced volume of the olfactory bulb was observed in 1.5 Tesla MRI, compared to patients with other types of headache [139], and might be more pronounced on the left, in comparison with controls [140]. In migraineurs with reported hypersensitivity to odours, regional blood flow in a study using $H_2{}^{15}O$-positron emission tomography was found to be increased in areas of the left piriform cortex and antero-superior temporal gyrus, as compared to controls, both with and without multiple odour stimuli [141]. During odour stimulation, blood flow was found to be decreased in bilateral fronto-temporo-parietal regions, as well as the posterior cingulate gyrus and right locus coeruleus [141]. Another study using fMRI to compare responses to the smell of roses found higher blood oxygen level-dependent activity in the amygdala and insular cortices of the amygdala and also in the midbrain, particularly the rostral pons. However, the smell of roses did not show significant interictal differences compared to the controls [142]. Activation of the amygdala and orbitofrontal cortex might be related, respectively, with the intensity and valence of the smell emotional experience [143]. The amygdala and cingulate cortex also showed abnormal activation in patients with multiple chemical sensitivity [144,145], which is associated with a high prevalence of headache [146] and was observed in up to 20% of migraineurs [147].

Olfactory hallucinations or phantosmia is a hallmark of temporal lobe epilepsy, and currently a no man's land when it presents in the form of aura. It is a rare symptom, with a reported prevalence of 0.66% in a headache center [148]. The majority of reported cases had normal electroencephalograms that were, however, taken during the interictal period, and usually respond to antiepileptic drugs.

The reported cases showed that the episodes have an average duration of less than 10 min and the onset occurs prior to the migraine attack [148,149]. Patients with symptoms of phantosmia scanned with FLASH and eco-planar imaging MRI techniques showed increased activation of different brain areas associated with the process of the sense of smell, such as the prefrontal, cingulate, temporal or insular cortex MRI activation was inhibited by typical antipsychotics that perform its activity through a wide range of binding receptors [150]. Peripheral blocking activities can alleviate phantosmia [151].

4. Neuro-Otological Manifestations

In 1984, Kayan and Hood described how vestibulocochlear symptoms were frequently reported, in up to 60% of patients with migraine, and these can be important or disabling enough for the patient to be the primary reason for referral to a specialist. The incidence of neuro-otological symptoms for migraineurs seemed homogeneous throughout all ages in males, but had a peculiar distribution in females. For women who reported audiovestibular symptoms only when asked during the study, a positive skew distribution could be observed, with the peak situated in the 3rd decade. However, the female patients whose reason of referral was the presence of disabling audio-vestibular symptoms had a peak in the peri-menopausal 5th and 6th decades. This group with disabling symptoms had a higher incidence in males [28]. They compared 80 patients referred for vestibulocochlear symptoms with 500 patients with multiple sclerosis for benign positional vertigo and Méniere's [28]. Only migraineurs described cochlear sensations, such as tinnitus, distortion of pitch, or hearing loss [28].

The frequency of migraine in Méniere's disease is higher than in normal subjects, and phonophobia has a high prevalence in these patients, independently of the presence of migraine headache [152].

4.1. *Phonophobia*

Phonophobia, along with photophobia, is one of the associated symptoms that define a migraine attack, according to the ICHD-3. As an asset for differential diagnosis, the presence of phonophobia may be able to exclude secondary headache types, such as cardiac cephalgia or sleep apnea headache; however, phonophobia is also reported in other headaches, such as a "tension-type headache", if it is not accompanied by photophobia in the episodic categories, or a "cervicogenic headache", which may make the clinician hesitate if the patient has a migrainous background [2]. This complication is simplified by using the appendix criteria for tension-type headaches that exclude both photophobia and phonophobia; and are clinically preferable.

In 1984, up to 81% of patients with migraine reported phonophobia, in comparison with only 12.1% of patients with a non-migrainous headache, and the combination of phonophobia and hearing loss was reported by some patients [28]. A recent meta-analysis showed that migraineurs may have a higher risk of developing sensorineural hearing loss [153]; therefore, the exclusion of migraine patients with hearing loss from the majority of the trials may lead to biased conclusions. In 1985, Blau and Solomon reported noise as a migraine trigger in 4/50 patients with migraine [133] and the potential measurability of phonophobia was suggested. Recently, it has been reported that annoying sounds, as well as other usually reported migraine triggers, may just represent early manifestations of migraine premonitory symptoms, as they demonstrate significant agreement with premonitory spontaneous phonophobia [154]. In studies that assessed sound discomfort using a range of Hertz stimuli, ictal [155,156] and interictal hearing discomfort thresholds were lower in migraineurs, as compared with healthy participants [156–158], with a low positive correlation with age [157]. Women may have a lower threshold than men [159]. Among migraineurs, ictal thresholds are lower than interictal ones [158]. Differences in monaural and binaural thresholds do not relate to the side of headache [156], and only a small proportion of participants with chronic migraine (5/48) report unilateral phonophobia, which was nonexistent in 54 participants with episodic migraine [160]. Similar to photophobia, unilaterality of phonophobia can be more specific to trigeminal autonomic cephalalgias [160].

The use of close-ended questions can be useful in increasing sensitivity for phonophobia during the neurological anamnesis [161].

Several electrophysiological studies have evaluated the hearing pathway in migraineurs with phonophobia. Phonophobia does not seem to be related with a recruitment phenomenon [155], which is commonly associated with cochlear damage.

The function of the cochlear efferents can be assessed by otoacoustic emission tests, which evaluates the suppression in the amplitude of transiently evoked signals from the olivary complex when a sound is produced on the contralateral ear [162,163]. It has been reported that for healthy controls, these amplitudes are significantly decreased, whereas in migraineurs, they are not suppressed [162,164]. This was specially observed in low-to-middle frequencies of 1–1.5 kHz, in a cohort of female phonophobic migraineurs during the interictal period [165]. However, this was not replicated in another study in patients with prominent vestibular symptoms, and phonophobia was not significantly associated with lack of suppression [163]. Neurotransmission in the outer hair cells of the cochlea may be mediated by CGRP [166], and increased CGRP activity in the inner ear has been hypothesized to be the cause of an insufficient suppression of the auditory pathway [165].

Another abnormality leading the patient to find sounds uncomfortable may lay in the cortical processing of auditory stimuli. Whereas latencies are similar, healthy participants experience a decrease in the amplitude of the auditory N1–P2 component following sequential blocks of stimuli in cortical-evoked auditory-evoked potentials, whereas participants with migraine experienced an increase, which could be considered a potentiation, instead of habituation. Intensity dependence of auditory-evoked potentials, which is measured as a slope after stimulation at increasing intensities, was also greater in migraineurs [159,167,168], and these may have a lower amplitude in the first blocks of stimuli, which may mean a decreased pre-activation of the sensory cortex [167,169]. The slope does not correlate with migraine frequency or duration, or with changes in visually evoked potentials [169], but may correlate with age [168], and has been associated with serotonergic activity [159,170–172] and response to preventive treatments [173].

Several studies have used brainstem auditory-evoked potentials. Interictal migraine patients have similar latency results to those of controls [174]. Podoshin et al. showed a significant impairment in interpeak latency differences in a group of patients during the migraine attack, when the rate of click sound stimuli was increased to 55 per second, in comparison with the same group between attacks [175]. Some studies found no differences between the side of the headache [175], but differences between sides were found by Schlake et al. in peak latencies at 10 clicks per second [174]. Peak latencies were delayed in 6/38 migraine patients, 2 of them with so-called basilar migraine [174], which can be normal [176] or abnormal during the ictal period [177]. Sand and Vingen showed that the discomfort threshold for low sound inversely correlated with low levels of habituation in wave IV-V, which corresponds with the lateral lemniscus in the pons and inferior colliculus in the midbrain [178]. Latency in waves III to V, corresponding to the tract between the cochlear nuclei to colliculus, has been correlated with migraine and attack duration [171]. In a recent study, participants with migraines showed that hearing threshold was inversely correlated with the severity of photophobia, and paradoxically, not with phonophobia, and was higher in patients on prophylactic medication or those who had taken a non-steroidal anti-inflammatory drug on the day of the test, and had higher wave amplitude in comparison with the controls [179].

There is increased blood flow in the auditory association cortex during an acute attack in patients with migraine and phonophobia [180].

Patients with episodic migraine that present with cranial and extracranial cutaneous allodynia have lower thresholds for auditory stimuli either between or during the attacks [181].

4.2. Vertigo

Vertigo is more frequent in people with migraine and vice versa [28,182–187].

Vestibular migraine (VM) is possibly the most frequent cause of recurrent vertigo [188]. It has received many names in the past [186,189,190], and recently, more conditions have been found to fall possibly under the current umbrella of what is considered today VM [191], as well as some diagnoses classified as functional disorders today that may, in the near future, be included. The mere fact of having a diagnosis has proven to be a positive predictor for the improvement of dizziness [192]. However, currently, VM still remains largely underdiagnosed [193]. Despite the consensus diagnostic criteria involving balance and headache societies [2,194], there are several mechanistic questions that remain unanswered, such as the controversy of whether migraine and VM are a continuum along the same spectrum or different entities, as well as important classification queries, such as whether there is a chronic form [195]. The current term of VM may not be well received by the patient, especially those examined outside a headache clinic environment, who usually do not report headaches as the main reason for referral [196], and a source of frustration for the clinician giving a diagnosis to patients who repeatedly report that they do not suffer from headaches.

The features of the attack of VM have been studied mainly retrospectively [186,189,190,197–200], and during the acute episode [201]. There may be a relationship between VM and Méniere disease (MD) [202]. Aural fullness may be an anamnestic key to differentiate VM from MD [203]. Patients with VM may have a high incidence of endolymphatic hydrops, although smaller than that of MD [204]; however, no anatomical differences were found between VM patients and healthy subjects with 3D-SPACE MRI [205]. A correlation between dizziness severity and cognitive dysfunction has been found [206].

Migraine and vestibular migraine: Similarities between migraine and VM are abundant. The majority (72/118) of patients with vestibular symptoms were considered in the 1980s as patients with "non-classical" migraine. Among those without vestibular symptoms, 59 out of 82 were given a diagnosis of "classical migraine". The incidence of "classical migraine" was therefore 11% higher among those without vestibular symptoms [28]. Vertigo can be triggered with nitroglycerin in up to 84% of migraineurs reporting vertigo during spontaneous attacks [207]. Patients with migraines exhibit greater visual and vestibular functional impairment, as well as lower results in the sensory organization test [208]. VM patients may be more sensitive to moving scenes and find it harder to maintain their posture [209–212], as they may tend to rely more on visual cues [213], whereas changes in the position of the head or posture could also trigger vestibular symptoms in some migrainous patients [28].

Patients with definite vestibular migraines demonstrated some changes in videonystagmography, but not canal paresis [214]. Spontaneous nystagmus can be triggered in migraineurs following supraorbital nociceptive inputs, which did not occur following extracephalic stimulation of the median nerve [215].

Pathophysiology: The pathophysiological research that has used neuroimaging approaches has contributed enormously to understanding the central anatomical structures with altered function in VM. In a small study using ^{18}F-deoxyglucose position-emission tomography, patients showed activation of the cerebellum, frontal cortices, thalami, dorsal pons and midbrain, right and insula and temporal cortex, and a deactivation of the posterior parietal and occipito-temporal areas during the attacks [216]. By using imaging-based voxel-based morphometry, patients with definite vestibular migraine showed a reduction in grey matter volume in several cortical areas, including the insula, parieto-occipital, dorsolateral prefrontal, cingulate cortex and the cingulate gyrus, and the volume of areas associated with vestibular and pain processing was negatively correlated with disease duration [217]. During caloric tests, patients with vestibular migraine exhibited increased thalamic activation, as observed in blood oxygenation level-dependent (BOLD) MRIs, which correlated with the attack frequency [218] and was proposed to hold right dominance [219]. A peripheral, vestibular alteration that involves serotonergic axons has also been suggested [220–223].

Treatment: Patient's treatment remains a grey zone, where the therapeutic choice is dependent on observational studies, as there are only a few randomized, placebo-controlled trials in this field for preventive [224,225] and acute medication [29,226]. A recent meta-analysis identified an improvement in the outcomes selected for several therapeutic agents, most of them migraine preventives, such as tricyclics and beta-blockers [224]. Vestibular rehabilitation can also be of help [227].

Recently, the inhibition of CGRP receptors has been shown to improve the vestibular function in animal models of chronic migraines [228], and retrospective studies in humans show a potential benefit when targeting the CGRP pathway [229]. Half of the patients were reported to respond to one prophylactic, 17% responded to a combination of two, and 10% did not have a response [203]. Predictors of poor response have been reported to be female sex, interictal imbalance, anxiety or depression, and our next topic, cutaneous allodynia [203].

4.3. *Allodynia*

Scalp tenderness was reported by 65% of the 500 patients characterized by Selby and Lance in 1960, and they described that this sensitivity could not be correlated with any trigeminal or cervical radicular innervation [230]. Cutaneous allodynia can be quantified in humans objectively [231,232] or by assessing the subjective patient's experience, by questionnaires [233,234]. A similar prevalence to that reported by Selby and Lance was found in large surveys of headache patients, slightly higher in those with the now obsolete term "transformed migraine", and was associated with female sex, high body mass index or depression [235]. Up to one-fifth of patients report severe allodynic symptoms [234]. When specifically measured, the prevalence increases up to 80% [236] and can be higher in patients with another concomitant pain syndrome, such as temporomandibular disorders [237]. Patients with chronification of attacks and migraine with aura may also have a higher prevalence of cutaneous allodynia during the attack [238], although other studies have not found an association with age or headache frequency of years having migraine in migraineurs reporting spontaneous attacks [231,239].

Pathophysiology: The mechanisms that predispose a patient to allodynia may represent a risk for other forms of sensory dysfunction [164,203]. Migraineurs have, in general, lower pressure-pain and heat thresholds than the general population [232]. The majority of cutaneous allodynia symptoms are focused on the cranial regions, but a proportion can also experience the symptoms in extracranial regions [231]. In contrast to patients with migraine, patients with trigeminal autonomic cephalalgias, such as cluster headaches, do not report cutaneous allodynia, unless they have a personal or family history of migraine, and have higher pain threshold both interictally and during the attack [240]. Allodynia can be triggered experimentally in humans [239], and the clinical sequence of onset and anatomical spread has been described [241].

Allodynia was initially reported to be an ictal marker of a "no-return point" that divides triptan efficacy [242]; however, triptans can treat spontaneous [243] and nitroglycerine-induced allodynia associated with migraine in humans [239], and the association appeared to be, instead, time-dependent [244,245]. In a similar way to low pain intensity, which can also increase as the attack progresses, lower allodynia may be an independent predictor for the efficacy of over-the-counter acute treatments [246], and recently, allodynia has been shown to be an independent risk factor for the worsening of migraine associated with the utilization of masks during the COVID-19 pandemic [247].

A complex network of peripheral and central structures is involved in allodynia. In 1994, reduced efficacy in the spinal inhibitory circuits, mediated by GABA-A, was proposed as a potential cause of allodynia in preclinical models of pain [248]. Two years later, it was shown that trigeminal afferents could be sensitized with a variety of chemical substances applied in the dural regions [249]. However, it is unlikely that the simple sensitization of peripheral afferents accounts for the single cause of allodynia. The periaqueductal grey holds inhibitory control over trigeminal afferent neurons [250,251] and also has a regulatory effect on the trigeminocervical nucleus (Figure 1), facilitated by CGRP [252]. Another neuropeptide, pituitary adenylate cyclase-activating peptide 38 (PACAP-38) can cause sensitization and delayed activation of trigemino-cervical neurons [253]. Under the bases of the role of the trigemino-cervical complex as a convergence center for afferent inputs [254], and its diencephalic connections, an increased response in central neurons could bring a reduced pain threshold in extracranial regions [255].

The diencephalon may be, indeed, strongly involved in the process of allodynia. Stress-related hypothalamic dysregulation of prolactin has recently been associated with allodynia in females [256]. Activation in posterior thalamic areas was demonstrated in rodents and also in migraine patients with extracephalic allodynia, with functional MRI BOLD techniques [257]. A first-line treatment in the prevention of migraine, propranolol, exerts part of its mechanisms upon these thalamic areas [258]. Thalamic projections are widely spread to many areas of the cortex, and have been traced from posterior and lateral nuclei to several cortical regions, including the auditory, entorhinal or visual cortex [259]. The medial area of the temporal lobe, for example, may be hyperexcitable in migraineurs, both during ictal and interictal moments, when applying painful heat stimuli to the forehead, as detected with diffusion tensor imaging in functional MRI [260]. An hyperexcitable state has also been suggested in subcortical regions in migraineurs [261].

Somatosensory-evoked potentials have not found significant abnormalities in migraineurs [262]. However, cortical thickness may be different in the associated temporo-parietal areas of migraineurs, and there is a positive correlation with pain threshold, contrary to healthy controls [263,264]. Activity is also increased in primary sensory areas, and between the pons and insula, implying a role in the patient's emotional response [265].

In preclinical models of allodynia, nitroglycerine is capable of increasing the firing of trigeminal neurons and dural-evoked action potentials, in addition to creating hypersensitive responses to facial stimulation with innocuous brush or noxious pinch. These responses were reversible with naratriptan [239] and also ibuprofen, suggesting both a serotonin and an inflammatory-mediated mechanism [266,267].

Allodynia has not been directly related to levels of amylin or CGRP [268]; however, it can be modulated to target CGRP [269–271], which may have a glial site of action [272] and stronger activity in females [271]. Nitroglycerine was able to trigger allodynia in 17/53 patients with migraine; among them, 14 responded to acute treatment with aspirin or sumatriptan, and those who reported allodynia in their usual attacks were more likely to experience it during the triggering session [239].

Finally, TRP channels are an interesting area in the understanding and treatment of migraine [273]. Migraineurs have less tolerance to heat during the interictal period [274]. Recent studies did not find an association between thermal quantitative sensory testing (QST) and allodynia. However, preclinical models suggest a potential genetic predisposition to mechanical allodynia, involving the non-selective cold-sensitive cation channel transient receptor potential melastatin 8 (TRPM8), the activation of which causes cranial and extracranial allodynia [275]. Fibres that express TRPM8 were progressively reduced in postnatal mice, in contrary to the fibres that express CGRP. Paradoxically, the use of the TRPM8 agonist menthol can reduce behavioural responses to meningeal chemical stimulation [276]. These channels may have potential hypothalamic modulation, as orexins may play a part in the emotional response to heat [277]. It may be speculated that these differences could potentially translate to different phenotypes of migraineurs, which find relief either with fresh air or a heated pad.

5. Conclusions

This article summarizes the literature regarding the associated symptoms in migraineurs. Knowledge concerning migraines and their associated symptoms continues to grow and is evolving into a concept that might not be as clinically simple as once imagined [278], with a wide spectrum of presentations of the same migrainous biology. Trials that have reported the most bothersome associated symptoms, together with pain, represent a more holistic approach to migraine research.

Associated symptoms of migraine are varied, extremely prevalent, and contribute to the disabling nature of migraines. Acknowledging the associated symptoms could contribute to a better outcome for the patient, and should never be forgotten in the anamnesis of the migraineur. Treatment should be focused on correct acute, preventive and anti-emetic migraine treatments, where needed.

The relationship between the central and peripheral sensitization processes with the associated symptoms of migraines is evident, and is comparable to the question of what was first to come, the chicken or the egg.

Author Contributions: Conceptualization, M.D.V.-M. and P.J.G.; methodology, M.D.V.-M. and P.J.G.; writing—original draft preparation, M.D.V.-M.; writing—review and editing, M.D.V.-M. and P.J.G. All authors have read and agreed to the published version of the manuscript.

Funding: The research was funded, in part, by the National Institute for Health and Social Care Research (NIHR203970). The APC was funded by King's College London.

Institutional Review Board Statement: Not applicable.

Informed Consent Statement: Not applicable.

Data Availability Statement: Not applicable.

Acknowledgments: We would like to thank Michael Murray and Andria Panteli for their etymological guidance, and Kostis Christoforou for his etymological guidance and continuous clinical support.

Conflicts of Interest: M.D.V.-M. has no conflict of interest. P.J.G. reports, over the last 36 months, grants and personal fees from Eli-Lilly and Company, a grant from Celgene, and personal fees from Aeon Biopharma, Allergan/Abbvie, Amgen, Biodelivery Sciences Intern, Biohaven Pharmaceuticals Inc., CoolTech LLC, Reddys, Epalex, Impel Neuropharma, Lundbeck, Novartis, Praxis, Sanofi, Satsuma and Teva Pharmaceuticals, and personal fees for advice through the Gerson Lehrman Group, Guidepoint, SAI Med Partners, Vector Metric, and fees for educational materials from CME Outfitters, Omnia Education, WebMD, and publishing royalties or fees from the Massachusetts Medical Society, Oxford University Press, UptoDate and Wolters Kluwer, and for medicolegal advice on headaches, and a patent magnetic stimulation for headaches (No. WO2016090333 A1) assigned to eNeura without fees.

References

1. Willis, T.; Pordage, S. *Two Discourses Concerning the Soul of Brutes, Which Is That of the Vital and Sensitive of Man: The First Is Physiological, Shewing the Nature, Parts, Powers, and Affections of the Same; and the Other Is Pathological, Which Unfolds the Diseases which Affect It and Its Primary Seat, to Wit, the Brain and Nervous Stock, and Treats of Their Cures: With Copper Cuts*; Thomas Dring, Ch. Harper and John Leigh: London, UK, 1683.
2. Headache Classification Committee of the International Headache Society (IHS) The International Classification of Headache Disorders, 3rd edition. *Cephalalgia* **2018**, *38*, 1–211. [CrossRef] [PubMed]
3. Gowers, W.R. *A Manual of Diseases of the Nervous System*, 3rd ed.; P. Blakiston, Son & Co.: Philadelphia, PA, USA, 1899; p. 1357.
4. Goadsby, P.; Holland, P.; Martins-Oliveira, M.; Hoffmann, J.; Schankin, C.; Akerman, S. Pathophysiology of Migraine: A Disorder of Sensory Processing. *Physiol. Rev.* **2017**, *97*, 553–622. [CrossRef] [PubMed]
5. Suzuki, K.; Suzuki, S.; Shiina, T.; Okamura, M.; Haruyama, Y.; Tatsumoto, M.; Hirata, K. Investigating the relationships between the burden of multiple sensory hypersensitivity symptoms and headache-related disability in patents with migraine. *J. Headache Pain* **2021**, *22*, 77. [CrossRef] [PubMed]

6. Lévêque, Y.; Masson, R.; Fornoni, L.; Moulin, A.; Bidet-Caulet, A.; Caclin, A.; Demarquay, G. Self-perceived attention difficulties are associated with sensory hypersensitivity in migraine. *Rev. Neurol.* **2020**, *176*, 829–838. [CrossRef] [PubMed]
7. Danno, D.; Wolf, J.; Ishizaki, K.; Kikui, S.; Hirata, K.; Takeshima, T. Cranial autonomic symptoms in migraine are related to central sensitization: A prospective study of 164 migraine patients at a tertiary headache center. *BMC Neurol.* **2022**, *22*, 89. [CrossRef] [PubMed]
8. Pan, L.-L.H.; Wang, Y.-F.; Ling, Y.-H.; Lai, K.-L.; Chen, S.-P.; Chen, W.-T.; Treede, R.-D.; Wang, S.-J. Pain sensitivities predict prophylactic treatment outcomes of flunarizine in chronic migraine patients: A prospective study. *Cephalalgia* **2022**, *11*, 899–909. [CrossRef] [PubMed]
9. Stubberud, A.; Buse, D.C.; Kristoffersen, E.S.; Linde, M.; Tronvik, E. Is there a causal relationship between stress and migraine? Current evidence and implications for management. *J. Headache Pain* **2021**, *22*, 155. [CrossRef]
10. Westgate, C.S.J.; Israelsen, I.M.E.; Jensen, R.H.; Eftekhari, S. Understanding the link between obesity and headache- with focus on migraine and idiopathic intracranial hypertension. *J. Headache Pain* **2021**, *22*, 123. [CrossRef]
11. Kang, L.; Tang, W.; Zhang, Y.; Zhang, M.; Liu, J.; Li, Y.; Kong, S.; Zhao, D.; Yu, S. The gut microbiome modulates nitroglycerin-induced migraine-related hyperalgesia in mice. *Cephalalgia* **2022**, *42*, 490–499. [CrossRef]
12. Raibin, K.; Markus, T.E. Cutaneous allodynia in pediatric and adolescent patients and their mothers: A comparative study. *Cephalalgia* **2022**, *42*, 579–589. [CrossRef]
13. Villar-Martinez, M.D.; Goadsby, P.J. Dim the Lights: A Narrative Review of Photophobia in Migraine. *TouchREVIEWS Neurol.* **2022**, *18*, 14–21. [CrossRef]
14. Lipton, R.B.; Buse, D.C.; Saiers, J.; Serrano, D.; Reed, M.L. Healthcare resource utilization and direct costs associated with frequent nausea in episodic migraine: Results from the American Migraine Prevalence and Prevention (AMPP) Study. *J. Med. Econ.* **2013**, *16*, 490–499. [CrossRef] [PubMed]
15. Gajria, K.; Lee, L.K.; Flores, N.M.; Aycardi, E.; Gandhi, S.K. Humanistic and economic burden of nausea and vomiting among migraine sufferers. *J. Pain Res.* **2017**, *10*, 689–698. [CrossRef]
16. Munjal, S.; Singh, P.; Reed, M.L.; Fanning, K.; Schwedt, T.J.; Dodick, D.W.; Buse, D.C.; Lipton, R.B. Most Bothersome Symptom in Persons with Migraine: Results from the Migraine in America Symptoms and Treatment (MAST) Study. *Headache* **2020**, *60*, 416–429. [CrossRef] [PubMed]
17. Lipton, R.B.; Buse, D.C.; Saiers, J.; Fanning, K.M.; Serrano, D.; Reed, M.L. Frequency and burden of headache-related nausea: Results from the American Migraine Prevalence and Prevention (AMPP) study. *Headache* **2013**, *53*, 93–103. [CrossRef]
18. Reed, M.L.; Fanning, K.M.; Serrano, D.; Buse, D.C.; Lipton, R.B. Persistent frequent nausea is associated with progression to chronic migraine: AMPP study results. *Headache* **2015**, *55*, 76–87. [CrossRef]
19. Kovacic, K.; Li, B.U.K. Cyclic vomiting syndrome: A narrative review and guide to management. *Headache* **2021**, *61*, 231–243. [CrossRef]
20. Li, B.U.; Murray, R.D.; Heitlinger, L.A.; Robbins, J.L.; Hayes, J.R. Is cyclic vomiting syndrome related to migraine? *J. Pediatr.* **1999**, *134*, 567–572. [CrossRef]
21. Abell, T.L.; Adams, K.A.; Boles, R.G.; Bousvaros, A.; Chong, S.K.F.; Fleisher, D.R.; Hasler, W.L.; Hyman, P.E.; Issenman, R.M.; Li, B.U.K.; et al. Cyclic vomiting syndrome in adults. *Neurogastroenterol. Motil.* **2008**, *20*, 269–284. [CrossRef]
22. Ellingsen, D.-M.; Garcia, R.G.; Lee, J.; Lin, R.L.; Kim, J.; Thurler, A.H.; Castel, S.; Dimisko, L.; Rosen, B.R.; Hadjikhani, N.; et al. Cyclic Vomiting Syndrome is characterized by altered functional brain connectivity of the insular cortex: A cross-comparison with migraine and healthy adults. *Neurogastroenterol. Motil.* **2017**, *29*, e13004. [CrossRef] [PubMed]
23. Sticht, M.A.; Limebeer, C.L.; Rafla, B.R.; Abdullah, R.A.; Poklis, J.L.; Ho, W.; Niphakis, M.J.; Cravatt, B.F.; Sharkey, K.A.; Lichtman, A.H.; et al. Endocannabinoid regulation of nausea is mediated by 2-arachidonoylglycerol (2-AG) in the rat visceral insular cortex. *Neuropharmacology* **2016**, *102*, 92–102. [CrossRef] [PubMed]
24. Perisetti, A.; Goyal, H. Endocannabinoid system and cannabis hyperemesis syndrome: A narrative update. *Eur. J. Gastroenterol. Hepatol.* **2022**, *34*, 1–8. [CrossRef] [PubMed]
25. Sorensen, C.J.; DeSanto, K.; Borgelt, L.; Phillips, K.T.; Monte, A.A. Cannabinoid Hyperemesis Syndrome: Diagnosis, Pathophysiology, and Treatment-a Systematic Review. *J. Med. Toxicol.* **2017**, *13*, 71–87. [CrossRef] [PubMed]
26. Venkatesan, T.; Levinthal, D.; Tarbell, S.E.; Jaradeh, S.S.; Hasler, W.L.; Issenman, R.M.; Adams, K.A.; Sarosiek, I.; Stave, C.; Sharaf, R.N.; et al. Guidelines on management of cyclic vomiting syndrome in adults by the American Neurogastroenterology and Motility Society and the Cyclic Vomiting Syndrome Association. *Neurogastroenterol. Motil.* **2019**, 2 (Suppl. S2), e13604. [CrossRef]
27. Abouzari, M.; Cheung, D.; Pham, T.; Goshtasbi, K.; Sarna, B.; Tajran, S.; Sahyouni, S.; Lin, H.W.; Djalilian, H.R. The Relationship between Vestibular Migraine and Motion Sickness Susceptibility. *Otol. Neurotol.* **2020**, *41*, 1116–1121. [CrossRef]
28. Kayan, A.; Hood, J.D. Neuro-otological manifestations of migraine. *Brain* **1984**, *107 Pt 4*, 1123–1142. [CrossRef]
29. Marcus, D.A.; Furman, J.M. Prevention of motion sickness with rizatriptan: A double-blind, placebo-controlled pilot study. *Med. Sci. Monit.* **2006**, *12*, PI1-7.
30. Drummond, P.D.; Granston, A. Facial pain increases nausea and headache during motion sickness in migraine sufferers. *Brain* **2004**, *127 Pt 3*, 526–534. [CrossRef]
31. Drummond, P.D.; Granston, A. Painful stimulation of the temple induces nausea, headache and extracranial vasodilation in migraine sufferers. *Cephalalgia* **2005**, *25*, 16–22. [CrossRef]

32. Stadler, M.; Bardiau, F.; Seidel, L.; Albert, A.; Boogaerts, J.G. Difference in risk factors for postoperative nausea and vomiting. *Anesthesiology* **2003**, *98*, 46–52. [CrossRef]
33. Khoronenko, V.E.; Baskakov, D.S. Role of migraine history in the development of postoperative nausea and vomiting in patients undergoing general and combined general-epidural anesthesia. *Anesteziol. Reanimatol.* **2014**, *2*, 41–44.
34. Apfel, C.C.; Kranke, P.; Eberhart, L.H. Comparison of surgical site and patient's history with a simplified risk score for the prediction of postoperative nausea and vomiting. *Anaesthesia* **2004**, *59*, 1078–1082. [CrossRef] [PubMed]
35. Heinrichs, L. Linking olfaction with nausea and vomiting of pregnancy, recurrent abortion, hyperemesis gravidarum, and migraine headache. *Am. J. Obstet. Gynecol.* **2002**, *186* (Suppl. S5), 123053. [CrossRef] [PubMed]
36. Hirst, B.; Noble, B. Migraine as a cause of persistent nausea or vomiting in palliative care: A case series. *J. Pain Symptom Manag.* **2009**, *37*, 918–922. [CrossRef] [PubMed]
37. Sicuteri, F. Dopamine, the second putative protagonist in headache. *Headache* **1977**, *17*, 129–131. [CrossRef]
38. Bes, A.; Dupui, P.; Guell, A.; Bessoles, G.; Geraud, G. Pharmacological exploration of dopamine hypersensitivity in migraine patients. *Int. J. Clin. Pharmacol. Res.* **1986**, *6*, 189–192.
39. Calabresi, P.; Silvestrini, M.; Stratta, F.; Cupini, L.M.; Argiro, G.; Atzei, G.P.; Bernardi, G. l-deprenyl test in migraine: Neuroendocrinological aspects. *Cephalalgia* **1993**, *13*, 406–409. [CrossRef]
40. Cerbo, R.; Barbanti, P.; Buzzi, M.G.; Fabbrini, G.; Brusa, L.; Roberti, C.; Zanette, E.; Lenzi, G.L. Dopamine hypersensitivity in migraine: Role of the apomorphine test. *Clin. Neuropharmacol.* **1997**, *20*, 36–41. [CrossRef]
41. Del Zompo, M.; Cherchi, A.; Palmas, M.A.; Ponti, M.; Bocchetta, A.; Gessa, G.L.; Piccardi, M.P. Association between dopamine receptor genes and migraine without aura in a Sardinian sample. *Neurology* **1998**, *51*, 781–786. [CrossRef]
42. Charbit, A.R.; Akerman, S.; Goadsby, P.J. Dopamine: What's new in migraine? *Curr. Opin. Neurol.* **2010**, *23*, 275–281. [CrossRef]
43. Walstab, J.; Krüger, D.; Stark, T.; Hofmann, T.; Demir, I.E.; Ceyhan, G.O.; Feistel, B.; Schemann, M.; Niesler, B. Ginger and its pungent constituents non-competitively inhibit activation of human recombinant and native 5-HT_3 receptors of enteric neurons. *Neurogastroenterol. Motil.* **2013**, *25*, 439–447, e302. [CrossRef] [PubMed]
44. Malick, A.; Jakubowski, M.; Elmquist, J.K.; Saper, C.B.; Burstein, R. A neurohistochemical blueprint for pain-induced loss of appetite. *Proc. Natl. Acad. Sci. USA* **2001**, *98*, 9930–9935. [CrossRef] [PubMed]
45. Giffin, N.; Ruggiero, L.; Lipton, R.; Silberstein, S.; Tvedskov, J.F.; Olesen, J.; Altman, J.; Goadsby, P.; Macrae, A. Premonitory symptoms in migraine: An electronic diary study. *Neurology* **2003**, *60*, 935–940. [CrossRef] [PubMed]
46. Karsan, N.; Bose, P.R.; Thompson, C.; Newman, J.; Goadsby, P.J. Headache and non-headache symptoms provoked by nitroglycerin in migraineurs: A human pharmacological triggering study. *Cephalalgia* **2020**, *40*, 828–841. [CrossRef] [PubMed]
47. Watson, J.; Gonsalves, S.; Fossa, A.; McLean, S.; Seeger, T.; Obach, S.; Andrews, P. The anti-emetic effects of CP-99,994 in the ferret and the dog: Role of the NK1 receptor. *Br. J. Pharmacol.* **1995**, *115*, 84–94. [CrossRef] [PubMed]
48. Saria, A. The tachykinin NK1 receptor in the brain: Pharmacology and putative functions. *Eur. J. Pharmacol.* **1999**, *375*, 51–60. [CrossRef]
49. McLean, S.; Ganong, A.H.; Seeger, T.F.; Bryce, D.K.; Pratt, K.G.; Reynolds, L.S.; Siok, C.J.; Lowe, J.A.; Heym, J. Activity and distribution of binding sites in brain of a nonpeptide substance P (NK1) receptor antagonist. *Science* **1991**, *251*, 437–439. [CrossRef]
50. Robinson, B.W.; Mishkin, M. Alimentary responses to forebrain stimulation in monkeys. *Exp. Brain Res.* **1968**, *4*, 330–366. [CrossRef]
51. Grélot, L.; Milano, S.; Portillo, F.; Miller, A.D.; Bianchi, A.L. Membrane potential changes of phrenic motoneurons during fictive vomiting, coughing, and swallowing in the decerebrate cat. *J. Neurophysiol.* **1992**, *68*, 2110–2119. [CrossRef]
52. Miller, A.D.; Nonaka, S.; Jakus, J. Brain areas essential or non-essential for emesis. *Brain Res.* **1994**, *647*, 255–264. [CrossRef]
53. Maniyar, F.H.; Sprenger, T.; Schankin, C.; Goadsby, P.J. The origin of nausea in migraine-a PET study. *J. Headache Pain* **2014**, *15*, 84. [CrossRef] [PubMed]
54. Goadsby, P.J.; Holland, P.R. An Update: Pathophysiology of Migraine. *Neurol. Clin.* **2019**, *37*, 651–671. [CrossRef] [PubMed]
55. Pereira, E.A.; Lu, G.; Wang, S.; Schweder, P.M.; Hyam, J.A.; Stein, J.F.; Paterson, D.J.; Aziz, T.Z.; Green, A.L. Ventral periaqueductal grey stimulation alters heart rate variability in humans with chronic pain. *Exp. Neurol.* **2010**, *223*, 574–581. [CrossRef] [PubMed]
56. Green, A.L.; Hyam, J.A.; Williams, C.; Wang, S.; Shlugman, D.; Stein, J.F.; Paterson, D.J.; Aziz, T.Z. Intra-operative deep brain stimulation of the periaqueductal grey matter modulates blood pressure and heart rate variability in humans. *Neuromodulation* **2010**, *13*, 174–181. [CrossRef] [PubMed]
57. Luo, H.; Huang, Y.; Green, A.L.; Aziz, T.Z.; Xiao, X.; Wang, S. Neurophysiological characteristics in the periventricular/periaqueductal gray correlate with pain perception, sensation, and affect in neuropathic pain patients. *Neuroimage Clin.* **2021**, *32*, 102876. [CrossRef]
58. Aguiar, D.C.; Almeida-Santos, A.F.; Moreira, F.A.; Guimaraes, F.S. Involvement of TRPV1 channels in the periaqueductal grey on the modulation of innate fear responses. *Acta Neuropsychiatr.* **2015**, *27*, 97–105. [CrossRef]
59. McGovern, A.E.; Ajayi, I.E.; Farrell, M.J.; Mazzone, S.B. A neuroanatomical framework for the central modulation of respiratory sensory processing and cough by the periaqueductal grey. *J. Thorac. Dis.* **2017**, *9*, 4098–4107. [CrossRef]
60. Subramanian, H.H.; Balnave, R.J.; Holstege, G. The midbrain periaqueductal gray control of respiration. *J. Neurosci.* **2008**, *28*, 12274–12283. [CrossRef]
61. Oliveras, J.L.; Besson, J.M.; Guilbaud, G.; Liebeskind, J.C. Behavioral and electrophysiological evidence of pain inhibition from midbrain stimulation in the cat. *Exp. Brain Res.* **1974**, *20*, 32–44. [CrossRef]

62. Makovac, E.; Venezia, A.; Hohenschurz-Schmidt, D.; Dipasquale, O.; Jackson, J.B.; Medina, S.; O'Daly, O.; Williams, S.C.R.; McMahon, S.B.; Howard, M.A. The association between pain-induced autonomic reactivity and descending pain control is mediated by the periaqueductal grey. *J. Physiol.* **2021**, *599*, 5243–5260. [CrossRef]
63. Wu, D.; Wang, S.; Stein, J.F.; Aziz, T.Z.; Green, A.L. Reciprocal interactions between the human thalamus and periaqueductal gray may be important for pain perception. *Exp. Brain Res.* **2014**, *232*, 527–534. [CrossRef]
64. Lau, B.K.; Winters, B.L.; Vaughan, C.W. Opioid presynaptic disinhibition of the midbrain periaqueductal grey descending analgesic pathway. *Br. J. Pharmacol.* **2020**, *177*, 2320–2332. [CrossRef] [PubMed]
65. Hyland, N.P.; Abrahams, T.P.; Fuchs, K.; Burmeister, M.A.; Hornby, P.J. Organization and neurochemistry of vagal preganglionic neurons innervating the lower esophageal sphincter in ferrets. *J. Comp. Neurol.* **2001**, *430*, 222–234. [CrossRef]
66. Kannan, H.; Yamashita, H. Connections of neurons in the region of the nucleus tractus solitarius with the hypothalamic paraventricular nucleus: Their possible involvement in neural control of the cardiovascular system in rats. *Brain Res.* **1985**, *329*, 205–212. [CrossRef]
67. Miller, A.D.; Leslie, R.A. The area postrema and vomiting. *Front. Neuroendocrinol.* **1994**, *15*, 301–320. [CrossRef]
68. Hyde, T.M.; Knable, M.B.; Murray, A.M. Distribution of dopamine D1–D4 receptor subtypes in human dorsal vagal complex. *Synapse* **1996**, *24*, 224–232. [CrossRef]
69. Jeong, J.K.; Dow, S.A.; Young, C.N. Sensory Circumventricular Organs, Neuroendocrine Control, and Metabolic Regulation. *Metabolites* **2021**, *11*, 494. [CrossRef]
70. Stafanini, E.; Clement-Cormier, Y. Detection of dopamine receptors in the area postrema. *Eur. J. Pharmacol.* **1981**, *74*, 257–260. [CrossRef]
71. Amin, A.H.; Crawford, T.B.; Gaddum, J.H. The distribution of substance P and 5-hydroxytryptamine in the central nervous system of the dog. *J. Physiol.* **1954**, *126*, 596–618. [CrossRef]
72. Lipton, R.B.; Schmidt, P.; Diener, H.C. Post Hoc Subanalysis of Two Randomized, Controlled Phase 3 Trials Evaluating Diclofenac Potassium for Oral Solution: Impact of Migraine-Associated Nausea and Prior Triptan Use on Efficacy. *Headache* **2017**, *57*, 756–765. [CrossRef]
73. Chan, T.L.H.; Cowan, R.P.; Woldeamanuel, Y.W. Calcitonin Gene-Related Peptide Receptor Antagonists (Gepants) for the Acute Treatment of Nausea in Episodic Migraine: A Systematic Review and Meta-Analysis. *Headache* **2020**, *60*, 1489–1499. [CrossRef] [PubMed]
74. Duthie, D.J.; Nimmo, W.S. Adverse effects of opioid analgesic drugs. *Br. J. Anaesth.* **1987**, *59*, 61–77. [CrossRef] [PubMed]
75. Bigal, M.E.; Serrano, D.; Buse, D.; Scher, A.; Stewart, W.F.; Lipton, R.B. Acute migraine medications and evolution from episodic to chronic migraine: A longitudinal population-based study. *Headache* **2008**, *48*, 1157–1168. [CrossRef] [PubMed]
76. Ho, T.W.; Rodgers, A.; Bigal, M.E. Impact of recent prior opioid use on rizatriptan efficacy. A post hoc pooled analysis. *Headache* **2009**, *49*, 395–403. [CrossRef]
77. Bonafede, M.; Wilson, K.; Xue, F. Long-term treatment patterns of prophylactic and acute migraine medications and incidence of opioid-related adverse events in patients with migraine. *Cephalalgia* **2019**, *39*, 1086–1098. [CrossRef]
78. Lipton, R.B.; Pascual, J.; Goadsby, P.J.; Massiou, H.; McCarroll, K.A.; Vandormael, K.; Jiang, K.; Lines, C.R. Effect of rizatriptan and other triptans on the nausea symptom of migraine: A post hoc analysis. *Headache* **2001**, *41*, 754–763. [CrossRef]
79. Freitag, F.; Taylor, F.R.; Hamid, M.A.; Ms, A.R.; Hustad, C.M.; Ramsey, K.E.; Skobieranda, F. Elimination of migraine-associated nausea in patients treated with rizatriptan orally disintegrating tablet (ODT): A randomized, double-blind, placebo-controlled study. *Headache* **2008**, *48*, 368–377. [CrossRef]
80. Pierce, M. Oral triptans and nausea: Treatment considerations in migraine. *Headache* **2013**, *1*, 17–20. [CrossRef]
81. Rowat, B.M.; Merrill, C.F.; Davis, A.; South, V. A double-blind comparison of granisetron and placebo for the treatment of acute migraine in the emergency department. *Cephalalgia* **1991**, *11*, 207–213. [CrossRef]
82. Abiri, S.; Chegin, M.; Soleimani, R.; Hatami, N.; Kalani, N.; Rayatdoost, E. Propofol + Granisetron vs. Propofol + Metoclopramide in Symptom Management of Acute Migraine Headache; a Double-Blind Randomized Clinical Trial. *Arch. Acad. Emerg. Med.* **2022**, *10*, e19.
83. Chen, L.; Cai, Z. The efficacy of ginger for the treatment of migraine: A meta-analysis of randomized controlled studies. *Am. J. Emerg. Med.* **2021**, *46*, 567–571. [CrossRef]
84. Maghbooli, M.; Golipour, F.; Moghimi Esfandabadi, A.; Yousefi, M. Comparison between the efficacy of ginger and sumatriptan in the ablative treatment of the common migraine. *Phytother. Res.* **2014**, *28*, 412–415. [CrossRef]
85. Sweetman, S.C. *Martindale: The Complete Drug Reference*, 36th ed.; Pharmaceutical Press: London, UK, 2009.
86. Jolliet, P.; Nion, S.; Allain-Veyrac, G.; Tilloy-Fenart, L.; Vanuxeem, D.; Berezowski, V.; Cecchelli, R. Evidence of lowest brain penetration of an antiemetic drug, metopimazine, compared to domperidone, metoclopramide and chlorpromazine, using an in vitro model of the blood-brain barrier. *Pharmacol. Res.* **2007**, *56*, 11–17. [CrossRef]
87. Volans, G.N. The effect of metoclopramide on the absorption of effervescent aspirin in migraine. *Br. J. Clin. Pharmacol.* **1975**, *2*, 57–63. [CrossRef]
88. Eken, C. Critical reappraisal of intravenous metoclopramide in migraine attack: A systematic review and meta-analysis. *Am. J. Emerg. Med.* **2015**, *33*, 331–337. [CrossRef]
89. Aydin, H.D.; Vuralli, D.; Akçali, D.T.; Bolay, H. Metoclopramide inhibits trigeminovascular activation:evidence for effective acute attack treatment in migraine. *Turk. J. Med. Sci.* **2017**, *47*, 343–347. [CrossRef]

90. Soltani, K.M.; Motamed, H.; Eslami, K.; Majdinasab, N.; Kouti, L. Randomised trial of IV metoclopramide vs. IV ketorolac in treatment of acute primary headaches. *Am. J. Emerg. Med.* **2021**, *50*, 376–380. [CrossRef]
91. Doğan, N.; Pekdemir, M.; Yılmaz, S.; Yaka, E.; Karadaş, A.; Durmuş, U.; Avcu, N.; Koçkan, E. Intravenous metoclopramide in the treatment of acute migraines: A randomized, placebo-controlled trial. *Acta Neurol. Scand.* **2019**, *139*, 334–339. [CrossRef]
92. Lapierre, J.; Amin, M.; Hattangadi, S. Prochlorperazine—A review of the literature since 1956. *Can. Psychiatr. Assoc. J.* **1969**, *14*, 267–274. [CrossRef]
93. Golikhatir, I.; Cheraghmakani, H.; Bozorgi, F.; Jahanian, F.; Sazgar, M.; Montazer, S.H. The Efficacy and Safety of Prochlorperazine in Patients with Acute Migraine: A Systematic Review and Meta-Analysis. *Headache* **2019**, *59*, 682–700. [CrossRef]
94. Bigal, M.E.; Bordini, C.A.; Speciali, J.G. Intravenous chlorpromazine in the emergency department treatment of migraines: A randomized controlled trial. *J. Emerg. Med.* **2002**, *23*, 141–148. [CrossRef]
95. Chou, D.E.; Tso, A.R.; Goadsby, P.J. Aprepitant for the management of nausea with inpatient IV dihydroergotamine. *Neurology* **2016**, *87*, 1613–1616. [CrossRef] [PubMed]
96. Dos Santos, L.V.; Souza, F.H.; Brunetto, A.T.; Sasse, A.D.; da Silveira Nogueira Lima, J.P. Neurokinin-1 receptor antagonists for chemotherapy-induced nausea and vomiting: A systematic review. *J. Natl. Cancer Inst.* **2012**, *104*, 1280–1292. [CrossRef]
97. Douek, E.E. Smell: Recent theories and their clinical application. *J. Laryngol. Otol.* **1967**, *81*, 431–439. [CrossRef]
98. Courteau, J.P.; Cushman, R.; Bouchard, F.; Quevillon, M.; Chartrand, A.; Bherer, L. Survey of construction workers repeatedly exposed to chlorine over a three to six month period in a pulpmill: I. Exposure and symptomatology. *Occup. Environ. Med.* **1994**, *51*, 219–224. [CrossRef] [PubMed]
99. Peatfield, R.C. Relationships between food, wine, and beer-precipitated migrainous headaches. *Headache* **1995**, *35*, 355–357. [CrossRef] [PubMed]
100. Wantke, F.; Focke, M.; Hemmer, W.; Bracun, R.; Wolf-Abdolvahab, S.; Gotz, M.; Jarisch, R.; Tschabitscher, M.; Gann, M.; Tappler, P.; et al. Exposure to formaldehyde and phenol during an anatomy dissecting course: Sensitizing potency of formaldehyde in medical students. *Allergy* **2000**, *55*, 84–87. [CrossRef]
101. Kelman, L. The triggers or precipitants of the acute migraine attack. *Cephalalgia* **2007**, *27*, 394–402. [CrossRef]
102. Barrett, S.A.; Gifford, E.W. Miwok Material Culture. *Bull. Public Mus. City Milwaukee* **1933**, *2*, 117–376.
103. Benemei, S.; Appendino, G.; Geppetti, P. Pleasant natural scent with unpleasant effects: Cluster headache-like attacks triggered by Umbellularia californica. *Cephalalgia* **2010**, *30*, 744–746. [CrossRef]
104. Nassini, R.; Materazzi, S.; Vriens, J.; Prenen, J.; Benemei, S.; De Siena, G.; La Marca, G.; Andrè, E.; Preti, D.; Avonto, C.; et al. The 'headache tree' via umbellulone and TRPA1 activates the trigeminovascular system. *Brain* **2012**, *135 Pt 2*, 376–390. [CrossRef]
105. Zhong, J.; Minassi, A.; Prenen, J.; Taglialatela-Scafati, O.; Appendino, G.; Nilius, B. Umbellulone modulates TRP channels. *Pflugers Arch.* **2011**, *462*, 861–870. [CrossRef]
106. Strecker, T.; Dux, M.; Messlinger, K. Nitric oxide releases calcitonin-gene-related peptide from rat dura mater encephali promoting increases in meningeal blood flow. *J. Vasc. Res.* **2002**, *39*, 489–496. [CrossRef]
107. Nicoletti, P.; Trevisani, M.; Manconi, M.; Gatti, R.; De Siena, G.; Zagli, G.; Benemei, S.; Capone, J.A.; Geppetti, P.; Pini, L.A. Ethanol causes neurogenic vasodilation by TRPV1 activation and CGRP release in the trigeminovascular system of the guinea pig. *Cephalalgia* **2008**, *28*, 9–17. [CrossRef]
108. Trevisani, M.; Smart, D.; Gunthorpe, M.J.; Tognetto, M.; Barbieri, M.; Campi, B.; Amadesi, S.; Gray, J.; Jerman, J.C.; Brough, S.J.; et al. Ethanol elicits and potentiates nociceptor responses via the vanilloid receptor-1. *Nat. Neurosci.* **2002**, *5*, 546–551. [CrossRef]
109. Andre, E.; Campi, B.; Materazzi, S.; Trevisani, M.; Amadesi, S.; Massi, D.; Creminon, C.; Vaksman, N.; Nassini, R.; Civelli, M.; et al. Cigarette smoke-induced neurogenic inflammation is mediated by alpha, beta-unsaturated aldehydes and the TRPA1 receptor in rodents. *J. Clin. Investig.* **2008**, *118*, 2574–2582.
110. Fujita, F.; Uchida, K.; Moriyama, T.; Shima, A.; Shibasaki, K.; Inada, H.; Sokabe, T.; Tominaga, M. Intracellular alkalization causes pain sensation through activation of TRPA1 in mice. *J. Clin. Investig.* **2008**, *118*, 4049–4057. [CrossRef]
111. Bessac, B.F.; Sivula, M.; von Hehn, C.A.; Escalera, J.; Cohn, L.; Jordt, S.E. TRPA1 is a major oxidant sensor in murine airway sensory neurons. *J. Clin. Investig.* **2008**, *118*, 1899–1910. [CrossRef]
112. McNamara, C.R.; Mandel-Brehm, J.; Bautista, D.M.; Siemens, J.; Deranian, K.L.; Zhao, M.; Hayward, N.J.; Chong, J.A.; Julius, D.; Moran, M.M.; et al. TRPA1 mediates formalin-induced pain. *Proc. Natl. Acad. Sci. USA* **2007**, *104*, 13525–13530. [CrossRef]
113. Silva-Néto, R.P.; Peres, M.F.; Valença, M.M. Odorant substances that trigger headaches in migraine patients. *Cephalalgia* **2014**, *34*, 14–21. [CrossRef]
114. Martins, B.; Costa, A. Migraine Improvement during COVID-19 Pandemic—A Case Report on the Wonders of a Mask. *Headache* **2020**, *60*, 2608–2609. [CrossRef]
115. Schulte, L.H.; Jürgens, T.P.; May, A. Photo-, osmo- and phonophobia in the premonitory phase of migraine: Mistaking symptoms for triggers? *J. Headache Pain* **2015**, *16*, 015–0495. [CrossRef]
116. Hoffmann, J.; Recober, A. Migraine and triggers: Post hoc ergo propter hoc? *Curr. Pain Headache Rep.* **2013**, *17*, 370. [CrossRef]
117. Tanik, N.; Bektas, M. Development of quality of life assessment questionnaire associated with osmophobia in people with migraine. *Pain Med.* **2021**, *21*, 1006–1014. [CrossRef]
118. Baldacci, F.; Lucchesi, C.; Ulivi, M.; Cafalli, M.; Vedovello, M.; Vergallo, A.; Prete, E.D.; Nuti, A.; Bonuccelli, U.; Gori, S. Clinical features associated with ictal osmophobia in migraine. *Neurol. Sci.* **2015**, *36*, 43–46. [CrossRef]

119. Albanês Oliveira Bernardo, A.; Lys Medeiros, F.; Sampaio Rocha-Filho, P.A. Osmophobia and Odor-Triggered Headaches in Children and Adolescents: Prevalence, Associated Factors, and Importance in the Diagnosis of Migraine. *Headache* **2020**, *60*, 954–966. [CrossRef]
120. Lovati, C.; Giani, L.; Capiluppi, E.; Preziosa, G.; D'Amico, D.; Mariani, C. O067. Osmophobia in allodynic migraine: Role of frequency of attacks and headache duration. *J. Headache Pain* **2015**, *16* (Suppl. S1), 1129–2377. [CrossRef]
121. Lovati, C.; Lombardo, D.; Peruzzo, S.; Bellotti, A.; Capogrosso, C.A.; Pantoni, L. Osmophobia in migraine: Multifactorial investigation and population-based survey. *Neurol. Sci.* **2020**, *41* (Suppl. S2), 453–454. [CrossRef]
122. Delussi, M.; Laporta, A.; Fraccalvieri, I.; de Tommaso, M. Osmophobia in primary headache patients: Associated symptoms and response to preventive treatments. *J. Headache Pain* **2021**, *22*, 109. [CrossRef]
123. Lovati, C.; Giani, L.; Castoldi, D.; Mariotti D'Alessandro, C.; DeAngeli, F.; Capiluppi, E.; D'Amico, D.; Mariani, C. Osmophobia in allodynic migraineurs: Cause or consequence of central sensitization? *Neurol. Sci.* **2015**, *1*, 145–147. [CrossRef]
124. Wang, Y.F.; Fuh, J.L.; Chen, S.P.; Wu, J.C.; Wang, S.J. Clinical correlates and diagnostic utility of osmophobia in migraine. *Cephalalgia* **2012**, *32*, 1180–1188. [CrossRef]
125. Park, S.P.; Seo, J.G.; Lee, W.K. Osmophobia and allodynia are critical factors for suicidality in patients with migraine. *J. Headache Pain* **2015**, *16*, 44. [CrossRef] [PubMed]
126. Saçmacı, H.; Cengiz, G.F.; Aktürk, T. Impact of dissociative experiences in migraine and its close relationship with osmophobia. *Neurol. Res.* **2020**, *42*, 529–536. [CrossRef] [PubMed]
127. Zanchin, G.; Dainese, F.; Mainardi, F.; Mampreso, E.; Perin, C.; Maggioni, F. Osmophobia in primary headaches. *J. Headache Pain* **2005**, *6*, 213–215. [CrossRef] [PubMed]
128. De Carlo, D.; Toldo, I.; Dal Zotto, L.; Perissinotto, E.; Sartori, S.; Gatta, M.; Balottin, U.; Mazzotta, G.; Moscato, D.; Raieli, V.; et al. Osmophobia as an early marker of migraine: A follow-up study in juvenile patients. *Cephalalgia* **2012**, *32*, 401–406. [CrossRef]
129. Rocha-Filho, P.A.; Marques, K.S.; Torres, R.C.; Leal, K.N. Osmophobia and Headaches in Primary Care: Prevalence, Associated Factors, and Importance in Diagnosing Migraine. *Headache* **2015**, *55*, 840–845. [CrossRef]
130. Silva-Néto, R.P.; Rodrigues, Â.B.; Cavalcante, D.C.; Ferreira, P.H.; Nasi, E.P.; Sousa, K.M.; Peres, M.F.; Valença, M.M. May headache triggered by odors be regarded as a differentiating factor between migraine and other primary headaches? *Cephalalgia* **2017**, *37*, 20–28. [CrossRef]
131. Chalmer, M.A.; Hansen, T.F.; Olesen, J. Nosographic analysis of osmophobia and field testing of diagnostic criteria including osmophobia. *Cephalalgia* **2019**, *39*, 38–43. [CrossRef]
132. Terrin, A.; Mainardi, F.; Lisotto, C.; Mampreso, E.; Fuccaro, M.; Maggioni, F.; Zanchin, G. A prospective study on osmophobia in migraine versus tension-type headache in a large series of attacks. *Cephalalgia* **2020**, *40*, 337–346. [CrossRef]
133. Blau, J.N.; Solomon, F. Smell and other sensory disturbances in migraine. *J. Neurol.* **1985**, *232*, 275–276. [CrossRef]
134. Porta-Etessam, J.; Casanova, I.; García-Cobos, R.; Lapeña, T.; Fernández, M.J.; García-Ramos, R.; Serna, C. Osmophobia analysis in primary headache. *Neurologia* **2009**, *24*, 315–317. [PubMed]
135. Kayabaşoglu, G.; Altundag, A.; Kotan, D.; Dizdar, D.; Kaymaz, R. Osmophobia and olfactory functions in patients with migraine. *Eur. Arch. Otorhinolaryngol.* **2017**, *274*, 817–821. [CrossRef]
136. Kandemir, S.; Pamuk, A.E.; Habipoğlu, Y.; Özel, G.; Bayar Muluk, N.; Kılıç, R. Olfactory acuity based on Brief Smell Identification Test (BSIT®) in migraine patients with and without aura: A cross-sectional, controlled study. *Auris Nasus Larynx* **2021**, *49*, 613–617. [CrossRef] [PubMed]
137. Marmura, M.J.; Monteith, T.S.; Anjum, W.; Doty, R.L.; Hegarty, S.E.; Keith, S.W. Olfactory function in migraine both during and between attacks. *Cephalalgia* **2014**, *34*, 977–985. [CrossRef]
138. Hirsch, A.R. Osmophobia and taste abnormality in migraineurs: A tertiary care study. *Headache* **2005**, *45*, 763–764. [CrossRef]
139. Doğan, A.; Bayar Muluk, N.; Şahan, M.H.; Asal, N.; Inal, M.; Ergün, U. Olfactory bulbus volume and olfactory sulcus depth in migraine patients: An MRI evaluation. *Eur. Arch. Otorhinolaryngol.* **2018**, *275*, 2005–2011. [CrossRef] [PubMed]
140. Aktürk, T.; Tanık, N.; Serin, H.; Saçmacı, H.; İnan, L.E. Olfactory bulb atrophy in migraine patients. *Neurol. Sci.* **2019**, *40*, 127–132. [CrossRef]
141. Demarquay, G.; Royet, J.P.; Mick, G.; Ryvlin, P. Olfactory hypersensitivity in migraineurs: A H(2)(15)O-PET study. *Cephalalgia* **2008**, *28*, 1069–1080. [CrossRef]
142. Stankewitz, A.; May, A. Increased limbic and brainstem activity during migraine attacks following olfactory stimulation. *Neurology* **2011**, *77*, 476–482. [CrossRef]
143. Anderson, A.K.; Christoff, K.; Stappen, I.; Panitz, D.; Ghahremani, D.G.; Glover, G.; Gabrieli, J.D.; Sobel, N. Dissociated neural representations of intensity and valence in human olfaction. *Nat. Neurosci.* **2003**, *6*, 196–202. [CrossRef] [PubMed]
144. Hillert, L.; Jovanovic, H.; Ahs, F.; Savic, I. Women with multiple chemical sensitivity have increased harm avoidance and reduced 5-HT(1A) receptor binding potential in the anterior cingulate and amygdala. *PLoS ONE* **2013**, *8*, e54781. [CrossRef] [PubMed]
145. Hillert, L.; Musabasic, V.; Berglund, H.; Ciumas, C.; Savic, I. Odor processing in multiple chemical sensitivity. *Hum. Brain Mapp.* **2007**, *28*, 172–182. [CrossRef] [PubMed]
146. Del Casale, A.; Ferracuti, S.; Mosca, A.; Pomes, L.M.; Fiaschè, F.; Bonanni, L.; Borro, M.; Gentile, G.; Martelletti, P.; Simmaco, M. Multiple Chemical Sensitivity Syndrome: A Principal Component Analysis of Symptoms. *Int. J. Environ. Res. Public Health* **2020**, *17*, 6551. [CrossRef] [PubMed]

147. Suzuki, K.; Okamura, M.; Haruyama, Y.; Suzuki, S.; Shiina, T.; Kobashi, G.; Hirata, K. Exploring the contributing factors to multiple chemical sensitivity in patients with migraine. *J. Occup. Health* **2022**, *64*, 1348–9585. [CrossRef]
148. Coleman, E.R.; Grosberg, B.M.; Robbins, M.S. Olfactory hallucinations in primary headache disorders: Case series and literature review. *Cephalalgia* **2011**, *31*, 1477–1489. [CrossRef]
149. Mainardi, F.; Rapoport, A.; Zanchin, G.; Maggioni, F. Scent of aura? Clinical features of olfactory hallucinations during a migraine attack (OHM). *Cephalalgia* **2017**, *37*, 154–160. [CrossRef]
150. Henkin, R.I.; Levy, L.M.; Lin, C.S. Taste and smell phantoms revealed by brain functional MRI (fMRI). *J. Comput. Assist. Tomogr.* **2000**, *24*, 106–123. [CrossRef]
151. Henkin, R.I.; Potolicchio, S.J.; Levy, L.M. Olfactory Hallucinations without Clinical Motor Activity: A Comparison of Unirhinal with Birhinal Phantosmia. *Brain Sci.* **2013**, *3*, 1483–1553. [CrossRef]
152. Saberi, A.; Nemati, S.; Amlashi, T.T.; Tohidi, S.; Bakhshi, F. Phonophobia and migraine features in patients with definite meniere's disease: Pentad or triad/tetrad? *Acta Otolaryngol.* **2020**, *140*, 548–552. [CrossRef]
153. Mohammadi, M.; Taziki Balajelini, M.H.; Rajabi, A. Migraine and risk of sudden sensorineural hearing loss: A systematic review and meta-analysis. *Laryngoscope Investig. Otolaryngol.* **2020**, *5*, 1089–1095. [CrossRef]
154. Karsan, N.; Bose, P.; Newman, J.; Goadsby, P.J. Are some patient-perceived migraine triggers simply early manifestations of the attack? *J. Neurol.* **2021**, *268*, 1885–1893. [CrossRef] [PubMed]
155. Woodhouse, A.; Drummond, P.D. Mechanisms of increased sensitivity to noise and light in migraine headache. *Cephalalgia* **1993**, *13*, 417–421. [CrossRef] [PubMed]
156. Vingen, J.V.; Pareja, J.A.; Støren, O.; White, L.R.; Stovner, L.J. Phonophobia in migraine. *Cephalalgia* **1998**, *18*, 243–249. [CrossRef] [PubMed]
157. Main, A.; Dowson, A.; Gross, M. Photophobia and phonophobia in migraineurs between attacks. *Headache* **1997**, *37*, 492–495. [CrossRef]
158. Ashkenazi, A.; Mushtaq, A.; Yang, I.; Oshinsky, M.L. Ictal and interictal phonophobia in migraine-a quantitative controlled study. *Cephalalgia* **2009**, *29*, 1042–1048. [CrossRef]
159. Wang, W.; Timsit-Berthier, M.; Schoenen, J. Intensity dependence of auditory evoked potentials is pronounced in migraine: An indication of cortical potentiation and low serotonergic neurotransmission? *Neurology* **1996**, *46*, 1404–1409. [CrossRef]
160. Irimia, P.; Cittadini, E.; Paemeleire, K.; Cohen, A.S.; Goadsby, P.J. Unilateral photophobia or phonophobia in migraine compared with trigeminal autonomic cephalalgias. *Cephalalgia* **2008**, *28*, 626–630. [CrossRef] [PubMed]
161. Evans, R.W.; Seifert, T.; Kailasam, J.; Mathew, N.T. The use of questions to determine the presence of photophobia and phonophobia during migraine. *Headache* **2008**, *48*, 395–397. [CrossRef]
162. Bolay, H.; Bayazit, Y.A.; Gunduz, B.; Ugur, A.K.; Akcali, D.; Altunyay, S.; Ilica, S.; Babacan, A. Subclinical dysfunction of cochlea and cochlear efferents in migraine: An otoacoustic emission study. *Cephalalgia* **2008**, *28*, 309–317. [CrossRef]
163. Murdin, L.; Premachandra, P.; Davies, R. Sensory dysmodulation in vestibular migraine: An otoacoustic emission suppression study. *Laryngoscope* **2010**, *120*, 1632–1636. [CrossRef]
164. Albanese, M.; Di Girolamo, S.; Silvani, L.; Ciaschi, E.; Chiaramonte, B.; Conti, M.; Passali, F.M.; Di Gioia, B.; Mercuri, N.B.; Di Stadio, A. Distortion Product Otoacoustic Emissions and Their Suppression as Predictors of Peripheral Auditory Damage in Migraine: A Case-Control Study. *J. Clin. Med.* **2021**, *10*, 5007. [CrossRef]
165. Joffily, L.; de Melo Tavares de Lima, M.A.; Vincent, M.B.; Frota, S.M. Assessment of otoacoustic emission suppression in women with migraine and phonophobia. *Neurol. Sci.* **2016**, *37*, 703–709. [CrossRef] [PubMed]
166. Cabanillas, L.A.; Luebke, A.E. CGRP- and cholinergic-containing fibers project to guinea pig outer hair cells. *Hear. Res.* **2002**, *172*, 14–17. [CrossRef]
167. Ambrosini, A.; Rossi, P.; De Pasqua, V.; Pierelli, F.; Schoenen, J. Lack of habituation causes high intensity dependence of auditory evoked cortical potentials in migraine. *Brain* **2003**, *126 Pt 9*, 2009–2015. [CrossRef]
168. Siniatchkin, M.; Kropp, P.; Neumann, M.; Gerber, W.; Stephani, U. Intensity dependence of auditory evoked cortical potentials in migraine families. *Pain* **2000**, *85*, 247–254. [CrossRef]
169. Afra, J.; Proietti Cecchini, A.; Sándor, P.S.; Schoenen, J. Comparison of visual and auditory evoked cortical potentials in migraine patients between attacks. *Clin. Neurophysiol.* **2000**, *111*, 1124–1129. [CrossRef]
170. Hegerl, U.; Juckel, G. Intensity dependence of auditory evoked potentials as an indicator of central serotonergic neurotransmission: A new hypothesis. *Biol. Psychiatry* **1993**, *33*, 173–187. [CrossRef]
171. Sand, T.; Zhitniy, N.; White, L.R.; Stovner, L.J. Brainstem auditory-evoked potential habituation and intensity-dependence related to serotonin metabolism in migraine: A longitudinal study. *Clin. Neurophysiol.* **2008**, *119*, 1190–1200. [CrossRef]
172. Viganò, A.; Torrieri, M.C.; Toscano, M.; Puledda, F.; Petolicchio, B.; Sasso D'Elia, T.; Verzina, A.; Ruggiero, S.; Altieri, M.; Vicenzini, E.; et al. Neurophysiological correlates of clinical improvement after greater occipital nerve (GON) block in chronic migraine: Relevance for chronic migraine pathophysiology. *J. Headache Pain* **2018**, *19*, 73. [CrossRef]
173. Sándor, P.S.; Afra, J.; Ambrosini, A.; Schoenen, J. Prophylactic treatment of migraine with beta-blockers and riboflavin: Differential effects on the intensity dependence of auditory evoked cortical potentials. *Headache* **2000**, *40*, 30–35. [CrossRef]
174. Schlake, H.P.; Grotemeyer, K.H.; Hofferberth, B.; Husstedt, I.W.; Wiesner, S. Brainstem auditory evoked potentials in migraine-evidence of increased side differences during the pain-free interval. *Headache* **1990**, *30*, 129–132. [CrossRef] [PubMed]

175. Podoshin, L.; Ben-David, J.; Pratt, H.; Fradis, M.; Sharf, B.; Weller, B.; Wajsbort, J.; Zellinger, M. Auditory brainstem evoked potentials in patients with migraine. *Headache* **1987**, *27*, 27–29. [CrossRef] [PubMed]
176. Ganji, S. Basilar artery migraine: EEG and evoked potential patterns during acute stage. *Headache* **1986**, *26*, 220–223. [CrossRef] [PubMed]
177. Ganji, S.; Hellman, S.; Stagg, S.; Furlow, J. Episodic coma due to acute basilar artery migraine: Correlation of EEG and brainstem auditory evoked potential patterns. *Clin. Electroencephalogr.* **1993**, *24*, 44–48. [CrossRef]
178. Benna, P.; Bianco, C.; Costa, P.; Piazza, D.; Bergamasco, B. Visual evoked potentials and brainstem auditory evoked potentials in migraine and transient ischemic attacks. *Cephalalgia* **1985**, *5* (Suppl. 2), 53–58. [CrossRef]
179. Kalita, J.; Misra, U.K.; Bansal, R. Phonophobia and brainstem excitability in migraine. *Eur. J. Neurosci.* **2021**, *53*, 1988–1997. [CrossRef]
180. Weiller, C.; May, A.; Limmroth, V.; Juptner, M.; Kaube, H.; Schayck, R.V.; Coenen, H.H.; Diener, H.C. Brain stem activation in spontaneous human migraine attacks. *Nat. Med.* **1995**, *1*, 658–660. [CrossRef]
181. Ashkenazi, A.; Yang, I.; Mushtaq, A.; Oshinsky, M.L. Is phonophobia associated with cutaneous allodynia in migraine? *J. Neurol. Neurosurg. Psychiatry* **2010**, *81*, 1256–1260. [CrossRef]
182. Akdal, G.; Ozge, A.; Ergor, G. The prevalence of vestibular symptoms in migraine or tension-type headache. *J. Vestib. Res.* **2013**, *23*, 101–106. [CrossRef]
183. Kuritzky, A.; Ziegler, D.K.; Hassanein, R. Vertigo, motion sickness and migraine. *Headache* **1981**, *21*, 227–231. [CrossRef]
184. Lee, H.; Sohn, S.I.; Jung, D.K.; Cho, Y.W.; Lim, J.G.; Yi, S.D.; Yi, H.A. Migraine and isolated recurrent vertigo of unknown cause. *Neurol. Res.* **2002**, *24*, 663–665. [CrossRef] [PubMed]
185. Moretti, G.; Manzoni, G.C.; Caffarra, P.; Parma, M. "Benign recurrent vertigo" and its connection with migraine. *Headache* **1980**, *20*, 344–346. [CrossRef] [PubMed]
186. Reploeg, M.D.; Goebel, J.A. Migraine-associated dizziness: Patient characteristics and management options. *Otol. Neurotol.* **2002**, *23*, 364–371. [CrossRef]
187. Vukovic, V.; Plavec, D.; Galinovic, I.; Lovrencic-Huzjan, A.; Budisic, M.; Demarin, V. Prevalence of vertigo, dizziness, and migrainous vertigo in patients with migraine. *Headache* **2007**, *47*, 1427–1435. [CrossRef]
188. Dieterich, M.; Obermann, M.; Celebisoy, N. Vestibular migraine: The most frequent entity of episodic vertigo. *J. Neurol.* **2016**, *263* (Suppl. 1), S82–S89. [CrossRef] [PubMed]
189. Cass, S.P.; Furman, J.M.; Ankerstjerne, K.; Balaban, C.; Yetiser, S.; Aydogan, B. Migraine-related vestibulopathy. *Ann. Otol. Rhinol. Laryngol.* **1997**, *106*, 182–189. [CrossRef]
190. Cutrer, F.M.; Baloh, R.W. Migraine-associated dizziness. *Headache* **1992**, *32*, 300–304. [CrossRef]
191. Van Leeuwen, R.B.; Colijn, C.; van Esch, B.F.; Schermer, T.R. Benign Recurrent Vertigo: The Course of Vertigo Attacks Compared to Patients with Menière's Disease and Vestibular Migraine. *Front. Neurol.* **2022**, *13*, 817812. [CrossRef]
192. Prell, T.; Finn, S.; Zipprich, H.M.; Axer, H. What Predicts Improvement of Dizziness after Multimodal and Interdisciplinary Day Care Treatment? *J. Clin. Med.* **2022**, *11*, 2005. [CrossRef] [PubMed]
193. Formeister, E.J.; Rizk, H.G.; Kohn, M.A.; Sharon, J.D. The Epidemiology of Vestibular Migraine: A Population-based Survey Study. *Otol. Neurotol.* **2018**, *39*, 1037–1044. [CrossRef]
194. Lempert, T.; Olesen, J.; Furman, J.; Waterston, J.; Seemungal, B.; Carey, J.; Bisdorff, A.; Versino, M.; Evers, S.; Newman-Toker, D. Vestibular migraine: Diagnostic criteria. *J. Vestib. Res.* **2012**, *22*, 167–172. [CrossRef] [PubMed]
195. Chae, R.; Krauter, R.; Pasquesi, L.L.; Sharon, J.D. Broadening the vestibular migraine diagnosis criteria: A prospective cohort study on vestibular migraine subtypes. *J. Vestib. Res.* **2022**, *2*, VES-210117. [CrossRef] [PubMed]
196. Neuhauser, H.K.; Radtke, A.; von Brevern, M.; Feldmann, M.; Lezius, F.; Ziese, T.; Lempert, T. Migrainous vertigo: Prevalence and impact on quality of life. *Neurology* **2006**, *67*, 1028–1033. [CrossRef] [PubMed]
197. Dieterich, M.; Brandt, T. Episodic vertigo related to migraine (90 cases): Vestibular migraine? *J. Neurol.* **1999**, *246*, 883–892. [CrossRef]
198. Eggers, S.D.; Staab, J.P.; Neff, B.A.; Goulson, A.M.; Carlson, M.L.; Shepard, N.T. Investigation of the coherence of definite and probable vestibular migraine as distinct clinical entities. *Otol. Neurotol.* **2011**, *32*, 1144–1151. [CrossRef] [PubMed]
199. Teggi, R.; Colombo, B.; Albera, R.; Asprella Libonati, G.; Balzanelli, C.; Batuecas Caletrio, A.; Casani, A.; Espinoza-Sanchez, J.M.; Gamba, P.; Lopez-Escamez, J.A.; et al. Clinical Features, Familial History, and Migraine Precursors in Patients with Definite Vestibular Migraine: The VM-Phenotypes Projects. *Headache* **2018**, *58*, 534–544. [CrossRef]
200. Li, Z.Y.; Shen, B.; Si, L.H.; Ling, X.; Li, K.Z.; Yang, X. Clinical characteristics of definite vestibular migraine diagnosed according to criteria jointly formulated by the Bárány Society and the International Headache Society. *Braz. J. Otorhinolaryngol.* **2022**, *21*, 1–5. [CrossRef]
201. Von Brevern, M.; Zeise, D.; Neuhauser, H.; Clarke, A.H.; Lempert, T. Acute migrainous vertigo: Clinical and oculographic findings. *Brain* **2005**, *128 Pt 2*, 365–374. [CrossRef]
202. Çelebisoy, N.; Kısabay Ak, A.; Özdemir, H.N.; Gökçay, F.; Durmaz, G.S.; Kartı, D.T.; Toydemir, H.E.; Yayla, V.; Çolpak Işıkay, A.; Erkent, İ.; et al. Vestibular migraine, demographic and clinical features of 415 patients: A multicenter study. *Clin. Neurol. Neurosurg.* **2022**, *215*, 107201. [CrossRef]

203. Kısabay Ak, A.; Çelebisoy, N.; Özdemir, H.N.; Gökçay, F.; Saruhan Durmaz, G.; Top Kartı, D.; Ertaşoğlu Toydemir, H.; Yayla, V.; Çolpak Işıkay, A.; Erkent, İ.; et al. Factors determining the response to treatment in patients with vestibular migraine. *Neurol. Res.* **2022**, *29*, 1–8. [CrossRef]
204. Seo, T.; Okano, Y.; Aomi, M.; Yamada, Z.; Kasugai, S.; Nakamura, M.; Miyamoto, Y.; Koizuka, I. Endolymphatic hydrops presumption tests for patients with vestibular migraine. *Acta Otolaryngol.* **2022**, *142*, 406–409. [CrossRef] [PubMed]
205. Leng, Y.; Lei, P.; Chen, C.; Liu, Y.; Xia, K.; Liu, B. Non-contrast MRI of Inner Ear Detected Differences of Endolymphatic Drainage System between Vestibular Migraine and Unilateral Ménière's Disease. *Front. Neurol.* **2022**, *13*, 814518. [CrossRef] [PubMed]
206. Donaldson, L.B.; Yan, F.; Liu, Y.F.; Nguyen, S.A.; Rizk, H.G. Does cognitive dysfunction correlate with dizziness severity in patients with vestibular migraine? *Am. J. Otolaryngol.* **2021**, *42*, 103124. [CrossRef] [PubMed]
207. Karsan, N.; Perez-Rodriguez, A.; Nagaraj, K.; Bose, P.R.; Goadsby, P.J. The migraine postdrome: Spontaneous and triggered phenotypes. *Cephalalgia* **2021**, *41*, 721–730. [CrossRef]
208. Carvalho, G.F.; Luedtke, K.; Pinheiro, C.F.; Moraes, R.; Lemos, T.W.; Carneiro, C.G.; Bigal, M.E.; Dach, F.; Bevilaqua-Grossi, D. Migraine and balance impairment: Influence of subdiagnosis, otoneurological function, falls, and psychosocial factors. *Headache* **2022**, *62*, 548–557. [CrossRef]
209. Nocini, R.; Baraldi, C.; Apa, E.; Ciorba, A.; Monzani, D.; Palma, S. Visually Evoked Postural Responses (VEPRs) in Children with Vestibular Migraine. *Children* **2021**, *9*, 14. [CrossRef]
210. Lim, Y.H.; Kim, J.S.; Lee, H.W.; Kim, S.H. Postural Instability Induced by Visual Motion Stimuli in Patients With Vestibular Migraine. *Front. Neurol.* **2018**, *9*, 433. [CrossRef]
211. So, C.W.; Bent, L.R. Increased vestibular contribution to posture control in individuals with chronic headache. *J. Vestib. Res.* **2009**, *19*, 49–58.
212. Ishizaki, K.; Mori, N.; Takeshima, T.; Fukuhara, Y.; Ijiri, T.; Kusumi, M.; Yasui, K.; Kowa, H.; Nakashima, K. Static stabilometry in patients with migraine and tension-type headache during a headache-free period. *Psychiatry Clin. Neurosci.* **2002**, *56*, 85–90. [CrossRef]
213. Teggi, R.; Colombo, B.; Bernasconi, L.; Bellini, C.; Comi, G.; Bussi, M. Migrainous vertigo: Results of caloric testing and stabilometric findings. *Headache* **2009**, *49*, 435–444. [CrossRef]
214. Janiak-Kiszka, J.; Nowaczewska, M.; Wierzbiński, R.; Kaźmierczak, W.; Kaźmierczak, H. The visual-ocular and vestibulo-ocular reflexes in vestibular migraine. *Otolaryngol. Pol.* **2021**, *76*, 21–28. [CrossRef] [PubMed]
215. Marano, E.; Marcelli, V.; Di Stasio, E.; Bonuso, S.; Vacca, G.; Manganelli, F.; Marciano, E.; Perretti, A. Trigeminal stimulation elicits a peripheral vestibular imbalance in migraine patients. *Headache* **2005**, *45*, 325–331. [CrossRef] [PubMed]
216. Shin, J.H.; Kim, Y.K.; Kim, H.J.; Kim, J.S. Altered brain metabolism in vestibular migraine: Comparison of interictal and ictal findings. *Cephalalgia* **2014**, *34*, 58–67. [CrossRef] [PubMed]
217. Obermann, M.; Wurthmann, S.; Steinberg, B.S.; Theysohn, N.; Diener, H.C.; Naegel, S. Central vestibular system modulation in vestibular migraine. *Cephalalgia* **2014**, *34*, 1053–1061. [CrossRef] [PubMed]
218. Russo, A.; Marcelli, V.; Esposito, F.; Corvino, V.; Marcuccio, L.; Giannone, A.; Conforti, R.; Marciano, E.; Tedeschi, G.; Tessitore, A. Abnormal thalamic function in patients with vestibular migraine. *Neurology* **2014**, *82*, 2120–2126. [CrossRef] [PubMed]
219. Fasold, O.; von Brevern, M.; Kuhberg, M.; Ploner, C.J.; Villringer, A.; Lempert, T.; Wenzel, R. Human vestibular cortex as identified with caloric stimulation in functional magnetic resonance imaging. *Neuroimage* **2002**, *17*, 1384–1393. [CrossRef]
220. Koo, J.W.; Balaban, C.D. Serotonin-induced plasma extravasation in the murine inner ear: Possible mechanism of migraine-associated inner ear dysfunction. *Cephalalgia* **2006**, *26*, 1310–1319. [CrossRef]
221. Ahn, S.K.; Balaban, C.D. Distribution of 5-HT1B and 5-HT1D receptors in the inner ear. *Brain Res.* **2010**, *1346*, 92–101. [CrossRef]
222. Vass, Z.; Steyger, P.S.; Hordichok, A.J.; Trune, D.R.; Jancso, G.; Nuttall, A.L. Capsaicin stimulation of the cochlea and electric stimulation of the trigeminal ganglion mediate vascular permeability in cochlear and vertebro-basilar arteries: A potential cause of inner ear dysfunction in headache. *Neuroscience* **2001**, *103*, 189–201. [CrossRef]
223. Halberstadt, A.L.; Balaban, C.D. Organization of projections from the raphe nuclei to the vestibular nuclei in rats. *Neuroscience* **2003**, *120*, 573–594. [CrossRef]
224. Byun, Y.J.; Levy, D.A.; Nguyen, S.A.; Brennan, E.; Rizk, H.G. Treatment of Vestibular Migraine: A Systematic Review and Meta-analysis. *Laryngoscope* **2021**, *131*, 186–194. [CrossRef] [PubMed]
225. Bayer, O.; Adrion, C.; Al Tawil, A.; Mansmann, U.; Strupp, M. Results and lessons learnt from a randomized controlled trial: Prophylactic treatment of vestibular migraine with metoprolol (PROVEMIG). *Trials* **2019**, *20*, 813. [CrossRef]
226. Neuhauser, H.; Radtke, A.; von Brevern, M.; Lempert, T. Zolmitriptan for treatment of migrainous vertigo: A pilot randomized placebo-controlled trial. *Neurology* **2003**, *60*, 882–883. [CrossRef]
227. Stancel-Lewis, J.; Lau, J.W.L.; Male, A.; Korres, G.; Rogel-Salazar, J.; Pavlou, M.; Bamiou, D.E. Vestibular Rehabilitation Therapy for the Treatment of Vestibular Migraine, and the Impact of Traumatic Brain Injury on Outcome: A Retrospective Study. *Otol. Neurotol.* **2022**, *43*, 359–367. [CrossRef] [PubMed]
228. Tian, R.; Zhang, Y.; Pan, Q.; Wang, Y.; Wen, Q.; Fan, X.; Qin, G.; Zhang, D.; Chen, L.; Zhang, Y.; et al. Calcitonin gene-related peptide receptor antagonist BIBN4096BS regulates synaptic transmission in the vestibular nucleus and improves vestibular function via PKC/ERK/CREB pathway in an experimental chronic migraine rat model. *J. Headache Pain* **2022**, *23*, 35. [CrossRef] [PubMed]

229. Hoskin, J.L.; Fife, T.D. New Anti-CGRP Medications in the Treatment of Vestibular Migraine. *Front. Neurol.* **2022**, *12*, 799002. [CrossRef]
230. Selby, G.; Lance, J.W. Observations on 500 cases of migraine and allied vascular headache. *J. Neurol. Neurosurg. Psychiatry* **1960**, *23*, 23–32. [CrossRef]
231. Ashkenazi, A.; Sholtzow, M.; Shaw, J.W.; Burstein, R.; Young, W.B. Identifying cutaneous allodynia in chronic migraine using a practical clinical method. *Cephalalgia* **2007**, *27*, 111–117. [CrossRef]
232. Nahman-Averbuch, H.; Shefi, T.; Schneider, V.J.; Li, D., 2nd; Ding, L.; King, C.D.; Coghill, R.C. Quantitative sensory testing in patients with migraine: A systematic review and meta-analysis. *Pain* **2018**, *159*, 1202–1223. [CrossRef]
233. Ashkenazi, A.; Silberstein, S.; Jakubowski, M.; Burstein, R. Improved identification of allodynic migraine patients using a questionnaire. *Cephalalgia* **2007**, *27*, 325–329. [CrossRef]
234. Lipton, R.B.; Bigal, M.E.; Ashina, S.; Burstein, R.; Silberstein, S.; Reed, M.L.; Serrano, D.; Stewart, W.F. Cutaneous allodynia in the migraine population. *Ann. Neurol.* **2008**, *63*, 148–158. [CrossRef] [PubMed]
235. Bigal, M.E.; Ashina, S.; Burstein, R.; Reed, M.L.; Buse, D.; Serrano, D.; Lipton, R.B.; Group, A. Prevalence and characteristics of allodynia in headache sufferers: A population study. *Neurology* **2008**, *70*, 1525–1533. [CrossRef] [PubMed]
236. Burstein, R.; Yarnitsky, D.; Goor-Aryeh, I.; Ransil, B.J.; Bajwa, Z.H. An association between migraine and cutaneous allodynia. *Ann. Neurol.* **2000**, *47*, 614–624. [CrossRef]
237. Bevilaqua-Grossi, D.; Lipton, R.B.; Napchan, U.; Grosberg, B.; Ashina, S.; Bigal, M.E. Temporomandibular disorders and cutaneous allodynia are associated in individuals with migraine. *Cephalalgia* **2010**, *30*, 425–432. [CrossRef]
238. Lovati, C.; D'Amico, D.; Rosa, S.; Suardelli, M.; Mailland, E.; Bertora, P.; Pomati, S.; Mariani, C.; Bussone, G. Allodynia in different forms of migraine. *Neurol. Sci.* **2007**, *28* (Suppl. S2), S220–S221. [CrossRef]
239. Akerman, S.; Karsan, N.; Bose, P.; Hoffmann, J.R.; Holland, P.R.; Romero-Reyes, M.; Goadsby, P.J. Nitroglycerine triggers triptan-responsive cranial allodynia and trigeminal neuronal hypersensitivity. *Brain* **2019**, *142*, 103–119. [CrossRef]
240. Ladda, J.; Straube, A.; Forderreuther, S.; Krause, P.; Eggert, T. Quantitative sensory testing in cluster headache: Increased sensory thresholds. *Cephalalgia* **2006**, *26*, 1043–1050. [CrossRef]
241. Burstein, R.; Cutrer, M.F.; Yarnitsky, D. The development of cutaneous allodynia during a migraine attack clinical evidence for the sequential recruitment of spinal and supraspinal nociceptive neurons in migraine. *Brain* **2000**, *123 Pt 8*, 1703–1709. [CrossRef]
242. Burstein, R.; Collins, B.; Jakubowski, M. Defeating migraine pain with triptans: A race against the development of cutaneous allodynia. *Ann. Neurol.* **2004**, *55*, 19–26. [CrossRef] [PubMed]
243. Cady, R.; Martin, V.; Mauskop, A.; Rodgers, A.; Hustad, C.; Ramsey, K.; Skobieranda, F. Symptoms of cutaneous sensitivity pre-treatment and post-treatment: Results from the rizatriptan TAME studies. *Cephalalgia* **2007**, *27*, 1055–1060. [CrossRef]
244. Goadsby, P.J.; Zanchin, G.; Geraud, G.; de Klippel, N.; Diaz-Insa, S.; Gobel, H.; Cunha, L.; Ivanoff, N.; Falques, M.; Fortea, J. Early vs. non-early intervention in acute migraine—'Act when Mild (AwM)'. A double-blind, placebo-controlled trial of almotriptan. *Cephalalgia* **2008**, *28*, 383–391. [CrossRef] [PubMed]
245. Cady, R.K.; Freitag, F.G.; Mathew, N.T.; Elkind, A.H.; Mao, L.; Fisher, A.C.; Biondi, D.M.; Finlayson, G.; Greenberg, S.J.; Hulihan, J.F. Allodynia-associated symptoms, pain intensity and time to treatment: Predicting treatment response in acute migraine intervention. *Headache* **2009**, *49*, 350–363. [CrossRef]
246. Ezzati, A.; Fanning, K.M.; Buse, D.C.; Pavlovic, J.M.; Armand, C.E.; Reed, M.L.; Martin, V.T.; Lipton, R.B. Predictive models for determining treatment response to nonprescription acute medications in migraine: Results from the American Migraine Prevalence and Prevention Study. *Headache* **2022**, *11*, 14312. [CrossRef] [PubMed]
247. Yuksel, H.; Kenar, S.G.; Gursoy, G.T.; Bektas, H. The impacts of masks and disinfectants on migraine patients in the COVID-19 pandemic. *J. Clin. Neurosci.* **2022**, *97*, 87–92. [CrossRef] [PubMed]
248. Sivilotti, L.; Woolf, C.J. The contribution of GABAA and glycine receptors to central sensitization: Disinhibition and touch-evoked allodynia in the spinal cord. *J. Neurophysiol.* **1994**, *72*, 169–179. [CrossRef] [PubMed]
249. Strassman, A.M.; Raymond, S.A.; Burstein, R. Sensitization of meningeal sensory neurons and the origin of headaches. *Nature* **1996**, *384*, 560–564. [CrossRef]
250. Hoskin, K.L.; Bulmer, D.C.; Lasalandra, M.; Jonkman, A.; Goadsby, P.J. Fos expression in the midbrain periaqueductal grey after trigeminovascular stimulation. *J. Anat.* **2001**, *198 Pt 1*, 29–35. [CrossRef]
251. Knight, Y.E.; Goadsby, P.J. The periaqueductal grey matter modulates trigeminovascular input: A role in migraine? *Neuroscience* **2001**, *106*, 793–800. [CrossRef]
252. Pozo-Rosich, P.; Storer, R.J.; Charbit, A.R.; Goadsby, P.J. Periaqueductal gray calcitonin gene-related peptide modulates trigeminovascular neurons. *Cephalalgia* **2015**, *35*, 1298–1307. [CrossRef]
253. Akerman, S.; Goadsby, P.J. Neuronal PAC1 receptors mediate delayed activation and sensitization of trigeminocervical neurons: Relevance to migraine. *Sci. Transl. Med.* **2015**, *7*, 308ra157. [CrossRef]
254. Sessle, B.J.; Hu, J.W.; Amano, N.; Zhong, G. Convergence of cutaneous, tooth pulp, visceral, neck and muscle afferents onto nociceptive and non-nociceptive neurones in trigeminal subnucleus caudalis (medullary dorsal horn) and its implications for referred pain. *Pain* **1986**, *27*, 219–235. [CrossRef]
255. Burstein, R.; Yamamura, H.; Malick, A.; Strassman, A.M. Chemical stimulation of the intracranial dura induces enhanced responses to facial stimulation in brain stem trigeminal neurons. *J. Neurophysiol.* **1998**, *79*, 964–982. [CrossRef] [PubMed]

256. Watanabe, M.; Kopruszinski, C.M.; Moutal, A.; Ikegami, D.; Khanna, R.; Chen, Y.; Ross, S.; Mackenzie, K.; Stratton, J.; Dodick, D.W.; et al. Dysregulation of serum prolactin links the hypothalamus with female nociceptors to promote migraine. *Brain* **2022**, *145*, 2894–2909. [CrossRef] [PubMed]
257. Burstein, R.; Jakubowski, M.; Garcia-Nicas, E.; Kainz, V.; Bajwa, Z.; Hargreaves, R.; Becerra, L.; Borsook, D. Thalamic sensitization transforms localized pain into widespread allodynia. *Ann. Neurol.* **2010**, *68*, 81–91. [CrossRef] [PubMed]
258. Shields, K.G.; Goadsby, P.J. Propranolol modulates trigeminovascular responses in thalamic ventroposteromedial nucleus: A role in migraine? *Brain* **2005**, *128 Pt 1*, 86–97. [CrossRef]
259. Noseda, R.; Jakubowski, M.; Kainz, V.; Borsook, D.; Burstein, R. Cortical projections of functionally identified thalamic trigeminovascular neurons: Implications for migraine headache and its associated symptoms. *J. Neurosci.* **2011**, *31*, 14204–14217. [CrossRef]
260. Moulton, E.A.; Becerra, L.; Maleki, N.; Pendse, G.; Tully, S.; Hargreaves, R.; Burstein, R.; Borsook, D. Painful heat reveals hyperexcitability of the temporal pole in interictal and ictal migraine States. *Cereb. Cortex* **2011**, *21*, 435–448. [CrossRef]
261. Lai, K.L.; Liao, K.K.; Fuh, J.L.; Wang, S.J. Subcortical hyperexcitability in migraineurs: A high-frequency oscillation study. *Can. J. Neurol. Sci.* **2011**, *38*, 309–316. [CrossRef]
262. Ambrosini, A.; Schoenen, J. Electrophysiological response patterns of primary sensory cortices in migraine. *J. Headache Pain* **2006**, *7*, 377–388. [CrossRef]
263. Schwedt, T.J.; Chong, C.D. Correlations between brain cortical thickness and cutaneous pain thresholds are atypical in adults with migraine. *PLoS ONE* **2014**, *9*, e99791.
264. Schwedt, T.J.; Berisha, V.; Chong, C.D. Temporal lobe cortical thickness correlations differentiate the migraine brain from the healthy brain. *PLoS ONE* **2015**, *10*, e0116687. [CrossRef] [PubMed]
265. Tso, A.R.; Trujillo, A.; Guo, C.C.; Goadsby, P.J.; Seeley, W.W. The anterior insula shows heightened interictal intrinsic connectivity in migraine without aura. *Neurology* **2015**, *84*, 1043–1050. [CrossRef] [PubMed]
266. Wen, Q.; Wang, Y.; Pan, Q.; Tian, R.; Zhang, D.; Qin, G.; Zhou, J.; Chen, L. MicroRNA-155-5p promotes neuroinflammation and central sensitization via inhibiting SIRT1 in a nitroglycerin-induced chronic migraine mouse model. *J. Neuroinflamm.* **2021**, *18*, 287. [CrossRef] [PubMed]
267. Chen, H.; Tang, X.; Li, J.; Hu, B.; Yang, W.; Zhan, M.; Ma, T.; Xu, S. IL-17 crosses the blood-brain barrier to trigger neuroinflammation: A novel mechanism in nitroglycerin-induced chronic migraine. *J. Headache Pain* **2022**, *23*, 1. [CrossRef]
268. Irimia, P.; Martinez-Valbuena, I.; Minguez-Olaondo, A.; Dominguez-Vivero, C.; Sanchez-Arias, J.A.; Martinez-Vila, E.; Luquin, M.-R.; Leira, R. Interictal amylin levels in chronic migraine patients: A case-control study. *Cephalalgia* **2021**, *41*, 604–612. [CrossRef]
269. De Luca, C.; Baldacci, F.; Mazzucchi, S.; Lombardo, I.; Curto, L.; Ulivi, M.; Chico, L.; Papa, M.; Siciliano, G.; Gori, S. CGRP Inhibitors and Oxidative Stress Biomarkers in Resistant Migraine: A Real-Life Study with Erenumab, Fremanezumab, and Galcanezumab. *J. Clin. Med.* **2021**, *10*, 4586. [CrossRef]
270. Greco, R.; Demartini, C.; Francavilla, M.; Zanaboni, A.M.; Tassorelli, C. Dual Inhibition of FAAH and MAGL Counteracts Migraine-like Pain and Behavior in an Animal Model of Migraine. *Cells* **2021**, *10*, 2543. [CrossRef]
271. Paige, C.; Plasencia-Fernandez, I.; Kume, M.; Papalampropoulou-Tsiridou, M.; Lorenzo, L.E.; David, E.T.; He, L.; Mejia, G.L.; Driskill, C.; Ferrini, F.; et al. A Female-Specific Role for Calcitonin Gene-Related Peptide (CGRP) in Rodent Pain Models. *J. Neurosci.* **2022**, *42*, 1930–1944. [CrossRef]
272. De Logu, F.; Nassini, R.; Hegron, A.; Landini, L.; Jensen, D.D.; Latorre, R.; Ding, J.; Marini, M.; de Araujo, D.S.M.; Ramírez-Garcia, P.; et al. Schwann cell endosome CGRP signals elicit periorbital mechanical allodynia in mice. *Nat. Commun.* **2022**, *13*, 646. [CrossRef]
273. Iannone, L.F.; De Logu, F.; Geppetti, P.; De Cesaris, F. The role of TRP ion channels in migraine and headache. *Neurosci. Lett.* **2022**, *768*, 136380. [CrossRef]
274. Schwedt, T.J.; Krauss, M.J.; Frey, K.; Gereau, I.R.W. Episodic and chronic migraineurs are hypersensitive to thermal stimuli between migraine attacks. *Cephalalgia* **2011**, *31*, 6–12. [CrossRef] [PubMed]
275. Burgos-Vega, C.C.; Ahn, D.D.; Bischoff, C.; Wang, W.; Horne, D.; Wang, J.; Gavva, N.; Dussor, G. Meningeal transient receptor potential channel M8 activation causes cutaneous facial and hindpaw allodynia in a preclinical rodent model of headache. *Cephalalgia* **2016**, *36*, 185–193. [CrossRef]
276. Ren, L.; Dhaka, A.; Cao, Y.Q. Function and postnatal changes of dural afferent fibers expressing TRPM8 channels. *Mol. Pain* **2015**, *11*, 37. [CrossRef] [PubMed]
277. Askari-Zahabi, K.; Abbasnejad, M.; Kooshki, R.; Raoof, M.; Esmaeili-Mahani, S.; Pourrahimi, A.M.; Zamyad, M. The role of basolateral amygdala orexin 1 receptors on the modulation of pain and psychosocial deficits in nitroglycerin-induced migraine model in adult male rats. *Korean J. Pain* **2022**, *35*, 22–32. [CrossRef]
278. Lance, J.W. Impact commentaries. Observations on 500 cases of migraine and allied vascular headache. *J. Neurol. Neurosurg. Psychiatry* **2012**, *83*, 673–674. [CrossRef] [PubMed]

cells

Article

Antagonism of CGRP Receptor: Central and Peripheral Mechanisms and Mediators in an Animal Model of Chronic Migraine

Rosaria Greco [1,*], Chiara Demartini [1,2], Miriam Francavilla [1], Anna Maria Zanaboni [1,2] and Cristina Tassorelli [1,2]

1 Unit of Translational Neurovascular Research, IRCCS Mondino Foundation, 27100 Pavia, Italy
2 Department of Brain and Behavioral Sciences, University of Pavia, 27100 Pavia, Italy
* Correspondence: rosaria.greco@mondino.it; Tel.: +39-(0382)-380255

Abstract: Calcitonin-gene-related peptide (CGRP) plays a key role in migraine pathophysiology and more specifically in the mechanisms underlying peripheral and central sensitization. Here, we explored the interaction of CGRP with other pain mediators relevant for neuronal sensitization in an animal model of chronic migraine. Male Sprague-Dawley rats were exposed to nitroglycerin (NTG, 5 mg/kg, i.p.) or vehicle co-administered with the CGRP receptor antagonist olcegepant (2 mg/kg i.p.), or its vehicle, every other day over a 9-day period. Twenty-four hours after the last injection of NTG (or vehicle), behavioral test and ex vivo analysis were performed. Olcegepant attenuated NTG-induced trigeminal hyperalgesia in the second phase of the orofacial formalin test. Interestingly, it also reduced gene expression and protein levels of CGRP, pro-inflammatory cytokines, inflammatory-associated miRNAs (miR-155-5p, miR-382-5p, and miR-34a-5p), and transient receptor potential ankyrin channels in the medulla–pons area, cervical spinal cord, and trigeminal ganglia. Similarly, olcegepant reduced the NTG-induced increase in CGRP and inflammatory cytokines in serum. The findings show that the activation of the CGRP pathway in a migraine animal model was associated to the persistent activation of inflammatory pathways, which was paralleled by a condition of hyperalgesia. These molecular events are relevant for informing us about the mechanisms underlying chronic migraine.

Keywords: olcegepant; chronic migraine; inflammation

Citation: Greco, R.; Demartini, C.; Francavilla, M.; Zanaboni, A.M.; Tassorelli, C. Antagonism of CGRP Receptor: Central and Peripheral Mechanisms and Mediators in an Animal Model of Chronic Migraine. *Cells* **2022**, *11*, 3092. https://doi.org/10.3390/cells11193092

Academic Editor: Alessandro Castorina

Received: 6 September 2022
Accepted: 28 September 2022
Published: 30 September 2022

1. Introduction

Migraine is associated with the activation of the trigeminovascular system, which induces peripheral and central sensitization [1]. The role of central sensitization, and the associated increase in the transmission of pain signals along the second and third neuron, seems particularly important for chronic migraine.

A huge number of studies support the involvement of calcitonin gene-related peptide (CGRP) in the development of sensitization and enhanced pain sensibility that characterize migraine pain [2]. CGRP exists in two forms: α-CGRP predominates in the peripheral and central nervous system, whereas β-CGRP is mostly distributed in the enteric nervous system [3]. α-CGRP is released from trigeminal nerves during migraine attacks. The activation of the trigeminal nerve causes an antidromic release of α-CGRP to induce non-endothelium-mediated vasodilatation. In the trigeminal nucleus caudalis (TNC), CGRP acts on second-order neurons to transmit pain signals through the brainstem and midbrain to higher levels [4]. The CGRP receptors are highly distributed throughout the trigeminovascular system, particularly, but not limited to, on neurons and glial cells [5]. CGRP is also involved in mast cell degranulation and thus in neurogenic inflammation [6]. In vitro, CGRP enhances pro-inflammatory cytokine' expression and release from rodent satellite glial cells of the trigeminal ganglia (TGs) [7], as well as from human peripheral blood mononuclear cells [8]. Circulating cytokines are known to be involved in the inflammatory pain phenomenon, and indirect evidence suggests that the release of these mediators from the trigeminovascular endings is involved in the initiation and persistence of pain [9]. The CGRP inflammatory mediator relationship may be bidirectional, since inflammatory mediators released during a migraine attack lead to the activation of transient receptor potential ankyrin 1 channels (TRPA1) [10,11], located on the trigeminal afferents, which in turn favor the release of CGRP [12,13]. The presence of inflammation in patients with migraine disease has been confirmed; for instance, several plasma cytokines, including tumor necrosis factor alpha (TNF-alpha), interleukin (IL)-1beta, and IL-6, increase during migraine attacks and in attack-free intervals [14]. Thus, the activation of the trigeminovascular system and the release of CGRP and neurogenic inflammation are key factors in migraine pathogenesis, together with the maintenance of a sensitized neuronal state [15–18].

Clinical studies confirmed the therapeutic potential of blocking CGRP in migraine pain by different CGRP receptor antagonists and monoclonal antibodies targeting either the CGRP or its receptor in the acute and preventive treatment of migraine with positive results [19–21]. Besides pain transmission, the blockade of CGRP signaling may exert beneficial effects also in relation to the inflammatory pathway in migraine, although the mechanisms are still not fully understood [22,23].

In the multifaceted scenario of migraine pathophysiology, there are also other elements that appear to be involved. MicroRNAs (miRNAs) are short non-coding RNAs that regulate a multiplicity of cellular processes, including immune/inflammatory responses and migraine pain [24,25]. Recently, we demonstrated that in patients with migraine disease, CGRP plasma levels and miR-382-5p and miR-34a-5p gene expression in peripheral cells are associated with each other at the individual level across the migraine spectrum [26]. In addition, CGRP plasma levels were positively correlated with the peripheral levels of both miRNAs, suggesting an interaction between CGRP and these signaling molecules. CGRP may induce IL-6 gene expression in macrophages by the upregulation of circular RNA_007893, a modulator of miR-485-5p [27].

The aim of this study was to explore in more depth the interplay between the neuropeptide CGRP and the inflammatory mediators within the mechanisms of neuronal sensitization in an animal model of chronic migraine, using olcegepant as a pharmacological probe. Specifically, we used the experimental animal model based on chronic systemic administration of nitroglycerin (NTG) that produces cephalic and extracephalic hypersensitivity [28–30] and evaluated ongoing changes of signaling molecules linked to inflammation in specific areas.

2. Materials and Methods

2.1. Animals and Experimental Design

Adult male Sprague-Dawley rats, weighing 250–270 g, were used in this study and randomly allocated in the different experimental groups, as reported in Table 1.

The IASP's guidelines for pain research in animals were followed [31]. All procedures were conducted in accordance with the European Convention for Care and Use of Laboratory Animals, and the experimental protocols were approved by the Italian Ministry of Health (no. 691/2020-PR). The animals were housed in the animal facility of the University of Pavia (Pavia, Italy) in groups of two per cages under controlled conditions (i.e., temperature 21–22 °C, 60–50% relative humidity) and 12/12 h light cycle (with lights on at 7.00 a.m.). Food and water were available ad libitum. Upon arrival, animals were habituated to the housing conditions for one week before the experimental testing. The experiments were performed in a randomized manner by an experimenter blinded to treatments.

Table 1. Experimental groups.

	Experimental Groups	I SET: OFT (*n*)	II SET: rt-PCR; ELISA (*n*)
Acute migraine model	CT (NTG vehicle + olcegepant vehicle)	8	-
	NTG (NTG 10 mg/kg + olcegepant vehicle)	9	-
	NTG + olcegepant 1 mg (NTG 10 mg/kg + olcegepant 1 mg/kg)	6	-
	olcegepant 1 mg (NTG vehicle + olcegepant 1 mg/kg)	6	-
	NTG + olcegepant 2 mg (NTG 10 mg/kg + olcegepant 2 mg/kg)	6	-
	olcegepant 2 mg (NTG vehicle + olcegepant 2 mg/kg)	6	-
Chronic migraine model	CT (NTG vehicle + olcegepant vehicle)	6	6
	NTG (NTG 5 mg/kg + olcegepant vehicle)	6	6
	NTG + olcegepant 2 mg (NTG 5 mg/kg + olcegepant 2 mg/kg)	6	6
	olcegepant 2 mg (NTG vehicle + olcegepant 2 mg/kg)	6	6

CT: control; ELISA: enzyme-linked immunosorbent assay; *n*: number of animals per group; NTG: nitroglycerin; OFT: orofacial formalin test; rt-PCR: real time polymerase chain reaction.

NTG (Bioindustria L.I.M., Novi Ligure (AL), Italy) was prepared from a stock solution of 5.0 mg/1.5 mL dissolved in 27% alcohol and 73% propylene glycol. For injections, NTG was further diluted in saline (0.9% NaCl) to reach the final concentration of 6% alcohol and 16% propylene glycol.

To test the appropriate dose of olcegepant able to counteract NTG-induced trigeminal hyperalgesia, we first tested two doses of the CGRP receptor antagonist in the acute animal model of migraine on the basis of a single NTG (10 mg/kg, i.p.) administration. For this purpose, groups of 6–9 rats were treated with olcegepant (1 or 2 mg/kg, i.p.) or vehicle (PEG 200/Tween-80/saline 1:1:18, 1 mL/kg, i.p.) 3 h after NTG (10 mg/kg, i.p.) or vehicle administration. Four hours after NTG (or vehicle) injection, the animals underwent the orofacial formalin test (Figure 1). The latency of h from NTG injection and the administration timing and doses of olcegepant were selected in agreement with previous observations [28,29,32–34].

The optimal dose identified with the above-described set of experiments was used within the animal model of chronic migraine, where 4 groups of 6 rats each received NTG (5 mg/kg, i.p.) or vehicle injection co-administered with olcegepant 2 mg/kg i.p. or its vehicle (1 mL/kg, i.p.) every 2 days over a 9-day period (see Table 1 for experimental group assignments).

Twenty-four hours after the last injection of NTG (or vehicle), a first set of rats underwent the orofacial formalin test. A second set of animals (not exposed to the orofacial formalin test) was used to evaluate CGRP, miRNAs, pro-inflammatory cytokines, and *TRPA1* gene expression in medulla-pons, cervical spinal cord (CSC), and TGs. In the same areas, we also evaluated the protein levels of the same cytokines. Additionally, we assayed CGRP serum levels (Table 1 and Figure 1).

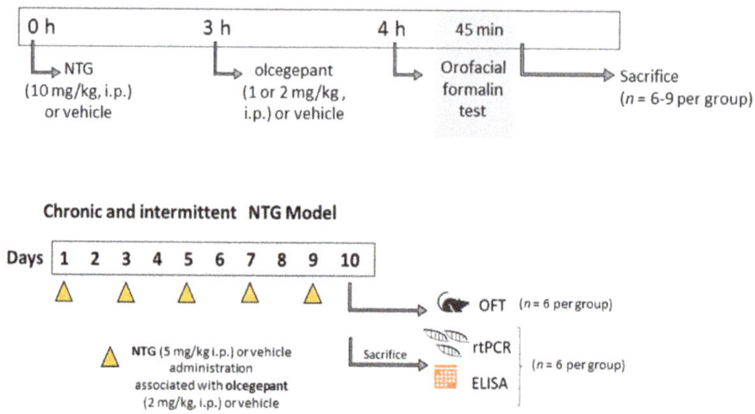

Figure 1. Experimental design of the acute and chronic models with specific procedures. Acute migraine model: after treatments, the animals underwent the orofacial formalin test; at the end of the behavioral test, all animals were sacrificed. Chronic migraine model: after treatments, the animals were divided into two different experimental sets. In the first set, the animals underwent the orofacial formalin test and then they were sacrificed. Within the second set, after sacrifice, the samples were collected to be used for rt-PCR and ELISA analysis. ELISA: enzyme-linked immunosorbent assay; *n*: number of animals per group; NTG: nitroglycerin; OFT: orofacial formalin test; rt-PCR: real time polymerase chain reaction.

2.2. *Orofacial Formalin Test*

Since NTG is systemically administered and its sensitization properties can induce cephalic and extra-cephalic changes, the orofacial formalin test was used in association with the NTG model to specifically activate the trigeminal system. Such a combination allows for the study of NTG-induced hyperalgesia at the trigeminal level, reflecting the orofacial/cephalic hypersensitivity observed in migraine patients.

The procedures applied for the behavioral test were those extensively described elsewhere [35]. Briefly, after rats' acclimatization (20 min) to the test chamber, they were injected subcutaneously with 50 µL of formalin 1.5% into the right upper lip. Face rubbing was measured by a researcher blind to treatments, counting the seconds the animal spent grooming the injected area with the ipsilateral forepaw or hindpaw 0–3 min (Phase I) and 12–45 min (Phase II) after formalin injection.

2.3. *Rt-PCR and Enzyme-Linked Immunosorbent Assay (ELISA)*

Rats of the second experimental set of the chronic model were euthanized under deep anesthesia (sodium thiopental) after the last NTG of vehicle injection. After decapitation, medulla–pons (bregma, −13.30 to −14.60 mm), CSC (C1–C2), and TGs were quickly dissected out, divided into right and left parts, rinsed in cold sterile 0.9% NaCl solution, placed in cryogenic tubes, and immediately frozen in liquid nitrogen. They were subsequently kept at −80 °C until rt-PCR processing for TNF-alpha, IL-1beta protein and gene expression, α-CGRP and TRPA1 gene expression, and miRNA evaluation. We selected medulla–pons and CSC because they contain several nuclei that play an important role in the mediation of migraine pain, e.g., TNC, locus coeruleus, and periacqueductal gray [36,37].

For mRNA, all procedures were performed under RNase-free conditions; after RNA extraction, the absorbance ratios (260/280 nm) ranged from 1.9 to 2.0 in all RNA samples, indicating no significant protein (including of blood origin) contamination. mRNA levels were measured by rt-PCR [38]. Primer sequences obtained from the Primer3 (https://primer3.ut.ee/, accessed on 14 January 2021) are reported in Table 2. Glyceraldehyde 3-phosphate dehydrogenase (GAPDH), whose expression remained constant in all experimental groups, was used for normalization.

Table 2. Primer sequences used in rt-PCR analysis.

Gene	Forward Primer	Reverse Primer
GAPDH	AACCTGCCAAGTATGATGAC	GGAGTTGCTGTTGAAGTCA
TNF-alpha	CCTCACACTCAGATCATCTTCTC	CGCTTGGTGGTTTGCTAC
IL-1beta	TCTTCCTTGTGCAAGTGTCTG	CAGGTCATTCTCCTCACTGTC
Calca (α-CGRP)	CAGTCTCAGCTCCAAGTCATC	TTCCAAGGTTGACCTCAAAG
TRPA1	CTCCCCGAGTGCATGAAAGT	TGCATATACGCGGGGATGTC
U6	TGCGGGTGCTCGCTTCGGCAGC	CCAGTGCAGGGTCCGAGGT
miR-155-5p	TTGAATTCTAACACCTTCGTGGCTACAGAG	TTAGATCTCATTTATCGAGGGAAGGATTG
miR-382-5p	GGCTGTGAGTAATTCTTTGGCAG	GGCAGTATACTTGCTGATTGCT
miR-34a-5p	GCAGTGTCTTAGCTGGTTGTTG	TGCAGCACTTCTAGGGCAGT

GAPDH: glyceraldehyde 3-phosphate dehydrogenase; *TNF-alpha*: tumor necrosis factor alpha; *IL-1beta:* interleukin 1beta; *Calca*: calcitonin related polypeptide alpha; *TRPA1*: transient receptor potential ankyrin 1 channels.

The same RNA was used for miRNA extraction in the same areas. Synthesis of cDNA was performed by using MirXMirna First Strand Synthesis (Takara-Diatech Labline, Jesi-Ancona, Italy), and TB Green q-Rt PCR was used (Takara-Diatech, Labline Jesi-Ancona, Italy) to determine expression levels of miR-155-5p, miR-34a-5p, and miR-382-5p. miRNAs expression was normalized with U6 (a type of small nuclear RNA), used as a housekeeping gene. The primers of miRNAs were selected from the Primer3 software (https://primer3.ut.ee/, accessed on 26 May 2021) and synthesized by Sigma Aldrich (Milan, Italy) (Table 2). Triplicate reactions were averaged for each mRNA and miRNA. The amount of mRNA was normalized to GAPDH or U6 using the 2–ΔΔCT method.

Pro-inflammatory cytokines in medulla–pons, CSC, and TG were evaluated using the ELISA procedure. All tissues were weighed and homogenized using a Precellys homogenizer, and central and peripheral cytokine levels were analyzed using a commercial ELISA kit (Diaclone Co, Besançon, France), adhering to the manufacturer's instructions.

For CGRP, TNF-alpha, and IL-1beta serum levels, the blood samples were collected in a clot activator with gel separator serum tubes and centrifuged for 15 min at 1000× *g* at 2–8 °C. Protein levels were measured using commercial ELISA kits (α-CGRP: Elabsciences, Houston, TX, USA; TNF-alpha and IL-1beta: Diaclone Co, Besançon, France). The measured absorbance of the samples in a microplate reader (Biotek, Santa Clara, CA, USA) was compared with a standard curve, and the concentrations were calculated.

2.4. Statistical Analysis

An a priori power analysis was conducted to determine the required sample size needed to obtain a statistical power of 0.80 at an alpha level of 0.05 (GPower version 3.1.9.4, Franz Faul, University Kiel, Kiel, Germany). We hypothesized a difference in total nociceptive response in Phase II of the orofacial formalin test (face rubbing time, in the acute and the chronic models) between rats injected with NTG and rats injected with NTG + olcegepant of that at least equal to the control condition (NTG = 160 ± 14 SEM; CT = 135 ± 18 SEM), and thus we estimated a sample size of at least 6 rats in each experimental group with an effect size of 1.55. The data were tested for normality using the Shapiro–Wilk normality test and considered normal. For nociceptive responses, gene expression, and protein levels, the statistical differences between groups were determined using the one-way ANOVA followed by post hoc Tukey's multiple comparisons test.

A probability level of <5% was considered significant.

3. Results

3.1. Acute Migraine Model

Orofacial Formalin Test

As illustrated in Figure 2, NTG administration induced a hyperalgesic state that was detectable as an increase in nocifensive behavior (total face rubbing time) during Phase II of the orofacial formalin test. Only 2 mg/kg of olcegepant significantly reduced NTG-induced nocifensive behavior in Phase II compared with NTG group, demonstrating a dose-dependent effect on trigeminal hyperalgesia. No significant effect was observed when olcegepant was injected with NTG vehicle compared with the CT group. No significant differences among groups were seen during Phase I of the test. The 2 mg/kg dose of olcegepant was then adopted to perform all the evaluations within the chronic migraine model.

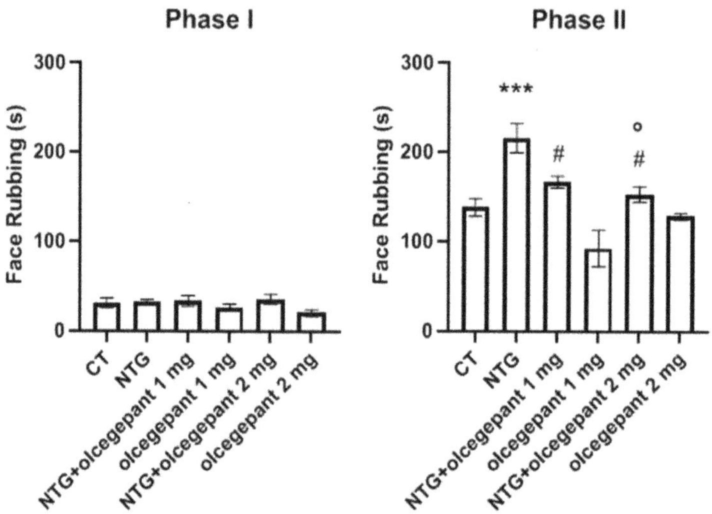

Figure 2. Time of face rubbing (expressed in seconds) during Phase I and II of the orofacial formalin test following acute systemic adminsitration of nitroglycerin (NTG)/vehicle and olcegepant/vehicle. Data are expressed as mean ± SEM. One-way ANOVA followed by Tukey's multiple comparisons test: *** $p < 0.001$ vs. control (CT), olcegepant 1 mg and olcegepant 2 mg; ° $p < 0.05$ vs. NTG; # $p < 0.05$ vs. olcegepant 1 mg.

3.2. *Chronic Migraine Model*

3.2.1. Orofacial Formalin Test

As illustrated in Figure 3, NTG administration induced a persistent hyperalgesic state, which was detectable as an increase in nocifensive behavior (total face rubbing time) during Phase II of the orofacial formalin test. This hyperalgesic state was detected 24 h after the last NTG treatment. The NTG-induced increase in face rubbing time was prevented by the chronic administration of olcegepant, confirming a key role of CGRP in the central and peripheral sensitization. Olcegepant did not induce any significant change in the orofacial formalin test when administered to the rats treated with NTG vehicle. No significant differences among groups were seen during Phase I of the test.

3.2.2. CGRP

Chronic NTG treatment increased CGRP gene expression in the central areas (CSC and medulla–pons, Figure 4A,B, respectively) and in the TG (Figure 4C) when compared with the control (CT) group. It also significantly increased CGRP serum levels (Figure 4D). These changes were significantly inhibited by chronic olcegepant treatment (Figure 4). No effect on CGRP gene expression and serum levels was observed when olcegepant was given with NTG vehicle (olcegepant 2 mg group).

The data suggest that olcegepant treatment reduces CGRP levels by blocking its receptor, probably via a negative feedback loop on NTG-induced inflammation.

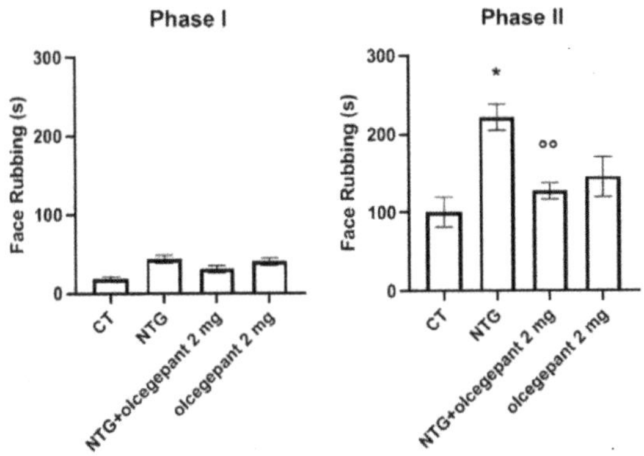

Figure 3. Time of face rubbing (expressed in seconds) during Phase I and II of the orofacial formalin test following chronic systemic administration of nitroglycerin (NTG)/vehicle and olcegepant/vehicle. Data are expressed as mean ± SEM. One-way ANOVA followed by Tukey's multiple comparisons test: * $p < 0.05$ vs. control (CT) and olcegepant 2 mg; °° $p < 0.01$ vs. NTG.

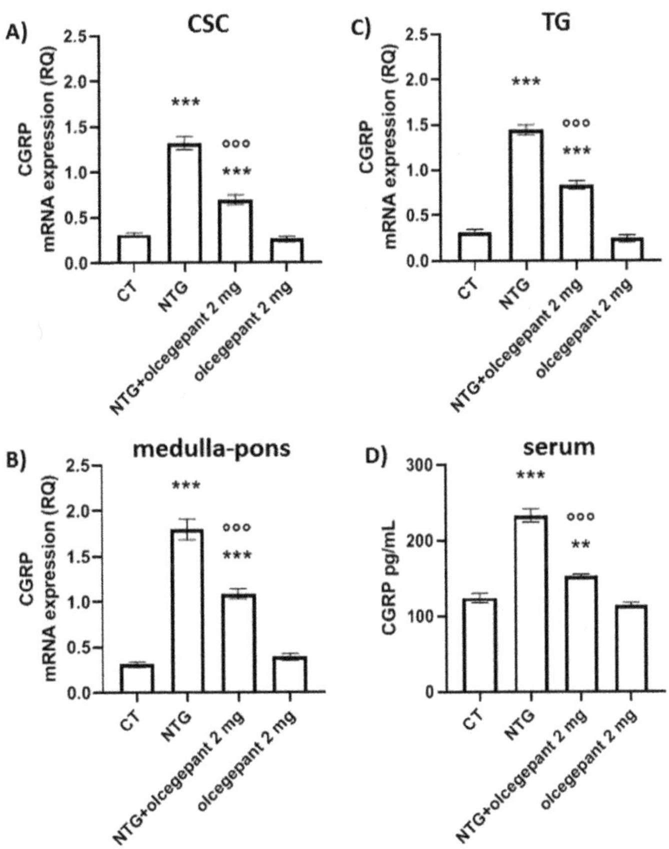

Figure 4. CGRP mRNA levels expressed as relative quantification (RQ) in cervical spinal cord (CSC) (**A**), medulla–pons (**B**), and trigeminal ganglia (TG) (**C**); CGRP serum levels expressed as pg/mL (**D**). Data are expressed as mean ± SEM; one-way ANOVA followed by Tukey's multiple comparisons test. ** $p < 0.01$ and *** $p < 0.001$, vs. control (CT) and olcegepant 2 mg; °°° $p < 0.001$ vs. nitroglycerin (NTG).

3.2.3. Cytokines

Chronic NTG treatment increased gene expression and protein levels of TNF-alpha (Figure 5) and IL-1-beta (Figure 6) in medulla–pons, CSC, and TG compared to the CT group; moreover, the NTG challenge significantly increased TNF-alpha (Figure 5G) and IL-1-beta (Figure 6G) serum levels compared to the CT group. Chronic treatment with olcegepant induced a significant decrease in NTG-induced mRNA TNF-alpha in all the areas under evaluation (Figure 5A–C). Olcegepant administration also reduced TNF-alpha protein levels in the medulla–pons and TG (Figure 5D–F). Olcegepant reduced the NTG-induced increase in IL-1beta mRNA in all the areas under investigation. It also reduced IL-1beta protein levels in the CSC and medulla–pons and in the serum (Figure 6A–F). Olcegepant did not induce any significant change in the parameters under evaluation when administered to the rats treated with NTG vehicle. Data are shown in Figures 5 and 6.

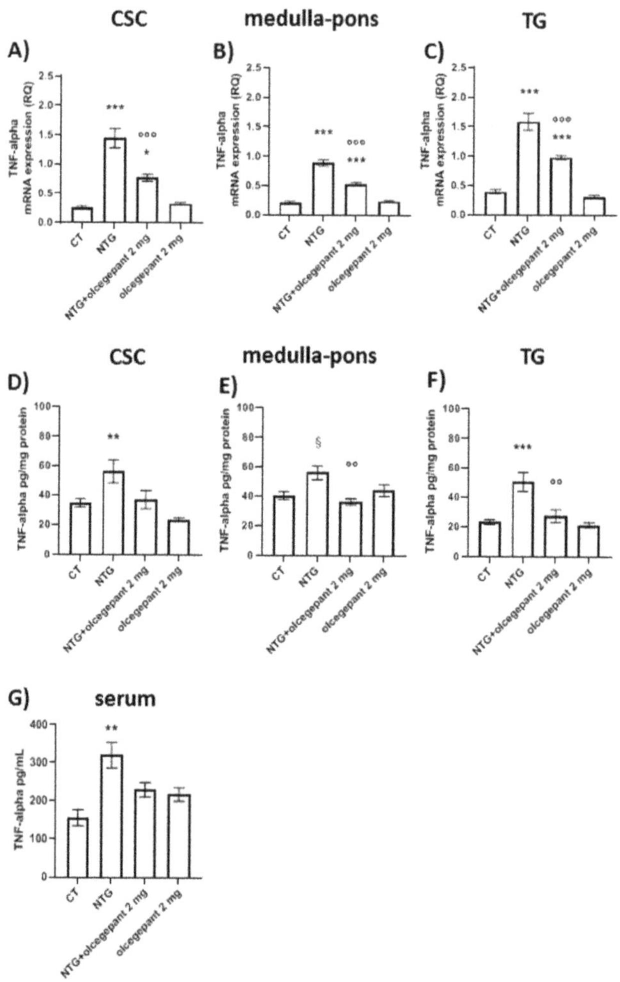

Figure 5. *TNF-alpha* gene expression levels (**A–C**) and protein levels (**D–F**) in cervical spinal cord (CSC), medulla–pons and trigeminal ganglion (TG), and TNF-alpha serum levels (**G**). mRNA levels are expressed as relative quantification (RQ); protein levels in tissues are expressed as pg/mg of protein and in serum are expressed as pg/mL. Data are expressed as mean ± SEM; one-way ANOVA followed by Tukey's multiple comparisons test. * $p < 0.05$, ** $p < 0.01$, and *** $p < 0.001$ vs. control (CT) and olcegepant 2 mg; °° $p < 0.01$ and °°° $p < 0.001$ vs. nitroglycerin (NTG); § $p < 0.05$ vs. CT.

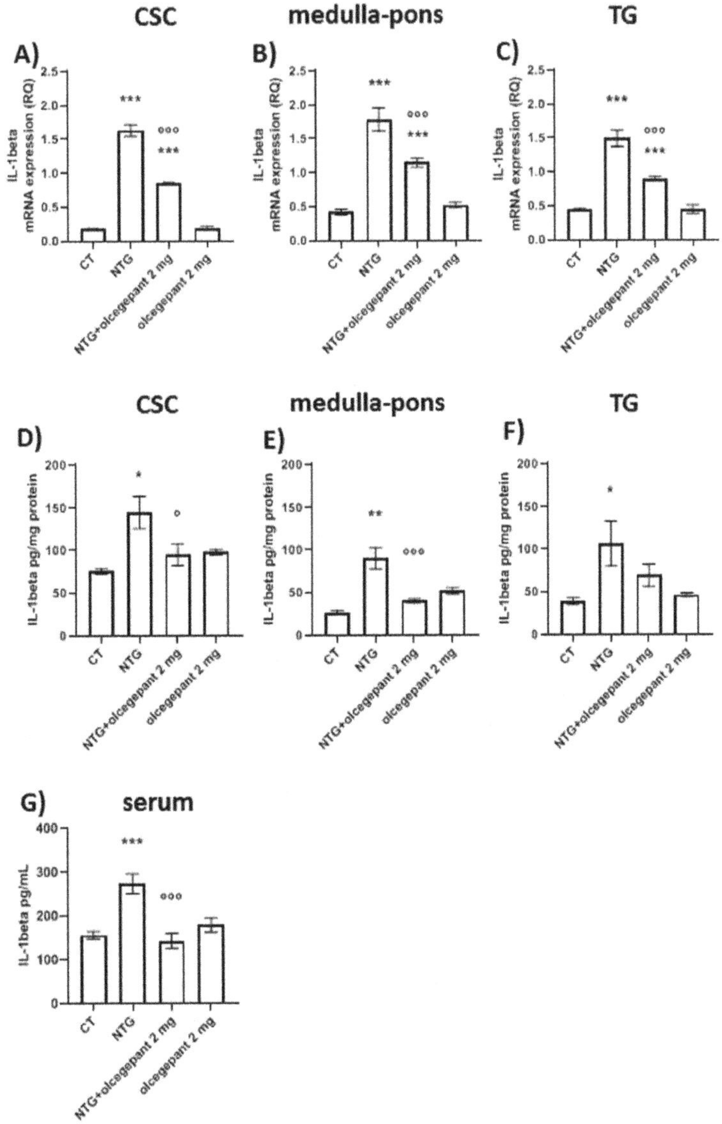

Figure 6. *IL-1beta* gene expression levels (**A–C**) and protein levels (**D–F**) in cervical spinal cord (CSC), medulla–pons and trigeminal ganglion (TG), and IL-1beta serum levels (**G**). mRNA levels are expressed as relative quantification (RQ); protein levels in tissues are expressed as pg/mg of protein and in serum are expressed as pg/mL. Data are expressed as mean ± SEM; one-way ANOVA followed by Tukey's multiple comparisons test. * $p < 0.05$, ** $p < 0.01$, and *** $p < 0.001$ vs. control (CT) and olcegepant 2 mg; ° $p < 0.05$ and °°° $p < 0.001$ vs. nitroglycerin (NTG).

These findings indicate that the release of cytokines is mediated by CGRP probably from the activated glial cells, which is responsible for the persistence of trigeminal pain.

3.2.4. microRNAs

Chronic NTG-treated animals showed a significant increase in miR-155-5p (Figure 7A–C), miR-34a-5p (Figure 7D–F), and miR-382-5p (Figure 7G–I) expression in CSC, medulla–pons, and TG compared to the CT group. These changes were significantly attenuated in all areas by olcegepant treatment (Figure 7). No effect on microRNA levels was observed when olcegepant was given with NTG vehicle (olcegepant 2 mg group) (Figure 7). The findings suggest that CGRP may interfere by still unknown mechanisms in the modulation of inflammatory-related miRNAs.

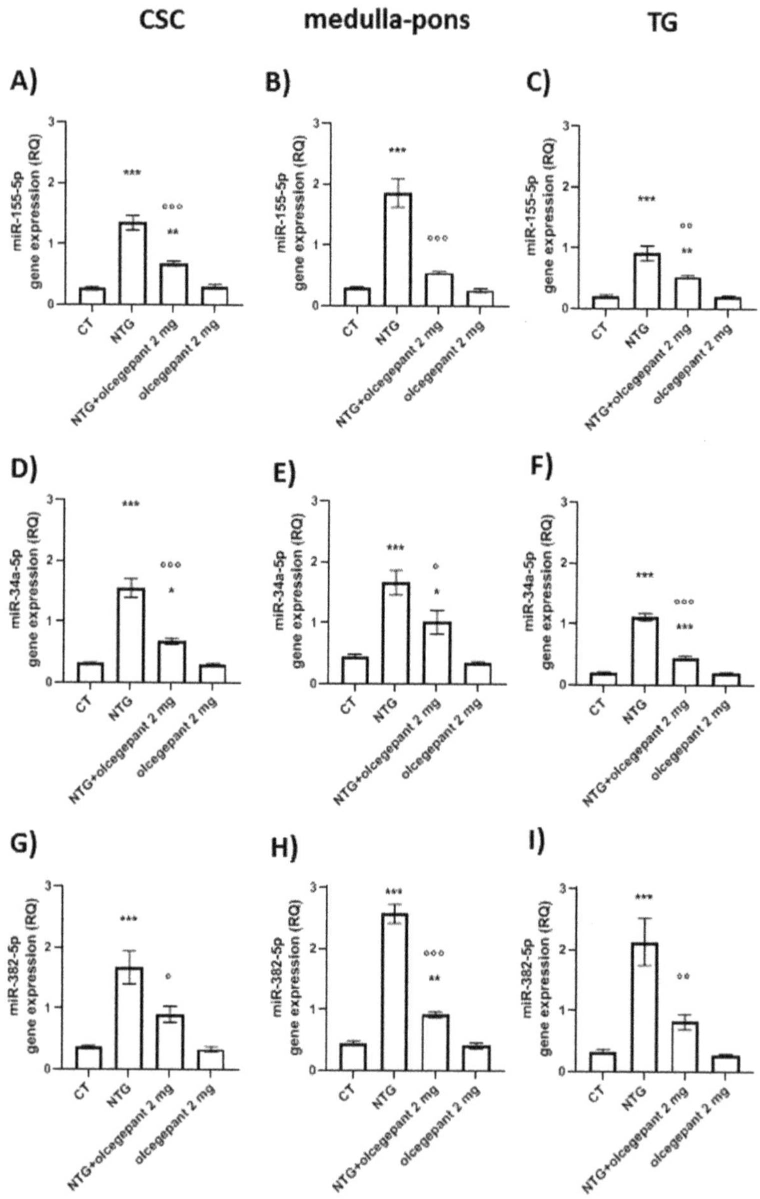

Figure 7. Expression levels of miR-155-5p (**A–C**), miR-34a-5p (**D–F**), and miR-382-5p (**G–I**) in the cervical spinal cord (CSC), medulla–pons, and trigeminal ganglion (TG). microRNA expression is expressed as relative quantification (RQ). Data are expressed as mean ± SEM; one-way ANOVA followed by Tukey's multiple comparisons test. * $p < 0.05$, ** $p < 0.01$, and *** $p < 0.001$ vs. control (CT) and olcegepant 2 mg; ° $p < 0.05$, °° $p < 0.01$, and °°° $p < 0.001$ vs. nitroglycerin (NTG).

3.2.5. TRPA1 Gene Expression

Chronic NTG administration caused a significant increase in *TRPA1* mRNA expression levels in all the areas investigated compared to the CT group. These changes were significantly attenuated by olcegepant treatment in all the three areas. Data are reported in Figure 8. These results suggest that the blockade of CGRP receptor, probably by inhibition of pro-inflammatory pathways, interrupts the processes that lead to the neuropeptide release following TRPA1 activation.

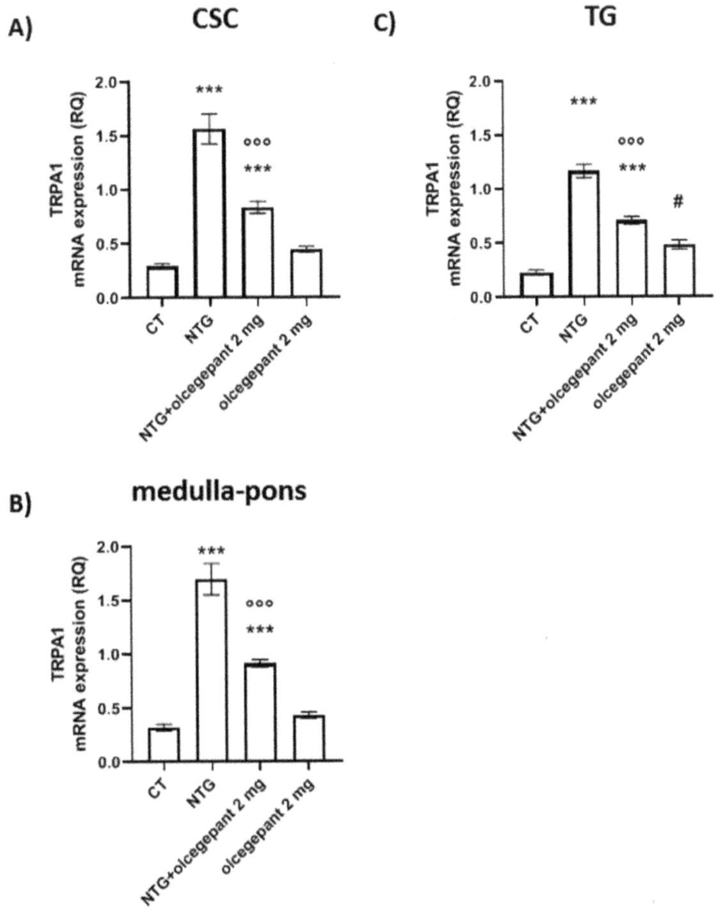

Figure 8. *TRPA1* gene expression levels in (**A**) cervical spinal cord (CSC), (**B**) medulla–pons, and (**C**) trigeminal ganglion (TG). mRNA levels are expressed as relative quantification (RQ). Data are expressed as mean ± SEM; one-way ANOVA followed by Tukey's multiple comparisons test. *** $p < 0.001$ vs. control (CT) and olcegepant 2 mg; °°° $p < 0.001$ vs. nitroglycerin (NTG); # $p < 0.05$ vs. CT.

4. Discussion

In migraine, peripheral and central sensitization play an important pathophysiological role, since they augment pain signal transmission, causing an altered processing of sensory stimuli. In this context, a major role belongs to CGRP, which is expressed in sensory afferents innervating the cranial vasculature and it exerts vasodilatory and neuroinflammatory action, contributing to the development of peripheral and central sensitization.

Here, we explored in depth the interaction of CGRP with other pain mediators relevant for neuronal sensitization in a migraine animal model of chronic migraine, using olcegepant as a pharmacological probe. In this context, our study yielded multiple important pieces of information. First, we confirmed the pivotal role of CGRP in migraine pathophysiology, as suggested by the increase in CGRP gene expression in medulla–pons, CSC, and TGs, as well as in the serum. In addition, we reported that the above-mentioned changes affecting the CGRP pathway are paralleled by an increase in (i) inflammatory cytokines in central and peripheral areas of the nervous system that are relevant in migraine circuitry and in the serum, (ii) miRNA associated to inflammation in the same areas, and (iii) TRPA1 gene expression in the same areas. These effects were all significantly counteracted by olcegepant administration in order to confirm the role of the CGRP pathway.

In agreement, mRNA and protein levels of CGRP were found to be significantly increased in the periaqueductal gray and TNC after NTG administration, supporting an involvement of CGRP in the endogenous pain modulatory system [39,40]. As noted, higher expression levels of CGRP in the brainstem are found in the locus coeruleus [36], a mainly noradrenergic nucleus involved in the regulation of autonomic [41], stress [42], and nociceptive functions [43]. The increase in CGRP serum levels likely reflects a peripherally restricted phenomenon, where CGRP is released from trigeminal fibers outside the blood–brain barrier (dura mater) and drained via intra- and extracranial venous vessels into the jugular blood [44–46].

Thus, the released CGRP would increase sensory activity at multiple levels, peripheral and central, a feature of migraine pain [47,48]. In the present study, NTG-induced increase in CGRP gene expression was significantly reversed after treatment with olcegepant, as was NTG-induced increase in CGRP serum levels. One of the possible ways through which CGRP is released after NTG is the activation of the TRPA1 channels [49], whose gene expression was increased by the NTG challenge, in agreement with previous data showing that TRPA1 channel activation in trigeminal nociceptive fibers leads to the release of the neuropeptide [12,13].

The increase in cytokines observed in the nervous tissues and in the serum in our experiments was inhibited by olcegepant. These findings indicate that the cytokine expression and release is mediated by CGRP. This latter was indeed reported to enhance the expression and release of pro-inflammatory cytokines from human peripheral blood mononuclear cells [8] and lymphocytes [50]. Interestingly, patients with an ongoing migraine attack have higher serum levels of pro-inflammatory cytokines compared to healthy controls, and their serum CGRP levels positively correlate with cytokine's levels [51]. It is tempting to hypothesize that NTG-activated glial cells have a role in the increased cytokine expression and release in the TGs, CSC, and medulla–pons [52]. In line with this idea, previous data suggest that CGRP stimulates satellite glial cells within the TGs, subsequently to sensory neurons' sensitization [7,53,54], probably also through the potentiation of the P2Y purinergic receptors [55]. Thus, we can speculate that the blockade of CGRP receptors by olcegepant led to a reduction in CGRP release and, hence, of pro-inflammatory mediators, thus breaking down the pathways that give rise to CGRP release. In line with this hypothesis, olcegepant inhibited TRPA1-mediated dilation [56] and blood flow [57] of the meningeal artery, thus suggesting a reduction of CGRP release [56,57]. Olcegepant-mediated block of pain and inflammation may in turn represent the signal for the system to further reduce CGRP release, again also via TRPA1, thus returning to a physiological level of CGRP receptor stimulation.

Mounting evidence suggests that several miRNAs expressed in the nervous system play important roles in neuroinflammation and chronic pain [58]. In particular, miR-155 has been shown to be deeply involved in regulating inflammation-associated diseases [5], including migraine [59,60]. Indeed, miR-155 was found to be upregulated in migraine patients in pain-free periods [59]; the same was also reported for miR-382-5p [61]. Intriguingly, the serum upregulation of miR-382-5p is also associated, together with miR-34a-5p, with migraine attacks [61].

In line with the above-mentioned observations, our study shows that the three miRNAs investigated here were upregulated in the model of chronic migraine, while they were downregulated by olcegepant treatment in the central and peripheral areas of the nervous system relevant for migraine pain. This downregulation was associated with a significant reduction of pro-inflammatory cytokines (gene and protein expression) in both peripheral and central nervous system areas, suggesting a relationship between CGRP, miRNAs, and cytokine pathways.

Recently, Chen et al. [39] reported an increase in TNF-alpha and IL-1beta levels and polarized microglia in TNC after chronic NTG; this suggests the possibility that NTG induces an upstream mechanism in which neuroinflammation is involved [39] and that blockade of the CGRP signaling is able to modulate neuroinflammation and pain. NTG may induce an inflammatory response within the dura mater and TG, potentiated by CGRP release, with an increase in pro-inflammatory cytokines IL-1beta and TNF-alpha levels by nuclear factor kappa B (NF-κB) activation [62,63]. This response may induce an upregulation of miRNAs, not only at the peripheral level but indirectly at the central level, confirming a link with pathways associated with pain transmission/modulation [25]. MiR-382-5p acts as a negative modulator of the interleukin 10 receptor alpha subunit (IL-10RA), an endogenous inhibitor of pro-inflammatory IL-1beta signaling gene expression [61], and thus its reduction is in keeping with the inhibition of CGRP pro-inflammatory activity after olcegepant treatment. Increased miR-382-5p expression in pain conditions may also be related to NF-κB signaling [64], which is a transcription factor for many miRNAs [65], including miR-382 [66]. MiR-34a-5p negatively modulates the GABAergic signaling and thus the reduced expression of miR-34a-5p after olcegepant may reflect a more active GABAergic transmission [61]. In agreement, we previously found that expression of miR-382-5p and miR-34a-5p in peripheral cells of patients with chronic migraine was significantly reduced after erenumab [67], the first-in-class fully human monoclonal antibody targeting the CGRP receptor. Moreover, in another study, we observed that CGRP plasma levels were positively correlated with miR-382-5p and miR-34a-5p in peripheral cells of chronic and episodic subjects, confirming a potential interaction [26] with the CGRP pathway. Here, we also demonstrated a significant upregulation of miR-155-5p, involved in the polarization of microglia and inflammatory processes in a variety of neurological diseases. MiR-155 overexpression promotes a pro-inflammatory phenotype in monocytes/macrophages and triggers the spontaneous production of several pro-inflammatory cytokines, including TNF-alpha, IL-6, and IL-1beta [68,69]. This finding is confirmed by a recent study, where changes in miR-155-5p expression were associated with microglial activation in the TNC after chronic NTG in rats, indicating that miR-155-5p may be involved in trigeminal hyperalgesia [60].

Altogether, these observations suggest the possibility that peripheral changes in miRNAs levels, previously reported by us in migraine patients, may likely reflect changes in central/peripheral areas of the nervous system involved in migraine pain modulation. Additionally, the findings also support a role for miR-382-5p, miR-34a-5p, and miR-155-5p in chronic migraine, which is partly, but probably not entirely, linked to the interaction with the CGRP pathway. The potential interplay between CGRP and the inflammatory mediators within the mechanisms of neuronal sensitization is reported in Figure 9.

Possible Limitations of the Study

A repeated use of acute migraine drugs can induce a condition of medication overuse headache [70]. In light of the experimental settings used in the present study, in which an acute migraine drug was used chronically, it must be noted that we did not detect any change in nociceptive and hyperalgesic behavior in the animals treated chronically with olcegepant and NTG vehicle (Figure 3), nor did we find an increase in the expression of inflammatory cytokines or of CGRP plasma levels, when comparing them with the control group. This is in agreement with previous literature showing that gepants (olcegepant and ubrogepant) did not induce central sensitization upon chronic administration [71,72]. Our data are also in accordance with clinical reports, in which neither gepants nor CGRP-targeting monoclonal antibodies were associated with the development of medication overuse headache [73,74].

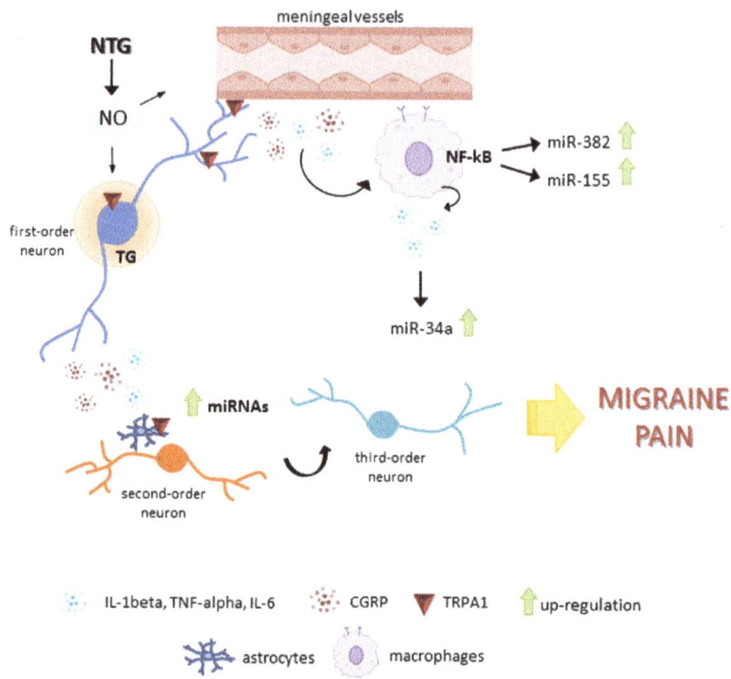

Figure 9. Potential interaction of CGRP with other pain/inflammatory mediators relevant for neuronal sensitization in an animal model of chronic migraine. Nitroglycerin (NTG), by releasing nitric oxide (NO), may induce an inflammatory response within the dura mater and the trigeminal ganglion (TG), potentiated by CGRP release, with an increase in pro-inflammatory cytokines interleukin (IL)-1beta and tumor necrosis factor alpha (TNF-alpha) levels by nuclear factor kappa B (NF-κB) activation [62,63]. In addition, NTG may activate transient receptor potential ankyrin 1 channels (TRPA1) receptors contributing to a further release of CGRP. This response may induce an upregulation of miRNAs, not only at the peripheral level but indirectly at the central level.

In this study, we show that chronic CGRP antagonism attenuates NTG-induced trigeminal hyperalgesia. In the clinical setting, olcegepant failed to prevent NTG-induced migraine attacks in subjects with migraine disease [75]. A comparison of methodology between pre-clinical and clinical experiments is of course not possible. However, it is worth noting that our finding is in accordance with a recent study, in which migraine-like intracranial and extracranial neuronal hypersensitivity, as well as central trigeminocervical neurons' activation stimulated by NTG, were reduced after olgecepant treatment [34]. Additionally, olcegepant was also effective in reducing the NTG-induced plantar allodynia [76] in another pre-clinical study. Another point that must be noted is that, differently from our study in animals and that of Juahsz et al. [45] in humans, Tvedskov and colleagues did not report changes in CGRP levels after NTG in migraine subjects [77].

The findings of the present study suggest a consistent modulation of the biomarkers under investigation in peripheral and central sites, which may be surprising when considering that olcegepant poorly penetrates the blood–brain barrier [78]. Thus, a possible limitation of the present study is that we tested olcegepant following systemic but not intracerebroventricular administration, thus not allowing a more comprehensive analysis of its site of action. It should, however, be noted that in the mouse model of chronic NTG, the intracerebroventricular administration of olcegepant was not effective compared with the systemic administration, thus suggesting that olcegepant has only peripheral effects [79]. Despite this observation, olcegepant was also reported to act in the TNC where it reduced the activation of second order neurons [80]. One possible interpretation is that olcegepant possesses a direct action in the periphery, which subsequently results in an indirect effect at the central level. This effect would in turn modulate the serum levels of the neuropeptide and also its gene expression in peripheral and central areas. An alternative hypothesis is that NTG-associated activation of neuroinflammation and CGRP release may alter blood–brain barrier permeability [81], thus allowing a direct central effect of olcegepant. Specifically targeted studies will be required to confirm or refute these alternative hypotheses.

5. Conclusions

The key role of CGRP in migraine pathogenesis and its chronicization is universally accepted. Here, we report that the changes in the CGRP pathway are paralleled by the activation of the neuroinflammation cascade and of miRNAs involved in the inflammatory pathway. To the best of our knowledge, this is the first report to evaluate these downstream molecular mechanisms following ligand-receptor blockade by olcegepant and demonstrating that the CGRP receptor antagonist reduces the mediators of sensitization in peripheral and central areas of the nervous system that are important in the circuitry of migraine pain.

Altogether, the present data contribute important additional pieces of knowledge in the understanding the complexity of the mechanisms related to CGRP in migraine pathogenesis; it seems reasonable to hypothesize that the miRNAs and cytokines investigated, together with the increased TRPA1 gene expression, may represent only a part of a multi-biomarker panel signature of the migraine disease.

Author Contributions: Conceptualization, R.G.; methodology, C.D., M.F. and A.M.Z.; validation, R.G.; formal analysis, R.G.; investigation, R.G., C.D., M.F. and A.M.Z.; resources, R.G., C.D., M.F. and A.M.Z.; data curation, R.G.; writing—original draft preparation, R.G. and C.D.; writing—review and editing, C.T.; visualization C.D.; supervision, R.G.; project administration, R.G.; Funding acquisition, C.T. All authors have read and agreed to the published version of the manuscript.

Funding: This study was supported by the European Funding for Neuroscience Research (Biomiga project, ERA-NET NEURON, Joint Transnational Research Projects 2019) to C.T., IRCCS Mondino Foundation, Pavia, Italy.

Institutional Review Board Statement: All procedures were approved by the Italian Ministry of Health (no. 691/2020-PR) and performed in agreement with the guidelines of the European Community Directive 2010/63/EU of 22 September 2010.

Data Availability Statement: The data presented in this study are available from the ZENODO repository (doi: 10.5281/zenodo.6794106).

Conflicts of Interest: The authors declare no conflict of interest.

References

1. Goadsby, P.J.; Holland, P.R.; Martins-Oliveira, M.; Hoffmann, J.; Schankin, C.; Akerman, S. Pathophysiology of Migraine: A Disorder of Sensory Processing. *Physiol. Rev.* **2017**, *97*, 553–622. [CrossRef]
2. Edvinsson, L. CGRP and migraine: From bench to bedside. *Rev. Neurol.* **2021**, *177*, 785–790. [CrossRef]
3. Mulderry, P.K.; Ghatei, M.A.; Spokes, R.A.; Jones, P.M.; Pierson, A.M.; Hamid, Q.A.; Kanse, S.; Amara, S.G.; Burrin, J.M.; Legon, S.; et al. Differential expression of alpha-CGRP and beta-CGRP by primary sensory neurons and enteric autonomic neurons of the rat. *Neuroscience* **1988**, *25*, 195–205. [CrossRef]
4. Iyengar, S.; Johnson, K.W.; Ossipov, M.H.; Aurora, S.K. CGRP and the Trigeminal System in Migraine. *Headache* **2019**, *59*, 659–681. [CrossRef]
5. Lennerz, J.K.; Rühle, V.; Ceppa, E.P.; Neuhuber, W.L.; Bunnett, N.W.; Grady, E.F.; Messlinger, K. Calcitonin receptor-like receptor (CLR), receptor activity-modifying protein 1 (RAMP1), and calcitonin gene-related peptide (CGRP) immunoreactivity in the rat trigeminovascular system: Differences between peripheral and central CGRP receptor distribution. *J. Comp. Neurol.* **2008**, *507*, 1277–1299. [CrossRef]
6. Theoharides, T.C.; Donelan, J.; Kandere-Grzybowska, K.; Konstantinidou, A. The role of mast cells in migraine pathophysiology. *Brain Res. Brain Res. Rev.* **2005**, *49*, 65–76. [CrossRef]
7. Afroz, S.; Arakaki, R.; Iwasa, T.; Oshima, M.; Hosoki, M.; Inoue, M.; Baba, O.; Okayama, Y.; Matsuka, Y. CGRP Induces Differential Regulation of Cytokines from Satellite Glial Cells in Trigeminal Ganglia and Orofacial Nociception. *Int. J. Mol. Sci.* **2019**, *20*, 711. [CrossRef]

8. Cuesta, M.C.; Quintero, L.; Pons, H.; Suarez-Roca, H. Substance P and calcitonin gene-related peptide increase IL-1 beta, IL-6 and TNF alpha secretion from human peripheral blood mononuclear cells. *Neurochem. Int.* **2002**, *40*, 301–306. [CrossRef]
9. Souza, G.R.; Talbot, J.; Lotufo, C.M.; Cunha, F.Q.; Cunha, T.M.; Ferreira, S.H. Fractalkine mediates inflammatory pain through activation of satellite glial cells. *Proc. Natl. Acad. Sci. USA* **2013**, *110*, 11193–11198. [CrossRef]
10. Diogenes, A.; Akopian, A.N.; Hargreaves, K.M. NGF up-regulates TRPA1: Implications for orofacial pain. *J. Dent. Res.* **2007**, *86*, 550–555. [CrossRef]
11. Hatano, N.; Itoh, Y.; Suzuki, H.; Muraki, Y.; Hayashi, H.; Onozaki, K.; Wood, I.C.; Beech, D.J.; Muraki, K. Hypoxia-inducible factor-1α (HIF1α) switches on transient receptor potential ankyrin repeat 1 (TRPA1) gene expression via a hypoxia response element-like motif to modulate cytokine release. *J. Biol. Chem.* **2012**, *287*, 31962–31972. [CrossRef]
12. Eberhardt, M.; Dux, M.; Namer, B.; Miljkovic, J.; Cordasic, N.; Will, C.; Kichko, T.I.; de la Roche, J.; Fischer, M.; Suárez, S.A.; et al. H2S and NO cooperatively regulate vascular tone by activating a neuroendocrine HNO-TRPA1-CGRP signalling pathway. *Nat. Commun.* **2014**, *5*, 4381. [CrossRef]
13. Kunkler, P.E.; Ballard, C.J.; Oxford, G.S.; Hurley, J.H. TRPA1 receptors mediate environmental irritant-induced meningeal vasodilatation. *Pain* **2011**, *152*, 38–44. [CrossRef]
14. Edvinsson, L.; Haanes, K.A.; Warfvinge, K. Does inflammation have a role in migraine? *Nat. Rev. Neurol.* **2019**, *15*, 483–490. [CrossRef]
15. Buzzi, M.G.; Pellegrino, M.G.; Bellantonio, P. Causes and mechanisms of primary headaches: Toward a bio-behavioral model. *Ital. J. Neurol. Sci.* **1995**, *16*, 15–19. [CrossRef] [PubMed]
16. Reuter, U.; Bolay, H.; Jansen-Olesen, I.; Chiarugi, A.; Sanchez del Rio, M.; Letourneau, R.; Theoharides, T.C.; Waeber, C.; Moskowitz, M.A. Delayed inflammation in rat meninges: Implications for migraine pathophysiology. *Brain* **2001**, *124 Pt 12*, 2490–2502. [CrossRef]
17. Bolay, H.; Reuter, U.; Dunn, A.K.; Huang, Z.; Boas, D.A.; Moskowitz, M.A. Intrinsic brain activity triggers trigeminal meningeal afferents in a migraine model. *Nat. Med.* **2002**, *8*, 136–142. [CrossRef]
18. Parsons, A.A.; Strijbos, P.J. The neuronal versus vascular hypothesis of migraine and cortical spreading depression. *Curr. Opin. Pharmacol.* **2003**, *3*, 73–77. [CrossRef]
19. Hong, P.; Tan, T.; Liu, Y.; Xiao, J. Gepants for abortive treatment of migraine: A network meta-analysis. *Brain Behav.* **2020**, *10*, e01701. [CrossRef] [PubMed]
20. Chiang, C.C.; Schwedt, T.J. Calcitonin gene-related peptide (CGRP)-targeted therapies as preventive and acute treatments for migraine-The monoclonal antibodies and gepants. *Prog. Brain Res.* **2020**, *255*, 143–170. [CrossRef]
21. Vandervorst, F.; Van Deun, L.; Van Dycke, A.; Paemeleire, K.; Reuter, U.; Schoenen, J.; Versijpt, J. CGRP monoclonal antibodies in migraine: An efficacy and tolerability comparison with standard prophylactic drugs. *J. Headache Pain* **2021**, *22*, 128. [CrossRef] [PubMed]
22. Cavestro, C.; Ferrero, M.; Mandrino, S.; Di Tavi, M.; Rota, E. Novelty in Inflammation and Immunomodulation in Migraine. *Curr. Pharm. Des.* **2019**, *25*, 2919–2936. [CrossRef]
23. Biscetti, L.; De Vanna, G.; Cresta, E.; Bellotti, A.; Corbelli, I.; Cupini, M.L.; Calabresi, P.; Sarchielli, P. Immunological findings in patients with migraine and other primary headaches: A narrative review. *Clin. Exp. Immunol.* **2022**, *207*, 11–26. [CrossRef] [PubMed]
24. Sayed, D.; Abdellatif, M. MicroRNAs in development and disease. *Physiol. Rev.* **2011**, *91*, 827–887. [CrossRef] [PubMed]
25. Ahmad, L.; Demartini, C.; Corrado, M.; Vaghi, G.; Piella, E.M.; Allena, M.; Zanaboni, A.M.; Greco, R.; Tassorelli, C.; De Icco, R. Expression of Selected microRNAs in Migraine: A New Class of Possible Biomarkers of Disease? *Processes* **2021**, *9*, 2199. [CrossRef]
26. Greco, R.; De Icco, R.; Demartini, C.; Zanaboni, A.M.; Tumelero, E.; Sances, G.; Allena, M.; Tassorelli, C. Plasma levels of CGRP and expression of specific microRNAs in blood cells of episodic and chronic migraine subjects: Towards the identification of a panel of peripheral biomarkers of migraine? *J. Headache Pain* **2020**, *21*, 122. [CrossRef] [PubMed]
27. Deng, T.; Yang, L.; Zheng, Z.; Li, Y.; Ren, W.; Wu, C.; Guo, L. Calcitonin gene-related peptide induces IL-6 expression in RAW264.7 macrophages mediated by mmu_circRNA_007893. *Mol. Med. Rep.* **2017**, *16*, 9367–9374. [CrossRef]
28. Greco, R.; Ferrigno, A.; Demartini, C.; Zanaboni, A.; Mangione, A.S.; Blandini, F.; Nappi, G.; Vairetti, M.; Tassorelli, C. Evaluation of ADMA-DDAH-NOS axis in specific brain areas following nitroglycerin administration: Study in an animal model of migraine. *J. Headache Pain* **2015**, *16*, 560. [CrossRef]
29. Greco, R.; Demartini, C.; Zanaboni, A.; Casini, I.; De Icco, R.; Reggiani, A.; Misto, A.; Piomelli, D.; Tassorelli, C. Characterization of the peripheral FAAH inhibitor, URB937, in animal models of acute and chronic migraine. *Neurobiol. Dis.* **2021**, *147*, 105157. [CrossRef]
30. Harriott, A.M.; Strother, L.C.; Vila-Pueyo, M.; Holland, P.R. Animal models of migraine and experimental techniques used to examine trigeminal sensory processing. *J. Headache Pain* **2019**, *20*, 91. [CrossRef]
31. Zimmermann, M. Ethical guidelines for investigations of experimental pain in conscious animals. *Pain* **1983**, *16*, 109–110. [CrossRef]
32. Ramachandran, R.; Bhatt, D.K.; Ploug, K.B.; Hay-Schmidt, A.; Jansen-Olesen, I.; Gupta, S.; Olesen, J. Nitric oxide synthase, calcitonin gene-related peptide and NK-1 receptor mechanisms are involved in GTN-induced neuronal activation. *Cephalalgia* **2014**, *34*, 136–147. [CrossRef] [PubMed]

33. Munro, G.; Petersen, S.; Jansen-Olesen, I.; Olesen, J. A unique inbred rat strain with sustained cephalic hypersensitivity as a model of chronic migraine-like pain. *Sci. Rep.* **2018**, *8*, 1836. [CrossRef] [PubMed]
34. Akerman, S.; Romero-Reyes, M.; Karsan, N.; Bose, P.; Hoffmann, J.R.; Holland, P.R.; Goadsby, P.J. Therapeutic targeting of nitroglycerin-mediated trigeminovascular neuronal hypersensitivity predicts clinical outcomes of migraine abortives. *Pain* **2021**, *162*, 1567–1577. [CrossRef] [PubMed]
35. Greco, R.; Demartini, C.; Zanaboni, A.M.; Tumelero, E.; Reggiani, A.; Misto, A.; Piomelli, D.; Tassorelli, C. FAAH inhibition as a preventive treatment for migraine: A pre-clinical study. *Neurobiol. Dis.* **2020**, *134*, 104624. [CrossRef] [PubMed]
36. Tajti, J.; Uddman, R.; Edvinsson, L. Neuropeptide localization in the "migraine generator" region of the human brainstem. *Cephalalgia* **2001**, *21*, 96–101. [CrossRef]
37. Vila-Pueyo, M.; Strother, L.C.; Kefel, M.; Goadsby, P.J.; Holland, P.R. Divergent influences of the locus coeruleus on migraine pathophysiology. *Pain* **2019**, *160*, 385–394. [CrossRef]
38. Greco, R.; Demartini, C.; Zanaboni, A.M.; Redavide, E.; Pampalone, S.; Toldi, J.; Fülöp, F.; Blandini, F.; Nappi, G.; Sandrini, G.; et al. Effects of kynurenic acid analogue 1 (KYNA-A1) in nitroglycerin-induced hyperalgesia: Targets and anti-migraine mechanisms. *Cephalalgia* **2017**, *37*, 1272–1284. [CrossRef]
39. Chen, H.; Tang, X.; Li, J.; Hu, B.; Yang, W.; Zhan, M.; Ma, T.; Xu, S. IL-17 crosses the blood-brain barrier to trigger neuroinflammation: A novel mechanism in nitroglycerin-induced chronic migraine. *J. Headache Pain* **2022**, *23*, 1. [CrossRef]
40. Yao, G.; Han, X.; Hao, T.; Huang, Q.; Yu, T. Effects of rizatriptan on the expression of calcitonin gene-related peptide and cholecystokinin in the periaqueductal gray of a rat migraine model. *Neurosci. Lett.* **2015**, *587*, 29–34. [CrossRef]
41. Samuels, E.R.; Szabadi, E. Functional neuroanatomy of the noradrenergic locus coeruleus: Its roles in the regulation of arousal and autonomic function part II: Physiological and pharmacological manipulations and pathological alterations of locus coeruleus activity in humans. *Curr. Neuropharmacol.* **2008**, *6*, 254–285. [CrossRef] [PubMed]
42. McCall, J.G.; Al-Hasani, R.; Siuda, E.R.; Hong, D.Y.; Norris, A.J.; Ford, C.P.; Bruchas, M.R. CRH Engagement of the Locus Coeruleus Noradrenergic System Mediates Stress-Induced Anxiety. *Neuron* **2015**, *87*, 605–620. [CrossRef] [PubMed]
43. Llorca-Torralba, M.; Borges, G.; Neto, F.; Mico, J.A.; Berrocoso, E. Noradrenergic Locus Coeruleus pathways in pain modulation. *Neuroscience* **2016**, *338*, 93–113. [CrossRef]
44. Goadsby, P.J.; Edvinsson, L.; Ekman, R. Vasoactive peptide release in the extracerebral circulation of humans during migraine headache. *Ann. Neurol.* **1990**, *28*, 183–187. [CrossRef] [PubMed]
45. Juhasz, G.; Zsombok, T.; Modos, E.A.; Olajos, S.; Jakab, B.; Nemeth, J.; Szolcsanyi, J.; Vitrai, J.; Bagdy, G. NO-induced migraine attack: Strong increase in plasma calcitonin gene-related peptide (CGRP) concentration and negative correlation with platelet serotonin release. *Pain* **2003**, *106*, 461–470. [CrossRef]
46. Juhasz, G.; Zsombok, T.; Jakab, B.; Nemeth, J.; Szolcsanyi, J.; Bagdy, G. Sumatriptan causes parallel decrease in plasma calcitonin gene-related peptide (CGRP) concentration and migraine headache during nitroglycerin induced migraine attack. *Cephalalgia* **2005**, *25*, 179–183. [CrossRef]
47. O'Connor, T.P.; van der Kooy, D. Enrichment of a vasoactive neuropeptide (calcitonin gene related peptide) in the trigeminal sensory projection to the intracranial arteries. *J. Neurosci.* **1988**, *8*, 2468–2476. [CrossRef]
48. Raddant, A.C.; Russo, A.F. Calcitonin gene-related peptide in migraine: Intersection of peripheral inflammation and central modulation. *Expert Rev. Mol. Med.* **2011**, *13*, e36. [CrossRef]
49. Demartini, C.; Tassorelli, C.; Zanaboni, A.M.; Tonsi, G.; Francesconi, O.; Nativi, C.; Greco, R. The role of the transient receptor potential ankyrin type-1 (TRPA1) channel in migraine pain: Evaluation in an animal model. *J. Headache Pain* **2017**, *18*, 94. [CrossRef]
50. Holzmann, B. Modulation of immune responses by the neuropeptide CGRP. *Amino. Acids* **2013**, *45*, 1–7. [CrossRef]
51. Han, D. Association of Serum Levels of Calcitonin Gene-related Peptide and Cytokines during Migraine Attacks. *Ann. Indian Acad. Neurol.* **2019**, *22*, 277–281. [CrossRef] [PubMed]
52. Conti, P.; D'Ovidio, C.; Conti, C.; Gallenga, C.E.; Lauritano, D.; Caraffa, A.; Kritas, S.K.; Ronconi, G. Progression in migraine: Role of mast cells and pro-inflammatory and anti-inflammatory cytokines. *Eur. J. Pharmacol.* **2019**, *844*, 87–94. [CrossRef] [PubMed]
53. Cady, R.J.; Glenn, J.R.; Smith, K.M.; Durham, P.L. Calcitonin gene-related peptide promotes cellular changes in trigeminal neurons and glia implicated in peripheral and central sensitization. *Mol. Pain* **2011**, *7*, 94. [CrossRef] [PubMed]
54. Vause, C.V.; Durham, P.L. Calcitonin gene-related peptide differentially regulates gene and protein expression in trigeminal glia cells: Findings from array analysis. *Neurosci. Lett.* **2010**, *473*, 163–167. [CrossRef]
55. Ceruti, S.; Villa, G.; Fumagalli, M.; Colombo, L.; Magni, G.; Zanardelli, M.; Fabbretti, E.; Verderio, C.; van den Maagdenberg, A.M.; Nistri, A.; et al. Calcitonin gene-related peptide-mediated enhancement of purinergic neuron/glia communication by the algogenic factor bradykinin in mouse trigeminal ganglia from wild-type and R192Q Cav2.1 Knock-in mice: Implications for basic mechanisms of migraine pain. *J. Neurosci.* **2011**, *31*, 3638–3649. [CrossRef]
56. Hansted, A.K.; Bhatt, D.K.; Olesen, J.; Jensen, L.J.; Jansen-Olesen, I. Effect of TRPA1 activator allyl isothiocyanate (AITC) on rat dural and pial arteries. *Pharmacol. Rep.* **2019**, *71*, 565–572. [CrossRef]
57. Nassini, R.; Materazzi, S.; Vriens, J.; Prenen, J.; Benemei, S.; De Siena, G.; la Marca, G.; Andrè, E.; Preti, D.; Avonto, C.; et al. The 'headache tree' via umbellulone and TRPA1 activates the trigeminovascular system. *Brain* **2012**, *135 Pt 2*, 376–390. [CrossRef]

58. Imai, S.; Saeki, M.; Yanase, M.; Horiuchi, H.; Abe, M.; Narita, M.; Kuzumaki, N.; Suzuki, T. Change in microRNAs associated with neuronal adaptive responses in the nucleus accumbens under neuropathic pain. *J. Neurosci.* **2011**, *31*, 15294–15299. [CrossRef] [PubMed]
59. Cheng, C.Y.; Chen, S.P.; Liao, Y.C.; Fuh, J.L.; Wang, Y.F.; Wang, S.J. Elevated circulating endothelial-specific microRNAs in migraine patients: A pilot study. *Cephalalgia* **2018**, *38*, 1585–1591. [CrossRef] [PubMed]
60. Wen, Q.; Wang, Y.; Pan, Q.; Tian, R.; Zhang, D.; Qin, G.; Zhou, J.; Chen, L. MicroRNA-155-5p promotes neuroinflammation and central sensitization via inhibiting SIRT1 in a nitroglycerin-induced chronic migraine mouse model. *J. Neuroinflammation* **2021**, *18*, 287. [CrossRef] [PubMed]
61. Andersen, H.H.; Duroux, M.; Gazerani, P. Serum MicroRNA Signatures in Migraineurs During Attacks and in Pain-Free Periods. *Mol. Neurobiol.* **2016**, *53*, 1494–1500. [CrossRef]
62. Reuter, U.; Chiarugi, A.; Bolay, H.; Moskowitz, M.A. Nuclear factor-kappaB as a molecular target for migraine therapy. *Ann. Neurol.* **2002**, *51*, 507–516. [CrossRef]
63. Greco, R.; Tassorelli, C.; Cappelletti, D.; Sandrini, G.; Nappi, G. Activation of the transcription factor NF-kappaB in the nucleus trigeminalis caudalis in an animal model of migraine. *Neurotoxicology* **2005**, *26*, 795–800. [CrossRef] [PubMed]
64. Sarchielli, P.; Floridi, A.; Mancini, M.L.; Rossi, C.; Coppola, F.; Baldi, A.; Pini, L.A.; Calabresi, P. NF-κB activity and iNOS expression in monocytes from internal jugular blood of migraine without aura patients during attacks. *Cephalalgia* **2006**, *26*, 1071–1079. [CrossRef] [PubMed]
65. Zhai, Y.; Zhu, Y.Y. MiR-30a relieves migraine by degrading CALCA. *Eur. Rev. Med. Pharmacol. Sci.* **2018**, *22*, 2022–2028. [CrossRef]
66. Tafuri, E.; Santovito, D.; De Nardis, V.; Marcantonio, P.; Paganelli, C.; Affaitati, G.; Bucci, M.; Mezzetti, A.; Giamberardino, M.A.; Cipollone, F. MicroRNA profiling in migraine without aura: Pilot study. *Ann. Med.* **2015**, *47*, 468–473. [CrossRef] [PubMed]
67. De Icco, R.; Fiamingo, G.; Greco, R.; Bottiroli, S.; Demartini, C.; Zanaboni, A.M.; Allena, M.; Guaschino, E.; Martinelli, D.; Putortì, A.; et al. Neurophysiological and biomolecular effects of erenumab in chronic migraine: An open label study. *Cephalalgia* **2020**, *40*, 1336–1345. [CrossRef] [PubMed]
68. Kurowska-Stolarska, M.; Alivernini, S.; Ballantine, L.E.; Asquith, D.L.; Millar, N.L.; Gilchrist, D.S.; Reilly, J.; Ierna, M.; Fraser, A.R.; Stolarski, B.; et al. MicroRNA-155 as a proinflammatory regulator in clinical and experimental arthritis. *Proc. Natl. Acad. Sci. USA* **2011**, *108*, 11193–11198. [CrossRef]
69. Elmesmari, A.; Fraser, A.R.; Wood, C.; Gilchrist, D.; Vaughan, D.; Stewart, L.; McSharry, C.; McInnes, I.B.; Kurowska-Stolarska, M. MicroRNA-155 regulates monocyte chemokine and chemokine receptor expression in Rheumatoid Arthritis. *Rheumatology* **2016**, *55*, 2056–2065. [CrossRef] [PubMed]
70. Thorlund, K.; Sun-Edelstein, C.; Druyts, E.; Kanters, S.; Ebrahim, S.; Bhambri, R.; Ramos, E.; Mills, E.J.; Lanteri-Minet, M.; Tepper, S. Risk of medication overuse headache across classes of treatments for acute migraine. *J. Headache Pain* **2016**, *17*, 107. [CrossRef]
71. Saengjaroentham, C.; Strother, L.C.; Dripps, I.; Sultan Jabir, M.R.; Pradhan, A.; Goadsby, P.J.; Holland, P.R. Differential medication overuse risk of novel anti-migraine therapeutics. *Brain* **2020**, *143*, 2681–2688. [CrossRef]
72. Navratilova, E.; Behravesh, S.; Oyarzo, J.; Dodick, D.W.; Banerjee, P.; Porreca, F. Ubrogepant does not induce latent sensitization in a preclinical model of medication overuse headache. *Cephalalgia* **2020**, *40*, 892–902. [CrossRef]
73. Tepper, S.J.; Diener, H.C.; Ashina, M.; Brandes, J.L.; Friedman, D.I.; Reuter, U.; Cheng, S.; Nilsen, J.; Leonardi, D.K.; Lenz, R.A.; et al. Erenumab in chronic migraine with medication overuse: Subgroup analysis of a randomized trial. *Neurology* **2019**, *92*, e2309–e2320. [CrossRef]
74. Moreno-Ajona, D.; Villar-Martínez, M.D.; Goadsby, P.J. New Generation Gepants: Migraine Acute and Preventive Medications. *J. Clin. Med.* **2022**, *11*, 1656. [CrossRef] [PubMed]
75. Olesen, J.; Diener, H.C.; Husstedt, I.W.; Goadsby, P.J.; Hall, D.; Meier, U.; Pollentier, S.; Lesko, L.M.; BIBN 4096 BS Clinical Proof of Concept Study Group. Calcitonin gene-related peptide receptor antagonist BIBN 4096 BS for the acute treatment of migraine. *N. Engl. J. Med.* **2004**, *350*, 1104–1110. [CrossRef]
76. Ernstsen, C.; Christensen, S.L.; Olesen, J.; Kristensen, D.M. No additive effect of combining sumatriptan and olcegepant in the GTN mouse model of migraine. *Cephalalgia* **2021**, *41*, 329–339. [CrossRef]
77. Tvedskov, J.F.; Tfelt-Hansen, P.; Petersen, K.A.; Jensen, L.T.; Olesen, J. CGRP receptor antagonist olcegepant (BIBN4096BS) does not prevent glyceryl trinitrate-induced migraine. *Cephalalgia* **2010**, *30*, 1346–1353. [CrossRef]
78. Edvinsson, L.; Warfvinge, K. Recognizing the role of CGRP and CGRP receptors in migraine and its treatment. *Cephalalgia* **2019**, *39*, 366–373. [CrossRef]
79. Christensen, S.L.; Ernstsen, C.; Olesen, J.; Kristensen, D.M. No central action of CGRP antagonising drugs in the GTN mouse model of migraine. *Cephalalgia* **2020**, *40*, 924–934. [CrossRef]
80. Sixt, M.L.; Messlinger, K.; Fischer, M.J. Calcitonin gene-related peptide receptor antagonist olcegepant acts in the spinal trigeminal nucleus. *Brain* **2009**, *132 Pt 11*, 3134–3341. [CrossRef]
81. Long, T.; He, W.; Pan, Q.; Zhang, S.; Zhang, D.; Qin, G.; Chen, L.; Zhou, J. Microglia P2X4R-BDNF signalling contributes to central sensitization in a recurrent nitroglycerin-induced chronic migraine model. *J. Headache Pain* **2020**, *21*, 4. [CrossRef] [PubMed]

cells

MDPI

Article

Src Family Kinases Facilitate the Crosstalk between CGRP and Cytokines in Sensitizing Trigeminal Ganglion via Transmitting CGRP Receptor/PKA Pathway

Lingdi Nie [1,2], Kai Sun [1], Ziyang Gong [1], Haoyang Li [1], John P. Quinn [2] and Minyan Wang [1,2,*]

1 Centre for Neuroscience, Department of Biological Sciences, Xi'an Jiaotong-Liverpool University (XJTLU), Suzhou 215123, China
2 Department of Pharmacology and Therapeutics, Institute of Systems, Molecular and Integrative Biology, Liverpool L69 7ZB, UK
* Correspondence: minyan.wang@xjtlu.edu.cn

Abstract: The communication between calcitonin gene-related peptide (CGRP) and cytokines plays a prominent role in maintaining trigeminal ganglion (TG) and trigeminovascular sensitization. However, the underlying regulatory mechanism is elusive. In this study, we explored the hypothesis that Src family kinases (SFKs) activity facilitates the crosstalk between CGRP and cytokines in sensitizing TG. Mouse TG tissue culture was performed to study CGRP release by enzyme-linked immunosorbent assay, cytokine release by multiplex assay, cytokine gene expression by quantitative polymerase chain reaction, and phosphorylated SFKs level by western blot. The results demonstrated that a SFKs activator, pYEEI (YGRKKRRQRRREPQY(PO3H2)EEIPIYL) alone, did not alter CGRP release or the inflammatory cytokine interleukin-1β (IL-1β) gene expression in the mouse TG. In contrast, a SFKs inhibitor, saracatinib, restored CGRP release, the inflammatory cytokines IL-1β, C-X-C motif ligand 1, C-C motif ligand 2 (CCL2) release, and IL-1β, CCL2 gene expression when the mouse TG was pre-sensitized with hydrogen peroxide and CGRP respectively. Consistently with this, the phosphorylated SFKs level was increased by both hydrogen peroxide and CGRP in the mouse TG, which was reduced by a CGRP receptor inhibitor BIBN4096 and a protein kinase A (PKA) inhibitor PKI (14–22) Amide. The present study demonstrates that SFKs activity plays a pivotal role in facilitating the crosstalk between CGRP and cytokines by transmitting CGRP receptor/PKA signaling to potentiate TG sensitization and ultimately trigeminovascular sensitization.

Keywords: Src family kinases; calcitonin gene-related peptide; interleukin-1β; C-C motif ligand 2; C-X-C motif ligand 1; protein kinase A; trigeminal ganglion; migraine

Citation: Nie, L.; Sun, K.; Gong, Z.; Li, H.; Quinn, J.P.; Wang, M. Src Family Kinases Facilitate the Crosstalk between CGRP and Cytokines in Sensitizing Trigeminal Ganglion via Transmitting CGRP Receptor/PKA Pathway. *Cells* **2022**, *11*, 3498. https://doi.org/10.3390/cells11213498

Academic Editors: László Vécsei, Bernadett Tuka and Masaru Tanaka

Received: 30 August 2022
Accepted: 1 November 2022
Published: 4 November 2022

1. Introduction

Migraine is a recurrent primary headache disorder that afflicts approximately 15% of the population worldwide [1]. A key mechanism by which nearly all migraine triggers induce migraine attacks is the activation and sensitization of the trigeminovascular pathway [2–4]. As an important peripheral component of the trigeminovascular pathway, trigeminal ganglion (TG) contains the cell bodies of meningeal nociceptors, the activation of which initiate trigeminovascular activation [5–7]. Active signaling mediated mainly by neuropeptides and inflammatory mediators occurs within the TG, among which calcitonin gene-related peptide (CGRP), the key drug target of migraine prevention and therapy, is a key player [8]. In the TG, released CGRP binds to CGRP receptor to facilitate neuronal excitability [9–11] and neuroinflammation, including elevated release and expression of inflammatory cytokines [11–14]. Importantly, cytokines can signal back to neurons, which promotes CGRP synthesis and release [15,16], thereby inducing a positive feedback loop of sensitization. Thus, the communication between CGRP and cytokines plays a prominent role in maintaining TG activation and sensitization as well as trigeminovascular sensitization [17–19], although the underlying regulatory mechanism is elusive.

Src family kinases (SFKs) activity has been previously found to mediate CGRP release in dorsal root ganglion neurons [20] and TG [21]. SFKs activity also mediates inflammatory cytokine release and expression in primary glial cells [22–25] and mediates interleukin-1 β (IL-1β) gene expression in the mouse TG [21]. Importantly, SFKs are known to play a key role in migraine pathogenesis. In an inflammatory soup-induced chronic migraine model, central inhibition of SFKs attenuates mechanical allodynia and synaptic plasticity [26]. In a genetic mouse migraine with aura model familial hemiplegic migraine type 2 (FHM2), deactivation of SFKs reduces the Ca^{2+} sensitivity and contraction of the cerebral arteries, which contributes to vascular tone and brain perfusion abnormalities [27]. Similarly, systemic deactivation of SFKs reduces cortical spreading depression (CSD), a migraine with aura model, and CSD-induced cerebral cortical inflammatory cytokines interleukin 1 beta (IL-1β) and tumor necrosis factor alpha (TNFα) gene expression [28]. Taken together, it is likely that SFKs activity facilitates the communication between CGRP and cytokines to activate and sensitize TG, which requires clarification.

In the present study, we examined whether SFKs activity facilitates the crosstalk between CGRP release and cytokines release and gene expression to activate and sensitize the mouse TG. How SFKs activity mediates the communication between CGRP and cytokines in TG is also explored by investigating the involvement of CGRP receptor/protein kinase A (PKA) pathway.

2. Materials and Methods

2.1. Animals

A total of 143 adult male C57BL/6J mice (21.4 ± 0.17 g) were used and purchased from Shanghai SLAC Laboratory Animal Corporation Ltd. (Shanghai, China). All studies in this paper were carried out in male rodents so that the effect of hormonal fluctuation in females is minimized. Mice were housed in the Experimental Animal Centre of Soochow University for at least one week to be acclimated to the housing room before use. Animal procedures were approved by the Ethical Review Panels of Xi'an Jiaotong–Liverpool University (XJTLU) under the agreement with Soochow University and performed in accordance with relevant China national and provincial guidelines. For each experiment, randomization of experimental groups was performed to reduce bias. All animals used were randomly allocated to different experimental groups.

2.2. Mouse TG Tissue Culture

Isolated TG culture is a commonly used model to study TG molecular and neurophysiological properties. Signaling molecules produced in TG cell bodies are delivered to the peripheral and central terminals via axonal transport to give rise to sensory transduction and neurotransmission [6,29–31]. Therefore, isolated TG culture is commonly used as a model of its peripheral or central endings to study meningeal nociceptors and trigeminal nociceptive transmission [17,19,32]. The method of TG tissue culture was established as reported previously [21]. Mice were sacrificed by rapid cervical dislocation. Both the left and right TG of each mouse were collected, and the merged TG were used for one individual experiment. The TG were recovered in 300 μL pre-oxygenated Kreb's solution (composition in mM: 126 NaCl, 2.5 KCl, 2.4 $CaCl_2·2H_2O$, 1.3 $MgCl_2·6H_2O$, 18 $NaHCO_3$, 1.2 NaH_2PO_4, 10 glucose; pH 7.4) for 30 min at 37 °C and then washed with pre-oxygenated Kreb's solution every 5 min for 30 min. Subsequently, the TG were incubated with each drug for 20 min or 1 h at 37 °C.

In order to explore whether SFKs activity mediates the communication between CGRP and cytokines in TG, three series of experiments were designed. Series 1: whether activation of SFKs increases CGRP release and IL-1β gene expression was examined in cultured mouse TG. This model has been validated in our previous publication by using KCl, a known trigger for neuronal activation and CGRP release, which successfully induces CGRP release [21]. A SFKs activator, pYEEI (YGRKKRRQRRREPQY(PO3H2)EEIPIYL) [33,34], or its negative control, the non-phosphorylated peptide, YEEI (YGRKKRRQRRREPQY-EEIPIYL), was applied at 1 mM [35,36] for 20 min. pYEEI binds to the SH2 domains of SFKs which hampers their closing conformation at the inactive state and induces their open active conformation [34,37,38]. A cell-penetrating peptide TAT (YGRKKRRQRRR) [39] was conjugated to both pYEEI and YEEI to make them cell permeable. These peptides were customized by A^+ Peptide (Shanghai, China). Two groups were designed: (i) 1 mM pYEEI; (ii) 1 mM YEEI (n = 8 for each). The level of CGRP released into the culture medium was detected by enzyme-linked immunosorbent assay (ELISA); the mRNA level of IL-1β in the TG was detected by quantitative polymerase chain reaction (qPCR). Series 2: whether deactivation of SFKs reduces CGRP release was examined in cultured mouse TG pre-sensitized by reactive oxygen species (ROS). Stress is the top trigger for migraineurs [40], and ROS is a common trigger of oxidative stress [41]. Since hydrogen peroxide (H_2O_2), a type of ROS, can induce CGRP release in dorsal root ganglion neurons [42], the present study used H_2O_2 to induce CGRP release in the TG. To inhibit SFKs activity, a SFKs inhibitor saracatinib (S1006, Selleckchem, Houston, TX, USA), which binds to the kinase (SH1) domains of SFKs [43], was used because saracatinib has been tested for treating different types of cancer [44–46] and Alzheimer's disease [47,48] in clinical trials and showed good tolerability and safety in patients. The cultured TG was treated with Kreb's, 1 mM H_2O_2 [42], 1.5 μM, 4 μM, or 10 μM saracatinib [49,50] in the presence of 1 mM H_2O_2 for 20 min. For this series, five groups were designed: (i) Kreb's (n = 8); (ii) 1 mM H_2O_2 (n = 8); (iii) 1 mM H_2O_2 + 1.5 μM saracatinib (n = 8); (iv) 1 mM H_2O_2 + 4 μM saracatinib (n = 7); (v) 1 mM H_2O_2 + 10 μM saracatinib (n = 7). Series 3: whether deactivation of SFKs reduces the release and gene expression of inflammatory cytokines induced by CGRP was examined in cultured mouse TG. CGRP (SCNTATCVTHRLAGLLSRSGGVVKDNFVPTNVGSEAF-NH2, disulfide bridge: Cys2-Cys7, A^+ Peptide, Shanghai, China), the potent neuroinflammatory mediator [51], was used to induce the release and gene expression of inflammatory cytokines in the TG, and the effect of the SFKs inhibitor saracatinib (S1006, Selleckchem, Houston, TX, USA) on these phenomena was studied. For studying cytokine release, the cultured TG was treated with Kreb's, 3 μM CGRP [9], or 1.5 μM saracatinib [49,50] in the presence of 3 μM CGRP for 20 min; for studying cytokine gene expression, the treatments were the same except for that 4 μM saracatinib in the presence of 3 μM CGRP was applied as an additional group, and the treatment time for each group was 60 min. Four groups were therefore designed: (i) Kreb's; (ii) 3 μM CGRP; (iii) 3 μM CGRP + 1.5 μM saracatinib; (iv) 3 μM CGRP + 4 μM saracatinib (n = 8 for each). To detect the release of multiple cytokines into the culture medium, a multiplex immunoassay was used to measure the levels of 12 pro-inflammatory cytokines, C-C motif ligand 2 (CCL2), C-C motif ligand 5 (CCL5), C-X-C motif ligand 1 (CXCL1), C-X-C motif ligand 10 (CXCL10), granulocyte-macrophage colony-stimulating factor (GM-CSF), interferon alpha (IFN-α), interferon beta (IFN-β), interferon gamma (IFN-γ), IL-1β, interleukin 6 (IL-6), interleukin 12 (IL-12), TNFα, and one anti-inflammatory cytokine, interleukin 10 (IL-10). Only the cytokines whose levels in the TG culture medium were significantly altered by both CGRP and saracatinib were selected to detect their mRNA levels by qPCR.

In order to explore how SFKs activity mediates the communication between CGRP and cytokines in TG, the signaling pathway that SFKs transmit during these processes is investigated, for which three series of experiments were designed. Series 4: to ensure that SFKs activity is increased by H_2O_2 in the mouse TG in Series 2, the TG treated with Kreb's and 1 mM H_2O_2 were collected to measure SFKs activity represented by the level of phosphorylated SFKs at Y416 using western blot in order to minimize animal use. How SFKs activity is enhanced by H_2O_2 was then investigated by examining whether inhibition of CGRP receptor reduces H_2O_2-enhanced SFKs activity in cultured mouse TG. To inhibit CGRP receptor, a CGRP receptor inhibitor, BIBN4096 (4561, Tocris, Bristol, UK), was used. The cultured TG was treated with 10 μM BIBN4096 in the presence of 1 mM H_2O_2 for 20 min. One additional group was designed: 1 mM H_2O_2 + 10 μM BIBN4096 (n = 7).

Series 5: to ensure that SFKs activity is increased by CGRP in the mouse TG in Series 3, the TG treated with Kreb's and 3 μM CGRP for 20 min were collected to measure the level of phosphorylated SFKs at Y416 using Western blot. Next, how SFKs activity is enhanced by CGRP was investigated by examining whether SFKs activity transmits CGRP receptor/PKA pathway as PKA is known to transmit signaling downstream CGRP [52–54] and activate SFKs in several models [36,55,56]. Specifically, whether inhibition of CGRP receptor and deactivation of PKA reduce CGRP-enhanced SFKs activity was examined in cultured mouse TG. To deactivate PKA, a PKA inhibitor, PKI (14-22) Amide (476485, Sigma-Aldrich, St. Louis, MO, USA), was used. The cultured TG was treated with 3 μM BIBN4096 [57] or 30 μM PKI (14-22) Amide [58] in the presence of 3 μM CGRP for 20 min. Two additional groups were designed: 3 μM CGRP + 3 μM BIBN4096, 3 μM CGRP + 30 μM PKI (14-22) Amide (n = 8 for each). Series 6: whether SFKs co-localize with CGRP or receptor activity modifying protein 1 (RAMP1), the unique and essential functional CGRP receptor subunit [59,60], was also examined in mouse TG using immunohistochemistry.

2.3. ELISA

After TG tissue culture, the level of CGRP released into the culture medium was measured using a mouse CGRP ELISA kit (CSB-EQ027706MO, CUSABIO, Houston, TX, USA). Briefly, 100 μL medium and each of 8 serially diluted standard solutions were added into an assay plate pre-coated with CGRP antibody, which was then incubated at 37 °C for 2 h. Next, after removing the remaining liquid in the wells, 100 μL 1 × biotin-conjugated antibody specific for CGRP was added to each well followed by incubating at 37 °C for 1 h. The wells were then aspirated and washed, after which each well was added with 100 μL 1 × avidin conjugated horseradish peroxidase (HRP) and incubated at 37 °C for 1 h. Following further wash to remove any unbound substances, each well was added with 90 μL TMB substrate and incubated at 37 °C for 30 min in the dark. The reaction was stopped by adding 50 μL stop solution to each well and the OD of the wells was read at 450 nm, 540 nm, and 570 nm using a colorimetric microplate reader (BioTek, Winooski, VT, USA). The mean reading at 540 nm and 570 nm were subtracted from that at 450 nm, which corrected for optical imperfections. A standard curve relating the OD values to the concentration of CGRP (pg/mL) in the standard solutions was plotted and an equation of the curve was obtained. The OD values of the media were used to calculate their CGRP concentration (pg/mL) using the equation.

2.4. Multiplex Immunoassay

A multi-analyte flow assay kit (740621, Biolegend, San Diego, CA, USA) was used to detect the release of 12 pro-inflammatory cytokines, CCL2, CCL5, CXCL1, CXCL10, GM-CSF, IFN-α, IFN-β, IFN-γ, IL-1β, IL-6, IL-12, TNFα, and one anti-inflammatory cytokine—IL-10—into the TG culture medium. First, 25 μL medium and each of 8 serially diluted standard solutions were added into an assay plate followed by adding 25 μL assay buffer and 25 μL mixed beads, which was then shaken at 500 rpm at room temperature for 2 h in the dark. After aspirating and washing the plate, 25 μL detection antibodies was added into the plate followed by shaking at 500 rpm at room temperature for 1 h. Next, 25 μL streptavidin-phycoerythrin (SA-PE) was added, and the plate was shaken at 500 rpm at room temperature for 30 min. After aspirating and washing the plate again, the beads in the plate were resuspended, and the plate was read on CytoFLEX S Flow Cytometer (C01161, Beckman Coulter, Brea, CA, USA). Each of the mixed beads was conjugated with a type of allophycocyanin (APC) fluorescence and an antibody specific to one of the 13 cytokines so that each cytokine in the medium was captured by its specific bead. The APC fluorescence conjugated to each bead had a differing level, which could be recognized by the flow cytometer at 660 nm to distinguish among different beads and identify the corresponding cytokine of each bead. The SA-PE bound to the detection antibody provided fluorescent signal in proportion to the amount of a certain cytokine bound to each bead, which was read as PE signal fluorescence intensity by the flow cytometer at 585 nm. Using

LEGENDplexTM Data Analysis Software 8.0 (Biolegend, San Diego, CA, USA), a standard curve relating the PE signal fluorescence intensities to the concentration of each of the 13 cytokines (pg/mL) in the standard solutions was plotted and an equation of the curve was obtained. The PE signal fluorescence intensities of the media were used to calculate the concentration (pg/mL) of each of the 13 cytokines using the equation.

2.5. qPCR

After 60 min of TG tissue culture, total RNA of mouse TG was extracted using TRIZOL reagent (T9424 Sigma-Aldrich, St. Louis, MO, USA) and was reverse transcribed to cDNA by a GoScript Reverse Transcription System (A5001 Promega, Madison, WI, USA). The mRNA levels of specific genes were detected by qPCR using GoTaq qPCR Master Mix (A6002, Promega, Madison, WI, USA). The qPCR reaction was performed in QuantStudio 5 Real-Time PCR System (Applied Biosystems, Waltham, MA, USA) under the following thermal cycling conditions: 95 °C for 2 min, 95 °C for 15 s, 60 °C for 1 min, and 60–95 °C for 1 min. The mRNA level of each target gene was presented as relative fold change by normalizing the individual mRNA level of the gene to the geometric mean of the mRNA levels of two housekeeping genes, β-actin, and peptidylprolyl isomerase A (PPIA). Primers specific to the target genes were shown as follows: IL-1β forward 5′ACTACAGGCTCCGAGATGAACAAC3′, reverse 5′CCCAAGGCCACAGGTATTTT3′; CCL2 forward 5′CACTCACCTGCTGCTACTCA3′, reverse 5′ GCTTGGTGACAAAAACTACAGC3′; ACTB forward 5′CTGTCCACCTTCCAGCAGAT3′, reverse 5′CGCAGCTCAGTAACAGTCCG3′; PPIA forward, 5′TTGCTGCAGACATGGTCAAC3′, reverse 5′TGTCTGCAAACAGCTCGAAG3′.

2.6. Western Blot

Total protein of mouse TG was extracted using sodium dodecyl sulfate (SDS, 74255, Sigma-Aldrich, St. Louis, MO, USA), as described previously [21]. The concentration of the extracted protein was measured using Bicinchoninic Acid Protein Assay Kit (P0010, Beyotime, Shanghai, China). The protein levels of phosphorylated SFKs at Y416 and SFKs were analyzed by Western blot. Except for that, the protein level of β-actin was also analyzed, which was used as an internal control to calculate the relative expression levels of phosphorylated SFKs at Y416 and SFKs. Protein samples were denatured with SDS polyacrylamide (SDS-PAGE) sample loading buffer (P0015, Beyotime, Shanghai, China) at 100 °C for 5 min. The protein samples were separated on a 10% SDS-PAGE gel followed by transfer onto nitrocellulose membranes (66485, Pall, Pensacola, FL, USA). The membranes were incubated in 5% milk at room temperature for 1 h, followed by incubation with anti-phospho-Y416 SFKs antibody (1:200, 6943, CST, Beverly, MA, USA) and anti-β-actin antibody (1:2000, 4970, CST, Beverly, MA, USA) at 4 °C overnight. Subsequently, the membranes were incubated with IRDye 680RD donkey anti-rabbit secondary antibody (1:5000, 925-68073, LI-COR, Lincoln, NE, USA) for 1 h in the dark. Odyssey Near-Infrared Fluorescent Imaging System (LI-COR, Lincoln, NE, USA) was used to detect the protein levels of phosphorylated SFKs at Y416 and β-actin on the membranes by scanning fluorescent signals at 700 nm. Next, the anti-phospho-Y416 SFK antibody on the membranes was stripped off using 0.2 M NaOH (134070010, Acros Organics, Geel, Belgium) for 15 min. After incubating in 5% milk, the membranes were incubated with anti-SFK antibody (1:1000, 2109, CST) at 4 °C overnight followed by incubating with the anti-rabbit secondary antibody and imaging to detect the protein level of SFKs. The mean gray value of protein band intensity was quantified using Image Studio Lite 5.0 (LI-COR, Lincoln, NE, USA). The level of phosphorylated SFKs at Y416 was presented as absolute ratio in the band intensities between phosphorylated SFKs at Y416 and β-actin, phosphorylated SFKs at Y416 and SFKs, and SFKs and β-actin.

2.7. Immunohistochemistry

As CGRP and CGRP receptor distribute differently in TG neurons and nerve fibers [10,60], we then detected SFKs distribution pattern in mouse TG and whether SFKs co-localize with CGRP or RAMP1. One C57BL6/J mouse was anesthetized in depth in 5% isoflurane with O_2:N_2O (1:2) and transcardially perfused with phosphate buffer saline (09–8912-100, Medicago, Uppsala, Sweden) and 4% paraformaldehyde (P804537, Macklin, Shanghai, China). The TG were collected and post-fixed in 4% paraformaldehyde at 4 °C overnight. The tissues were dehydrated in 10%, 20%, and 30% sucrose (V900116, Sigma-Aldrich, St. Louis, MO, USA) solutions at 4 °C overnight. Before sectioning, the tissues were embedded in Tissue-Tek O.C.T. Compound (4583, Sakura, Flemingweg, The Netherlands). Coronal sections at 20 μm of the TG tissues were prepared using a cryostat (CM1950, Leica, Tokyo, Japan) and fixed on glass slides. The TG slice was permeabilized in 0.25% Triton X-100 (V90050210, Sigma-Aldrich, St. Louis, MO, USA) for 15 min and blocking in 2% donkey serum (D9663, Sigma-Aldrich) with 2% bovine serum albumin (V900933, Sigma-Aldrich) and 0.1% Tween20 (P1379, Sigma-Aldrich) for 1.5 h at room temperature. Subsequently, the TG slice was incubated with anti-SFKs antibody (1:40, AF3389, R&D Systems, Minneapolis, MN, USA) with anti-CGRP antibody (1:50, ab81887, Abcam, Cambridge, UK) or anti-RAMP1 antibody (1:50, ARR-021, Alomone Labs, Jerusalem, Israel) respectively at 4 °C overnight. The next day, the TG slice was incubated with Alexa fluor 488 donkey anti-goat secondary antibody (1:500, A11055, Invitrogen, Carlsbad, CA, USA) with 568 goat anti-mouse secondary antibody (1:500, A11004, Invitrogen) or 568 donkey anti-rabbit secondary antibody (1:500, A10042, Invitrogen) at room temperature for 1 h in the dark, after which they were incubated in 4′,6-diamidino-2-phenylindole (DAPI, 1:5000, D8417, Sigma-Aldrich) for 5 min. After mounting in mounting solution (S36936, Invitrogen), the expressions of SFKs, CGRP, and RAMP1 in the TG slice were imaged using a Confocal Laser Scanning Microscope (LSM880, Zeiss, Jena, Germany). The co-localization of SFKs and CGRP or RAMP1 in the acquired images was analyzed qualitatively.

2.8. Statistical Analysis

For quantitative studies, all raw data generated in experiments were statistically analyzed using GraphPad Prism 7.0 (San Diego, CA, USA) for testing if each dataset followed normal distribution and if significant difference existed between the data of two comparable experimental groups. In order to choose a proper test for analyzing significant statistical difference between two groups, a Shapiro–Wilk test was performed for all the datasets to determine if they followed normal distribution. If the normality test was passed, the data were presented as mean ± standard error of the mean, and the significance of intergroup statistical difference was analyzed by two-tailed unpaired t-test; if not, the data were presented as median (interquartile range), and the significance of intergroup statistical difference was analyzed by two-tailed Mann–Whitney test. Four types of significant intergroup statistical difference were used: * $p < 0.05$, ** $p < 0.01$, *** $p < 0.001$, or **** $p < 0.0001$. Detailed data presentation and statistical analysis for each quantitative study was described in the respective figure legend.

3. Results

3.1. pYEEI Alone Did Not Increase CGRP Release and IL-1β Gene Expression in the Mouse TG

We examined whether activation of SFKs increases CGRP release and IL-1β gene expression in the TG. When treated the TG with 1 mM pYEEI, the SFKs activator, the CGRP level was 32.8 ± 1 pg/mL, which was not significantly different from the CGRP level at 35.1 ± 1.9 pg/mL in the YEEI group (n = 8 per group, Figure 1A). Similarly, 1 mM pYEEI did not affect the IL-1β mRNA level either, which was 1 ± 0.1 (vs. 1 ± 0.1 in the YEEI group, Figure 1B) (n = 8 per group).

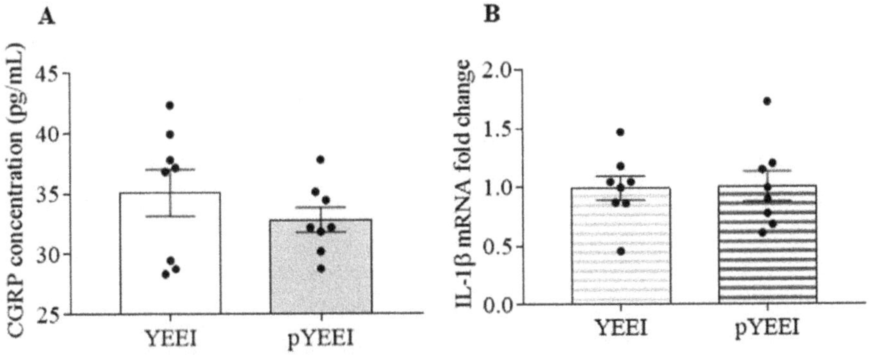

Figure 1. pYEEI alone did not alter CGRP release and IL-1β gene expression in the mouse TG. (**A**,**B**) Effects of 1 mM pYEEI or 1 mM YEEI ($n = 8$ per group) on CGRP release (pg/mL) and IL-1β mRNA level at 20 min post treatment. IL-1β mRNA level was present in the fold change relative to the geometric mean of β-actin and PPIA mRNA levels. Two-tailed unpaired t-test was used for the comparison in CGRP release and IL-1β mRNA level between the YEEI group and the pYEEI group.

3.2. Saracatinib Reduced CGRP Release Induced by H_2O_2 in the Mouse TG

This section determined whether deactivation of SFKs reduces CGRP release induced by ROS in the TG. H_2O_2 at 1 mM increased the level of CGRP in the TG culture medium to 28.4 ± 4.4 pg/mL in comparison with that at 13.7 ± 2.2 pg/mL in the Kreb's group ($n = 8$ per group, $p = 0.0126$, Figure 2). In the presence of 1 mM H_2O_2, 1.5 μM saracatinib ($n = 7$), the SFKs inhibitor, slightly reduced the CGRP level from the TG to 19.8 ± 4 pg/mL, which was not significantly different from that in the H_2O_2 group (Figure 2). When saracatinib was applied at 4 μM ($n = 7$), a significant reduction in the CGRP level to 15.9 ± 1 pg/mL was seen compared to that in the H_2O_2 group ($p = 0.024$, Figure 2). Saracatinib at 10 μM ($n = 7$) also significantly reduced the level of CGRP to 13.4 ± 0.9 compared to that in the H_2O_2 group ($p = 0.0105$, Figure 2). These data supported a concentration-response effect of saracatinib on CGRP release from the TG primed by H_2O_2.

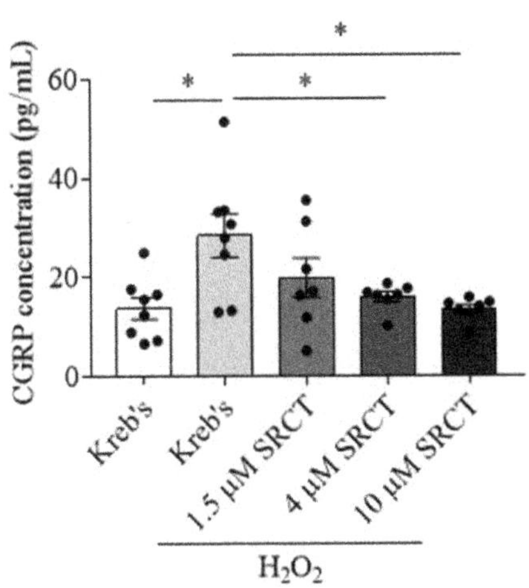

Figure 2. Saracatinib reduced CGRP release induced by H_2O_2 in the mouse TG. Effects of Kreb's ($n = 8$); 1 mM H_2O_2 ($n = 8$); 1.5 μM ($n = 7$), 4 μM ($n = 7$), or 10 μM ($n = 7$) saracatinib in the presence of 1 mM H_2O_2 at 20 min post treatment on CGRP release (pg/mL). Abbreviations: saracatinib (SRCT). Two-tailed unpaired t-test was used for the comparison in CGRP release between the H_2O_2 group and either the Kreb's group or the saracatinib in the presence of H_2O_2 group. Significant differences were labeled as * $p < 0.05$.

3.3. Saracatinib Reduced IL-1β, CCL2, and CXCL1 Release Induced by CGRP in the Mouse TG

We addressed whether deactivation of SFKs reduces inflammatory cytokine release induced by CGRP in the TG. Among the 12 pro-inflammatory cytokines (CCL2, CCL5, CXCL1, CXCL10, GM-CSF, IFN-α, IFN-β, IFN-γ, IL-1β, IL-6, IL-12, TNFα) and one anti-inflammatory cytokine (IL-10), 3 μM CGRP promoted the levels of IL-1β to 4.6 ± 0.4 pg/mL (vs. 2.9 ± 0.5 pg/mL in the Kreb's group, $p = 0.0288$), CCL2 to 8.9 ± 1.7 pg/mL (vs. 3.9 ± 0.7 pg/mL in the Kreb's group, $p = 0.0232$), and CXCL1 to 2.6 ± 0.7 pg/mL (vs. 0.9 ± 0.2 pg/mL in the Kreb's group, $p = 0.0303$) in the TG culture medium ($n = 8$ per group, Figure 3A–C). Differently, 3 μM CGRP decreased the level of IL-10 to 0.2 ± 0.1 pg/mL (vs. 3.7 ± 0.9 pg/mL in the Kreb's group, $p = 0.0017$, $n = 8$ per group, Figure 3D) in the medium. It is noted that 3 μM CGRP did not alter the levels of the other 9 cytokines (Supplementary Figure S1). As expected, in the presence of 3 μM CGRP, 1.5 μM saracatinib decreased the levels of IL-1β to 1.9 ± 0.5 pg/mL ($p = 0.001$), CCL2 to 2.5 ± 0.5 pg/mL ($p = 0.0069$), and CXCL1 to 1 ± 0.3 pg/mL ($p = 0.0476$) when compared to the respective data in the CGRP group ($n = 8$ per group, Figure 3A–C). However, 1.5 μM saracatinib did not significantly affect the reduced level of IL-10 elicited by 3 μM CGRP, which was 1.2 ± 0.5 ($n = 8$ per group, Figure 3D).

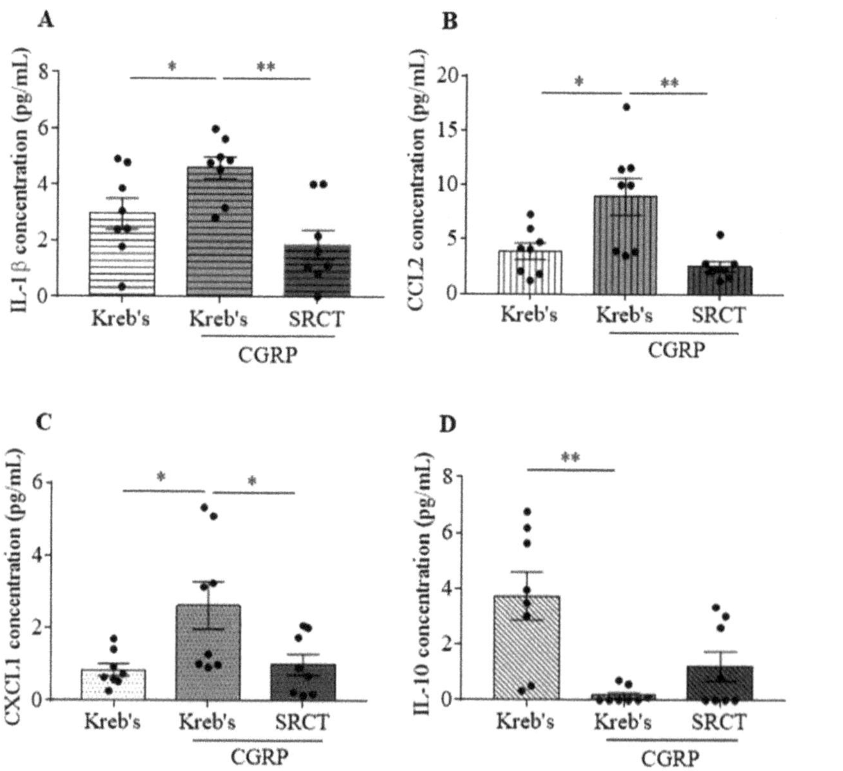

Figure 3. Saracatinib reduced IL-1β, CCL2, and CXCL1 release induced by CGRP in the mouse TG. (**A–D**) Effects of Kreb's, 3 μM CGRP, or 1.5 μM saracatinib in the presence of 3 μM CGRP ($n = 8$ per group) at 20 min post treatment on IL-1β, CCL2, CXCL1, and IL-10 release (pg/mL). Abbreviations: saracatinib (SRCT). Two-tailed unpaired t-test was used for the comparison in IL-1β, CCL2, CXCL1, and IL-10 release between the CGRP group and either the Kreb's group or the saracatinib in the presence of CGRP group. Significant differences were labeled as * $p < 0.05$ or ** $p < 0.01$.

3.4. Saracatinib Reduced IL-1β and CCL2 Gene Expression Induced by CGRP in the Mouse TG

Our data demonstrated that saracatinib reduced IL-1β, CCL2, and CXCL1 release promoted by CGRP in the TG (Figure 3A–C). All these proteins are associated with pain hypersensitivity and migraine [61–65]. We therefore further investigated whether SFKs activity promotes cytokines production machinery by examining the effect of the SFKs inhibitor saracatinib on their CGRP-induced gene expression. CGRP at 3 μM increased the mRNA levels of IL-1β to 2.1 (2.1) (vs. 1.1 (0.9) in the Kreb's group, $p = 0.0003$) and CCL2 to 1.5 (4.5) (vs. 0.9 (1.2) in the Kreb's group, $p = 0.0207$) in the TG ($n = 8$ per group, Figure 4A,B). In contrast, 3 μM CGRP did not affect the CXCL1 mRNA level, which was 1.1 ± 0.2 compared to that at 1 ± 0.1 in the Kreb's group ($n = 8$ per group, Figure 4C). When 1.5 μM saracatinib was applied in the presence of 3 μM CGRP, the IL-1β mRNA level was 4 (1.2), which was insignificantly different from that in the CGRP group (Figure 4A). Saracatinib at 4 μM, however, resulted in a pronounced reduction in the mRNA levels of IL-1β to 1 (1.1) ($p = 0.0148$) and CCL2 to 0.8 (0.3) ($p = 0.0002$) in comparison with that in the CGRP group ($n = 8$ per group, Figure 4A,B).

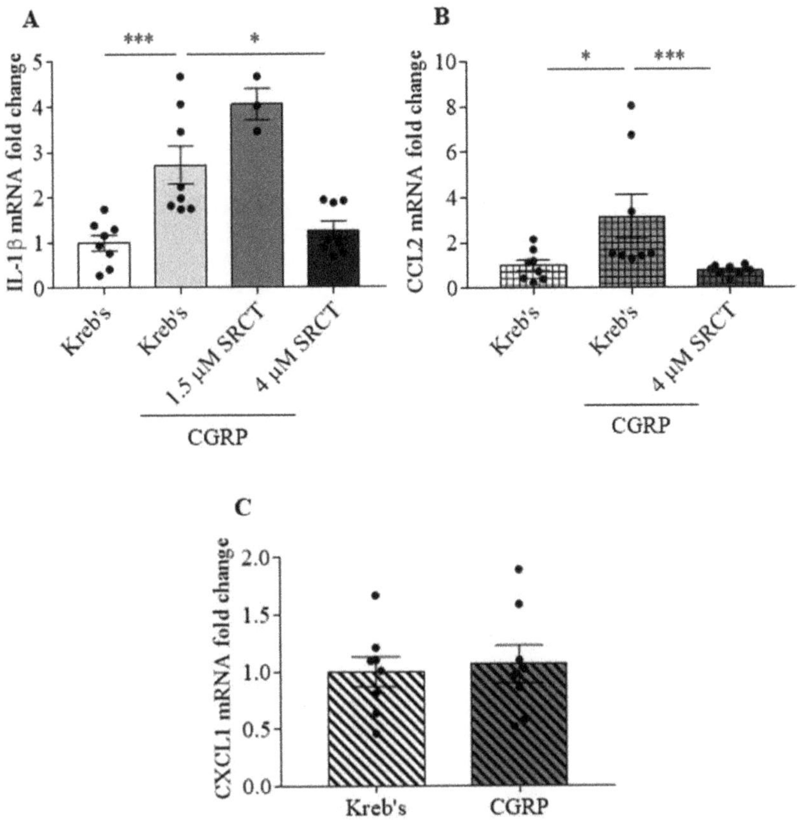

Figure 4. Saracatinib reduced IL-1β, CCL2 and CCL1 gene expression induced by CGRP in the mouse TG. (**A**–**C**) Effects of Kreb's, 3 μM CGRP, 1.5 μM, or 4 μM saracatinib in the presence of 3 μM CGRP ($n = 8$ per group) at 60 min post treatment on mRNA level of IL-1β, CCL2 and CCL1 in respective order. IL-1β, CCL2, and CCL1 mRNA levels were present in the fold change relative to the geometric mean of β-actin and PPIA mRNA levels. Abbreviations: saracatinib (SRCT). Two-tailed unpaired Mann–Whitney test was used for the comparison in IL-1β and CCL2 mRNA levels between the CGRP group and either the Kreb's group or the saracatinib in the presence of CGRP group. Significant differences were labeled as * $p < 0.05$ or *** $p < 0.001$.

3.5. The Protein Level of Phosphorylated SFKs at Y416 Was Increased by H_2O_2, Which Was Reduced by BIBN4096 in the Mouse TG

We examined whether SFKs activity is increased by H_2O_2 and whether such elevation can be reversed by inhibition of CGRP receptor using BIBN4096. When the protein level of phosphorylated SFKs at Y416 was normalized to that of β-actin, exposure to 1 mM H_2O_2 increased the protein level of phosphorylated SFKs at Y416 to 0.48 ± 0.06 compared to that at 0.13 ± 0.02 in the Kreb's group (n = 7 per group, p = 0.0010, Figure 5B). BIBN4096 at 10 μM reduced the H_2O_2-enhanced protein level of phosphorylated SFKs at Y416 to 0.24 ± 0.04 compared to that in the H_2O_2 group (n = 7 per group, p = 0.0082, Figure 5B). In contrast, the protein level of SFKs was unchanged among the three groups (Figure 5C), suggesting that the protein level of SFKs was insensitive to H_2O_2 or BIBN4096. When the protein level of phosphorylated SFKs at Y416 was normalized to that of SFKs, consistently, the protein level of phosphorylated SFKs at Y416 was increased to 0.32 ± 0.03 by H_2O_2 in comparison with that at 0.07 ± 0.01 in the Kreb's group (n = 7 per group, $p < 0.0001$, Figure 5D). BIBN4096 reduced H_2O_2-enhanced protein level of phosphorylated SFKs at Y416 to 0.19 ± 0.03 in comparison with that in the H_2O_2 group (n = 7 per group, p = 0.0117, Figure 5D).

3.6. The Protein Level of Phosphorylated SFKs at Y416 Was Increased by CGRP, Which Was Reduced by Both PKI (14-22) Amide and BIBN4096 in the Mouse TG

We next investigated whether SFKs activity can be elevated by CGRP and whether such elevation is sensitive to PKA or CGRP receptor inhibition using PKI (14-22) Amide and BIBN4096, respectively. When the protein level of phosphorylated SFKs at Y416 was normalized to that of β-actin, 3 μM CGRP markedly increased the protein level of phosphorylated SFKs at Y416 to 0.13 ± 0.01 compared to that at 0.06 ± 0.01 in the Kreb's group (n = 9 per group, p = 0.0001, Figure 6B). In the presence of 3 μM CGRP, both 30 μM PKI (14-22) Amide (n = 8) and 3 μM BIBN4096 (n = 8) reduced the protein level of phosphorylated SFKs at Y416 to 0.07 ± 0.01 compared to that in the CGRP group (p = 0.0023 and 0.002 respectively, Figure 6B). In contrast, the protein level of SFKs was unchanged among the four groups (Figure 6C). Consistently, when the protein level of phosphorylated SFKs at Y416 was normalized to that of SFKs, the protein level of phosphorylated SFKs at Y416 was increased to 0.11 ± 0.01 by CGRP in comparison with that at 0.06 ± 0.01 in the Kreb's group (n = 9 per group, p = 0.0158, Figure 6D). In the presence of CGRP, both PKI (14-22) Amide (n = 8) and BIBN4096 (n = 8) reduced the protein level of phosphorylated SFKs at Y416 to 0.06 ± 0.01 and 0.05 ± 0.01 in comparison with that in the CGRP alone group (p = 0.0283 and 0.0063, respectively, Figure 6D).

3.7. The Protein Levels of Phosphorylated SFKs at Y416 and Released Cytokines Induced by CGRP Were Positively Correlated in the Mouse TG

We then carried out further analysis to explore whether the SFKs activity and cytokine release enhanced by CGRP are correlated. A positive relationship between the increased levels of phosphorylated SFKs at Y416 and IL-1β release (r = 0.7261, p = 0.0014, Figure 7A), CCL2 release (r = 0.7462, p = 0.0009, Figure 7B), and CXCL1 release (r = 0.7768, p = 0.0004, Figure 7C) was seen in the TG treated by both Kreb's and 3 μM CGRP (n = 8 per group).

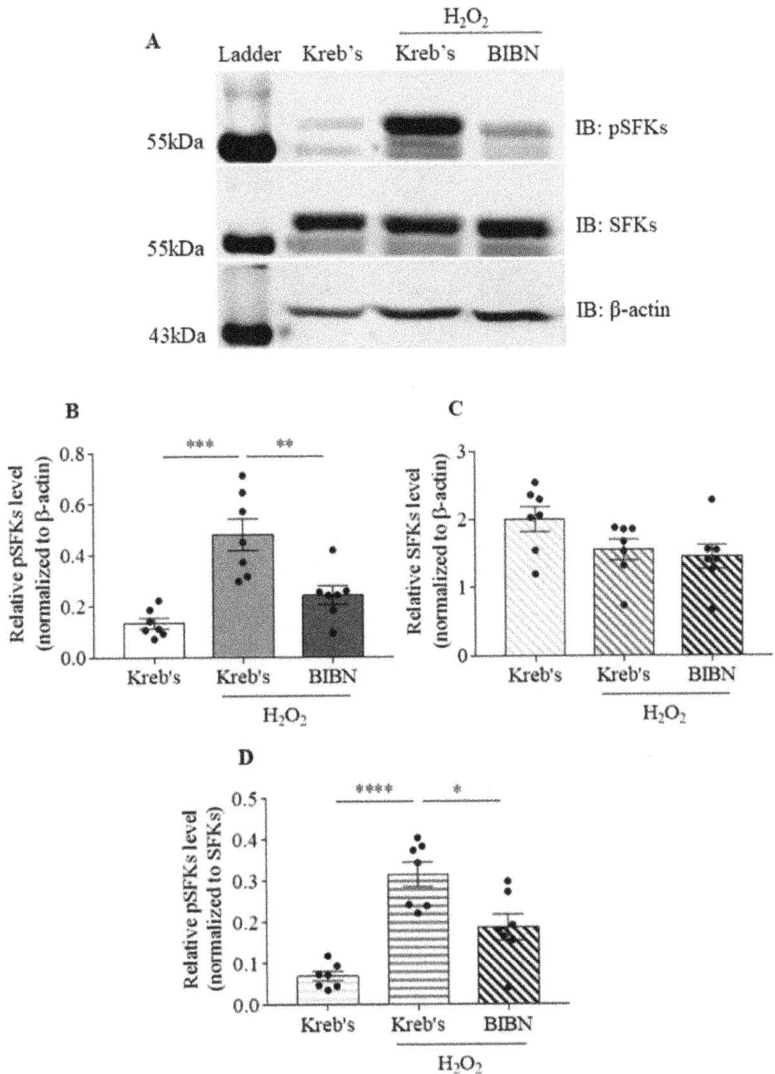

Figure 5. The protein level of phosphorylated SFKs at Y416 was increased by H_2O_2 in the mouse TG. (**A**) The representative Western blot bands of phosphorylated SFKs at Y416, SFKs, and β-actin subjected to the treatment with Kreb's, 1 mM H_2O_2, or 10 μM BIBN4096 in the presence of 1 mM H_2O_2. (**B–D**) Effects of Kreb's, 1 mM H_2O_2, or 10 μM BIBN4096 in the presence of 1 mM H_2O_2 ($n = 7$ per group) at 20 min post treatment on the protein levels of phosphorylated SFKs at Y416 and SFKs relative to that of β-actin and on the protein level of phosphorylated SFKs at Y416 relative to that of SFKs, all of which were presented in the absolute ratio. Abbreviations: BIBN4086 (BIBN), phosphorylated SFKs at Y416 (pSFKs). Two-tailed unpaired t-test was used for the comparison in the protein level of phosphorylated SFKs at Y416 between the H_2O_2 group and either the Kreb's group or the BIBN4096 in the presence of 1 mM H_2O_2 group. Significant differences were labeled as * $p < 0.05$, ** $p < 0.01$, *** $p < 0.001$, or **** $p < 0.0001$. Original western blot images for the representative images in Figure 5 was shown in Supplementary Figure S2.

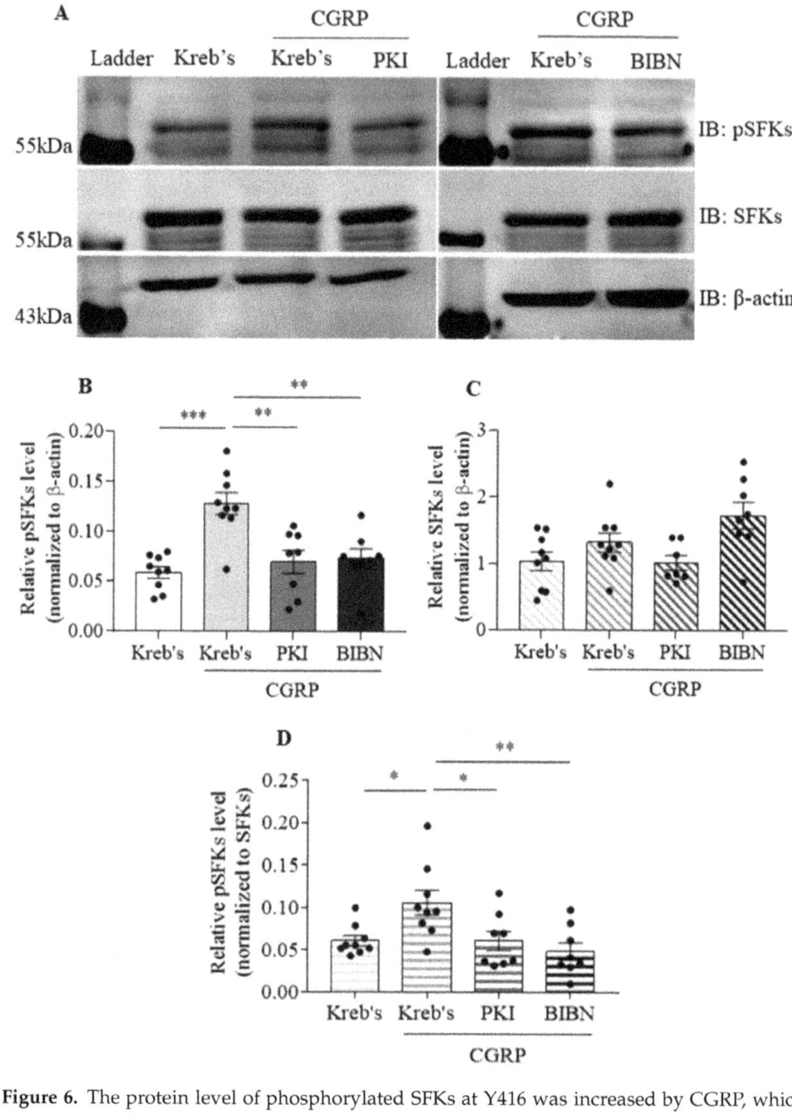

Figure 6. The protein level of phosphorylated SFKs at Y416 was increased by CGRP, which was reduced by both PKI (14-22) Amide and BIBN4096 in the mouse TG. (**A**) The representative Western blot bands of phosphorylated SFKs at Y416, SFKs, and β-actin subjected to the treatment with Kreb's, 3 μM CGRP, 30 μM PKI (14-22) Amide or 10 μM BIBN4096 in the presence of 3 μM CGRP for 20 min. (**B–D**) Effects of Kreb's ($n = 9$), 3 μM CGRP ($n = 9$), 30 μM PKI (14-22) Amide ($n = 8$) or 3 μM BIBN4096 ($n = 8$) in the presence of 3 μM CGRP on the protein levels of phosphorylated SFKs at Y416 and SFKs relative to that of β-actin and on the protein level of phosphorylated SFKs at Y416 relative to that of SFKs, all of which were presented in the absolute ratio. Abbreviations: PKI (14-22) Amide (PKI), BIBN4086 (BIBN), phosphorylated SFKs at Y416 (pSFKs). Two-tailed unpaired t-test was used for the comparison in the protein level of phosphorylated SFKs at Y416 between the CGRP group and either the Kreb's group, the PKI (14-22) Amide, or the BIBN4096 in the presence of CGRP groups. Significant differences were labeled as * $p < 0.05$, ** $p < 0.01$, or *** $p < 0.001$. Original western blot images for the representative images in Figure 6 were shown in Supplementary Figures S3 and S4.

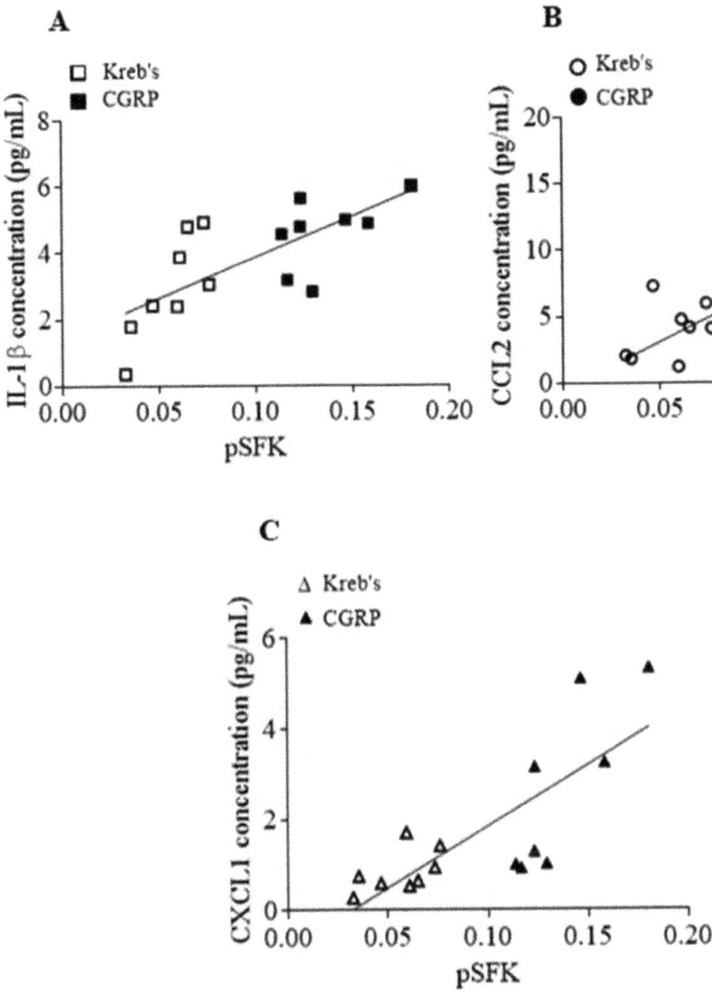

Figure 7. The levels of phosphorylated SFKs at Y416 and released cytokines induced by CGRP were correlated in the mouse TG. Correlation analysis between the levels of phosphorylated SFKs at Y416 and IL-1β (**A**), CCL2 (**B**), CXCL1 (**C**) release in the mouse TG treated by Kreb's and 3 μM CGRP (n = 8 per group) for 20 minutes.

3.8. SFKs Co-Localized with CGRP and RAMP1 in the Mouse TG

Consistent with the previous reports [10,66–68], we were able to demonstrate that CGRP immunoreactivity was present in small to medium-sized neurons and C fibers (Figure 8A) while RAMP1 immunoreactivity was present in large neurons and Aδ fibers (Figure 8B) in the TG. Further, SFKs immunoreactivity was present in the neurons of nearly all sizes and fibers in the TG (Figure 8A,B). Double staining with the anti-SFKs antibody and anti-CGRP antibody or anti-RAMP1 antibody demonstrated that SFKs co-localized with both CGRP and RAMP1 proteins in their respective cell types and fibers of the TG (Figure 8A,B).

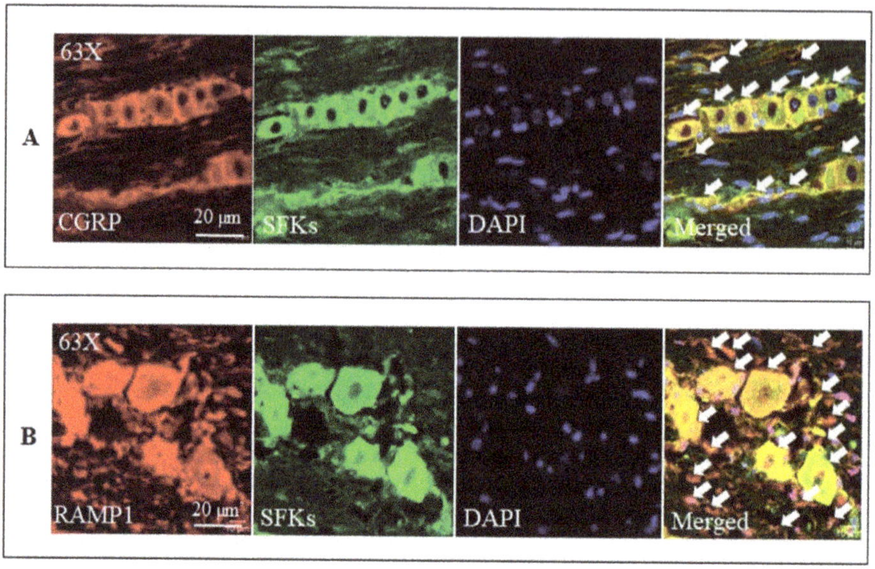

Figure 8. SFKs co-localized with both CGRP and RAMP1 in the mouse TG. (**A**) The representative images of double staining with the anti-CGRP antibody and anti-SFKs antibody in the TG. (**B**) The representative images of double staining with the anti-RAMP1 antibody and anti-SFKs antibody in the TG. CGRP and RAMP1 were stained with the anti-CGRP antibody and anti-RAMP1 antibody, respectively, and are shown in red; SFKs were stained with the anti-SFKs antibody and are shown in green; nucleus was stained with DAPI and is shown in blue. Co-localization of SFKs and CGRP or RAMP1 is shown in yellow and is indicated by the white arrows.

4. Discussion

Our data demonstrate that SFKs activity facilitates the crosstalk between CGRP and cytokines in sensitizing trigeminal ganglion by transmitting CGRP receptor/PKA signaling. These findings uncover an unprecedented role of SFKs in migraine pain transmission.

We have previously demonstrated that SFKs mediate CGRP release and IL-1β gene expression and SFKs inhibition by saracatinib reduces the stress-sensing cation channel transient receptor potential ankyrin 1 (TRPA1)-activated CGRP release and IL-1β gene expression in the mouse TG [21]. We therefore postulated that direct SFKs activation may induce CGRP release and neuroinflammation from the TG thus triggering TG activation. Unexpectedly in this communication, activation of SFKs using pYEEI does not alter CGRP release or IL-1β gene expression in the mouse TG (Figure 1). pYEEI, the SFKs activator, activates SFKs by binding to their SH2 domains, which disrupts the closing conformation of inactive SFKs ensuing their open active conformation [34,37,38]. Given that the concentration (1 mM) of pYEEI applied in this study is high enough to trigger neuronal activation [35,36], the current data suggest that SFKs activation by pYEEI alone is least likely to be a stimulus for TG activation, which is unlike other stimuli such as KCl or TRPA1 activator that can activate multiple pathways triggering mass activation and sensitization of TG [21].

We then explored whether modulating SFKs activity may be only effective when the TG is pre-primed. Indeed, in the mouse TG that is pre-sensitized by H_2O_2, the SFKs inhibitor, saracatinib, markedly reduces CGRP release from TG in a concentration-dependent manner (Figure 2). Consistent with this, SFKs activity is increased by H_2O_2 in the mouse TG (Figure 5). These data highlight a key role of SFKs activity in facilitating TG sensitization by mediating endogenous CGRP release. Moreover, these data extend the previous findings that deactivation of SFKs reverses CGRP release promoted by nerve growth factor and capsaicin in dorsal root ganglion neurons [20] and by a TRPA1 activator, umbellulone, in the TG [21].

Similar to the release of CGRP, the SFKs inhibitor saracatinib also reduces the release of IL-1β, CCL2, and CXCL1 from the mouse TG pre-sensitized by exogenous CGRP (Figure 3), which consistently increases SFKs activity (Figure 6). The levels of all these three cytokines released show positive correlations with the level of respective phosphorylated SFKs induced by CGRP (Figure 7), highlighting the importance of SFK activity in promoting TG neuroinflammation. Moreover, these data extend the previous finding that SFKs activity contributes to the release of inflammatory cytokines in astrocytes [25] and microglia [22–24]. Among these inflammatory cytokines, IL-1β potentiates the excitability of nociceptive neurons in the TG and directly causes the hypersensitivity to nociception ensuing the nociceptive behaviors, hyperalgesia, and allodynia [69]. While IL-1β has a well-identified role in migraine pathogenesis [65], CXCL1 [61,62,70,71] and CCL2 [63,64,72] are associated with pain hypersensitivity and more recently with migraine. It is therefore concluded that SFKs activity can synergistically elevate CGRP and cytokine release to reinforce TG sensitization and facilitate pain transmission. It is noted that, unlike the three inflammatory cytokines, the anti-inflammatory cytokine, IL-10, release is insensitive to SFK inhibition. This might suggest that SFKs activity plays a more prominent role in controlling inflammatory cytokine but not anti-inflammatory cytokine release.

One question to ask from our data is how the SFKs-mediated TG sensitization is sustained. In our study, besides cytokine release by CGRP being reduced by saracatinib at 20 min post treatment (Figure 3), the induction of IL-1β and CCL2 gene expression by exogenous CGRP was also reduced by SFKs deactivation at 1 h post treatment in the mouse TG (Figure 4). These data are consistent with our previous finding that SFKs activity contributes to IL-1β gene expression induced by the TRPA1 activator umbellulone in the mouse TG [21]. It is highly likely that the SFKs-induced transcriptional machinery activation of these cytokines is crucial to sustain the TG sensitization, especially in that IL-1β mRNA expression and protein expression are well correlated [73]. The possible mechanism by which SFKs mediate cytokines gene expression could be via transcription factors or histone modification. SFKs are known to activate the transcription factor NFκB in models of neurodegenerative diseases [25,74], which is important for inflammatory responses [75]. Interestingly, SFKs-mediated CCL2 gene expression and histone H3 acetylation at the CCL2 promoter are associated in macrophages [76]. Furthermore, CGRP is a potent neuroinflammatory mediator that induces the expression and release of cytokines in the TG [11–14], including IL-1β, CCL2, CXCL1, all of which in turn stimulate CGRP release [16,77], thereby inducing a positive feedback loop of TG sensitization. Similar to CGRP, IL-1β can activate SFKs in several cell lines [78–81], which suggests that released cytokines are highly likely to strengthen SFKs activity again to induce CGRP release and aggravate TG sensitization. This is consistent with the significant positive correlation between the elevated SFKs activity and cytokine release induced by CGRP, which supports the model that SFKs activity facilitates the crosstalk between CGRP and cytokines (Figure 7). Taken together, we propose a feedback mechanism by which SFKs activity facilitates the crosstalk and intraganglionic signaling between CGRP and cytokines in stress-primed TG to potentiate TG sensitization and ultimately trigeminovascular sensitization.

The molecular mechanism underlying the SFKs-mediated crosstalk between CGRP and cytokines in sensitizing TG has yet to be fully defined. Interestingly, SFKs activation induced by either H_2O_2 (Figure 5) or exogenous CGRP (Figure 6) can be reduced by the CGRP receptor inhibitor, BIBN4096, in the mouse TG, supporting CGRP receptor-dependent SFKs activation in TG. As SFKs co-localize with CGRP in small- to medium-sized neurons and C fibers, whilst RAMP1 in large neurons, Aδ fibers, and satellite glial cells of the mouse TG (Figure 8), we can conclude that the SFKs-mediated crosstalk between CGRP and cytokines is dependent on CGRP/CGRP receptor signaling. Notably, PKA is known to actively transmit CGRP/CGRP receptor signaling to initiate downstream signaling cascades, thus leading to TG activation [52–54]. PKA robustly increases SFKs activity in cell lines [55], spinal dorsal horn [56], and hypothalamic arcuate nucleus neurons [36], and PKA/SFKs pathway facilitates neuronal firing [56] and pain sensitivity [36]. This can be compared with our previous study which demonstrates that SFKs activity is elevated by PKA upon TRPA1 activation to promote CGRP release in the mouse TG [21]. Furthermore, in the present study, the enhanced phosphorylated SFKs at Y416 induced by CGRP is also reduced by the PKA inhibitor PKI (14-22) Amide, which is similar to that by the CGRP receptor inhibitor BIBN4096 (Figure 6). Taken together, these data pinpoint that SFKs activity is increased downstream of CGRP/CGRP receptor signaling via PKA activity in the TG, thereby contributing to CGRP-cytokines crosstalk and TG sensitization (Figure 9). Given that PKA promotes the phosphorylation of SFKs at the S17 site followed by autophosphorylation at their Y416 site [55], it is likely that SFKs are activated directly by PKA downstream of CGRP/CGRP receptor signaling in the TG, which awaits future validation. Future work should also examine whether CGRP co-localizes with PKA in the TG.

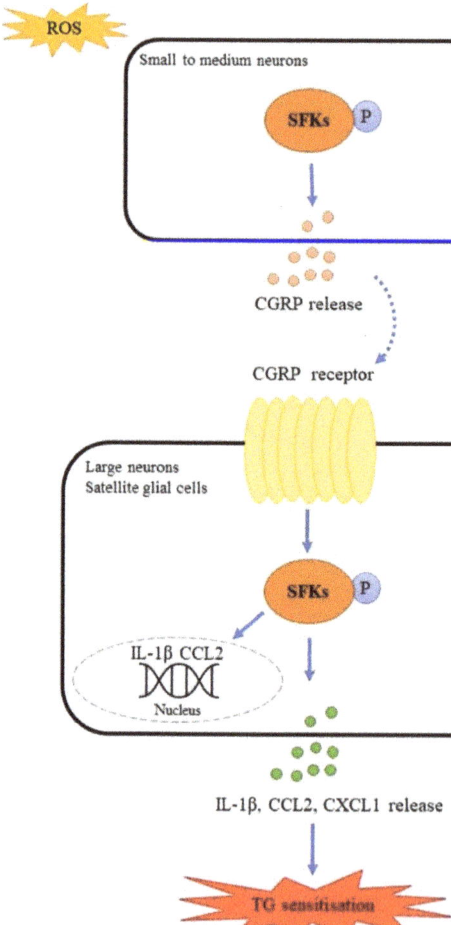

Figure 9. Model of SFKs activity facilitating the crosstalk between CGRP and cytokines by transmitting CGRP receptor signaling to potentiate TG sensitization. SFKs are activated in response to ROS to induce CGRP release in small to medium neurons; released CGRP binds to CGRP receptor (dotted line with arrow) to activate SFKs in large neurons and satellite glial cells, which causes IL-1β, CCL2, CXCL1 release and IL-1β, CCL2 gene expression, thus leading to TG sensitization.

In this study, only male mice are used to explore the role of SFKs in TG sensitization so that the effect of hormonal fluctuation in females can be minimized. Similarly, the previous studies on investigating the role of SFKs in the CSD-induced migraine with aura model [82] and the inflammatory soup-induced chronic migraine model [26] only use male rodents. Interestingly, SFKs deactivation does not show sex difference in brain perfusion abnormalities in the genetic migraine with aura model FHM2 [27]. Nevertheless, future work should explore whether SFKs mediate different migraine pathogenesis in females in order to understand if there are gender-specific effects of targeting SFKs.

5. Conclusions

The present study demonstrates that SFKs activity plays a pivotal role in facilitating the crosstalk between CGRP and cytokines by transmitting CGRP receptor/PKA signaling to potentiate TG sensitization and ultimately trigeminovascular sensitization. These findings shed light on the SFKs-mediated peripheral mechanism of migraine pathogenesis and support the promising efficacy of drugs targeting SFKs for migraine therapy.

Supplementary Materials: The following supporting information can be downloaded at: https://www.mdpi.com/article/10.3390/cells11213498/s1. Figure S1: CGRP did not alter the release of CCL5, CXCL10, GM-CSF, IFN-α, IFN-β, IFN-γ, IL-6, IL-12, or TNFα in the mouse TG; Figure S2: Original western blot images for the representative images presented in Figure 5; Figures S3 and S4: Original western blot images for the representative images presented in Figure 6.

Author Contributions: L.N. and M.W. designed the study and drafted the manuscript. L.N., K.S., Z.G. and H.L. performed the experiments. M.W. sponsored the study. M.W. and J.P.Q. edited the manuscript. All authors have read and agreed to the published version of the manuscript.

Funding: This study was supported by XJTLU PGR studentship (RDF-17-02-057) and Wangwenli Charitable Foundation (RDS-0006).

Institutional Review Board Statement: The study was conducted according to the guidelines of the ARRIVE, approved by the Research Ethics Committee of XJTLU (protocol code EXT 19-01-04 and date of approval 4 March 2019).

Informed Consent Statement: Not applicable.

Data Availability Statement: Not applicable.

Acknowledgments: We thank Xiaojuan Guo for her technical support; other lab members in the laboratory at XJTLU.

Conflicts of Interest: The authors declare no potential conflict of interest with respect to the research, authorship, and/or publication of this article.

References

1. Ashina, M.; Katsarava, Z.; Do, T.P.; Buse, D.C.; Pozo-Rosich, P.; Özge, A.; Krymchantowski, A.V.; Lebedeva, E.R.; Ravishankar, K.; Yu, S.; et al. Migraine: Epidemiology and systems of care. *Lancet* **2021**, *397*, 1485–1495. [CrossRef]
2. Bernstein, C.; Burstein, R. Sensitization of the Trigeminovascular Pathway: Perspective and Implications to Migraine Pathophysiology. *J. Clin. Neurol.* **2012**, *8*, 89–99. [CrossRef]
3. Noseda, R.; Burstein, R. Migraine pathophysiology: Anatomy of the trigeminovascular pathway and associated neurological symptoms, cortical spreading depression, sensitization, and modulation of pain. *Pain* **2013**, *154*, S44–S53. [CrossRef] [PubMed]
4. Ashina, M.; Hansen, J.M.; Do, T.P.; Melo-Carrillo, A.; Burstein, R.; Moskowitz, M.A. Migraine and the trigeminovascular system-40 years and counting. *Lancet Neurol.* **2019**, *18*, 795–804. [CrossRef]
5. Ray, B.S.; Wolff, H.G. Experimental studies on headache: Pain-sensitive structures of the head and their significance in heaadache. *Arch. Surg.* **1940**, *41*, 813–856. [CrossRef]
6. Moskowitz, M.A.; Macfarlane, R. Neurovascular and molecular mechanisms in migraine headaches. *Cerebrovasc. Brain Metab. Rev.* **1993**, *5*, 159–177. [PubMed]
7. Olesen, J.; Burstein R Fau-Ashina, M.; Ashina M Fau-Tfelt-Hansen, P.; Tfelt-Hansen, P. Origin of pain in migraine: Evidence for peripheral sensitisation. *Lancet Neurol.* **2009**, *8*, 679–690. [CrossRef]
8. Edvinsson, L.; Haanes, K.A.; Warfvinge, K.; Krause, D.N. CGRP as the target of new migraine therapies—successful translation from bench to clinic. *Nat. Rev. Neurol.* **2018**, *14*, 338–350. [CrossRef]

9. Chatchaisak, D.; Connor, M.; Srikiatkhachorn, A.; Chetsawang, B. The potentiating effect of calcitonin gene-related peptide on transient receptor potential vanilloid-1 activity and the electrophysiological responses of rat trigeminal neurons to nociceptive stimuli. *J. Physiol. Sci.* **2018**, *68*, 261–268. [CrossRef]
10. Edvinsson, J.C.A.; Warfvinge, K.; Krause, D.N.; Blixt, F.W.; Sheykhzade, M.; Edvinsson, L.; Haanes, K.A. C-fibers may modulate adjacent Aδ-fibers through axon-axon CGRP signaling at nodes of Ranvier in the trigeminal system. *J. Headache Pain* **2019**, *20*, 105. [CrossRef]
11. Afroz, S.; Arakaki, R.; Iwasa, T.; Oshima, M.; Hosoki, M.; Inoue, M.; Baba, O.; Okayama, Y.; Matsuka, Y. CGRP Induces Differential Regulation of Cytokines from Satellite Glial Cells in Trigeminal Ganglia and Orofacial Nociception. *Int. J. Mol. Sci.* **2019**, *20*, 711. [CrossRef] [PubMed]
12. Thalakoti, S.; Patil, V.V.; Damodaram, S.; Vause, C.V.; Langford, L.E.; Freeman, S.E.; Durham, P.L. Neuron–Glia Signaling in Trigeminal Ganglion: Implications for Migraine Pathology. *Headache* **2007**, *47*, 1008–1023. [CrossRef] [PubMed]
13. Vause, C.V.; Durham, P.L. Calcitonin gene-related peptide differentially regulates gene and protein expression in trigeminal glia cells: Findings from array analysis. *Neurosci. Lett.* **2010**, *473*, 163–167. [CrossRef]
14. Afroz, S.; Arakaki, R.; Iwasa, T.; Waskitho, A.; Oshima, M.; Matsuka, Y. Role of CGRP in Neuroimmune Interaction via NF-κB Signaling Genes in Glial Cells of Trigeminal Ganglia. *Int. J. Mol. Sci.* **2020**, *21*, 6005. [CrossRef]
15. Bowen, E.J.; Schmidt, T.W.; Firm, C.S.; Russo, A.F.; Durham, P.L. Tumor necrosis factor-α stimulation of calcitonin gene-related peptide expression and secretion from rat trigeminal ganglion neurons. *J. Neurochem.* **2006**, *96*, 65–77. [CrossRef] [PubMed]
16. Capuano, A.; De Corato, A.; Lisi, L.; Tringali, G.; Navarra, P.; Russo, C.D. Proinflammatory-Activated Trigeminal Satellite Cells Promote Neuronal Sensitization: Relevance for Migraine Pathology. *Mol. Pain* **2009**, *5*, 43. [CrossRef] [PubMed]
17. Messlinger, K.; Russo, A.F. Current understanding of trigeminal ganglion structure and function in headache. *Cephalalgia Int. J. Headache* **2018**, *39*, 1661–1674. [CrossRef]
18. Iyengar, S.; Johnson, K.W.; Ossipov, M.H.; Aurora, S.K. CGRP and the Trigeminal System in Migraine. *Headache* **2019**, *59*, 659–681. [CrossRef]
19. Messlinger, K.; Balcziak, L.K.; Russo, A.F. Cross-talk signaling in the trigeminal ganglion: Role of neuropeptides and other mediators. *J. Neural Transm.* **2020**, *127*, 431–444. [CrossRef]
20. Park, K.A.; Fehrenbacher, J.C.; Thompson, E.L.; Duarte, D.B.; Hingtgen, C.M.; Vasko, M.R. Signaling pathways that mediate nerve growth factor-induced increase in expression and release of calcitonin gene-related peptide from sensory neurons. *Neuroscience* **2010**, *171*, 910–923. [CrossRef]
21. Nie, L.; Jiang, L.; Quinn, J.P.; Grubb, B.D.; Wang, M. TRPA1-Mediated Src Family Kinases Activity Facilitates Cortical Spreading Depression Susceptibility and Trigeminovascular System Sensitization. *Int. J. Mol. Sci.* **2021**, *22*, 12273. [CrossRef] [PubMed]
22. Chen, C.-J.; Ou, Y.-C.; Chang, C.-Y.; Pan, H.-C.; Lin, S.-Y.; Liao, S.-L.; Raung, S.-L.; Chen, S.-Y.; Chang, C.-J. Src signaling involvement in Japanese encephalitis virus-induced cytokine production in microglia. *Neurochem. Int.* **2011**, *58*, 924–933. [CrossRef] [PubMed]
23. Portugal Camila, C.; Socodato, R.; Canedo, T.; Silva Cátia, M.; Martins, T.; Coreixas Vivian, S.M.; Loiola Erick, C.; Gess, B.; Röhr, D.; Santiago Ana, R.; et al. Caveolin-1–mediated internalization of the vitamin C transporter SVCT2 in microglia triggers an inflammatory phenotype. *Sci. Signal.* **2017**, *10*, eaal2005. [CrossRef] [PubMed]
24. Huang, Y.; Li, Y.; Zhong, X.; Hu, Y.; Liu, P.; Zhao, Y.; Deng, Z.; Liu, X.; Liu, S.; Zhong, Y. Src-family kinases activation in spinal microglia contributes to central sensitization and chronic pain after lumbar disc herniation. *Mol. Pain* **2017**, *13*, 1744806917733637. [CrossRef]
25. Lee, C.; Low, C.Y.B.; Wong, S.Y.; Lai, M.K.P.; Tan, M.G.K. Selective induction of alternatively spliced FynT isoform by TNF facilitates persistent inflammatory responses in astrocytes. *Sci. Rep.* **2017**, *7*, 43651. [CrossRef]
26. Wang, X.-Y.; Zhou, H.-R.; Wang, S.; Liu, C.-Y.; Qin, G.-C.; Fu, Q.-Q.; Zhou, J.-Y.; Chen, L.-X. NR2B-Tyr phosphorylation regulates synaptic plasticity in central sensitization in a chronic migraine rat model. *J. Headache Pain* **2018**, *19*, 102. [CrossRef]
27. Staehr, C.; Hangaard, L.; Bouzinova, E.V.; Kim, S.; Rajanathan, R.; Boegh Jessen, P.; Luque, N.; Xie, Z.; Lykke-Hartmann, K.; Sandow, S.L.; et al. Smooth muscle Ca2+ sensitization causes hypercontractility of middle cerebral arteries in mice bearing the familial hemiplegic migraine type 2 associated mutation. *J. Cereb. Blood Flow Metab.* **2018**, *39*, 1570–1587. [CrossRef]
28. Nie, L.; Ma, D.; Quinn, J.P.; Wang, M. Src family kinases activity is required for transmitting purinergic P2X7 receptor signaling in cortical spreading depression and neuroinflammation. *J. Headache Pain* **2021**, *22*, 146. [CrossRef]
29. Williamson, D.J.; Hargreaves, R.J.; Hill, R.G.; Shepheard, S.L. Sumatriptan Inhibits Neurogenic Vasodilation of Dural Blood Vessels in the Anaesthetized Rat—Intravital Microscope Studies. *Cephalalgia Int. J. Headache* **1997**, *17*, 525–531. [CrossRef]
30. Storer, R.J.; Akerman, S.; Goadsby, P.J. Calcitonin gene-related peptide (CGRP) modulates nociceptive trigeminovascular transmission in the cat. *Br. J. Pharmacol.* **2004**, *142*, 1171–1181. [CrossRef]
31. Coste, J.; Voisin, D.L.; Miraucourt, L.S.; Dallel, R.; Luccarini, P. Dorsal horn NK1-expressing neurons control windup of downstream trigeminal nociceptive neurons. *Pain* **2008**, *137*, 340–351. [CrossRef] [PubMed]
32. Dux, M.; Rosta, J.; Messlinger, K. TRP Channels in the Focus of Trigeminal Nociceptor Sensitization Contributing to Primary Headaches. *Int. J. Mol. Sci.* **2020**, *21*, 342. [CrossRef] [PubMed]
33. Zhou, S.; Steven, E.S.; Manas, C.; Gerald, G.; Tony, P.; Wayne, G.H.; Fred, K.; Tom, B.; Sheldon, R.; Robert, J.L.; et al. SH2 domains recognize specific phosphopeptide sequences. *Cell* **1993**, *72*, 767–778. [CrossRef]

34. Pellicena, P.; Stowell, K.R.; Miller, W.T. Enhanced Phosphorylation of Src Family Kinase Substrates Containing SH2 Domain Binding Sites*. *J. Biol. Chem.* **1998**, *273*, 15325–15328. [CrossRef]
35. Yu, X.-M.; Askalan, R.; Keil Gary, J.; Salter Michael, W. NMDA Channel Regulation by Channel-Associated Protein Tyrosine Kinase Src. *Science* **1997**, *275*, 674–678. [CrossRef]
36. Sun, X.-D.; Wang, A.; Ma, P.; Gong, S.; Tao, J.; Yu, X.-M.; Jiang, X. Regulation of the firing activity by PKA-PKC-Src family kinases in cultured neurons of hypothalamic arcuate nucleus. *J. Neurosci. Res.* **2020**, *98*, 384–403. [CrossRef]
37. Liu, X.; Brodeur Sr Fau-Gish, G.; Gish G Fau-Songyang, Z.; Songyang Z Fau-Cantley, L.C.; Cantley Lc Fau-Laudano, A.P.; Laudano Ap Fau-Pawson, T.; Pawson, T. Regulation of c-Src tyrosine kinase activity by the Src SH2 domain. *Oncogene* **1993**, *8*, 1119–1126.
38. Xu, W.; Doshi A Fau-Lei, M.; Lei M Fau-Eck, M.J.; Eck Mj Fau-Harrison, S.C.; Harrison, S.C. Crystal structures of c-Src reveal features of its autoinhibitory mechanism. *Mol. Cell* **1999**, *3*, 629–638. [CrossRef]
39. Xie, J.; Bi, Y.; Zhang, H.; Dong, S.; Teng, L.; Lee, R.J.; Yang, Z. Cell-Penetrating Peptides in Diagnosis and Treatment of Human Diseases: From Preclinical Research to Clinical Application. *Front. Pharmacol.* **2020**, *11*, 697. [CrossRef]
40. IHS, H.C.C.O.T. Headache Classification Committee of the International Headache Society (IHS) The International Classification of Headache Disorders, 3rd edition. *Cephalalgia Int. J. Headache* **2018**, *38*, 1–211.
41. Schieber, M.; Chandel, N.S. ROS function in redox signaling and oxidative stress. *Curr. Biol.* **2014**, *24*, R453–R462. [CrossRef] [PubMed]
42. Shatillo, A.; Koroleva, K.; Giniatullina, R.; Naumenko, N.; Slastnikova, A.A.; Aliev, R.R.; Bart, G.; Atalay, M.; Gu, C.; Khazipov, R.; et al. Cortical spreading depression induces oxidative stress in the trigeminal nociceptive system. *Neuroscience* **2013**, *253*, 341–349. [CrossRef] [PubMed]
43. Hennequin, L.F.; Allen, J.; Breed, J.; Curwen, J.; Fennell, M.; Green, T.P.; Lambert-van der Brempt, C.; Morgentin, R.; Norman, R.A.; Olivier, A.; et al. N-(5-Chloro-1,3-benzodioxol-4-yl)-7-[2-(4-methylpiperazin-1-yl)ethoxy]-5- (tetrahydro-2H-pyran-4-yloxy)quinazolin-4-amine, a Novel, Highly Selective, Orally Available, Dual-Specific c-Src/Abl Kinase Inhibitor. *J. Med. Chem.* **2006**, *49*, 6465–6488. [CrossRef] [PubMed]
44. Fury, M.G.; Baxi, S.; Shen, R.; Kelly, K.W.; Lipson, B.L.; Carlson, D.; Stambuk, H.; Haque, S.; Pfister, D.G. Phase II Study of Saracatinib (AZD0530) for Patients with Recurrent or Metastatic Head and Neck Squamous Cell Carcinoma (HNSCC). *Anticancer Res.* **2011**, *31*, 249. [PubMed]
45. Renouf, D.J.; Moore, M.J.; Hedley, D.; Gill, S.; Jonker, D.; Chen, E.; Walde, D.; Goel, R.; Southwood, B.; Gauthier, I.; et al. A phase I/II study of the Src inhibitor saracatinib (AZD0530) in combination with gemcitabine in advanced pancreatic cancer. *Investig. New Drugs* **2012**, *30*, 779–786. [CrossRef] [PubMed]
46. Fujisaka, Y.; Onozawa, Y.; Kurata, T.; Yasui, H.; Goto, I.; Yamazaki, K.; Machida, N.; Watanabe, J.; Shimada, H.; Shi, X.; et al. First report of the safety, tolerability, and pharmacokinetics of the Src kinase inhibitor saracatinib (AZD0530) in Japanese patients with advanced solid tumours. *Invest. New Drugs* **2013**, *31*, 108–114. [CrossRef] [PubMed]
47. Nygaard, H.B.; Wagner, A.F.; Bowen, G.S.; Good, S.P.; MacAvoy, M.G.; Strittmatter, K.A.; Kaufman, A.C.; Rosenberg, B.J.; Sekine-Konno, T.; Varma, P.; et al. A phase Ib multiple ascending dose study of the safety, tolerability, and central nervous system availability of AZD0530 (saracatinib) in Alzheimer's disease. *Alzheimers Res. Ther.* **2015**, *7*, 35. [CrossRef]
48. van Dyck, C.H.; Nygaard, H.B.; Chen, K.; Donohue, M.C.; Raman, R.; Rissman, R.A.; Brewer, J.B.; Koeppe, R.A.; Chow, T.W.; Rafii, M.S.; et al. Effect of AZD0530 on Cerebral Metabolic Decline in Alzheimer Disease: A Randomized Clinical Trial. *JAMA Neurol.* **2019**, *76*, 1219–1229. [CrossRef]
49. Kaufman, A.C.; Salazar, S.V.; Haas, L.T.; Yang, J.; Kostylev, M.A.; Jeng, A.T.; Robinson, S.A.; Gunther, E.C.; van Dyck, C.H.; Nygaard, H.B.; et al. Fyn inhibition rescues established memory and synapse loss in Alzheimer mice. *Ann. Neurol.* **2015**, *77*, 953–971. [CrossRef]
50. Wang, X.-J.; Yu, J.-P.; An, X.; Jia, Z.-W.; Zhang, J.; Su, Y.-X. Attenuation of cartilage pathogenesis in osteoarthritis by blocking the phosphorylation of tyrosine kinase Fyn to β-catenin, AZD0530. *Bone* **2022**, *154*, 116259. [CrossRef]
51. Benemei, S.; Appendino, G.; Geppetti, P. Pleasant natural scent with unpleasant effects: Cluster headache-like attacks triggered by Umbellularia californica. *Cephalalgia Int. J. Headache* **2009**, *30*, 744–746. [CrossRef] [PubMed]
52. Drake, W.M.; Lowe, S.R.; Mirtella, A.; Bartlett, T.J.; Clark, A.J. Desensitisation of calcitonin gene-related peptide responsiveness but not adrenomedullin responsiveness in vascular smooth muscle cells. *J. Endocrinol.* **2000**, *165*, 133–138. [CrossRef] [PubMed]
53. Sun, R.-Q.; Tu, Y.-J.; Lawand, N.B.; Yan, J.-Y.; Lin, Q.; Willis, W.D. Calcitonin Gene-Related Peptide Receptor Activation Produces PKA- and PKC-Dependent Mechanical Hyperalgesia and Central Sensitization. *J. Neurophysiol.* **2004**, *92*, 2859–2866. [CrossRef] [PubMed]
54. Walker, C.S.; Conner Ac Fau-Poyner, D.R.; Poyner Dr Fau-Hay, D.L.; Hay, D.L. Regulation of signal transduction by calcitonin gene-related peptide receptors. *Trends Pharmacol. Sci.* **2010**, *31*, 476–483. [CrossRef]
55. Obara, Y.; Labudda, K.; Dillon, T.J.; Stork, P.J.S. PKA phosphorylation of Src mediates Rap1 activation in NGF and cAMP signaling in PC12 cells. *J. Cell Sci.* **2004**, *117*, 6085–6094. [CrossRef]
56. Yang, H.-B.; Yang, X.; Cao, J.; Li, S.; Liu, Y.-N.; Suo, Z.-W.; Cui, H.-B.; Guo, Z.; Hu, X.-D. cAMP-dependent protein kinase activated Fyn in spinal dorsal horn to regulate NMDA receptor function during inflammatory pain. *J. Neurochem.* **2011**, *116*, 93–104. [CrossRef]

57. Tozzi, A.; de Iure, A.; Di Filippo, M.; Costa, C.; Caproni, S.; Pisani, A.; Bonsi, P.; Picconi, B.; Cupini, L.M.; Materazzi, S.; et al. Critical role of calcitonin gene-related peptide receptors in cortical spreading depression. *Proc. Natl. Acad. Sci. USA* **2012**, *109*, 18985–18990. [CrossRef]
58. Swierczewski, B.E.; Davies, S.J. Developmental regulation of protein kinase A expression and activity in Schistosoma mansoni. *Int. J. Parasitol.* **2010**, *40*, 929–935. [CrossRef]
59. Héroux, M.; Hogue, M.; Lemieux, S.; Bouvier, M. Functional Calcitonin Gene-related Peptide Receptors Are Formed by the Asymmetric Assembly of a Calcitonin Receptor-like Receptor Homo-oligomer and a Monomer of Receptor Activity-modifying Protein-1*. *J. Biol. Chem.* **2007**, *282*, 31610–31620. [CrossRef]
60. Edvinsson, L.; Grell, A.-S.; Warfvinge, K. Expression of the CGRP Family of Neuropeptides and their Receptors in the Trigeminal Ganglion. *J. Mol. Neurosci.* **2020**, *70*, 930–944. [CrossRef]
61. Silva, R.L.; Lopes, A.H.; Guimarães, R.M.; Cunha, T.M. CXCL1/CXCR2 signaling in pathological pain: Role in peripheral and central sensitization. *Neurobiol. Dis.* **2017**, *105*, 109–116. [CrossRef] [PubMed]
62. Khaiboullina, S.F.; Mendelevich, E.G.; Shigapova, L.H.; Shagimardanova, E.; Gazizova, G.; Nikitin, A.; Martynova, E.; Davidyuk, Y.N.; Bogdanov, E.I.; Gusev, O.; et al. Cerebellar Atrophy and Changes in Cytokines Associated with the CACNA1A R583Q Mutation in a Russian Familial Hemiplegic Migraine Type 1 Family. *Front. Cell Neurosci.* **2017**, *11*, 263. [CrossRef] [PubMed]
63. Ma, S.-B.; Xian, H.; Wu, W.-B.; Ma, S.-Y.; Liu, Y.-K.; Liang, Y.-T.; Guo, H.; Kang, J.-J.; Liu, Y.-Y.; Zhang, H.; et al. CCL2 facilitates spinal synaptic transmission and pain via interaction with presynaptic CCR2 in spinal nociceptor terminals. *Mol. Brain* **2020**, *13*, 161. [CrossRef] [PubMed]
64. Takizawa, T.; Qin, T.; Lopes de Morais, A.; Sugimoto, K.; Chung, J.Y.; Morsett, L.; Mulder, I.; Fischer, P.; Suzuki, T.; Anzabi, M.; et al. Non-invasively triggered spreading depolarizations induce a rapid pro-inflammatory response in cerebral cortex. *J. Cereb. Blood Flow Metab. Off. J. Int. Soc. Cereb. Blood Flow Metab.* **2020**, *40*, 1117–1131. [CrossRef] [PubMed]
65. Yamanaka, G.; Suzuki, S.; Morishita, N.; Takeshita, M.; Kanou, K.; Takamatsu, T.; Suzuki, S.; Morichi, S.; Watanabe, Y.; Ishida, Y.; et al. Role of Neuroinflammation and Blood-Brain Barrier Permutability on Migraine. *Int. J. Mol. Sci.* **2021**, *22*, 8929. [CrossRef] [PubMed]
66. Lennerz, J.K.; Rühle, V.; Ceppa, E.P.; Neuhuber, W.L.; Bunnett, N.W.; Grady, E.F.; Messlinger, K. Calcitonin receptor-like receptor (CLR), receptor activity-modifying protein 1 (RAMP1), and calcitonin gene-related peptide (CGRP) immunoreactivity in the rat trigeminovascular system: Differences between peripheral and central CGRP receptor distribution. *J. Comp. Neurol.* **2008**, *507*, 1277–1299. [CrossRef]
67. Eftekhari, S.; Salvatore, C.A.; Calamari, A.; Kane, S.A.; Tajti, J.; Edvinsson, L. Differential distribution of calcitonin gene-related peptide and its receptor components in the human trigeminal ganglion. *Neuroscience* **2010**, *169*, 683–696. [CrossRef]
68. Miller, S.; Liu, H.; Warfvinge, K.; Shi, L.; Dovlatyan, M.; Xu, C.; Edvinsson, L. Immunohistochemical localization of the calcitonin gene-related peptide binding site in the primate trigeminovascular system using functional antagonist antibodies. *Neuroscience* **2016**, *328*, 165–183. [CrossRef]
69. Takeda, M.; Tanimoto, T.; Kadoi, J.; Nasu, M.; Takahashi, M.; Kitagawa, J.; Matsumoto, S. Enhanced excitability of nociceptive trigeminal ganglion neurons by satellite glial cytokine following peripheral inflammation. *Pain* **2007**, *129*, 155–166. [CrossRef]
70. Cao, D.-L.; Zhang, Z.-J.; Xie, R.-G.; Jiang, B.-C.; Ji, R.-R.; Gao, Y.-J. Chemokine CXCL1 enhances inflammatory pain and increases NMDA receptor activity and COX-2 expression in spinal cord neurons via activation of CXCR2. *Exp. Neurol.* **2014**, *261*, 328–336. [CrossRef]
71. Chen, G.; Park, C.-K.; Xie, R.-G.; Berta, T.; Nedergaard, M.; Ji, R.-R. Connexin-43 induces chemokine release from spinal cord astrocytes to maintain late-phase neuropathic pain in mice. *Brain* **2014**, *137*, 2193–2209. [CrossRef] [PubMed]
72. Dansereau, M.-A.; Midavaine, É.; Bégin-Lavallée, V.; Belkouch, M.; Beaudet, N.; Longpré, J.-M.; Mélik-Parsadaniantz, S.; Sarret, P. Mechanistic insights into the role of the chemokine CCL2/CCR2 axis in dorsal root ganglia to peripheral inflammation and pain hypersensitivity. *J. Neuroinflamm.* **2021**, *18*, 79. [CrossRef] [PubMed]
73. Zhang, Y.; Saccani, S.; Shin, H.; Nikolajczyk, B.S. Dynamic Protein Associations Define Two Phases of IL-1β Transcriptional Activation. *J. Immunol.* **2008**, *181*, 503. [CrossRef] [PubMed]
74. Panicker, N.; Sarkar, S.; Harischandra, D.S.; Neal, M.; Kam, T.-I.; Jin, H.; Saminathan, H.; Langley, M.; Charli, A.; Samidurai, M.; et al. Fyn kinase regulates misfolded α-synuclein uptake and NLRP3 inflammasome activation in microglia. *J. Exp. Med.* **2019**, *216*, 1411–1430. [CrossRef] [PubMed]
75. Liu, T.; Zhang, L.; Joo, D.; Sun, S.-C. NF-κB signaling in inflammation. *Signal Transduct. Target. Ther.* **2017**, *2*, 17023. [CrossRef] [PubMed]
76. Veerasubramanian, P.K.; Shao, H.; Meli, V.S.; Phan, T.A.Q.; Luu, T.U.; Liu, W.F.; Downing, T.L. A Src-H3 acetylation signaling axis integrates macrophage mechanosensation with inflammatory response. *Biomaterials* **2021**, *279*, 121236. [CrossRef]
77. Qin, X.; Wan, Y.; Wang, X. CCL2 and CXCL1 trigger calcitonin gene-related peptide release by exciting primary nociceptive neurons. *J. Neurosci. Res.* **2005**, *82*, 51–62. [CrossRef]
78. Viviani, B.; Bartesaghi, S.; Gardoni, F.; Vezzani, A.; Behrens, M.M.; Bartfai, T.; Binaglia, M.; Corsini, E.; Di Luca, M.; Galli, C.L.; et al. Interleukin-1β Enhances NMDA Receptor-Mediated Intracellular Calcium Increase through Activation of the Src Family of Kinases. *J. Neurosci.* **2003**, *23*, 8692. [CrossRef]

79. Davis, N.C.; Mann, E.; Behrens, M.M.; Gaidarova, S.; Rebek, M.; Rebek, J.; Bartfai, T. MyD88-dependent and -independent signaling by IL-1 in neurons probed by bifunctional Toll/IL-1 receptor domain/BB-loop mimetics. *Proc. Natl. Acad. Sci. USA* **2006**, *103*, 2953–2958. [CrossRef]
80. Huang, Y.; Smith, D.E.; Ibáñez-Sandoval, O.; Sims, J.E.; Friedman, W.J. Neuron-Specific Effects of Interleukin-1β Are Mediated by a Novel Isoform of the IL-1 Receptor Accessory Protein. *J. Neurosci.* **2011**, *31*, 18048. [CrossRef]
81. Mon, N.N.; Senga, T.; Ito, S. Interleukin-1β activates focal adhesion kinase and Src to induce matrix metalloproteinase-9 production and invasion of MCF-7 breast cancer cells. *Oncol. Lett.* **2017**, *13*, 955–960. [CrossRef] [PubMed]
82. Bu, F.; Wang, Y.; Jiang, L.; Ma, D.; Quinn, J.P.; Wang, M. Sarcoma family kinase activity is required for cortical spreading depression. *Cephalalgia Int. J. Headache* **2018**, *38*, 1748–1758. [CrossRef] [PubMed]

cells

Review

Exploring the Tryptophan Metabolic Pathways in Migraine-Related Mechanisms

Tamás Körtési [1,2,†], Eleonóra Spekker [2,†] and László Vécsei [2,3,*]

1 Faculty of Health Sciences and Social Studies, University of Szeged, Temesvári krt. 31, H-6726 Szeged, Hungary
2 ELKH-SZTE Neuroscience Research Group, University of Szeged, H-6725 Szeged, Hungary
3 Department of Neurology, Faculty of Medicine, Albert Szent-Györgyi Clinical Center, University of Szeged, H-6725 Szeged, Hungary
* Correspondence: vecsei.laszlo@med.u-szeged.hu
† These authors contributed equally to this work.

Abstract: Migraine is a complex neurovascular disorder, which causes intense socioeconomic problems worldwide. The pathophysiology of disease is enigmatic; accordingly, therapy is not sufficient. In recent years, migraine research focused on tryptophan, which is metabolized via two main pathways, the serotonin and kynurenine pathways, both of which produce neuroactive molecules that influence pain processing and stress response by disturbing neural and brain hypersensitivity and by interacting with molecules that control vascular and inflammatory actions. Serotonin has a role in trigeminal pain processing, and melatonin, which is another product of this pathway, also has a role in these processes. One of the end products of the kynurenine pathway is kynurenic acid (KYNA), which can decrease the overexpression of migraine-related neuropeptides in experimental conditions. However, the ability of KYNA to cross the blood–brain barrier is minimal, necessitating the development of synthetic analogs with potentially better pharmacokinetic properties to exploit its therapeutic potential. This review summarizes the main translational and clinical findings on tryptophan metabolism and certain neuropeptides, as well as therapeutic options that may be useful in the prevention and treatment of migraine.

Keywords: primary headaches; migraine; tryptophan; serotonin pathway; kynurenic pathway; serotonin; melatonin; kynurenic acid; PACAP; CGRP

Citation: Körtési, T.; Spekker, E.; Vécsei, L. Exploring the Tryptophan Metabolic Pathways in Migraine-Related Mechanisms. *Cells* **2022**, *11*, 3795. https://doi.org/10.3390/cells11233795

Academic Editor: Alessandra Pacini

Received: 25 October 2022
Accepted: 24 November 2022
Published: 27 November 2022

1. Migraine

Migraine is one of the most common neurological conditions with a high prevalence and morbidity [1] and is associated with a high economic burden [2]. The estimation of the Migraine Impact Model projected approximately 60,000–686,000 annual workdays as being affected by lost productive time due to migraine and estimated annual indirect costs as totaling 6.2–8.5 times the annual direct costs in USA [3]. Clinically, migraine is characterized by a unilateral throbbing, pulsing headache, associated with various symptoms, such as allodynia, photophobia, and phonophobia, which lasts for hours to days, and the pain has a negative impact on daily activities [4].

Despite extensive research, there are still questions that have not been fully answered about the pathomechanism of migraine; however, translational and clinical trials suggest that activation and sensitization of the trigeminal system (TS) are important during the attacks [5]. The theory of TS constitutes neurovascular incidence, peripheral and central sensitization, and neurogenic inflammation in the dural vessels. According to the literature, the major contributing pathophysiological event thought to initiate migraine is cerebral and meningeal arterial vasodilation. Nevertheless, the role of vasodilation in migraine is not fully understood, and recent findings challenge its necessity. During the attacks, several mediators are released from blood vessels, such as growth factors, cytokines, adenosine triphosphate (ATP), and nitric oxide (NO), which induce local sterile meningeal inflammation [6,7].

Glutamate is an important excitatory neurotransmitter in the central nervous system (CNS), and it plays a role in pain transmission, central sensitization, and cortical spreading depolarization [8–10]. Increased glutamate levels have been noticed in blood and cerebrospinal fluid both interictally and ictally in migraine patients [11–13]; thus, they are involved in migraine pathophysiology.

In addition to glutamate, other neurotransmitters are involved in the development of attacks. Serotonin (5-HT) has a vasoconstrictor effect on blood vessels, thereby affecting nociceptive pain [14]. 5-HT receptors are present in the TS and cranial vessels [15,16], and their agonists, i.e., triptans, are effective for migraine relief [17,18]. Accordingly, neurotransmission mediated by 5-HT is also involved in migraine [19].

Tryptophan is an essential amino acid required for different metabolic reactions among others, such as 5-HT production [20]; however, only a small amount of 5-HT is formed during tryptophan metabolism. The kynurenine pathway (KP) is responsible for 95% of tryptophan metabolism, which is closely related to both glutamatergic and serotonergic mechanisms; thus, the catabolites of this pathway are the focus of migraine research (Figure 1).

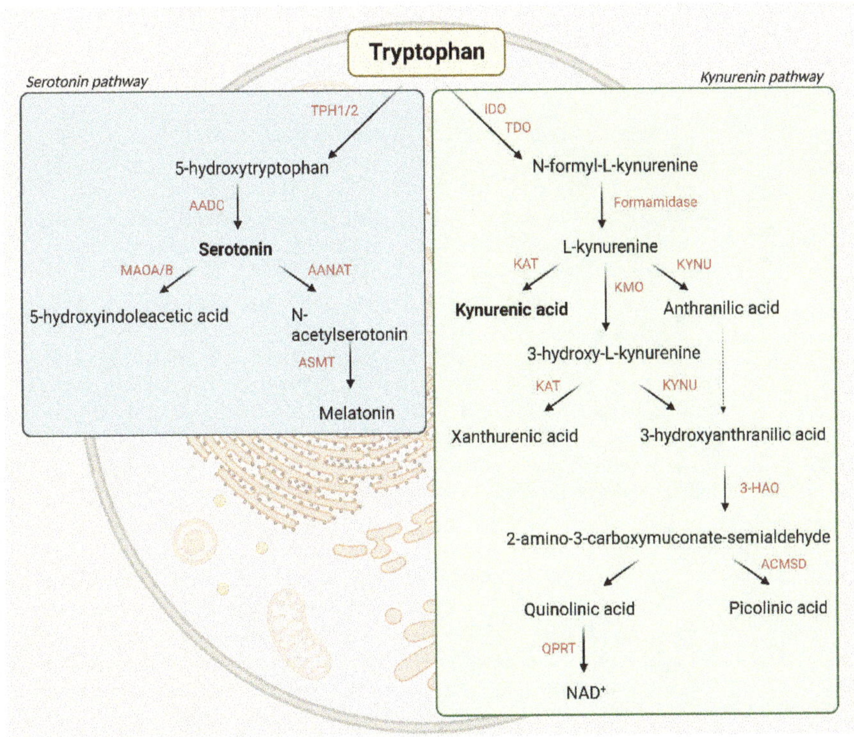

Figure 1. The two main pathways of tryptophan metabolism: serotonin and kynurenine pathways.

This review article summarizes the emerging evidence supporting the involvement of tryptophan metabolism in the pathophysiology of migraine, as well as presents the latest results of preclinical research and the therapeutic possibilities of the disease.

2. Tryptophan and Its Role in Migraine

Tryptophan is an essential amino acid needed to produce and maintain proteins, muscles, enzymes, and neurotransmitters. Changes in tryptophan levels can cause an imbalance in the synthesis of 5-HT and melatonin in the brain and may play a role in the pathophysiology of numerous neuropsychiatric and neurodegenerative disorders [21].

Some research groups observed decreased serum and plasma tryptophan levels in migraine sufferers compared to healthy controls [22,23]. Furthermore, other clinical investigations showed increased tryptophan levels in migraine, especially during the aura phase [24,25]. Similarly, increased serum tryptophan was reported in cluster headaches [26].

Several studies have confirmed a reduction in tryptophan level in the interictal period and an increase in the ictal phase of migraine patients [27,28]. Tryptophan depletion does not trigger migraine attacks but causes lower levels of 5-HT in the brain, which enhance symptoms of migraine [29–31]. In a study, tryptophan depletion induced headache in migraineurs and increased nausea and dizziness. Moreover, ratings of glare and light-induced pain were greater in the tryptophan depletion condition [32]. Consistent with the results above, Jahromi et al. demonstrated that increased tryptophan intake reduces migraine attacks [33,34].

The fact that tryptophan is the precursor of several components that are possibly involved in migraine pathogenesis (e.g., 5-HT and kynurenines) can explain the relationship between tryptophan and migraine (Figure 1).

3. Role of the Tryptophan/Serotonin Pathway in Migraine

5-HT was first identified as a vasoconstrictor present in the blood [35], which constricts blood vessels, thereby potentially modulating nociceptive pain [14,36]. 5-HT receptors can be classified into seven families, which can be further divided into 14 subtypes, all of which are members of the G-protein-coupled receptor family, except the 5-HT$_3$ receptor, which is a ligand-gated ion channel [37]. 5-HT receptors are widely distributed in the CNS, including several areas involved in migraine, such as the striatum, cortex, hippocampus, thalamus, cerebellum, and raphe nuclei [38,39].

Sicuteri et al. were the first to suggest the importance of 5-HT in migraine when they found that, during a migraine attack, the amount of 5-hydroxyindoleacetic acid (5-HIAA), considered the main metabolite of 5-HT, increased in the urine, while the platelet 5-HT concentration decreased [40,41]; these results were confirmed by Curran et al. [42]. Other studies have reported that 5-HT infusion can interrupt spontaneous [43] or reserpine-induced [44] headache. Ren et al. reported low levels of serum serotonin in migraine patients, which was consistent with previous studies [45,46]. Moreover, they also found low levels of tryptophan in these patients [22].

3.1. Serotonin Pathway

The biochemical pathway for 5-HT synthesis initially involves the transformation of L-tryptophan into 5-hydroxytryptophan (5-HTP) by the rate-limiting enzyme L-tryptophan hydroxylase (TPH). 5-HTP is then decarboxylated to become 5-HT via the action of the cytosolic enzyme L-aromatic amino acid decarboxylase (AADC) [47,48]. Extracellular 5-HT enters the cells using the serotonin transporter (5-HTT), and excess 5-HT is metabolized. The metabolism of serotonin is primarily carried out by the outer mitochondrial membrane enzyme, monoamine oxidase (MAO) [49,50]. Finally, with the help of an aldehyde dehydrogenase enzyme, it is converted into 5-HIAA, which is excreted in the urine [47] (Figure 2).

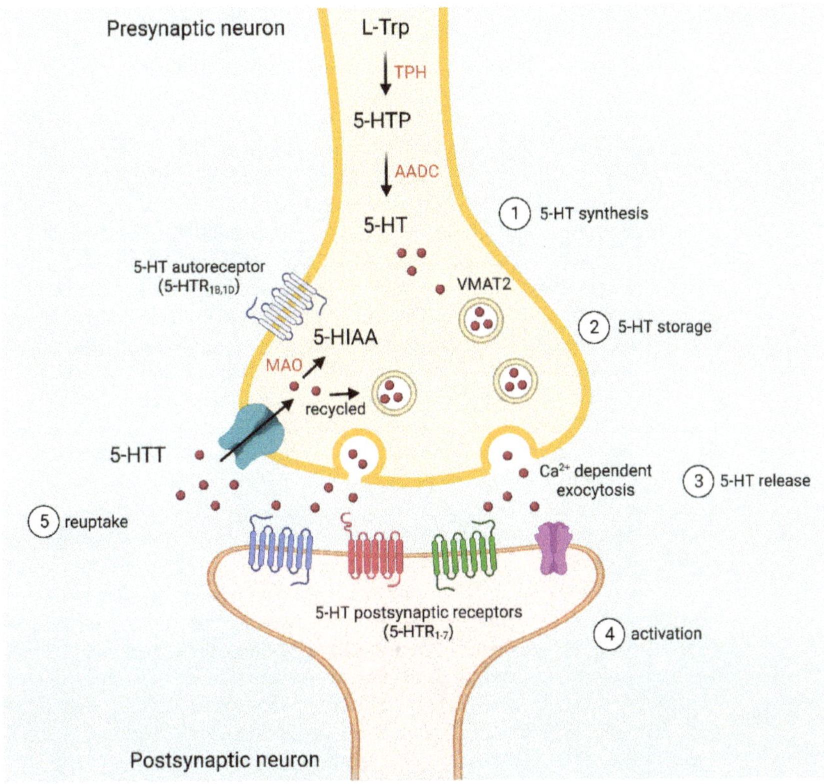

Figure 2. Serotonin pathway. (1) L-tryptophan is converted to 5-HT by TPH and AADC enzymes. (2) 5-HT is then taken up into vesicles in the axon terminal via VMAT2. (3) After an action potential, 5-HT is released into the synapse. 5-HT can also interact with presynaptic and postsynaptic receptors. (4) All 5-HT receptors are post-synaptically expressed on non-serotonergic neurons, and autoreceptors are located pre-synaptically on the serotonergic neurons. (5) Free 5-HT is removed from the synapse by 5-HTT, which controls the extent and duration of 5-HT receptor activation. Furthermore, 5-HT can be metabolized by MAO and aldehyde dehydrogenase into 5-HIAA, which is excreted in the urine. L-Trp: L-tryptophan, TPH: L-tryptophan hydroxylase, AADC: L-aromatic amino acid decarboxylase, 5-HTP: 5-hydroxytryptophan, 5-HT: serotonin, VMAT2: vesicular monoamine transporter isoform 2, 5-HTT: serotonin transporter, 5-HIAA: 5-hydroxyindoleacetic acid, MAO: monoamine oxidase.

3.2. Serotonin Transporter

5-HTT retakes 5-HT from the synaptic gap to the presynaptic terminals, thereby reducing the effect of 5-HT. The transport process is controlled by the $Na+/Cl^-$ ion gradient [51]. 5-HTT occurs mainly in the area of the raphe nuclei and serotonergic projection areas (e.g., cortical areas, thalamus, hippocampus CA3 region, and amygdala) [52]. Imaging studies have established that the distribution of 5-HTT in the brain stem area is greater in migraine patients [53]. It has been observed that familial hemiplegic migraine (FHM) patients have a low level of 5-HT in platelets, and it has also been described that the 5-HT transport capacity is low. In addition, reduced metabolite levels in cerebrospinal fluid were observed in these patients [54].

3.3. Serotonin Receptors

5-HT receptors are important in the regulation of serotonergic neurotransmission, and they play a distinguished role in several behavioral and physiological functions [55]. In previous studies, it was observed that the neurons of the dorsal raphe and the trigeminal ganglia (TG) are mostly serotonergic [16,56].

In humans, it has been demonstrated that both receptor $5HT_{1B}$ and $5HT_{1D}$ subtypes are present in trigeminal neurons [57,58], and both receptors have been detected at mRNA and protein levels in the TG [59] and colocalize with calcitonin gene-related peptide (CGRP), substance P (SP), and nitric oxide synthases (NOS) [58].

Triptans are $5\text{-}HT_{1B/1D}$ agonists with some affinity for the $5\text{-}HT_{1F}$ receptor subtype, and they are clinically effective anti-migraine drugs. They can have an inhibitory effect on the trigeminal sensory fibers, which is attributed to the inhibition of endogenous CGRP and SP release [60]. The efficacy of triptans also suggests that 5-HT may modulate the pathogenesis of migraine. Unfortunately, triptans are contraindicated in patients with high blood pressure and cardiovascular or cerebrovascular disease due to their vasoconstrictive effect. In addition, these drugs are not effective for everyone, often leading to excessive drug use, which eventually causes migraines to become chronic [61].

These facts led to the development of ditans, the new class of selective $5\text{-}HT_{1F}$ receptor agonists that do not have vasoconstrictive properties [62,63]. The $5\text{-}HT_{1F}$ receptor is expressed in several brain areas involved in migraine attacks, such as the cortex, the hypothalamus, the trigeminal ganglia, the trigeminal nucleus caudalis (TNC), the locus coeruleus, the middle cerebral artery, and the upper cervical cord [64,65]. Several selective $5\text{-}HT_{1F}$ receptor agonists have been developed in the past years; in preclinical studies, they could successfully inhibit dural extravasation after TG stimulation and hinder neuronal activation in the TNC following trigeminovascular activation [66–69]. However, only lasmiditan can currently be used as anti-migraine therapy, but it has no therapeutic gain over triptans. Lasmiditan can cross the BBB and, thus, exert its effects centrally on the trigeminovascular system; however, at the same time, it also has a peripheral effect, via $5\text{-}HT_{1F}$ receptors expressed on trigeminal afferents or TG [70]. Lasmiditan can probably moderate the activation of Sp5C second-order trigeminal neurons, which has an important role in the pathomechanism of migraine [71,72].

$5\text{-}HT_{2B}$ receptors can influence the release of 5-HT through the 5-HTT and are also involved in the normal physiological regulation of blood plasma 5-HT levels [73]. In rats, $5\text{-}HT_{2B}$ receptors are slightly expressed in neurons located in the cerebellum, the posterior hypothalamus, the lateral septum, the medial amygdala, the spinal cord, and the dorsal root ganglion (DRG). Unlike the $5\text{-}HT_1$ receptor, it seems that the $5\text{-}HT_{2B}$ receptors do not inhibit/decrease the release of neuropeptides involved in migraine (CGRP, glutamate) from trigeminal neurons [74]. Indeed, the $5\text{-}HT_{2B}$ receptor can activate NOS, which promotes the synthesis of NO [75], a potentially key component in the development of a migraine attack. In guinea pigs, acute activation of $5\text{-}HT_{2B}$ receptors by m-chlorophenylpiperazine (mCPP) led to NO-dependent plasma protein extravasation (PPE) in the dura mater and neuronal activation in the TNC, which could be inhibited by selective $5\text{-}HT_{2B}$ receptor antagonists [76–78]. In humans, mCPP, with $5\text{-}HT_{2B/2C}$ receptor affinity, leads to delayed migraine-like headaches in migraine sufferers and nonspecific headaches in healthy subjects [79]. Methysergide, a $5\text{-}HT_{2B}$ antagonist, can reduce the frequency of migraine, but it has to be used for a longer period to exert its therapeutic effect [80]. Johnson et al. reported that, after electrical stimulation of the TG, LY202146, a selective $5\text{-}HT_{2B}$ receptor antagonist, failed to inhibit protein extravasation [77], suggesting that the $5\text{-}HT_{2B}$ receptor may play a role in triggering the migraine attack, but is not related directly to the release of peptides from trigeminal neurons. These observations resemble the results obtained in clinical research where effective preventive agents, such as methysergide and pizotifen, could not inhibit the onset of a migraine attack.

On this basis, it was suggested that meningeal $5\text{-}HT_{2B}$ receptors may play a role in the onset of migraine attacks (Figure 3).

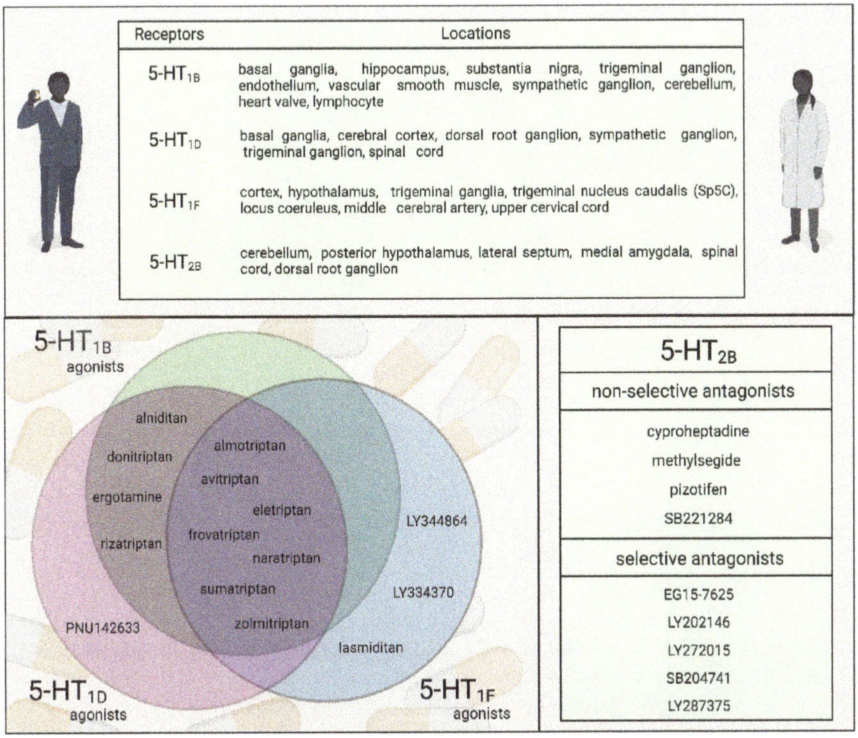

Figure 3. 5-HT receptors and their relevance in migraine therapy.

3.4. Melatonin

Melatonin is a tryptophan metabolite that plays a role in regulating circadian rhythms, and numerous studies have demonstrated that melatonin can exert its anti-migraine effect in several ways. Melatonin can regulate neurotransmitters and neural pathways; it can inhibit the synthesis of NO, as well as the release of CGRP and dopamine, and it can antagonize glutamate-induced excitotoxicity [81–84]. Furthermore, it has an anti-free radical effect and inhibits the release of inflammatory factors [85]. It is supported by many studies that melatonin has a role in pain transmission and sensitization [84,86–89]. Membrane melatonin receptors (MT1 and MT2) have been identified in the thalamus, dorsal horn of the spinal cord, trigeminal tract, and trigeminal nucleus, which are involved in nociceptive transmission [90,91].

Melatonin can increase the release of β-endorphin from the pituitary gland and interacts with opioidergic, muscarinic, nicotinic, serotonergic, and α1 and α2-adrenergic receptors located in the CNS and the dorsal horn of the spinal cord; thus, it may be able to exert an analgesic effect [92–94] (Figure 4). In fibromyalgia, inflammatory bowel syndrome, and migraine, melatonin was able to reduce pain [95–97]. In another study, melatonin treatment was able to modify the central level of brain-derived neurotrophic factor (BDNF) in rats submitted to acute and chronic inflammation [98].

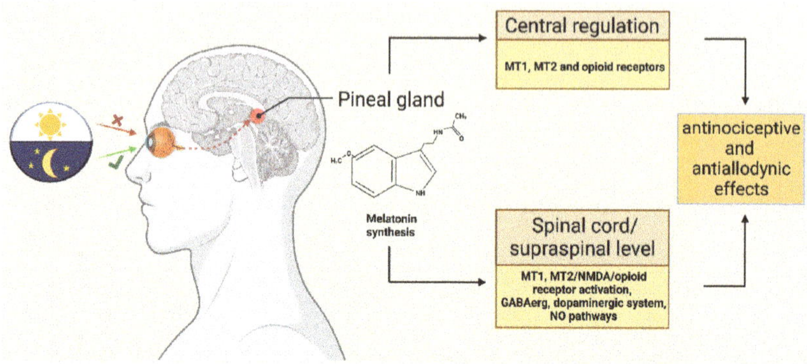

Figure 4. Melatonin synthesis and its anti-nociceptive and anti-allodynic effects. Melatonin can probably induce an anti-nociceptive effect through the regulation of MT1/MT2 receptors in the spinal cord and brain. It also interacts with other receptors such as NMDA, opioids, the dopaminergic system, the GABAergic system, and the NO pathway to exert anti-nociceptive and anti-allodynic effects. MT1/2: melatonin receptor 1/2, NMDA: N-methyl-D-aspartate, GABA: gamma-aminobutyric acid, NO: nitric oxide.

Masruha and colleagues found low levels of 6-sulfatoxymelatonin—a urinary metabolite of melatonin—in migraine patients [99]. Previous studies found low levels of melatonin in episodic [100] and chronic [101] migraine patients. Murialdo et al. found that, during the luteal phase, migraineurs showed a less pronounced change in melatonin levels than controls. Melatonin secretion was further decreased during migraine attack [102]. In line with these observations, Brun et al. found significantly lower melatonin levels in women with migraine during the cycle, while healthy participants showed a significant increase in melatonin secretion from the follicular to the luteal phase [103].

According to these data, it is possible that melatonin may be beneficial in migraine prophylaxis.

4. Role of Tryptophan/Kynurenine Pathway in Migraine

The role of the tryptophan/kynurenine metabolic pathway is receiving more attention in various illnesses including migraine [48]. In parallel to 5-HT synthesis, the central route of the tryptophan metabolism is the KP [104].

4.1. Kynurenine Pathway

The transformation process of tryptophan into N-formyl-L-kynurenine is carried out by two rate-limiting enzymes: tryptophan-2,3-dioxygenase (TDO) and indoleamine-2,3-dioxygenase (IDO). N-formyl-L-kynurenine is degraded by formamidase to L-kynurenine (L-KYN). L-KYN can be metabolized into kynurenic acid (KYNA), 3-hydroxy-L-kynurenine (3-HK), or anthranilic acid (AA) under the action of kynurenine aminotransferase (KAT), kynurenine-3-monooxygenase (KMO), and kynureninase (KYNU) enzymes. 3-HK can be further converted to xanthurenic acid (XA) by KAT or to 3-hydroxyanthranilic acid (3-HANA) by KYNU. 3-Hydroxyanthranilic acid is then metabolized by 3-hydroxyanthranilate oxidase (3-HAO) to 2-amino-3-carboxymuconate-semialdehyde, which is transformed into picolinic acid (PIC) or quinolinic acid (QUIN). In the last step of the KP, QUIN is converted into the coenzymes nicotinamide adenine dinucleotide (NAD) and nicotinamide adenine dinucleotide phosphate (NADP) [105] (Figure 5).

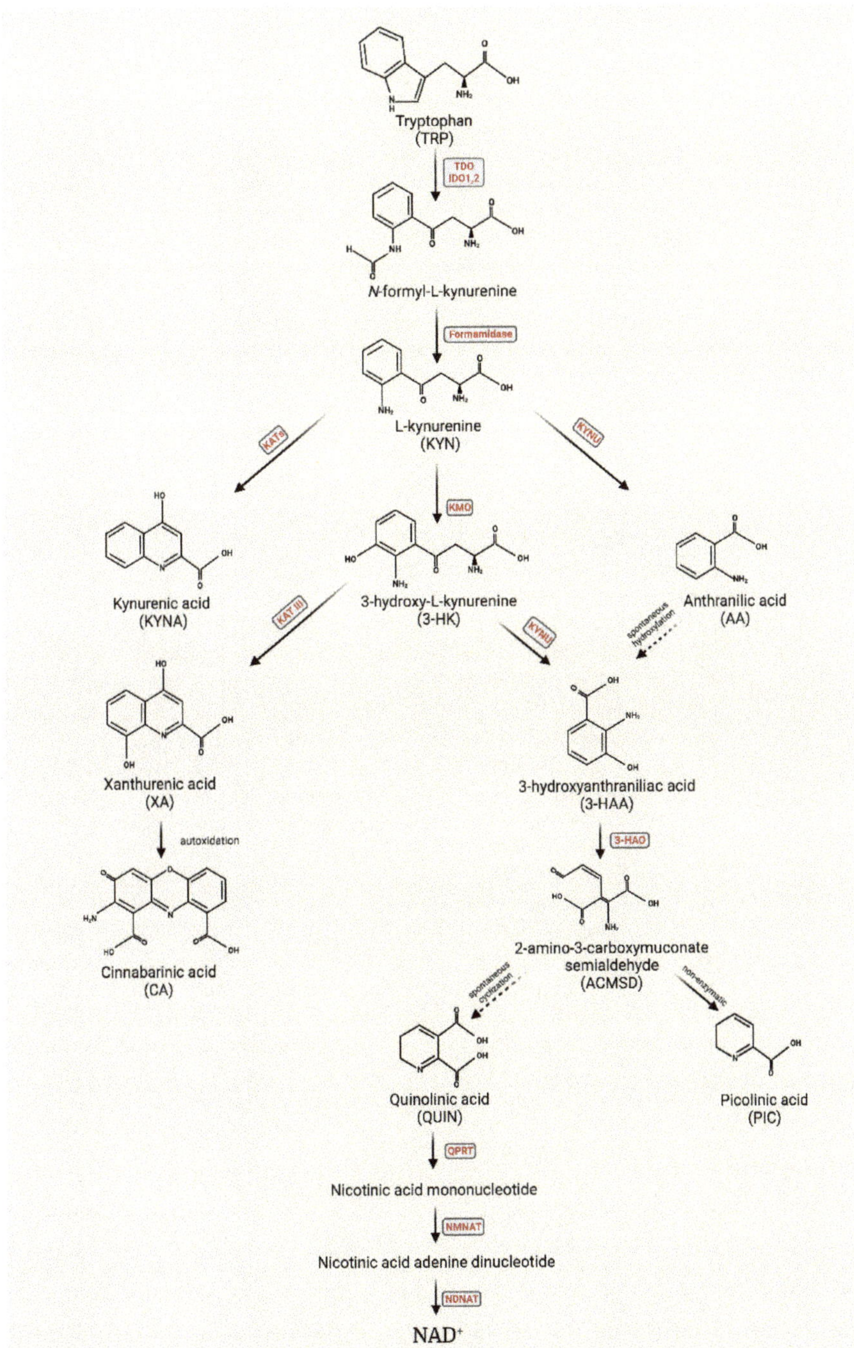

Figure 5. Kynurenine pathway.

4.2. Kynurenines

KP produces neuroactive metabolites which have a role in the modification of the trigemino-vascular activation processes and can interact with glutamate receptors in the CNS [106]; therefore, they may be involved in the pathophysiology of migraine.

Among the kynurenines, KYNA should be mentioned, which can act through N-methyl-D-aspartate (NMDA), α-amino-3-hydroxy-5-methyl-4-isoxazole propionic acid (AMPA), kainate receptors, and G-protein-coupled receptor 35 (GPR35), and these receptors have a major role in pain processing and neuroinflammation [105]. Experimental data suggest, that in the brain, an increased level of KYNA has neuroprotective effects [107,108]. Additionally, in an animal model of migraine, KYNA was able to inhibit trigemino-vascular activation [109,110]. Furthermore, KYNA can modulate the activation of migraine generators and inhibits cortical spreading depression (CSD) [19]. Oláh and colleagues reported that, in rats, peripherally administered KYNA was able to reduce the number of CSD waves; moreover, it decreased the permeability of the blood–brain barrier (BBB) during CSD [111]. Knyihár-Csillik et al. reported reduced KAT expression after the electrical stimulation of the TG [112]. Moreover, Spekker et al. found that inflammatory soup was able to cause sterile neurogenic inflammation in the dura mater and increased the area covered by CGRP and transient receptor potential vanilloid 1 (TRPV1) immunoreactive fibers, as well as the number of neuronal nNOS-positive cells in the caudal trigeminal nucleus, and pretreatment with KYNA was able to modulate the changes caused by inflammatory soup. KYNA probably inhibited the glutamate system, thereby preventing the sensitization processes which are key actors in migraine [113].

It has been reported that KYNA has anti-nociceptive effects in both the first- and second-order trigeminal nociceptors. Zhang et al. found that KYNA dose-dependently suppressed carrageenan-induced thermal hyperalgesia and significantly reduced c-fos expression in both the superficial and the deep laminae of the dorsal horn in rats [114]. In another study, after carrageenan injection into the tibio-tarsal joint, locally administered KYNA was able to abolish allodynia and cause anti-nociception [115].

The therapeutic use of KYNA is hampered by the fact that it is difficult to cross the BBB [116]. The development of KYNA analogs with retained or modified activity can be a solution to this problem. These compounds are promising because they are capable of selectively inhibiting NMDA receptors containing the NR2B subunit, which play a role in the modulation of pain perception.

L-KYN is the source of all the other kynurenine metabolites, and it is readily transported across the BBB [116]. L-KYN in combination with probenecid can prevent nitroglycerin (NTG)-induced changes in c-fos expression in rat TNC [109]. Peripheral treatment with L-KYN can dose-dependently enhance the concentration of KYNA in the brain; thus, it may provide a possible therapeutic solution for the treatment of several neurological disorders, including primary headaches. However, the physiological effect and safety of L-KYN in vivo in humans are still awaiting clarification.

5. Neuropeptides in Migraine

5.1. Pituitary Adenylate Cyclase-Activating Polypeptide

Some neuropeptides play a role in neurogenic inflammation, thereby activating TS. The pituitary adenylate cyclase-activating polypeptide (PACAP) is a member of the vasoactive intestinal peptide (VIP)/secretin/glucagon peptide family [117]. PACAP is widely expressed in the human body, with extensive effects [118]. Literature data prove that this peptide plays roles such as neuromodulation [119] and neuroprotection [120], in addition to antiapoptotic effects [121] and differentiation-inducing effects in the developing nervous system [122]. The peptide has two biologically active forms; PACAP1-38, which consists of 38 amino acids, and PACAP1-27, which contains 27 amino acids at its N-terminus. These are produced by alternative splicing from the PACAP precursor, preproPACAP [123,124]. The effects of PACAP are mediated through three G-protein-coupled receptors: VPAC1, VPAC2, and PAC1. The latter is a high-affinity and PACAP-selective receptor, while VPAC1 and VPAC2 receptors show a comparable affinity to PACAP and VIP [125]. The PAC1 receptor has been shown to play crucial roles in the functioning of the nervous system. The activation of this receptor induces numerous signal transduction cascades, including phospholipase C, adenylyl cyclase, MEK/extracellular signal-regulated kinase (ERK), and Akt pathways that regulate a number of physiological systems to maintain functional homeostasis [126,127]. Previous studies evidence that, through PACAP activation, PAC1 receptor-mediated pathways are implicated in a number of disorders including depression, posttraumatic stress disorder, metabolic abnormalities, chronic pain, and migraine [128].

In recent years, several clinical investigations have reflected the possible relevance of PACAP in migraine. In experimental conditions, intraperitoneal administration of PACAP1-38 evoked notable photophobia and meningeal vasodilatation, as well as increased the number of c-fos-positive activated neurons in the brainstem in wildtype, but not in PACAP1-38-deficient mice [129]. Elevation of PACAP1-38 concentration was also detected in the brainstem after the activation of the TS in different animal models. The intraperitoneal administration of NTG also provoked an increase in PACAP1-38 and PACAP1-27 expression 3 h after the treatment in the TNC [130]. Furthermore, electrical stimulation of trigeminal ganglion (ES-TG) resulted in significantly increased PACAP1-38 immunoreactivity 3 h after ES-TG of the plasma and PACAP1-38 and PACAP1-27 immunoreactivity in the TNC [130]. The endogenous antagonists of NMDA receptor, KYNA and its synthetic analog SZR-72, were able to inhibit overexpression of PACAP at both the proteome and transcriptome levels, suggesting that KYNA and SZR-72 is a potential new drug candidate for PACAP-targeted headache therapy in the future.

In patients suffering from migraines, the level of PACAP1-38 in the blood is increased during the migraine attack compared to the interictal period, suggesting a potential biomarker function of peptide in the disease [131]. Furthermore, intravenous administration of PACAP1-38 provoked headache and vasodilatation, in both healthy participants and migraine sufferers, whereas it delayed migraine-like attacks only in migraineurs [132,133]. In migraineurs without aura, the development of PACAP1-38-induced migraine-like attack was independent of the severity of the family load [134]. In the same study, 90 min after the PACAP treatment, the levels of numerous markers relevant to the disease (such as VIP, prolactin, S100B, and thyroid-stimulating hormone (TSH)) were increased in the plasma [134]. Correlation was shown between the microstructural integrity of the white matter and the interictal plasma PACAP1-38 immunoreactivity in migraineurs [135]. In addition, magnetic resonance imaging angiography examinations revealed that PACAP1-38 evoked headache was associated with prolonged vasodilatation of the middle meningeal artery (MMA), but not the middle cerebral artery (MCA) [136]. The anti-migraine drug sumatriptan was able to alleviate the headache, which mirrored the contraction of the MMA, but not the MCA, suggesting that PACAP1-38-induced headaches may arise from the extracerebral arteries.

5.2. Calcitonin Gene-Related Peptide

The "old warrior" CGRP is another pathogenic factor in the pathomechanism of migraine. A previous study confirmed that the expression of CGRP and SP was elevated during ES-TG of the external jugular vein of cats [137]. In addition to PACAP, CGRP can activate mast cells, leading to the secretion of vasoactive, proinflammatory, and neurosensitizing mediators, thereby contributing to the activation of TS [138,139] PACAP1-38 administration can cause increased CGRP expression in the brainstem, suggesting a possible link between CGRP and PACAP1-38 release [140]. CGRP and PACAP show co-expression; 23% of the neurons expressed both CGRP and PACAP1-38 in rat TRG, and CGRP (49%) was expressed in more neurons compared to PACAP1-38 (29%) [141]. In an experimental model of migraine, the simultaneous release of these neuropeptides was detected; a chronic NTG injection caused elevated concentrations of CGRP and PACAP in the plasma of rats, while the intervention resulted in mechanical and thermal hyperalgesia [142]. These data are consistent with our experimental results; orofacial complete Freund's adjuvant (CFA) treatment caused significant CGRP and PACAP release in the brainstem. This elevation showed correlation with the mechanical hyperalgesia of animals [143]. However, activation of the TS is possible with different CFA treatments, which eventuates pain-associated pathological states, including migraine, neuralgias, and temporomandibular joint (TMJ) disorders [144]. A recent study examined the effect of CFA on the mitogen-activated protein kinase (MAPK) expression, which has a major role in the pain-related process. Administration of CFA in the TMJ resulted in significant ERK1/2 and p38 MAPK elevation in the TG [145]. Dural administration of CFA increased the expression of ERK1/2, interleukin-1 (IL-1β), and CGRP in the TG. In addition, high glutamate and c-fos immunoreactivity was observed in TNC and cervical neurons [146]. Following orofacial CFA treatment in the TG and TNC, CGRP, ionized calcium-binding adapter molecule 1 (Iba1) and glial fibrillary acidic protein (GFAP) gene expression changes were revealed, reflecting that CFA-induced neuroinflammation induces elevated CGRP and PACAP1-38 levels. [147]. Despite the similarities between CGRP and PACAP, experimental investigations suggest that these neuropeptides act independently, increasing their future therapeutic potential [148,149].

5.3. The Relationship between Neuropeptides and the Kynurenine System

The main mediator of CGRP and PACAP gene expression is intracellular calcium homeostasis. In addition to the action of voltage-dependent calcium channels, the main inducer of the gene expression of these peptides is calcium influx through the NMDA receptors. KYNA and its analogs can block the NMDA receptors, thereby moderating the amount of calcium coming into the cell, which may result in decreased CGRP and PACAP gene expression. Since KYNA and its analogs can decrease migraine-related neuropeptides expression, targeting CGRP and PACAP with KYNA may have a therapeutic role in the future (Figure 6). In a previous experiment by Körtési et al., the expression levels of PACAP were significantly different between the uncompetitive antagonist of the NMDA glutamate receptor MK-801 and SZR-72 treatment groups, raising the possibility of the involvement of KYNA targets other than NMDA. In addition to NMDA receptor, KYNA has an effect on the AMPA, kainate, aryl hydrocarbon, GPR35, and opiate receptors. Regarding SZR-72, investigations are in process, but the exact targets and mechanisms of this analog have not yet been identified [127].

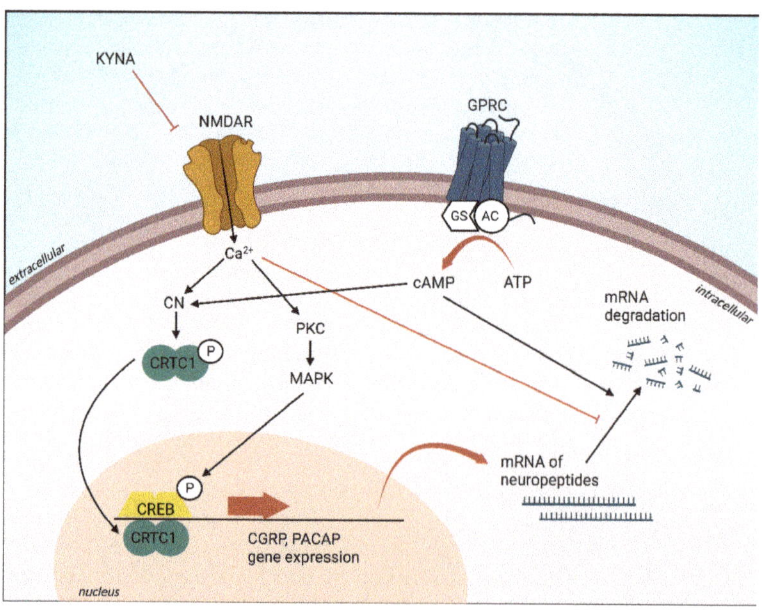

Figure 6. Proposed regulation of CGRP and PACAP gene expression. AC: adenylate cylase, ATP:

adenosine monophosphate, CaM: calmodulin, cAMP: cyclic adenosine monophosphate, CGRP: calcitonin gene-related peptide CN: calcineurin, CREB: cAMP response element-binding protein, CRTC1: CN/Cre-binding protein, GPCR: G-protein-coupled receptor, Gs: stimulatory G protein, KYNA: kynurenic acid, MAPK: mitogen-activated protein kinase, NMDAR: NMDA receptor, PACAP: pituitary adenylate cyclase activating polypeptide PKC: protein kinase C.

6. Clinical Studies

Several lines of evidence suggest that imbalance of the kynurenine pathway plays roles in several diseases [105]. Several preclinical studies reflect a link between the kynurenine pathway and migraine. Indeed, numerous studies have demonstrated that the NMDA receptor inhibitor KYNA and its analogs have anti-nociceptive effects at the levels of both first- and second-order sensory neurons [19]. KYNA and one of its derivatives both decreased the levels of several inflammatory mediators in the animal model of CFA-induced TS activation [145]. The effects of two KYNA analogs were tested in the orofacial formalin model, revealing that both were able to inhibit the formalin-induced behavioral and morphological changes, as well as increase the concentration of KYNA in the rat brainstem [150]. Notably, systemic administration of NTG decreased the expression of KAT II in the TS of rats, an enzyme catalyzing the transformation of L-KYN to KYNA [151]. In line with this, in another model of TS activation, decreased KAT immunoreactivity was observed in mast cells, Schwann cells, and dural macrophages [113]. In addition to preclinical studies, clinical results have provided evidence for the connection between the kynurenine system and various headache disorders, including migraine or cluster headache. Indeed, in patients suffering from primary headache disorders, alterations of the kynurenine pathway were observed, which, among others, manifested in the reduction in KYNA concentration in the serum [25,26]. A clinical study also proved that plasma concentrations of most tryptophan metabolites were remarkably decreased in the interictal period of migraineurs compared to healthy control subjects, especially in the migraine without aura subgroup (tryptophan, L-KYN, KYNA, ANA, PIC, 5-HIAA, and melatonin). In patients suffering from migraine without aura, several metabolites showed a tendency to elevate during the ictal phase, but this was significant only in the cases of ANA, 5-HIAA, and melatonin [152]. A clinical phase I investigation proved that intravenous administration of L-KYN is safe and well tolerated. The lack of change in kynurenine metabolites in plasma reflects a relatively slow metabolism of L-KYN and no or little feed-back effect of this metabolite on its synthesis [153].

7. Conclusions

The aim of the present work was to draw some attention to the role of different tryptophan catabolites; furthermore, we were able to gain insight into the role of various neuropeptides in the pathomechanism of migraine. The serotonin and kynurenine pathways are closely connected, and alterations in one arm of the pathway may influence the other. Tryptophan metabolites play an important role in primary headaches, and they can be potential therapeutic targets in the treatment of the migraine and other primary headaches.

Neuropeptides, CGRP and PACAP in particular, are implicated in trigeminal activation. The expression of CGRP and PACAP and its receptors, and their main effects and mechanisms in the nociceptive pathways suggest that these neuropeptides have a special role in migraine. Identification of their molecular mechanisms might open up future perspectives for the development of novel analgesic drugs.

Author Contributions: Conceptualization, T.K. and E.S.; writing—original draft preparation, E.S. and T.K.; writing—review and editing, E.S., T.K. and L.V.; visualization, E.S.; supervision, L.V.; project administration, L.V.; funding acquisition, L.V. All authors have read and agreed to the published version of the manuscript.

Funding: Tamás Körtési's was supported by the ÚNKP-22-4 New National Excellence Program of the Ministry for Culture and Innovation from the source of the National Research, Development, and Innovation Fund. This research was also funded by the National Research, Development and Innovation Office—NKFIH 138125.

Acknowledgments: All figures were created using BioRender.com.

Conflicts of Interest: The authors declare no conflict of interest.

Abbreviations

3-HANA	3-hydroxyanthranilic acid
3-HAO	3-hydroxyanthranilate oxidase
3-HK	3-hydroxy-L-kynurenine
5-HIAA	5-hydroxyindoleacetic acid
5-HT	serotonin
5-HTP	5-hydroxytryptophan
5-HTT	serotonin transporter
AA	anthranilic acid
AADC	L-aromatic amino acid decarboxylase
AMPA	α-amino-3-hydroxy-5-methyl-4-isoxazole propionic acid
ATP	adenosine triphosphate
BBB	blood–brain barrier
BDNF	brain-derived neurotrophic factor
CFA	complete Freund's adjuvant
CGRP	calcitonin gene-related peptide
CNS	central nervous system
CSD	cortical spreading depression
DRG	dorsal root ganglion
ERK	extracellular signal-regulated kinase
FHM	familial hemiplegic migraine
GABA	gamma-aminobutyric acid
GFAP	glial fibrillary acidic protein
GPR35	G-protein-coupled receptor 35
Iba1	ionized calcium-binding adapter molecule 1
IDO	indoleamine-2,3-dioxygenase
IL-1 β	interleukin-1
KAT	kynurenine aminotransferase
KMO	kynurenine-3-monooxygenase
KP	kynurenine pathway
KYNA	kynurenic acid
KYNU	kynureninase
L-KYN	L-kynurenine
MAO	monoamine oxidase
MAPK	mitogen-activated protein kinases
MCA	middle cerebral artery
mCPP	m-chlorophenylpiperazine
MMA	middle meningeal artery
MT1/2	melatonin receptors 1/2
NAD	nicotinamide adenine dinucleotide
NADP	nicotinamide adenine dinucleotide phosphate
NMDA	N-methyl-D-aspartate
NO	nitric oxide
NOS	nitric oxide synthases
NTG	nitroglycerin
PACAP	pituitary adenylate cyclase-activating polypeptide

PIC	picolinic acid
PPE	plasma protein extravasation
QUIN	quinolinic acid
SP	substance P
TDO	tryptophan-2,3-dioxygenase
TG	trigeminal ganglia
TMJ	temporomandibular joint
TNC	trigeminal nucleus caudalis
TPH	L-tryptophan-hydroxylase
TRP	tryptophan
TRPV1	transient receptor potential vanilloid 1
TS	trigeminal system
TSH	thyroid-stimulating hormone
VIP	vasoactive intestinal peptide
VMAT2	vesicular monoamine transporter isoform 2
XA	xanthurenic acid

References

1. Steiner, T.J.; Stovner, L.J.; Vos, T. GBD 2015: Migraine is the third cause of disability in under 50s. *J. Headache Pain* **2016**, *17*, 104. [CrossRef] [PubMed]
2. Cerbo, R.; Pesare, M.; Aurilia, C.; Rondelli, V.; Barbanti, P. Socio–economic costs of migraine. *J. Headache Pain* **2001**, 2 (Suppl. S1), s15–s19. [CrossRef]
3. Yucel, A.; Thach, A.; Kumar, S.; Loden, C.; Bensink, M.; Goldfarb, N. Estimating the Economic Burden of Migraine on US Employers. *Am. J. Manag. Care* **2020**, *26*, e403–e408. [CrossRef]
4. Headache Classification Committee of the International Headache Society (IHS). The International Classification of Headache Disorders, 3rd edition (beta version). *Cephalalgia* **2013**, *33*, 629–808. [CrossRef]
5. Goadsby, P.; Holland, P.; Martins-Oliveira, M.; Hoffmann, J.; Schankin, C.; Akerman, S. Pathophysiology of Migraine: A Disorder of Sensory Processing. *Physiol. Rev.* **2017**, *97*, 553–622. [CrossRef]
6. Edvinsson, L. Tracing neural connections to pain pathways with relevance to primary headaches. *Cephalalgia* **2011**, *31*, 737–747. [CrossRef]
7. Spekker, E.; Tanaka, M.; Szabó, A.; Vécsei, L. Neurogenic Inflammation: The Participant in Migraine and Recent Advancements in Translational Research. *Biomedicines* **2021**, *10*, 76. [CrossRef]
8. Lauritzen, M.; Hansen, A.J. The Effect of Glutamate Receptor Blockade on Anoxic Depolarization and Cortical Spreading Depression. *J. Cereb. Blood Flow Metab.* **1992**, *12*, 223–229. [CrossRef]
9. Kramer, D.R.; Fujii, T.; Ohiorhenuan, I.; Liu, C.Y. Cortical spreading depolarization: Pathophysiology, implications, and future directions. *J. Clin. Neurosci.* **2016**, *24*, 22–27. [CrossRef]
10. Hoffmann, J.; Charles, A. Glutamate and Its Receptors as Therapeutic Targets for Migraine. *Neurotherapeutics* **2018**, *15*, 361–370. [CrossRef] [PubMed]
11. Ferrari, M.D.; Odink, J.; Bos, K.D.; Malessy, M.; Bruyn, G.W. Neuroexcitatory plasma amino acids are elevated in migraine. *Neurology* **1990**, *40*, 1582–1586. [CrossRef] [PubMed]
12. Martínez, F.; Castillo, J.; Rodríguez, J.R.; Leira, R.; Noya, M. Neuroexcitatory Amino Acid Levels in Plasma and Cerebrospinal Fluid During Migraine Attacks. *Cephalalgia* **1993**, *13*, 89–93. [CrossRef] [PubMed]
13. Peres, M.; Zukerman, E.; Soares, C.S.; Alonso, E.; Santos, B.; Faulhaber, M. Cerebrospinal Fluid Glutamate Levels in Chronic Migraine. *Cephalalgia* **2004**, *24*, 735–739. [CrossRef] [PubMed]
14. Taylor, B.K.; Basbaum, A.I. Neurochemical Characterization of Extracellular Serotonin in the Rostral Ventromedial Medulla and Its Modulation by Noxious Stimuli. *J. Neurochem.* **1995**, *65*, 578–589. [CrossRef]
15. Rebeck, G.W.; Maynard, K.I.; Hyman, B.T.; Moskowitz, M.A. Selective 5-HT1Da serotonin receptor gene expression in trigeminal ganglia: Implications for antimigraine drug development. *Proc. Natl. Acad. Sci. USA* **1994**, *91*, 3666–3669. [CrossRef] [PubMed]
16. Berman, N.E.; Puri, V.; Chandrala, S.; Puri, S.; MacGregor, R.; Liverman, C.S.; Klein, R.M. Serotonin in Trigeminal Ganglia of Female Rodents: Relevance to Menstrual Migraine. *Headache* **2006**, *46*, 1230–1245. [CrossRef] [PubMed]
17. Saxena, P.; De Vries, P.; Villalón, C. 5-HT 1-like receptors: A time to bid goodbye. *Trends Pharmacol. Sci.* **1998**, *19*, 311–316. [CrossRef]
18. Cameron, C.; Kelly, S.; Hsieh, S.-C.; Murphy, M.; Chen, L.; Kotb, A.; Peterson, J.; Coyle, D.; Skidmore, B.; Gomes, T.; et al. Triptans in the Acute Treatment of Migraine: A Systematic Review and Network Meta-Analysis. *Headache* **2015**, *55* (Suppl. S4), 221–235. [CrossRef]
19. Párdutz, A.; Fejes, A.; Bohár, Z.; Tar, L.; Toldi, J.; Vécsei, L. Kynurenines and headache. *J. Neural Transm.* **2012**, *119*, 285–296. [CrossRef]
20. Stone, T.W.; Darlington, L.G. Endogenous kynurenines as targets for drug discovery and development. *Nat. Rev. Drug Discov.* **2002**, *1*, 609–620. [CrossRef]

21. Comai, S.; Bertazzo, A.; Brughera, M.; Crotti, S. Tryptophan in health and disease. *Adv. Clin. Chem.* **2020**, *95*, 165–218. [CrossRef] [PubMed]
22. Ren, C.; Liu, J.; Zhou, J.; Liang, H.; Wang, Y.; Sun, Y.; Ma, B.; Yin, Y. Low levels of serum serotonin and amino acids identified in migraine patients. *Biochem. Biophys. Res. Commun.* **2018**, *496*, 267–273. [CrossRef] [PubMed]
23. D'Andrea, G.; D'Amico, D.; Bussone, G.; Bolner, A.; Aguggia, M.; Saracco, M.G.; Galloni, E.; De Riva, V.; D'Arrigo, A.; Colavito, D.; et al. Tryptamine levels are low in plasma of chronic migraine and chronic tension-type headache. *Neurol. Sci.* **2014**, *35*, 1941–1945. [CrossRef]
24. Alam, Z.; Coombes, N.; Waring, R.H.; Williams, A.C.; Steventon, G.B. Plasma levels of neuroexcitatory amino acids in patients with migraine or tension headache. *J. Neurol. Sci.* **1998**, *156*, 102–106. [CrossRef]
25. Curto, M.; Lionetto, L.; Negro, A.; Capi, M.; Fazio, F.; Giamberardino, M.A.; Simmaco, M.; Nicoletti, F.; Martelletti, P. Altered kynurenine pathway metabolites in serum of chronic migraine patients. *J. Headache Pain* **2015**, *17*, 47. [CrossRef]
26. Curto, M.; Lionetto, L.; Negro, A.; Capi, M.; Perugino, F.; Fazio, F.; Giamberardino, M.A.; Simmaco, M.; Nicoletti, F.; Martelletti, P. Altered serum levels of kynurenine metabolites in patients affected by cluster headache. *J. Headache Pain* **2015**, *17*, 27. [CrossRef] [PubMed]
27. Young, S. Acute tryptophan depletion in humans: A review of theoretical, practical and ethical aspects. *J. Psychiatry Neurosci.* **2013**, *38*, 294–305. [CrossRef]
28. Deen, M.; Christensen, C.E.; Hougaard, A.; Hansen, H.D.; Knudsen, G.M.; Ashina, M. Serotonergic mechanisms in the migraine brain—A systematic review. *Cephalalgia* **2017**, *37*, 251–264. [CrossRef]
29. Carpenter, L.L.; Anderson, G.M.; Pelton, G.H.; Gudin, J.A.; Kirwin, P.D.S.; Price, L.H.; Heninger, G.R.; McDougle, C.J. Tryptophan Depletion During Continuous CSF Sampling in Healthy Human Subjects. *Neuropsychopharmacology* **1998**, *19*, 26–35. [CrossRef]
30. Williams, W.A.; Shoaf, S.E.; Hommer, D.; Rawlings, R.; Linnoila, M. Effects of acute tryptophan depletion on plasma and cerebrospinal fluid tryptophan and 5-hydroxyindoleacetic acid in normal volunteers. *J. Neurochem.* **1999**, *72*, 1641–1647. [CrossRef]
31. Van Der Stelt, H.M.; Broersen, L.M.; Olivier, B.; Westenberg, H.G.M. Effects of dietary tryptophan variations on extracellular serotonin in the dorsal hippocampus of rats. *Psychopharmacology* **2004**, *172*, 137–144. [CrossRef] [PubMed]
32. Drummond, P. Tryptophan Depletion Increases Nausea, Headache and Photophobia in Migraine Sufferers. *Cephalalgia* **2006**, *26*, 1225–1233. [CrossRef]
33. Jahromi, S.R.; Togha, M.; Ghorbani, Z.; Hekmatdoost, A.; Khorsha, F.; Rafiee, P.; Shirani, P.; Nourmohammadi, M.; Ansari, H. The association between dietary tryptophan intake and migraine. *Neurol. Sci.* **2019**, *40*, 2349–2355. [CrossRef] [PubMed]
34. Gecse, K.; Dobos, D.; Aranyi, C.S.; Galambos, A.; Baksa, D.; Kocsel, N.; Szabó, E.; Pap, D.; Virág, D.; Ludányi, K.; et al. Association of plasma tryptophan concentration with periaqueductal gray matter functional connectivity in migraine patients. *Sci. Rep.* **2022**, *12*, 739. [CrossRef]
35. Villalón, C.M.; vanDenBrink, A.M. The Role of 5-Hydroxytryptamine in the Pathophysiology of Migraine and its Relevance to the Design of Novel Treatments. *Med. Chem.* **2017**, *17*, 928–938. [CrossRef]
36. Comings, D.E. Serotonin: A Key to Migraine Disorders? *Health Fit. Mag.* **1994**.
37. Nichols, D.E.; Nichols, C.D. Serotonin Receptors. *Chem. Rev.* **2008**, *108*, 1614–1641. [CrossRef]
38. Varnäs, K.; Halldin, C.; Hall, H. Autoradiographic distribution of serotonin transporters and receptor subtypes in human brain. *Hum. Brain Mapp.* **2004**, *22*, 246–260. [CrossRef]
39. Barnes, N.M.; Sharp, T. A review of central 5-HT receptors and their function. *Neuropharmacology* **1999**, *38*, 1083–1152. [CrossRef]
40. Sicuteri, F.; Testi, A.; Anselmi, B. Biochemical investigations in headache: Increase in hydroxytryindoleacetic acid excretion during migraine attacks. *Int. Arch. Allergy* **1961**, *19*, 55–58. [CrossRef]
41. Anthony, M.; Hinterberger, H.; Lance, J.W. Plasma Serotonin in Migraine and Stress. *Arch. Neurol.* **1967**, *16*, 544–552. [CrossRef]
42. Curran, D.A.; Hinterberger, H.; Lance, J.W. Total plasma serotonin, 5-hydroxyindoleacetic acid and p-hydroxy-m-methoxymandelic acid excretion in normal and migrainous subjects. *Brain* **1965**, *88*, 997–1010. [CrossRef]
43. Lance, J.W.; Anthony, M.; Hinterberger, H. The control of cranial arteries by humoral mechanisms and its relation to the migraine syndrome. *Headache* **1967**, *7*, 93–102. [CrossRef] [PubMed]
44. Kimball, R.W.; Friedman, A.P.; Vallejo, E. Effect of serotonin in migraine patients. *Neurology* **1960**, *10*, 107–111. [CrossRef]
45. Ferrari, M.D.; Odink, J.; Tapparelli, C.; Van Kempen, G.; Pennings, E.J.; Bruyn, G.W. Serotonin metabolism in migraine. *Neurology* **1989**, *39*, 1239–1242. [CrossRef]
46. Rossi, C.; Pini, L.A.; Cupini, M.L.; Calabresi, P.; Sarchielli, P. Endocannabinoids in platelets of chronic migraine patients and medication-overuse headache patients: Relation with serotonin levels. *Eur. J. Clin. Pharmacol.* **2008**, *64*, 1–8. [CrossRef] [PubMed]
47. Best, J.; Nijhout, H.F.; Reed, M. Serotonin synthesis, release and reuptake in terminals: A mathematical model. *Theor. Biol. Med. Model.* **2010**, *7*, 34. [CrossRef]
48. Höglund, E.; Øverli, Ø.; Winberg, S. Tryptophan Metabolic Pathways and Brain Serotonergic Activity: A Comparative Review. *Front. Endocrinol.* **2019**, *10*, 158. [CrossRef]
49. Sandler, M. Changes in 5-hydroxytryptamine and its metabolites in neuropsychiatric disorders. *Psychopharmacol. Bull.* **1981**, *17*, 19–21. [PubMed]
50. Mohammad-Zadeh, L.F.; Moses, L.; Gwaltney-Brant, S.M. Serotonin: A review. *J. Vet.-Pharmacol. Ther.* **2008**, *31*, 187–199. [CrossRef]

51. Lesch, K.; Aulakh, C.S.; Wolozin, B.L.; Tolliver, T.J.; Hill, J.L.; Murphy, D.L. Regional brain expression of serotonin transporter mRNA and its regulation by reuptake inhibiting antidepressants. *Mol. Brain Res.* **1993**, *17*, 31–35. [CrossRef]
52. Murphy, D.L.; Mueller, E.A.; Aulakh, C.S.; Bagdy, G.; Garrick, N.A. Serotoninergic Function in Neuropsychiatric Disorders. In *Serotonin*; Mylechrane, E.J., Angus, J.A., de la Lande, I.S., Humphrey, P.P.A., Eds.; Macmillan Press: London, UK, 1989; pp. 257–264.
53. Schuh-Hofer, S.; Richter, M.; Geworski, L.; Villringer, A.; Israel, H.; Wenzel, R.; Munz, D.L.; Arnold, G. Increased serotonin transporter availability in the brainstem of migraineurs. *J. Neurol.* **2007**, *254*, 789–796. [CrossRef]
54. Horvath, G.A.; Selby, K.; Poskitt, K.; Hyland, K.; Waters, P.J.; Coulter-Mackie, M.; Stockler-Ipsiroglu, S.G. Hemiplegic migraine, seizures, progressive spastic paraparesis, mood disorder, and coma in siblings with low systemic serotonin. *Cephalalgia* **2011**, *31*, 1580–1586. [CrossRef]
55. Meneses, A. Physiological, pathophysiological and therapeutic roles of 5-HT systems in learning and memory. *Rev. Neurosci.* **1998**, *9*, 275–289. [CrossRef]
56. Lambert, G. The Lack of Peripheral Pathology in Migraine Headache. *Headache* **2010**, *50*, 895–908. [CrossRef] [PubMed]
57. Longmore, J.; Shaw, D.; Smith, D.; Hopkins, R.; McAlliste, G.; Pickard, J.; Sirinathsinghji, D.; Butler, A.; Hill, R. Differential Distribution of $5Ht_{1D}$-and $5HT_{1B}$-Immunoreactivity within the Human Trigemino-Cerebrovascular System: Implications for the Discovery of New Antimigraine Drugs. *Cephalalgia* **1997**, *17*, 833–842. [CrossRef]
58. Hou, M.; Kanje, M.; Longmore, J.; Tajti, J.; Uddman, R.; Edvinsson, L. 5-HT1B and 5-HT1D receptors in the human trigeminal ganglion: Co-localization with calcitonin gene-related peptide, substance P and nitric oxide synthase. *Brain Res.* **2001**, *909*, 112–120. [CrossRef]
59. Bouchelet, I.; Cohen, Z.; Case, B.; Séguéla, P.; Hamel, E. Differential expression of sumatriptan-sensitive 5-hydroxytryptamine receptors in human trigeminal ganglia and cerebral blood vessels. *Mol. Pharmacol.* **1996**, *50*, 219–223. [PubMed]
60. Longmore, J.; Dowson, A.J.; Hill, R.G. Advances in migraine therapy—5-HT receptor subtype-specific agonist drugs. *Curr. Opin. CPNS Investig. Drugs* **1999**, *1*, 39–53.
61. Dodick, D.; Lipton, R.B.; Martin, V.; Papademetriou, V.; Rosamond, W.; MaassenVanDenBrink, A.; Loutfi, H.; Welch, K.M.; Goadsby, P.J.; Hahn, S.; et al. Consensus Statement: Cardiovascular Safety Profile of Triptans (5-HT1B/1D Agonists) in the Acute Treatment of Migraine. *Headache* **2004**, *44*, 414–425. [CrossRef] [PubMed]
62. Cohen, M.L.; Schenck, K. Contractile responses to sumatriptan and ergotamine in the rabbit saphenous vein: Effect of selective 5-HT1F receptor agonists and PGF2α. *J. Cereb. Blood Flow Metab.* **2000**, *131*, 562–568. [CrossRef]
63. Ramadan, N.; Skljarevski, V.; Phebus, L.; Johnson, K. $5\text{-}HT_{1F}$ Receptor Agonists in Acute Migraine Treatment: A Hypothesis. *Cephalalgia* **2003**, *23*, 776–785. [CrossRef] [PubMed]
64. Adham, N.; Kao, H.T.; Schecter, L.E.; Bard, J.; Olsen, M.; Urquhart, D.; Durkin, M.; Hartig, P.R.; Weinshank, R.L.; Branchek, T.A. Cloning of another human serotonin receptor (5-HT1F): A fifth 5-HT1 receptor subtype coupled to the inhibition of adenylate cyclase. *Proc. Natl. Acad. Sci. USA* **1993**, *90*, 408–412. [CrossRef]
65. Vila-Pueyo, M. Targeted 5-HT1F Therapies for Migraine. *Neurotherapeutics* **2018**, *15*, 291–303. [CrossRef]
66. Johnson, K.W.; Schaus, J.M.; Durkin, M.M.; Audia, J.E.; Kaldor, S.W.; Flaugh, M.E.; Adham, N.; Zgombick, J.M.; Cohen, M.L.; Branchek, T.A.; et al. 5-HT1F receptor agonists inhibit neurogenic dural inflammation in guinea pigs. *NeuroReport* **1997**, *8*, 2237–2239. [CrossRef] [PubMed]
67. Mitsikostas, D.D.; del Rio, M.S.; Moskowitz, M.A.; Waeber, C. Both 5-HT1B and 5-HT1F receptors modulate c-fos expression within rat trigeminal nucleus caudalis. *Eur. J. Pharmacol.* **1999**, *369*, 271–277. [CrossRef]
68. Mitsikostas, D.; del Rio, M.S.; Waeber, C. $5\text{-}Hydroxytryptamine_{1B/1D}$ and $5\text{-}Hydroxytryptamine_{1F}$Receptors Inhibit Capsaicin-Induced *C-Fos* Immunoreactivity within Mouse Trigeminal Nucleus Caudalis. *Cephalalgia* **2002**, *22*, 384–394. [CrossRef] [PubMed]
69. Nelson, D.L.; Phebus, L.A.; Johnson, K.W.; Wainscott, D.B.; Cohen, M.L.; Calligaro, D.O.; Xu, Y.-C. Preclinical pharmacological profile of the selective $5\text{-}HT_{1F}$ receptor agonist lasmiditan. *Cephalalgia* **2010**, *30*, 1159–1169. [CrossRef]
70. Kovalchin, J.; Ghiglieri, A.; Zanelli, E.; Ings, R.; Mathers, T. Lasmiditan acts specifically on the 5-HT1F receptors in the central nervous system. *Cephalalgia* **2016**, *36*, 103.
71. Akerman, S.; Romero-Reyes, M.; Holland, P.R. Current and novel insights into the neurophysiology of migraine and its implications for therapeutics. *Pharmacol. Ther.* **2017**, *172*, 151–170. [CrossRef]
72. Huang, P.-C.; Yang, F.-C.; Chang, C.-M.; Yang, C.-P. Targeting the 5-HT1B/1D and 5-HT1F receptors for acute migraine treatment. *Prog. Brain Res.* **2020**, *255*, 99–121. [CrossRef] [PubMed]
73. Callebert, J.; Esteve, J.M.; Hervé, P.; Peoc'H, K.; Tournois, C.; Drouet, L.; Launay, J.M.; Maroteaux, L. Evidence for a Control of Plasma Serotonin Levels by 5-Hydroxytryptamine2B Receptors in Mice. *J. Pharmacol. Exp. Ther.* **2006**, *317*, 724–731. [CrossRef]
74. Xiao, Y.; Richter, J.A.; Hurley, J.H. Release of Glutamate and CGRP from Trigeminal Ganglion Neurons: Role of Calcium Channels and 5-HT1 Receptor Signaling. *Mol. Pain* **2008**, *4*, 12. [CrossRef] [PubMed]
75. Florian, J.A.; Watts, S.W. Integration of mitogen-activated protein kinase kinase activation in vascular 5-hydroxytryptamine2A receptor signal transduction. *J. Pharmacol. Exp. Ther.* **1998**, *284*, 346–355.
76. Martin, R.S.; Martin, G.R. Investigations into migraine pathogenesis: Time course for effects of m-CPP, BW723C86 or glyceryl trinitrate on appearance of Fos-like immunoreactivity in rat trigeminal nucleus caudalis (TNC). *Cephalalgia* **2001**, *21*, 46–52. [CrossRef]

77. Johnson, K.; Nelson, D.; Dieckman, D.; Wainscott, D.; Lucaites, V.; Audia, J.; Owton, W.; Phebus, L. Neurogenic Dural Protein Extravasation Induced by Meta-Chlorophenylpiperazine (mCPP) Involves Nitric Oxide and 5-HT_{2B} Receptor Activation. *Cephalalgia* **2003**, *23*, 117–123. [CrossRef]
78. Schmitz, B.; Ullmer, C.; Segelcke, D.; Gwarek, M.; Zhu, X.-R.; Lübbert, H. BF-1—A novel selective 5-HT2B receptor antagonist blocking neurogenic dural plasma protein extravasation in guinea pigs. *Eur. J. Pharmacol.* **2015**, *751*, 73–80. [CrossRef] [PubMed]
79. Brewerton, T.D.; Murphy, D.L.; Mueller, E.A.; Jimerson, D.C. Induction of migrainelike headaches by the serotonin agonist m-chlorophenylpiperazine. *Clin. Pharmacol. Ther.* **1988**, *43*, 605–609. [CrossRef] [PubMed]
80. Silberstein, S.D. Methysergide. *Cephalalgia* **1998**, *18*, 421–435. [CrossRef]
81. Bettahi, I.; Pozo, D.; Osuna, C.; Reiter, R.J.; Acuña-Castroviejo, D.; Guerrero, J.M. Melatonin reduces nitric oxide synthase activity in rat hypothalamus. *J. Pineal Res.* **1996**, *20*, 205–210. [CrossRef] [PubMed]
82. Lance, J.W.; Anthony, M.; Somerville, B. Comparative Trial of Serotonin Antagonists in the Management of Migraine. *Br. Med. J.* **1970**, *2*, 327–330. [CrossRef]
83. Reiter, R.J.; Calvo, J.R.; Karbownik, M.; Qi, W.; Tan, D.X. Melatonin and Its Relation to the Immune System and Inflammation. *Ann. N. Y. Acad. Sci.* **2000**, *917*, 376–386. [CrossRef] [PubMed]
84. Peres, M.F.P.; Masruha, M.R.; Zukerman, E.; Moreira-Filho, C.A.; Cavalheiro, E. Potential therapeutic use of melatonin in migraine and other headache disorders. *Expert Opin. Investig. Drugs* **2006**, *15*, 367–375. [CrossRef] [PubMed]
85. Adnyana, I.M.O.; Tertia, C.; Widyadharma, I.P.E.; Mahadewi, N.P.A.P. Melatonin as a treatment for migraine sufferers: A systematic review. *Egypt. J. Neurol. Psychiatry Neurosurg.* **2022**, *58*, 94. [CrossRef]
86. Scarabelot, V.L.; Medeiros, L.F.; de Oliveira, C.; Adachi, L.N.S.; de Macedo, I.C.; Cioato, S.G.; de Freitas, J.S.; de Souza, A.; Quevedo, A.; Caumo, W.; et al. Melatonin Alters the Mechanical and Thermal Hyperalgesia Induced by Orofacial Pain Model in Rats. *Inflammation* **2016**, *39*, 1649–1659. [CrossRef] [PubMed]
87. Torres, I.L.; Laste, G.; de Macedo, I.C.; Rozisky, J.R.; da Silva, F.R.; Caumo, W. Melatonin administration reduces inflammatory pain in rats. *J. Pain Res.* **2012**, *5*, 359–362. [CrossRef]
88. Noseda, R.; Hernández, A.; Valladares, L.; Mondaca, M.; Laurido, C.; Soto-Moyano, R. Melatonin-induced inhibition of spinal cord synaptic potentiation in rats is MT2 receptor-dependent. *Neurosci. Lett.* **2004**, *360*, 41–44. [CrossRef]
89. Danilov, A.; Kurganova, J. Melatonin in Chronic Pain Syndromes. *Pain Ther.* **2016**, *5*, 1–17. [CrossRef]
90. Weaver, D.; Rivkees, S.; Reppert, S. Localization and characterization of melatonin receptors in rodent brain by in vitro autoradiography. *J. Neurosci.* **1989**, *9*, 2581–2590. [CrossRef]
91. Williams, L.M.; Hannah, L.T.; Hastings, M.H.; Maywood, E.S. Melatonin receptors in the rat brain and pituitary. *J. Pineal Res.* **1995**, *19*, 173–177. [CrossRef]
92. Yu, C.-X.; Zhu, C.-B.; Xu, S.-F.; Cao, X.-D.; Wu, G.-C. The analgesic effects of peripheral and central administration of melatonin in rats. *Eur. J. Pharmacol.* **2000**, *403*, 49–53. [CrossRef]
93. Shavali, S.; Ho, B.; Govitrapong, P.; Sawlom, S.; Ajjimaporn, A.; Klongpanichapak, S.; Ebadi, M. Melatonin exerts its analgesic actions not by binding to opioid receptor subtypes but by increasing the release of β-endorphin an endogenous opioid. *Brain Res. Bull.* **2005**, *64*, 471–479. [CrossRef]
94. Srinivasan, V.; Lauterbach, E.C.; Ho, K.Y.; Acuña-Castroviejo, D.; Zakaria, R.; Brzezinski, A. Melatonin in Antinociception: Its Therapeutic Applications. *Curr. Neuropharmacol.* **2012**, *10*, 167–178. [CrossRef]
95. Mease, P. Fibromyalgia syndrome: Review of clinical presentation, pathogenesis, outcome measures, and treatment. *J. Rheumatol.* **2005**, *75*, 6–21.
96. Reiter, R.J.; Acuna-Castroviejo, D.; Tan, D.-X. Melatonin therapy in fibromyalgia. *Curr. Pain Headache Rep.* **2007**, *11*, 339–342. [CrossRef] [PubMed]
97. Song, G.H.; Leng, P.H.; Gwee, K.A.; Moochhala, S.M.; Ho, K.Y. Melatonin improves abdominal pain in irritable bowel syndrome patients who have sleep disturbances: A randomised, double blind, placebo controlled study. *Gut* **2005**, *54*, 1402–1407. [CrossRef] [PubMed]
98. Laste, G.; Rozisky, J.R.; Caumo, W.; Torres, I.L.D.S. Short- but Not Long-Term Melatonin Administration Reduces Central Levels of Brain-Derived Neurotrophic Factor in Rats with Inflammatory Pain. *Neuroimmunomodulation* **2015**, *22*, 358–364. [CrossRef] [PubMed]
99. Masruha, M.R.; Lin, J.; Vieira, D.S.D.S.; Minett, T.S.; Cipolla-Neto, J.; Zukerman, E.; Vilanova, L.C.; Peres, M.F. Urinary 6-Sulphatoxymelatonin Levels Are Depressed in Chronic Migraine and Several Comorbidities. *Headache* **2010**, *50*, 413–419. [CrossRef]
100. Claustrat, B.; Loisy, C.; Brun, J.; Beorchia, S.; Arnaud, J.L.; Chazot, G. Nocturnal Plasma Melatonin Levels in Migraine: A Preliminary Report. *Headache* **1989**, *29*, 242–245. [CrossRef] [PubMed]
101. Masruha, M.R.; Vieira, D.S.D.S.; Minett, T.S.C.; Cipolla-Neto, J.; Zukerman, E.; Vilanova, L.C.P.; Peres, M.F.P. Low urinary 6-sulphatoxymelatonin concentrations in acute migraine. *J. Headache Pain* **2008**, *9*, 221–224. [CrossRef]
102. Murialdo, G.; Fonzi, S.; Costelli, P.; Solinas, G.P.; Parodi, C.; Marabini, S.; Fanciullacci, M.; Polleri, A. Urinary Melatonin Excretion Throughout the Ovarian Cycle in Menstrually Related Migraine. *Cephalalgia* **1994**, *14*, 205–209. [CrossRef] [PubMed]
103. Brun, J.; Claustrat, B.; Saddier, P.; Chazot, G. Nocturnal Melatonin Excretion is Decreased in Patients with Migraine without Aura Attacks Associated with Menses. *Cephalalgia* **1995**, *15*, 136–139. [CrossRef] [PubMed]

104. Wolf, H. The effect of hormones and vitamin B6 on urinary excretion of metabolites of the kynurenine pathway. *Scand. J. Clin. Lab. Investig.* **1974**, *136*, 1–186.
105. Vécsei, L.; Szalárdy, L.; Fülöp, F.; Toldi, J. Kynurenines in the CNS: Recent advances and new questions. *Nat. Rev. Drug Discov.* **2013**, *12*, 64–82. [CrossRef] [PubMed]
106. Schwarcz, R.; Bruno, J.P.; Muchowski, P.J.; Wu, H.-Q. Kynurenines in the mammalian brain: When physiology meets pathology. *Nat. Rev. Neurosci.* **2012**, *13*, 465–477. [CrossRef] [PubMed]
107. Miranda, A.; Boegman, R.; Beninger, R.; Jhamandas, K. Protection against quinolinic acid-mediated excitotoxicity in nigrostriatal dopaminergic neurons by endogenous kynurenic acid. *Neuroscience* **1997**, *78*, 967–975. [CrossRef]
108. Silva-Adaya, D.; La Cruz, V.P.-D.; Villeda-Hernández, J.; Carrillo-Mora, P.; González-Herrera, I.G.; García, E.; Colín-Barenque, L.; Pedraza-Chaverrí, J.; Santamaría, A. Protective effect of l-kynurenine and probenecid on 6-hydroxydopamine-induced striatal toxicity in rats: Implications of modulating kynurenate as a protective strategy. *Neurotoxicol. Teratol.* **2011**, *33*, 303–312. [CrossRef]
109. Knyihár-Csillik, E.; Toldi, J.; Krisztin-Péva, B.; Chadaide, Z.; Németh, H.; Fenyő, R.; Vécsei, L. Prevention of electrical stimulation-induced increase of c-fos immunoreaction in the caudal trigeminal nucleus by kynurenine combined with probenecid. *Neurosci. Lett.* **2007**, *418*, 122–126. [CrossRef]
110. Vamos, E.; Párdutz, A.; Varga, H.; Bohár, Z.; Tajti, J.; Fülöp, F.; Toldi, J.; Vécsei, L. l-kynurenine combined with probenecid and the novel synthetic kynurenic acid derivative attenuate nitroglycerin-induced nNOS in the rat caudal trigeminal nucleus. *Neuropharmacology* **2009**, *57*, 425–429. [CrossRef]
111. Toldi, J.; Oláh, G.; Herédi, J.; Menyhárt, A.; Czinege, Z.; Nagy, D.; Fuzik, J.; Krucsó, E.; Kocsis, K.; Knapp, L.; et al. Unexpected effects of peripherally administered kynurenic acid on cortical spreading depression and related blood–brain barrier permeability. *Drug Des. Dev. Ther.* **2013**, *7*, 981–987. [CrossRef]
112. Knyihár-Csillik, E.; Chadaide, Z.; Okuno, E.; Krisztin-Péva, B.; Toldi, J.; Varga, C.; Molnár, A.; Csillik, B.; Vécsei, L. Kynurenine aminotransferase in the supratentorial dura mater of the rat: Effect of stimulation of the trigeminal ganglion. *Exp. Neurol.* **2004**, *186*, 242–247. [CrossRef]
113. Spekker, E.; Laborc, K.F.; Bohár, Z.; Nagy-Grócz, G.; Fejes-Szabó, A.; Szűcs, M.; Vécsei, L.; Párdutz, A. Effect of dural inflammatory soup application on activation and sensitization markers in the caudal trigeminal nucleus of the rat and the modulatory effects of sumatriptan and kynurenic acid. *J. Headache Pain* **2021**, *22*, 17. [CrossRef] [PubMed]
114. Zhang, Y.-Q.; Ji, G.-C.; Wu, G.-C.; Zhao, Z.-Q. Kynurenic acid enhances electroacupuncture analgesia in normal and carrageenan-injected rats. *Brain Res.* **2003**, *966*, 300–307. [CrossRef] [PubMed]
115. Mecs, L.; Tuboly, G.; Nagy, E.; Benedek, G.; Horvath, G. The Peripheral Antinociceptive Effects of Endomorphin-1 and Kynurenic Acid in the Rat Inflamed Joint Model. *Anesthesia Analg.* **2009**, *109*, 1297–1304. [CrossRef] [PubMed]
116. Fukui, S.; Schwarcz, R.; Rapoport, S.I.; Takada, Y.; Smith, Q.R. Blood?Brain Barrier Transport of Kynurenines: Implications for Brain Synthesis and Metabolism. *J. Neurochem.* **1991**, *56*, 2007–2017. [CrossRef]
117. Arimura, A. PACAP: The road to discovery. *Peptides* **2007**, *28*, 1617–1619. [CrossRef] [PubMed]
118. Vaudry, D.; Falluel-Morel, A.; Bourgault, S.; Basille, M.; Burel, D.; Wurtz, O.; Fournier, A.; Chow, B.K.C.; Hashimoto, H.; Galas, L.; et al. Pituitary Adenylate Cyclase-Activating Polypeptide and Its Receptors: 20 Years after the Discovery. *Pharmacol. Rev.* **2009**, *61*, 283–357. [CrossRef]
119. Hashimoto, H.; Shintani, N.; Tanida, M.; Hayata, A.; Hashimoto, R.; Baba, A. PACAP is implicated in the stress axes. *Curr. Pharm. Des.* **2011**, *17*, 985–989. [CrossRef]
120. Reglodi, D.; Kiss, P.; Lubics, A.; Tamas, A. Review on the Protective Effects of PACAP in Models of Neurodegenerative Diseases In Vitro and In Vivo. *Curr. Pharm. Des.* **2011**, *17*, 962–972. [CrossRef]
121. Seaborn, T.; Masmoudi-Kouli, O.; Fournier, A.; Vaudry, H.; Vaudry, D. Protective effects of pituitary adenylate cyclase-activating polypeptide (PACAP) against apoptosis. *Curr. Pharm. Des.* **2011**, *17*, 204–214. [CrossRef]
122. Ohtsuka, M.; Fukumitsu, H.; Furukawa, S. PACAP decides neuronal laminar fate via PKA signaling in the developing cerebral cortex. *Biochem. Biophys. Res. Commun.* **2008**, *369*, 1144–1149. [CrossRef]
123. Palkovits, M.; Somogyvári-Vigh, A.; Arimura, A. Concentrations of pituitary adenylate cyclase activating polypeptide (PACAP) in human brain nuclei. *Brain Res.* **1995**, *699*, 116–120. [CrossRef] [PubMed]
124. Vécsei, L.; Tuka, B.; Tajti, J. Role of PACAP in migraine headaches. *Brain* **2014**, *137 Pt 3*, 650–651. [CrossRef] [PubMed]
125. Laburthe, M.; Couvineau, A.; Tan, V. Class II G protein-coupled receptors for VIP and PACAP: Structure, models of activation and pharmacology. *Peptides* **2007**, *28*, 1631–1639. [CrossRef]
126. Fukuchi, M.; Kuwana, Y.; Tabuchi, A.; Tsuda, M. Balance between cAMP and Ca2+ signals regulates expression levels of pituitary adenylate cyclase-activating polypeptide gene in neurons. *Genes Cells* **2016**, *21*, 921–929. [CrossRef]
127. Körtési, T.; Tuka, B.; Tajti, J.; Bagoly, T.; Fülöp, F.; Helyes, Z.; Vécsei, L. Kynurenic Acid Inhibits the Electrical Stimulation Induced Elevated Pituitary Adenylate Cyclase-Activating Polypeptide Expression in the TNC. *Front. Neurol.* **2018**, *8*, 745. [CrossRef] [PubMed]
128. Liao, C.; De Molliens, M.P.; Schneebeli, S.T.; Brewer, M.; Song, G.; Chatenet, D.; Braas, K.M.; May, V.; Li, J. Targeting the PAC1 Receptor for Neurological and Metabolic Disorders. *Curr. Top. Med. Chem.* **2019**, *19*, 1399–1417. [CrossRef] [PubMed]
129. MMarkovics, A.; Kormos, V.; Gaszner, B.; Lashgarara, A.; Szoke, E.; Sandor, K.; Szabadfi, K.; Tuka, B.; Tajti, J.; Szolcsanyi, J.; et al. Pituitary adenylate cyclase-activating polypeptide plays a key role in nitroglycerol-induced trigeminovascular activation in mice. *Neurobiol. Dis.* **2012**, *45*, 633–644. [CrossRef]

130. Tuka, B.; Helyes, Z.; Markovics, A.; Bagoly, T.; Németh, J.; Márk, L.; Brubel, R.; Reglődi, D.; Párdutz, A.; Szolcsányi, J.; et al. Peripheral and central alterations of pituitary adenylate cyclase activating polypeptide-like immunoreactivity in the rat in response to activation of the trigeminovascular system. *Peptides* **2012**, *33*, 307–316. [CrossRef]
131. Tuka, B.; Helyes, Z.; Markovics, A.; Bagoly, T.; Szolcsányi, J.; Szabó, N.; Tóth, E.; Kincses, Z.T.; Vécsei, L.; Tajti, J. Alterations in PACAP-38-like immunoreactivity in the plasma during ictal and interictal periods of migraine patients. *Cephalalgia* **2013**, *33*, 1085–1095. [CrossRef]
132. Schytz, H.W.; Birk, S.; Wienecke, T.; Kruuse, C.; Olesen, J.; Ashina, M. PACAP38 induces migraine-like attacks in patients with migraine without aura. *Brain* **2009**, *132 Pt 1*, 16–25. [CrossRef] [PubMed]
133. Schytz, H.W.; Olesen, J.; Ashina, M. The PACAP receptor: A novel target for migraine treatment. *Neurotherapeutics* **2010**, *7*, 191–196. [CrossRef] [PubMed]
134. Guo, S.; Christensen, A.F.; Liu, M.L.; Janjooa, B.N.; Olesen, J.; Ashina, M. Calcitonin gene-related peptide induced migraine attacks in patients with and without familial aggregation of migraine. *Cephalalgia* **2017**, *37*, 114–124. [CrossRef]
135. VVeréb, D.; Szabó, N.; Tuka, B.; Tajti, J.; Király, A.; Faragó, P.; Kocsis, K.; Tóth, E.; Kincses, B.; Bagoly, T.; et al. Correlation of neurochemical and imaging markers in migraine: PACAP38 and DTI measures. *Neurology* **2018**, *91*, e1166–e1174. [CrossRef] [PubMed]
136. Amin, F.M.; Asghar, M.S.; Guo, S.; Hougaard, A.; Hansen, A.E.; Schytz, H.W.; van der Geest, R.; de Koning, P.J.; Larsson, H.B.; Olesen, J.; et al. Headache and prolonged dilatation of the middle meningeal artery by PACAP38 in healthy volunteers. *Cephalalgia* **2012**, *32*, 140–149. [CrossRef]
137. Goadsby, P.J.; Edvinsson, L.; Ekman, R. Release of vasoactive peptides in the extracerebral circulation of humans and the cat during activation of the trigeminovascular system. *Ann. Neurol.* **1988**, *23*, 193–196. [CrossRef] [PubMed]
138. Tajti, J.; Szok, D.; Majláth, Z.; Tuka, B.; Csáti, A.; Vécsei, L. Migraine and neuropeptides. *Neuropeptides* **2015**, *52*, 19–30. [CrossRef]
139. Romero-Reyes, M.; Pardi, V.; Akerman, S. A potent and selective calcitonin gene-related peptide (CGRP) receptor antagonist, MK-8825, inhibits responses to nociceptive trigeminal activation: Role of CGRP in orofacial pain. *Exp. Neurol.* **2015**, *271*, 95–103. [CrossRef]
140. Jansen-Olesen, I.; Baun, M.; Amrutkar, D.V.; Ramachandran, R.; Christophersen, D.V.; Olesen, J. PACAP-38 but not VIP induces release of CGRP from trigeminal nucleus caudalis via a receptor distinct from the PAC1 receptor. *Neuropeptides* **2014**, *48*, 53–64. [CrossRef]
141. Eftekhari, S.; Salvatore, C.A.; Johansson, S.; Chen, T.-B.; Zeng, Z.; Edvinsson, L. Localization of CGRP, CGRP receptor, PACAP and glutamate in trigeminal ganglion. Relation to the blood–brain barrier. *Brain Res.* **2015**, *1600*, 93–109. [CrossRef]
142. Mahmoudi, J.; Mohaddes, G.; Erfani, M.; Sadigh-Eteghad, S.; Karimi, P.; Rajabi, M.; Reyhani-Rad, S.; Farajdokht, F. Cerebrolysin attenuates hyperalgesia, photophobia, and neuroinflammation in a nitroglycerin-induced migraine model in rats. *Brain Res. Bull.* **2018**, *140*, 197–204. [CrossRef] [PubMed]
143. Körtési, T.; Tuka, B.; Nyári, A.; Vécsei, L.; Tajti, J. The effect of orofacial complete Freund's adjuvant treatment on the expression of migraine-related molecules. *J. Headache Pain* **2019**, *20*, 43. [CrossRef] [PubMed]
144. Romero-Reyes, M.; Uyanik, J.M. Orofacial pain management: Current perspectives. *J. Pain Res.* **2014**, *7*, 99–115. [CrossRef] [PubMed]
145. Csáti, A.; Edvinsson, L.; Vécsei, L.; Toldi, J.; Fulop, F.; Tajti, J.; Warfvinge, K. Kynurenic acid modulates experimentally induced inflammation in the trigeminal ganglion. *J. Headache Pain* **2015**, *16*, 99. [CrossRef] [PubMed]
146. Lukács, M.; Warfvinge, K.; Tajti, J.; Fülöp, F.; Toldi, J.; Vécsei, L.; Edvinsson, L. Topical dura mater application of CFA induces enhanced expression of c-fos and glutamate in rat trigeminal nucleus caudalis: Attenuated by KYNA derivate (SZR72). *J. Headache Pain* **2017**, *18*, 39. [CrossRef] [PubMed]
147. Aczél, T.; Kun, J.; Szőke, E.; Rauch, T.; Junttila, S.; Gyenesei, A.; Bölcskei, K.; Helyes, Z. Transcriptional Alterations in the Trigeminal Ganglia, Nucleus and Peripheral Blood Mononuclear Cells in a Rat Orofacial Pain Model. *Front. Mol. Neurosci.* **2018**, *11*, 219. [CrossRef] [PubMed]
148. Veres, G.; Fejes-Szabó, A.; Zádori, D.; Nagy-Grócz, G.; László, A.M.; Bajtai, A.; Mándity, I.; Szentirmai, M.; Bohár, Z.; Laborc, K.; et al. A comparative assessment of two kynurenic acid analogs in the formalin model of trigeminal activation: A behavioral, immunohistochemical and pharmacokinetic study. *J. Neural Transm.* **2017**, *124*, 99–112. [CrossRef]
149. Nagy-Grócz, G.; Tar, L.; Bohár, Z.; Fejes-Szabó, A.; Laborc, K.F.; Spekker, E.; Vécsei, L.; Párdutz, A. The modulatory effect of anandamide on nitroglycerin-induced sensitization in the trigeminal system of the rat. *Cephalalgia* **2016**, *36*, 849–861. [CrossRef]
150. Kuburas, A.; Mason, B.N.; Hing, B.; Wattiez, A.-S.; Reis, A.S.; Sowers, L.P.; Loomis, C.M.; Garcia-Martinez, L.F.; Russo, A.F. PACAP Induces Light Aversion in Mice by an Inheritable Mechanism Independent of CGRP. *J. Neurosci.* **2021**, *41*, 4697–4715. [CrossRef]
151. Ernstsen, C.; Christensen, S.L.; Rasmussen, R.H.; Nielsen, B.S.; Jansen-Olesen, I.; Olesen, J.; Kristensen, D.M. The PACAP pathway is independent of CGRP in mouse models of migraine: Possible new drug target? *Brain* **2022**, *145*, 2450–2460. [CrossRef]

152. Tuka, B.; Nyári, A.; Cseh, E.K.; Körtési, T.; Veréb, D.; Tömösi, F.; Kecskeméti, G.; Janáky, T.; Tajti, J.; Vécsei, L. Clinical relevance of depressed kynurenine pathway in episodic migraine patients: Potential prognostic markers in the peripheral plasma during the interictal period. *J. Headache Pain* **2021**, *22*, 60. [CrossRef] [PubMed]
153. Al-Karagholi, M.A.; Hansen, J.M.; Abou-Kassem, D.; Hansted, A.K.; Ubhayasekera, K.; Bergquist, J.; Vécsei, L.; Jansen-Olesen, I.; Ashina, M. Phase 1 study to access safety, tolerability, pharmacokinetics, and pharmacodynamics of kynurenine in healthy volunteers. *Pharmacol. Res. Perspect.* **2021**, *9*, e00741. [CrossRef] [PubMed]

cells

MDPI

Review

Monoclonal Antibodies against Calcitonin Gene-Related Peptide for Migraine Prophylaxis: A Systematic Review of Real-World Data

Antun R. Pavelic [1], Christian Wöber [2,3,*], Franz Riederer [1,4] and Karin Zebenholzer [2,3]

1 Department of Neurology, Hietzing Hospital, 1130 Vienna, Austria
2 Department of Neurology, Medical University of Vienna, 1090 Vienna, Austria
3 Comprehensive Center for Clinical Neurosciences & Mental Health, Medical University of Vienna, 1090 Vienna, Austria
4 Faculty of Medicine, University of Zurich, 8032 Zurich, Switzerland
* Correspondence: christian.woeber@meduniwien.ac.at

Abstract: Objective: To perform a systematic review of real-world outcomes for anti-CGRP-mAbs. Methods: Following the PRISMA guidelines, we searched PubMed for real-world data of erenumab, galcanezumab, fremanezumab, or eptinezumab in patients with migraines. Results: We identified 134 publications (89 retrospective), comprising 10 pharmaco-epidemiologic and 83 clinic-based studies, 38 case reports, and 3 other articles. None of the clinic-based studies provided follow-up data over more than one year in more than 200 patients. Findings suggest that there are reductions in health insurance claims and days with sick-leave as well as better treatment adherence with anti-CGRP-mAbs. Effectiveness, reported in 77 clinic-based studies, was comparable to randomized controlled trials. A treatment pause was associated with an increase in migraine frequency, and switching to another antibody resulted in a better response in some of the patients. Adverse events and safety issues were addressed in 86 papers, including 24 single case reports. Conclusion: Real-world data on anti-CGRP-mAbs are limited by retrospective data collection, small patient numbers, and short follow-up periods. The majority of papers seem to support good effectiveness and tolerability of anti-CGRP-mAbs in the real-world setting. There is an unmet need for large prospective real-world studies providing long-term follow-ups of patients treated with anti-CGRP-mAbs.

Keywords: real-world; erenumab; galcanezumab; fremanezumab; eptinezumab; pharmacoepidemiology; effectiveness; tolerability; safety; treatment pause; switching

Citation: Pavelic, A.R.; Wöber, C.; Riederer, F.; Zebenholzer, K. Monoclonal Antibodies against Calcitonin Gene-Related Peptide for Migraine Prophylaxis: A Systematic Review of Real-World Data. *Cells* **2023**, *11*, 143. https://doi.org/10.3390/cells12010143

Academic Editors: Bernadett Tuka, László Vécsei and Masaru Tanaka

Received: 5 October 2022
Revised: 22 December 2022
Accepted: 23 December 2022
Published: 29 December 2022

1. Introduction

For decades, the pharmacological prophylaxis of migraines has been based on medications that were non-specific for migraines, which led to low adherence rates due to limited efficacy and poor tolerability [1]. Monoclonal antibodies against the calcitonin gene-related peptide (CGRP) or its receptor (anti-CGRP-mAbs) have opened a new era for migraine prevention.

CGRP is a neuropeptide also acting as neurotransmitter that has, among others, a crucial role within the pathophysiology of migraines. Its release is increased during migraine attacks [2] and intravenous infusion of CGRP can trigger migraine-like attacks in migraine patients. CGRP is a very potent vasodilator and exerts its action not exclusively in the brain. It contributes to reactive vasodilation during myocardial infarction and vasospasms during subarachnoid hemorrhages. It is involved in the transmission of pain and sensory stimuli, in wound healing, and it has functions in the gastrointestinal system [3].

Phase 2 and phase 3 trials showed no signs of an increased incidence of vascular events or vascular complications in patients under therapy with an anti-CGRP-mAb. Moreover, package information leaflets do not list any vascular disease or risk factor as contraindications against these antibodies. Nonetheless, these leaflets contain warnings to be cautious in patients with a history of cardiovascular or cerebrovascular diseases.

Anti-CGRP-mAbs are effective in episodic [4–8] and chronic migraines [9–12], including difficult-to-treat patient groups with multiple treatment failures, psychiatric comorbidities [13–18], or medication overuse [19–22]. Outcome measures involve monthly days with migraines, any headache and use of acute medication, the 50% responder rate (i.e., the proportion of patients experiencing a reduction in monthly migraine days by 50% or more), as well as functional and patient-related outcomes [23–26].

The CGRP-antibodies fremanezumab and galcanezumab as well as the CGRP-receptor antibody erenumab, all of which are administered subcutaneously, have been licensed for migraine prevention since 2018. More recently, eptinezumab was licensed, another CGRP-antibody, which is administered intravenously. Instead of a daily intake of medication, as required for standard pharmacoprophylaxis, anti-CGRP-mAbs are administered once every four weeks, every month, or every three months.

Altogether, they are approved for episodic migraines with at least four migraine days per month, and chronic migraine. Reimbursement regulations differ from country to country. This leads to different uses in daily clinical practice, with respect to the number of previously prescribed prophylactic medications, necessity of therapy breaks, or switches from one antibody to another.

While some long-term studies, mostly open-label extensions of phase 2 or phase 3 studies in highly selected populations, are reassuring concerning safety [27–30], real-world evidence in unselected patient groups is of particular interest. Issues deserving further study in the real-world setting include long-term safety and effectiveness, impact on migraine auras, outcomes of pausing the treatment and of switching to another antibody, and data in special groups (such as elderly persons and patients with comorbidities).

Since the approval of anti-CGRP-mAbs, plenty of studies and case reports dealing with real-world experience and focusing on various aspects of these antibodies have been published. The aim of this article was to gather real-world data on anti-CGRP-mAbs and to review these data systematically with respect to pharmaco-epidemiological findings, headache diagnoses, general effectiveness, effectiveness in patients with previous treatment failures, differences in effectiveness of the antibodies, outcomes of pausing treatment, switching to another antibody, and discontinuing treatment, as well as tolerability and safety.

2. Methods

2.1. Search Methods

We performed a review of the literature using PubMed, concerning real-world studies of migraine patients treated with anti-CGRP-mAbs. Search terms included the following: erenumab, fremanezumab, galcanezumab, eptinezumab, CGRP, calcitonin, real, case, migraine, vertigo, cyclic vomiting, and visual snow. To focus the results, we conducted 8 individualized searches: 2 for each monoclonal antibody—one using the keyword real and one search using the keyword case.

2.2. Selection Criteria

Our selection criteria were language (English), primary headache type (migraine and migraine-related disorders), and study design (real-world data). The last search took place on 1 December 2022.

2.3. Review Preparation and Statistics

The systematic review was prepared according to the latest PRISMA (Preferred Reporting Items for Systematic Reviews and Meta-Analyses) guidelines [31], and study data were gathered into an Excel table. Descriptive statistics were conducted in IBM SPSS Statistics 21.

3. Study Characteristics

Our search yielded 251 results from the eight individual searches. After we applied selection criteria and excluded duplicates, 145 articles remained for hand-search to exclude additional nonrelevant publications. Finally, we included 134 articles in this review. An exact breakdown of the search results can be seen in Figure 1.

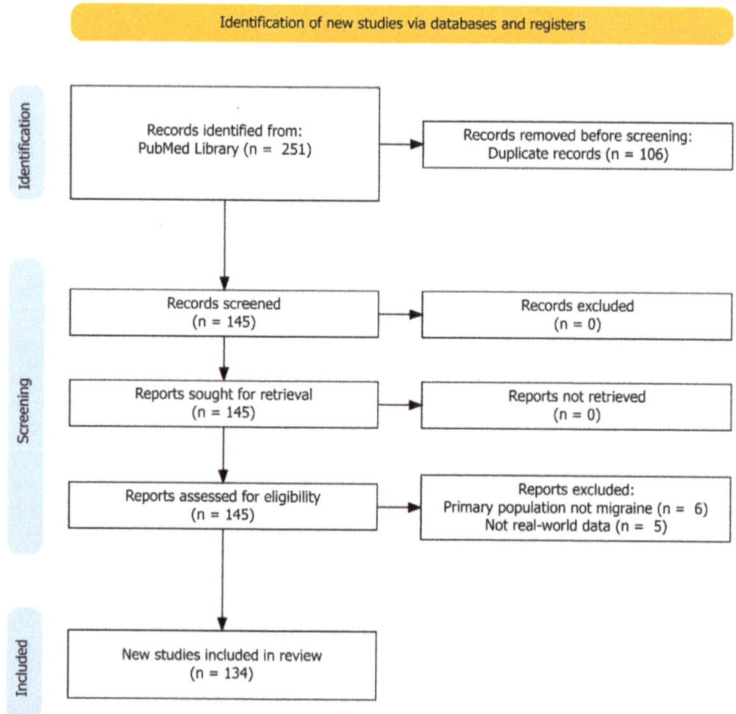

Figure 1. Identification of studies according to the PRISMA Guidelines (Preferred Reporting Items for Systematic Reviews and Meta-Analyses).

We classified these articles into pharmacoepidemiologic studies (n = 8) [32–39], clinic-based studies (n = 83) [40–122], case reports (n = 40) [123–162], and other articles (n = 5) [163–167]. Eighty-nine articles were retrospective [32–71,106,109,111,117,118,120,123–167] and 45 prospective [72–105,107,108,110,112–116,119,121,122]. Outcomes for erenumab, galcanezumab, and fremanezumab were reported in 113, 45, and 31 studies. Real-world data of eptinezumab were only available in one study [167].

4. Pharmacoepidemiologic Studies

Table 1 summarizes the pharmacoepidemiologic studies that looked at real-world prescription data. Due to the nature of such databases, clinical outcomes such as efficacy, adverse events, or days with acute medication use could not be collected. However, large insurance-based datasets allowed us to look at physicians' prescription patterns or claims made by the patients. Thus, the persistence of treatment and adherence could be assessed. Inferences on the efficacy of the therapies could only be made indirectly.

We grouped the main study results by outcome parameters and looked at prescriptions of acute and prophylactic migraine medications, treatment adherence, health care resource utilization (HCRU), days with sick-leave, and impact of migraine and adverse events.

Table 1. Pharmacoepidemiologic studies.

Reference	CGRP-mAb	Patients (n)	Women (%)	Mean Age (Years)	Migraine Diagnosis Available/Diagnosis According to	Inclusion of Patients with				
						Migraine with Aura	Chronic Migraine	Medication Overuse	Prior Treatment Failure	Other Prophylactic Medication
[32]	E	82	85.4	45	Yes/ICD-10	NA	NA	NA	Yes	Yes
[33]	E	4437	85.8	47	Yes/ICD-10	Yes	Yes	NA	NA	Yes
[34]	E	14,282	83.0	46	No	Yes	Yes	NA	Yes	NA
[35]	E	29,451	79.2	47	No	NA	NA	NA	Yes	Yes
[36]	F	172	83.7	46	No	NA	NA	NA	Yes	Yes
[37]	E, OBTA	2676	91.6	50	Yes/ICD	Yes	Yes	NA	Yes	Yes
[38]	E	3171	84.8	51	Yes/ICD	Yes	Yes	NA	Yes	Yes
[39]	E, F, G	3082	85.7	44	Yes/ICD-10	Yes	Yes	NA	Yes	Yes

Abbreviations: E erenumab, F fremanezumab, G galcanezumab, OBTA OnabotulinumtoxinA, ICD International Classification of Diseases, ICD-10 International Classification of Diseases, Tenth Revision, NA: information not available.

4.1. Acute Medication

Five studies assessed the prescription of acute migraine medications six to twelve months before and six to twelve months the after first administration of an anti-CGRP-mAb [32,33,35,37,38]. The different methods of data representations do not allow us to calculate direct comparisons or summaries of data. Comparing baseline to treatment with erenumab, the prescription of acute migraine medications decreased by 49% [35] and by 23% [38], respectively; and the proportion of patients using no prescription acute medication at all or only one type increased [33]. Analyzing specific acute migraine medications, the prescription of non-steroidal anti-inflammatory drugs decreased significantly [37]. In addition, there was a (numerical) decrease in the prescription of triptans [32,37] and barbiturate-containing acute medications [37]. Comparing erenumab to OnabotulinumtoxinA, reductions were stronger for erenumab [37].

4.2. Prophylactic Medication Apart from Anti-CGRP-mAbs

Four studies assessed the prescription of prophylactic medications before and after the first administration of an anti-CGRP-mAb. Three studies reported on erenumab [32,33,35] and one included erenumab, fremanezumab, and galcanezumab [39]. In the first, the prescription of other prophylactics decreased by roughly 30%. In addition, this study found that 50% percent of the patients with standard therapies stopped them within one month, but less than 20% of the patients on anti-CGRP-mAbs stopped their antibody-therapy within one month [32]. The second study [33] observed a shift to fewer prescriptions of preventive medications. The mean time until other ongoing preventive medications were stopped was 185 to 230 days, and 36% had stopped other prophylactics at twelve months in the third study [35]. In the study including three antibodies [39], patients received significantly less often other prophylactics during follow-up and 75% stopped other prophylactics during the twelve-month follow-up.

4.3. Adherence and Persistence

Three studies examined the adherence or persistence. [34,35,39] The adherence to anti-CGRP-mAbs was higher ($\geq$0.8) than to oral prophylactics but still not at the optimum [35]. In the Novartis Go Program [34] offering advice, injection training, and erenumab free of charge until the individual insurance was willing/able to pay for erenumab, the persistence of treatment was 71% at 360 days and 63% at 450 days, which is better than under oral preventives [1]. Varnado et al. [39] found a higher main persistence under anti-CGRP-mAbs than under standard prophylactics and a significantly higher adherence at six and twelve months (medication possession rate 58% vs. 37%, proportion of days covered 55% vs. 35%).

4.4. Health Care Resource Utilization

HCRU was analyzed in four studies [32,36–38]. During treatment with erenumab, migraine-specific office visits decreased statistically significantly from 86.2% to 77.6% [38], claims for health care utilization decreased by 10–19% [37], and health care visits decreased by 45% in the study of Autio et al. [32]. Similarly, treatment with [36] was associated with a significant reduction in HCRU. Emergency visits decreased by 25% and outpatient visits by 22%.

4.5. Sick-Leave and Impact of Migraine

Only one very small study [32] addressed the question of whether treatment with an anti-CGRP-mAb, namely erenumab, has an impact on sick-leave. The results suggest that erenumab may significantly reduce the number of headache-related sick-leave days in employed patients with migraines, managed in routine clinical practice. In detail, sick leave days per patient year decreased by 74%, i.e., from 4.9 to 1.3.

Another single study [36] found a reduction in self-reported headache frequency and migraine pain intensity during treatment with fremanezumab.

4.6. Summary

These pharmacoepidemiologic data indirectly hint to the real-world effectiveness of and adherence to anti-CGRP-mAbs. The biggest limitation is that clinical outcome data were not available. Most of these studies were carried out in the Unites States of America or Canada, only one in Europe, reflecting the insurance systems of these countries which cannot be generalized to other countries. The observation periods were limited to 6 to 12 months.

Such databases capture the prescription of medications and the dispensation to patients; however, they cannot capture if the medications are actually used by the patients, and they were not primarily made for research. Moreover, pharmacoepidemiologic data do not provide information on the reasons for stopping therapy with an anti-CGRP-mAb.

All but four pharmacoepidemiologic studies included only erenumab, which was marketed first [32–35,37,38]. All studies bear the risk of bias, as they were supported by pharmaceutical companies. The risk of bias is highest in the studies by Varnado et al. [39] and Gladstone et al [34]. The first, reporting claims data of erenumab, galcanezumab, and fremanezumab [39], was performed by Eli Lilly and focused on the switch to galcanezumab. The second [34] was biased because Novartis offered erenumab for free if the patient's insurance did not cover the costs.

5. Clinic-Based Studies

As of 1 December 2022, we found 83 clinically based, real-world studies involving all anti-CGRP-mAbs except for eptinezumab. Details of all studies are given in Supplementary Tables S1 and S2. Out of the 83 studies, 21 were supported by pharmaceutical companies.

5.1. Study Design

About half of the studies had a prospective study design (45/83), stating more often clear inclusion (77/83) and to a lesser degree, clear exclusion criteria (41/83). All but one study cited the latest ICHD-3 criteria [168], while about half of the studies stated whether the migraine patients had auras or not—26 did not make this distinction. Practically all studies recruited patients with chronic migraine (81/83) and just over half of these included patients with episodic migraine (49/83—no study focused solely on episodic migraine). Medication overuse headache (MOH) was clearly reported in sixty of these, three did not, while twenty studies did not explicitly state the presence of MOH patients.

5.2. Patients

On average, 180 patients (SD 269.6, median 100, IQR 52–160) were recruited. As can be expected, most of these patients were women (mean 149 patients, SD 220, median 85, IQR 41–132); however, four studies did not specify gender. The average age of the patients was 46.7 (median 47.1, IQR 45.7–49), although some studies opted to report median and IQR instead. All but four studies reported patients as having prior prophylactic treatment failure or refractory migraines. Unfortunately, many of these studies lost their patients during the study period. In 73 articles reporting patient numbers at baseline as well as at the last available follow-up, the total number of patients decreased by a mean of 18.9%. Driessen et al., in both of their papers [49,111], went on to lose over 90% of the initially recruited patients (1003 recruited and 92 patients analyzed at 6 months of treatment). Thus, the reported results must be considered critically.

5.3. Anti-CGRP-mAbs

Erenumab alone was studied in 48/83 articles, eight studied galcanezumab alone, three examined exclusively fremanezumab; meanwhile seven studies compared the effects of erenumab and galcanezumab, three studied examined patients treated with erenumab and fremanezumab, and 14/83 studies included all three.

5.4. Effectiveness

The effectiveness of anti-CGRP-mAb treatment was reported in 77 of the 83 clinic-based studies; however, the data is grossly heterogeneous. Only 16 studies reported both monthly migraine and monthly headache days. Baseline average monthly migraine days were not reported by 52 studies; instead, 13/52 studies opted to report median and IQR. The other 39 decided to split their results in terms of either anti-CGRP responders or non-responders, episodic migraine or chronic migraine, or did not report this data at all. The average number of monthly migraine days at 3 months of treatment with an anti-CGRP-mAb was reported by just thirteen studies, at 6 months by eleven studies, none reported at 9 months of treatment, while four reported average monthly migraine days after 12 months of treatment. Similarly, the average number of monthly headache days was also inconsistently reported. Only 29 studies reported baseline monthly headache days and this number dwindled with the respective 3-month, 6-month, 9-month, and 12-month follow-ups (thirteen, ten, one, and four studies, respectively). Another effectiveness metric, monthly acute medication use, was comparably inconsistently reported. Only twenty studies reported baseline data, which went on to be sparsely reported, with only five studies reporting 12-month data. Finally, 50% responder rates (≥50% reduction in monthly migraine/headache days compared to baseline) were reported in 71/83 articles—however, again with varying methodologic preference. The 50% responder rates in terms of monthly migraine days at specific time points, namely 3, 6, and 12 months were reported only in twenty-two, thirteen, and eight articles, respectively. The average proportion of 50% responders seemed to increase over time and was 44% (SD 20.1%, median 48.7%, IQR 27.5%–58.3%) at 3 months, 49.7% (SD 27.1%, median 53.3%, IQR 26.8%–67.1%) at 6 months, and 63.6% (SD 25.6%, median 61.1%, IQR 46.6%–91.1%) at 12 months. A similar trend could be seen in the 50% responder rate in terms of monthly headache days. A summary of the effectiveness of the different anti-CGRP-mAbs is provided in the Supplemental Tables S1 and S2.

The overall conclusion is that a significant treatment benefit is reported in the real-world longitudinal studies, just as in the Phase 3 open-label extensions [27–30]; however, these real-world results must be treated critically as many studies are limited by their short observation period and many lost patients to follow-up, which significantly affected the responder rates reported; i.e., non-responders are probably more likely to be lost during follow-up than responders, and thus the response rate will increase. Moreover, 34 of the 56 studies did not include baseline data and therefore, it was impossible to verify the authors' claimed observed effectiveness rates.

5.5. Concomitant Pharmacoprophylaxis

Around half of the studies (42/83) also tracked whether patients remained on previous migraine prophylaxis while undergoing treatment with an anti-CGRP-mAb. Thirteen studies conducted direct comparisons to treatment with OnabotulinumtoxinA, after which antidepressants were the next most common concomitant prophylactic reported. Patients treated concomitantly with OnabotulinumtoxinA showed significant reductions in migraine and headache days, displaying a possible synergistic benefit of the two treatments in patients with chronic migraine. None of the 42 articles clearly stated whether the concomitant prophylactic treatment was slowly titrated out or whether they were regular migraine therapies. Thus, the real-world data do not allow us to infer whether concomitant prophylactic migraine treatment works synergistically to relieve the burden of disease in migraine patients.

5.6. Treatment Break

Two studies described patients undergoing planned and unplanned treatment breaks [47,68], ten explicitly described a planned break in treatment with the anti-CGRP-mAb [52,62,73,80,83,84,92,95,101,122], and six reported an unplanned break in treatment [42,45,58,66,74,96]. In contrast, 65 of these 83 real-world studies did not have study periods that allowed for analysis of a treatment break or did not describe a treatment break at all. Most interestingly, all but one of the studies addressing planned treatment breaks made their primary endpoints the effect of pausation of treatment, which meant little was discussed about their treatment benefit leading up to the treatment break [52,62,73,80,83,84,92,95,101]. Nine studies reported the time to migraine return and the corresponding patient number [47,52,74,80,83,84,92,101,122]. Eight found that in a range from 4 to 12 weeks after pausing or interrupting treatment with anti-CGRP-mAbs, patients began to experience increased migraine frequency [47,52,74,80,83,84,92,122]. Vernieri et al. reported no worsening of migraine frequency within the first 3 months [101]. In this regard, the studies by Gantenbein et al. [52], Iannone et al. [85], and Nsaka et al. [122] give us the most relevant real-world data, as they shared the initial 12-month treatment benefit in addition to the effects of treatment pausation of 3 months and 1 month after re-initiation. Gantenbein et al. and Nsaka et al. reported that no participants experienced lasting effects (i.e., longer than 3 months) of their anti-CGRP therapy [52,122], while Iannone et al. reported that 12/44 patients did not meet criteria to restart anti-CGRP therapy [84].

5.7. Switching to Another Anti-CGRP-mAb

Of the 19 studies looking at ≥2 anti-CGRP-mAbs, 11 studies considered the effects of switching therapies. These studies examined a variety of questions without consistent reporting. The overarching aims were to reaffirm effectiveness and safety of the studied anti-CGRP-mAbs and to compare them against other prophylactic treatments (i.e., OnabotulinumtoxinA). In general, the clinical aspects of anti-CGRP-mAb treatment appear very heterogeneous. Two studies documented an improvement after switching to another anti-CGRP-mAb; 8/25 [62] and 8/15 [65] patients showed a ≥30% improvement in monthly migraine days after switching from anti-receptor-mAb to an anti-ligand-mAb.

5.8. Discontinuation of Antibody Treatment

Many studies discussed treatment discontinuation (57/83). Interestingly, 26 studies had no patients discontinue treatment. An often-cited reason for discontinuation was "perceived lack of effectiveness"; however, no paper went on to state the migraine or headache frequencies of these patients.

5.9. Adverse Events

Sixty-one studies reported adverse events, eighteen saw no adverse event in their patient populations, and four did not give any information on adverse events (Table 2). Studies mainly relied on patient reporting of adverse events (61/83), while one went further and used a structured questionnaire. Adverse event intensity and duration were rarely gathered (five and eight articles, respectively). Causality of the adverse event with anti-CGRP treatment was discussed in 55 articles and adverse event frequency (i.e., how many patients) was mentioned in 57/83 articles. Constipation was the most common side effect reported, while reaction at the site of injection was the next most common. A plethora of other adverse events was reported in the studies that are not part of the official list of side effects for anti-CGRP-mAbs. Among these, flu-like symptoms, arthralgia, gastric pain, and chest pain were more frequent. In addition, there were single observations of hypertension and hair loss. Forty-four of the eighty-three articles described the cessation of treatment due to adverse events. Generally, an average of 5.9% of the patients (SD 11.4%, median 1.2%, IQR: 0–5.9%) stopped treatment due to side effects.

Table 2. Most frequent adverse events reported in clinic-based studies.

Adverse Event	Inquired (Number of Studies)	Observed (Number of Patients)
Constipation	50	1251
Reaction at injection site	42	217
Dizziness	39	78
Muscle cramps	38	41
Pruritus	37	44
Pain at injection site	36	76
Skin rash	36	19
Urticaria	36	12

Recently, a prospective study from the Leiden Headache Center reported a small blood pressure increase in migraine patients after initiation of erenumab or fremanezumab [116]. In this study, the effect was more consistent after erenumab initiation, where systolic blood pressure was elevated in all follow-up visits, whereas only systolic blood pressure was elevated in the first follow-up visit on fremanezumab. No blood pressure increase was observed in a control group without CGRP treatment. However, this study contrasts with pivotal erenumab and fremanezumab phase-3 studies and open-label extension studies [4,9,11,13–15]. No blood pressure increase was observed in an open-label study over 5 years [27]. Methodological issues such as the standardization of blood pressure measurements and the balancing of investigational groups merit discussion. In summary, a subtle signal for the development of worsening of blood pressure after CGRP blockade is possible in the real-world setting, but further investigation is needed. Thus, repeated blood pressure measurements can be recommended for patients on anti-CGRP-mAb therapy.

5.10. *Severe Adverse Events*

We found 18 articles that reported one or multiple severe adverse events, 39 that found none, and 26 that did not make any mention of severe adverse events. The most common severe adverse event reported was severe constipation; no deaths were directly attributed to the therapy.

5.11. *Summary*

The results from clinically based, real-world studies are diverse and generally did not have reporting guidelines to refer to until recently [169–171]. This lack of reporting guidelines—or at least lack of awareness—has led to a variety of data to be published since the approval of anti-CGRP-mAbs. Nevertheless, clinic-based real-world studies seem to suggest that the monoclonal antibodies are similarly effective as seen in the clinical trials. Furthermore, their safety and tolerability profiles appear to be equally similar; except, for hypertension being added to the official list of possible side effects, even though the causal relation is disputed [163,164].

6. Case Reports

Among forty case reports, twenty-seven described a single patient, four reported on two, and five on three patients, and one paper each included four, five, eight, and ten patients [123–162]. In these 77 patients, the mean age was 44.6 (SD 9.63) and 76.6% were women, 30 had used erenumab, 12 had used fremanezumab, and 6 had used galcanezumab.

Case reports may give hints on rare adverse events in the clinical setting. Inherently, causal associations between single observations and a given drug can hardly be drawn, but collecting information is important to detect the possible clustering of events. Notably, beneficial effects of anti-CGRP-mAbs beyond their actual indication are also possible. The fact that most reports were on erenumab, the first anti-CGRP-mAb to be licensed, may give a biased view on effects or side effects. Furthermore, most reports were on observations in women, reflecting prescription practice and migraine epidemiology. Conceptually, case reports were found to cover the following situations:

i. Improvement of a symptom or comorbid condition;
ii. Effectiveness and no adverse events under special circumstances;
iii. Adverse events in otherwise healthy individuals;
iv. Adverse events because of possible drug interactions, or potentiation of side effects;
v. Deterioration of preexisting disorder.

Improvement of a symptom or comorbid condition with anti-CGRP-mAbs was reported for migraine aura [124], cluster headache [134,148], headache related to sexual activity [139], nummular headache [160], restless leg syndrome [159], sleep terrors [149], and stuttering [150]. In three patients, severe nausea induced by erenumab led to smoking cessation [140].

Single reports on effectiveness without adverse events covered the exposure to erenumab in the first weeks of pregnancy [129,157], during whole pregnancy [155], during breast feeding [133], and in myasthenia gravis treated with immunoglobulins [143]. A case-report of three pregnancies reported two full-term deliveries and one miscarriage after exposure to erenumab. The two full-term pregnancies administered one dose of erenumab during the pregnancy, immediately stopping treatment afterwards and not experiencing any complications, while the patient who had a miscarriage ceased treatment 1 month prior to learning she was pregnant. In the latter case, a rare intrauterine complication was found (gestational throphoplastic neoplasia), but the known risk factors for this complication do not seem to correlate with the mechanisms of erenumab [155]. According to a WHO pharmacovigilance database on erenumab, galcanezumab, and fremanezumab exposure during pregnancy and lactation, no specific risk for toxicity could be detected, but data were limited to 94 cases (more than half on erenumab). Although adverse events including spontaneous abortions or birth defects were reported, this was not increased in the exposed patients. Further data collection, for instance, in registries, seems mandatory before definite advice concerning the safety of CGRP-mAbs in pregnancy and lactation can be given [172].

In one patient each, erenumab and fremanezumab were effective in COVID-19-related migraine exacerbations [126,132], and in two patients, the use of rimegepant during treatment with erenumab was found effective and was well-tolerated [142]. Notably, no recommendation concerning the safety in these conditions can be given based on this anecdotal evidence. In contrast, a series of 10 patients treated with both erenumab and OnabotulinumtoxinA added to the pharmacoepidemiologic data on this combination [147]. In the absence of evidence from RCTs, patients with otherwise refractory migraine, may benefit form anti-CGRP-mAbs administered together with OnabotulinumtoxinA.

A possible anti-CGRP-mAb adverse event in an otherwise healthy individual was reported by Rozen et al. [145]. In summary, a 43-year-old woman developed a sexual headache and a thunderclap headache 2 days after the second dose of erenumab and after high-altitude exposure and triptan use in the week before. CT angiogram results showed narrowing of the left middle and anterior cerebral arteries, consistent with reversible cerebral vasoconstriction syndrome. Treatment with erenumab and triptans was stopped, and verapamil was initiated. The CT angiogram was normal 4 weeks after initial neuroimaging, supporting the diagnosis of reversible cerebral vasoconstriction syndrome (RCVS). [145]. The observation of cerebral vasospasms after a CGRP blockade is of considerable interest, given the vasodilatory effects of CGRP. However, it has been suggested that anti-CGRP-mAbs might not reach the abluminal compartment of cerebral blood vessels within the blood brain barrier and thus might be an unlikely cause of RCVS [173,174]. Another limitation of the hypothesis of a possible causal relationship between erenumab and RCVS is pharmacokinetics, as the maximum concentration of erenumab is reached later.

From a clinical point of view, this case report contrasts with the patient mentioned above who used erenumab for migraines and experienced improvement of headaches related to sexual activity.

An Australian-Irish collaboration found serious adverse events (SAEs) in eight patients from centers in Australia and Ireland, forcing all patients to cease their use of anti-CGRP-mAbs, related to inflammatory complications of CGRP monoclonal antibodies [161]. In this article, three of the eight patients had a pre-existing, well-controlled rheumatological or dermatological disease, which worsened significantly in eight patients after the anti-CGRP-mAb therapy was started. Six patients developed a de novo inflammatory disease after exposure. Causality was established based on the remission of symptoms after withdrawal of anti-CGRP-mAbs. Patient 1, for example, suffered from rheumatoid arthritis, dyslipidemia, and pulmonary fibrosis, finally experiencing autoimmune hepatitis after one injection of erenumab and ceasing therapy thereafter. One patient with fibromyalgia and chronic fatigue syndrome developed ocular Susac's syndrome after 12 months of erenumab treatment. In this series, one patient without significant comorbidities developed granulomatosis with polyangiitis after treatment with fremanezumab. Patients 6 and 7, on the other hand, experienced worsening of their psoriatic conditions, leading them to stop their therapies with galcanezumab and erenumab, respectively. The cases provided and explained by Ray and colleagues show that antagonism of CGRP should be carefully considered, especially in patients with pre-existing immunological diseases, as CGRP's role in inflammatory regulation should not be underestimated, and its inhibition can lead to serious, albeit rare, SAEs. The authors discussed possible effects of CGRP blockades on Langerhans cells, macrophages, and mast cells, as well as effects on cytokine production. In this real-life series with SAEs, causality could not be further undermined since re-exposure was not possible; thus, it is vital that such events be consistently reported. While these case series are extremely engaging, reports regarding the real-life, complication-free use of anti-CGRP-mAbs in patients with pre-existing autoimmune conditions should be reported—and are equally valuable.

A further interesting article by Wurthmann et al. reported skin lesions and impaired wounds in a previously healthy patient [151]. In essence, the patient using erenumab presented with crescent-shaped necroses on the inner surface of the left forearm that formed from a singular erythematous papular skin lesion, no bigger than 1 cm. The vessels supplying the upper cervicobrachial plexus were thrombosed, and the authors hypothesized that erenumab caused a decreased blood flow to small blood vessels, leading to necrosis. Whether remission of the symptoms following cessation of erenumab supports this hypothesis must remain open.

In a case report by Aradi et al. [125], it is less clear if the patient was otherwise healthy. The authors describe a 41-year-old woman with migraine without aura who developed a right thalamic infarction following a first dose of erenumab. The stroke developed 34 days after the first exposure to erenumab and 4 h after medication with rizatriptan, which the patient had taken before without complications. In addition, the patient was on a low-dose estrogen oral contraceptive. She had no other vascular risk factors. A CT angiography of the head and neck demonstrated a proximal right posterior cerebral artery stenosis in the P1 segment, which resolved after 2 months and was thus interpreted as a vasospasm. In this patient, blood tests for hypercoagulopathy were negative and transesophageal echocardiography revealed no source of embolus; however, long-term electrocardiograms to rule out atrial fibrillation were not reported. Thus, this case is potentially confounded by incomplete diagnostic work up and concomitant use of other substances potentially related to ischemic stroke. The authors discussed the possibility that CGRP blockades might impair vasodilatory mechanisms to compensate for triptan-induced vasoconstriction. However, triptans seem to reverse vasodilatation of intracranial arteries during the migraine attack rather than cause intracranial vasoconstriction [175] and have been safely used for migraine therapy for decades.

The case report from Lehman et al. describing deterioration of a pre-existing cerebrovascular disorders warrants serious scrutiny [138]. An anti-CGRP-mAb was prescribed to a migraine patient with cerebral proliferative angiopathy. Two days after the first subcutaneous administration of erenumab, the patient presented with status epilepticus and showed diffusion abnormalities in the MRI in vicinity to the cerebral proliferative angiopathy. The authors summarized that the patient had recurrent refractory epilepsy with lasting damage to his motor as well as visuospatial functions. This report serves to teach that anti-CGRP-mAbs should be prescribed with caution, weighing the risks and benefits of anti-CGRP-mAbs in certain comorbid conditions.

More case reports on adverse events are summarized in Table 3.

Table 3. Adverse events from case reports.

Adverse Events in Otherwise Healthy Individuals					
Ref.	**Age**	**Sex**	**Exposure**	**Adverse Event**	**Comment**
[123]	54	M	G	Erectile dysfunction	More than 2 months after start, reversible after 2 half-lives, rare use of metoprolol for palpitations
[128]	33	M	E	Raynaud's phenomenon	When in the cold cca. 1 h, had RP of all the fingers and toes bilateral with temperature change and numbness lasting about 1 h
[131]	38	F	E	Restless leg-like symptoms	De novo symptoms; erenumab continued despite symptoms
[131]	47	F	G	Restless leg-like symptoms	De novo symptoms; cessation of symptoms after erenumab discontinuation
[136]	61	F	G	Migraine aura	Unsuccessful with erenumab, 1 month after last injection switch to galcanezumab (240 mg loading dose, followed by a maintenance dose of 120 mg 28 days later), within 1 week after the first dose of 120 mg, experienced first visual aura
[137]	48	F	G	Skin lesions in fixed location	After several months, developed erythema and pruritus of left upper arm within 24 h of self-injection (lasting up to 3 days), evolved into a nonpruritic, non-painful, chronic, brown-to-blue patch. Each monthly injection of galcanezumab resulted in same clinical course (at identical site on the left arm), despite injecting different areas on body (incl. the abdomen and thighs), without reaction at injection site
[141]	52	F	F	Non-immediate rash	Causal relation confirmed with pinprick test
[144]	26	F	E	Stypsis	Exteroceptive suppression period of the temporalis muscle was assessed during a ten-day washout period, before starting erenumab and after 4 months of erenumab treatment
[146]	60	F	E	Xerostomia	After first injection, reported dry mouth in the next ten days; similar duration after 2nd injection
[151]	51	F	E	Impaired wound healing of trivial injury	Improvement after discontinuation of erenumab
[156]	57	F	E	Myocardial infarction	Former smoker, family history of cardiovascular disease
[162]	55	M	E	Myocardial infarction	BMI of 29, non-smoker, suffered from hypertension, dyslipidemia, and prior myocardial infarction in 2012
[154]	48	F	E	Symmetrical drug-related intertriginous and flexural exanthema	Erenumab discontinued and switched to fremanezumab

Table 3. *Cont.*

Ref.	Age	Sex	Exposure	Adverse event	Comment
Adverse events because of possible drug interactions, or potentiation of side effects					
Ref.	**Age**	**Sex**	**Exposure**	**Adverse event**	**Comment**
[127]	41	F	E + fish oil	Extreme ecchymoses	Improvement after discontinuation of fish oil
Deterioration of preexisting disorder					
Ref.	**Age**	**Sex**	**Exposure**	**Adverse event**	**Comment**
[128]	45	F	F	Raynaud's phenomenon	At 6-month follow-up, reported frequent and more severe RP (the thumb was not involved) including mild digital ulcers (which had healed by the time of the visit) for about 1 month after receiving galcanezumab.
[128]	65	M	G	Raynaud's phenomenon	Onset few weeks after fremanezumab injection, frequent episodes of RP involving all the fingers of both hands in cool temperatures
[130]	39	F	E	Paralytic ileus in a patient after undergone abdominal surgery	Paralytic ileus is a known complication of abdominal surgery
[146]	35	F	E	Xerostomia	Previous xerostomia, and patient was on amitriptyline

Abbreviations: E erenumab, F fremanezumab, G galcanezumab.

Summary

Based on anecdotal evidence from case reports, no definite conclusions can be drawn. Case reports included observations of contradictory findings, e.g., de novo appearance [136] or substantial improvement of auras [124] and de novo appearance or significant improvement [139] of headaches related to sexual activity [145]. This could be explained by differential effects based on unknown cofactors or reflect the report of mere coincidences. Based on current real-world data, no clustering of rare side effects was observed.

However, in our opinion, the observation of possible adverse events related to the blockade of the vasodilator CGRP deserves attention. One stroke related to vasoconstriction [163] and one case of RCVS [145] were reported. It must be emphasized that cryptogenic stroke is common in young individuals. In addition, new appearance or exacerbation of Raynaud´s phenomenon was observed [128]. The issue of the possible development or exacerbation of hypertension is not fully understood yet. Thus, we conclude that patients should be screened for high vascular risk before the initiation of CGRP-based therapies.

7. Other Articles

Finally, we want to review five articles: two related to hypertension [163,164], two articles focusing on adverse events (AEs) [165,166], and one reporting on Raynaud's phenomenon [167] in patients using anti-CGRP-mAbs.

Saely et al. summarized 57 reports of elevated blood pressure associated with the use of erenumab submitted to the FDA Adverse Event Reporting System [164]. In this case series, baseline blood pressure was reported in only half the patients, and reports of hypertension were based on single elevated blood pressure measurements, which precludes robust conclusions. Subsequently, Dodick et al. gathered information on all post-marketing adverse event reports of hypertension in erenumab users using the Amgen global safety database and summarized them into a single article containing 355 patient cases [163]. Adverse events of hypertension occurred, in part, in patients with pre-existing hypertension—one third of patients with serious hypertension had previous hypertension. Time of onset was not described in more than half of the reports, while about half of the cases with hypertension were reported after 1 week of the first administration. The authors conclude that adverse event rates of hypertension reported with erenumab in the post-marketing setting were generally low and that only with additional studies can this risk be properly characterized.

Two studies focused specifically on adverse events during real-world use of anti-CGRP-mAbs [165,166]. Overall, patients reported "migraine", "headache", and "drug ineffective", along with migraine-associated symptoms (i.e., nausea) and "injection-site" reactions as the most common AEs for all erenumab, galcanezumab, and fremanezumab. Cardiovascular events were outside of the top ten AEs for any of the three anti-CGRP-mAbs. "Constipation" was the second most commonly reported AE for erenumab; however, it found itself outside the top ten AEs for fremanezumab or galcanezumab. Serious AEs were infrequent across all three anti-CGRP-mAbs [166]. A particular topic of interest was Raynaud's phenomenon (RP), which is followed by the World Health Organization in its VigiBase® [165]. CGRP-targeting drugs were significantly associated with Raynaud's phenomenon. Erenumab was the most reported anti-CGRP-mAb (with 56/99 reports). The median time to RP onset was 84 days; however, it never led to fatality, with one patient suffering gangrene and extremity necrosis [165]. The authors could not, however, conclusively determine from the evidence in the database whether the occurrence of RP was de novo or a worsening of pre-existing RP. Nevertheless, consideration should be taken before prescribing anti-CGRP-mAbs to migraine patients with the potential to develop RP.

Breen et al. [167] examined a cohort of patients with Raynaud's phenomenon from a specialized clinic who were treated with CGRP antagonists for migraines. Most Raynaud patients (160/169) experienced no complications, and a minority (9/160) of patients experienced complications including microvascular complications (such as worsening facial telangiectasias or digital necrosis requiring surgery), all of whom had received anti-CGRP-mAbs (erenumab, galcanezumab, fremanezumab, and eptinezumab). Approximately half of patients with complications developed Raynaud's phenomenon de novo shortly after the first exposure. In this cohort, no significant difference in demographic or clinical variables was detected in patients with or without complications. The authors concluded that anti-CGRP-mAbs should be used with caution in patients with Raynaud's phenomenon.

8. Conclusions

With few exceptions, available real-world data are limited by retrospective data collection, small patient numbers, and short follow-up periods. For the time being, the majority of real-world papers seem to support good efficacy and tolerability of anti-CGRP-mAbs in the real-world setting. Furthermore, direct head-to-head comparisons between the anti-CGRP-mAbs are made difficult by the heterogeneity of results reported. Reports of rare adverse events must be carefully monitored, but causal relations may not be concluded from single case studies. Particular attention is given to vascular events related to anti-CGRP-mAbs, although no clear vascular safety signal has emerged yet. De novo appearance or worsening of Raynaud's phenomenon must be carefully monitored. There is an unmet need for large prospective real-world studies and registries providing long-term follow-ups of patients treated with anti-CGRP-mAbs.

Supplementary Materials: The following supporting information can be downloaded at: https://www.mdpi.com/article/10.3390/cells12010143/s1, Table S1: Characteristics reported in clinic-based studies; Table S2: Outcome variables reported in clinic-based studies.

Author Contributions: Conceptualization, A.R.P. and C.W.; methodology, A.R.P.; formal analysis, A.R.P., C.W., F.R. and K.Z.; data curation, A.R.P. and C.W.; writing—original draft preparation, A.R.P., C.W., F.R. and K.Z.; writing—review and editing, A.R.P., C.W., F.R. and K.Z.; visualization, A.R.P. and C.W.; supervision, C.W. All authors have read and agreed to the published version of the manuscript.

Funding: This research received no external funding.

Data Availability Statement: Not applicable.

Conflicts of Interest: A.R.P. has nothing to report; C.W. has received honoraria for consultancy/speaking from Apomedica, Curelator, Eli Lilly, Grünenthal, Hermes, Novartis, Pfizer, Ratiopharm/Teva; and

Stada, F.R. has received honoraria for consultancy/speaking from Teva, Novartis, Eli Lilly, Lundbeck, Abbvie, Grünenthal. K.Z. has received honoraria for consultancy/speaking from Teva Ratiopharm, Novartis, Eli Lilly, Abbvie, Lundbeck, Grünenthal, Pfizer, MedAhead.

References

1. Hepp, Z.; Bloudek, L.M.; Varon, S.F. Systematic Review of Migraine Prophylaxis Adherence and Persistence. *J. Manag. Care Pharm.* **2014**, *20*, 22–33. [CrossRef] [PubMed]
2. Ohlsson, L.; Haanes, K.A.; Kronvall, E.; Xu, C.; Snellman, J.; Edvinsson, L. Erenumab (AMG 334), a monoclonal antagonist antibody against the canonical CGRP receptor, does not impair vasodilatory or contractile responses to other vasoactive agents in human isolated cranial arteries. *Cephalalgia* **2019**, *39*, 1745–1752. [CrossRef] [PubMed]
3. Rubio-Beltrán, E.; Labastida-Ramírez, A.; Haanes, K.A.; Bogaerdt, A.V.D.; Bogers, A.J.; Dirven, C.; Danser, A.J.; Xu, C.; Snellman, J.; MaassenVanDenBrink, A. Characterisation of vasodilatory responses in the presence of the CGRP receptor antibody erenumab in human isolated arteries. *Cephalalgia* **2019**, *39*, 1735–1744. [CrossRef] [PubMed]
4. Dodick, D.W.; Ashina, M.; Brandes, J.L.; Kudrow, D.; Lanteri-Minet, M.; Osipova, V.; Palmer, K.; Picard, H.; Mikol, D.D.; Lenz, R.A. ARISE: A Phase 3 randomized trial of erenumab for episodic migraine. *Cephalalgia* **2018**, *38*, 1026–1037. [CrossRef] [PubMed]
5. Goadsby, P.J.; Reuter, U.; Hallström, Y.; Broessner, G.; Bonner, J.H.; Zhang, F.; Sapra, S.; Picard, H.; Mikol, D.D.; Lenz, R.A. A Controlled Trial of Erenumab for Episodic Migraine. *N. Engl. J. Med.* **2017**, *377*, 2123–2132. [CrossRef] [PubMed]
6. Stauffer, V.L.; Dodick, D.W.; Zhang, Q.; Carter, J.N.; Ailani, J.; Conley, R.R. Evaluation of Galcanezumab for the Prevention of Episodic Migraine: The EVOLVE-1 Randomized Clinical Trial. *JAMA Neurol.* **2018**, *75*, 1080–1088. [CrossRef] [PubMed]
7. Dodick, D.W.; Silberstein, S.D.; Bigal, M.E.; Yeung, P.P.; Goadsby, P.J.; Blankenbiller, T.; Grozinski-Wolff, M.; Yang, R.; Ma, Y.; Aycardi, E. Effect of Fremanezumab Compared With Placebo for Prevention of Episodic Migraine: A randomized clinical trial. *JAMA* **2018**, *319*, 1999–2008. [CrossRef]
8. Ashina, M.; Saper, J.; Cady, R.; Schaeffler, B.A.; Biondi, D.M.; Hirman, J.; Pederson, S.; Allan, B.; Smith, J. Eptinezumab in episodic migraine: A randomized, double-blind, placebo-controlled study (PROMISE-1). *Cephalalgia* **2020**, *40*, 241–254. [CrossRef] [PubMed]
9. Lipton, R.B.; Goadsby, P.J.; Smith, J.; Schaeffler, B.A.; Biondi, D.M.; Hirman, J.; Pederson, S.; Allan, B.; Cady, R. Efficacy and safety of eptinezumab in patients with chronic migraine: PROMISE-2. *Neurology* **2020**, *94*, e1365–e1377. [CrossRef]
10. Silberstein, S.D.; Dodick, D.W.; Bigal, M.E.; Yeung, P.P.; Goadsby, P.J.; Blankenbiller, T.; Grozinski-Wolff, M.; Yang, R.; Ma, Y.; Aycardi, E. Fremanezumab for the Preventive Treatment of Chronic Migraine. *N. Engl. J. Med.* **2017**, *377*, 2113–2122. [CrossRef]
11. Tepper, S.; Ashina, M.; Reuter, U.; Brandes, J.L.; Doležil, D.; Silberstein, S.; Winner, P.; Leonardi, D.; Mikol, D.; Lenz, R. Safety and efficacy of erenumab for preventive treatment of chronic migraine: A randomised, double-blind, placebo-controlled phase 2 trial. *Lancet Neurol.* **2017**, *16*, 425–434. [CrossRef] [PubMed]
12. Detke, H.C.; Goadsby, P.J.; Wang, S.; Friedman, D.I.; Selzler, K.J.; Aurora, S.K. Galcanezumab in chronic migraine: The randomized, double-blind, placebo-controlled REGAIN study. *Neurology* **2018**, *91*, e2211–e2221. [CrossRef] [PubMed]
13. Reuter, U.; Goadsby, P.J.; Lanteri-Minet, M.; Wen, S.; Hours-Zesiger, P.; Ferrari, M.D.; Klatt, J. Efficacy and tolerability of erenumab in patients with episodic migraine in whom two-to-four previous preventive treatments were unsuccessful: A randomised, double-blind, placebo-controlled, phase 3b study. *Lancet* **2018**, *392*, 2280–2287. [CrossRef] [PubMed]
14. Goadsby, P.J.; Paemeleire, K.; Broessner, G.; Brandes, J.; Klatt, J.; Zhang, F.; Picard, H.; Lenz, R.; Mikol, D.D. Efficacy and safety of erenumab (AMG334) in episodic migraine patients with prior preventive treatment failure: A subgroup analysis of a randomized, double-blind, placebo-controlled study. *Cephalalgia* **2019**, *39*, 817–826. [CrossRef]
15. Mulleners, W.M.; Kim, B.-K.; Láinez, M.J.A.; Lanteri-Minet, M.; Pozo-Rosich, P.; Wang, S.; Tockhorn-Heidenreich, A.; Aurora, S.K.; Nichols, R.M.; Yunes-Medina, L.; et al. Safety and efficacy of galcanezumab in patients for whom previous migraine preventive medication from two to four categories had failed (CONQUER): A multicentre, randomised, double-blind, placebo-controlled, phase 3b trial. *Lancet Neurol.* **2020**, *19*, 814–825. [CrossRef]
16. Ferrari, M.D.; Diener, H.C.; Ning, X.; Galic, M.; Cohen, J.M.; Yang, R.; Mueller, M.; Ahn, A.H.; Schwartz, Y.C.; Grozinski-Wolff, M.; et al. Fremanezumab versus placebo for migraine prevention in patients with documented failure to up to four migraine preventive medication classes (FOCUS): A randomised, double-blind, placebo-controlled, phase 3b trial. *Lancet* **2019**, *394*, 1030–1040. [CrossRef]
17. Smitherman, T.A.; Tietjen, G.E.; Schuh, K.; Skljarevski, V.; Lipsius, S.; D'Souza, D.N.; Pearlman, E.M. Efficacy of Galcanezumab for Migraine Prevention in Patients With a Medical History of Anxiety and/or Depression: A Post Hoc Analysis of the Phase 3, Randomized, Double-Blind, Placebo-Controlled REGAIN, and Pooled EVOLVE-1 and EVOLVE-2 Studies. *Headache* **2020**, *60*, 2202–2219. [CrossRef]
18. Lipton, R.B.; Cohen, J.M.; Galic, M.; Seminerio, M.J.; Yeung, P.P.; Aycardi, E.; Bigal, M.E.; Bibeau, K.; Buse, D.C. Effects of fremanezumab in patients with chronic migraine and comorbid depression: Subgroup analysis of the randomized HALO CM study. *Headache* **2021**, *61*, 662–672. [CrossRef]
19. Tepper, S.J.; Diener, H.-C.; Ashina, M.; Brandes, J.L.; Friedman, D.I.; Reuter, U.; Cheng, S.; Nilsen, J.; Leonardi, D.K.; Lenz, R.A.; et al. Erenumab in chronic migraine with medication overuse: Subgroup analysis of a randomized trial. *Neurology* **2019**, *92*, e2309–e2320. [CrossRef]

20. Dodick, D.W.; Doty, E.G.; Aurora, S.K.; Ruff, D.D.; Stauffer, V.L.; Jedynak, J.; Dong, Y.; Pearlman, E.M. Medication overuse in a subgroup analysis of phase 3 placebo-controlled studies of galcanezumab in the prevention of episodic and chronic migraine. *Cephalalgia* **2021**, *41*, 340–352. [CrossRef]
21. Silberstein, S.D.; Cohen, J.M.; Seminerio, M.J.; Yang, R.; Ashina, S.; Katsarava, Z. The impact of fremanezumab on medication overuse in patients with chronic migraine: Subgroup analysis of the HALO CM study. *J. Headache Pain* **2020**, *21*, 114. [CrossRef] [PubMed]
22. Diener, H.; Marmura, M.J.; Tepper, S.J.; Cowan, R.; Starling, A.J.; Diamond, M.L.; Hirman, J.; Mehta, L.; Brevig, T.; Sperling, B.; et al. Efficacy, tolerability, and safety of eptinezumab in patients with a dual diagnosis of chronic migraine and medication-overuse headache: Subgroup analysis of PROMISE-2. *Headache* **2021**, *61*, 125–136. [CrossRef] [PubMed]
23. Lipton, R.B.; Tepper, S.J.; Reuter, U.; Silberstein, S.; Stewart, W.F.; Nilsen, J.; Leonardi, D.K.; Desai, P.; Cheng, S.; Mikol, D.D.; et al. Erenumab in chronic migraine: Patient-reported outcomes in a randomized double-blind study. *Neurology* **2019**, *92*, e2250–e2260. [CrossRef] [PubMed]
24. Lipton, R.B.; Cohen, J.M.; Gandhi, S.K.; Yang, R.; Yeung, P.P.; Buse, D.C. Effect of fremanezumab on quality of life and productivity in patients with chronic migraine. *Neurology* **2020**, *95*, e878–e888. [CrossRef]
25. Ford, J.H.; Ayer, D.W.; Zhang, Q.; Carter, J.N.; Leroux, E.; Skljarevski, V.; Aurora, S.K.; Tockhorn-Heidenreich, A.; Lipton, R.B. Two randomized migraine studies of galcanezumab: Effects on patient functioning and disability. *Neurology* **2019**, *93*, e508–e517. [CrossRef]
26. Lipton, R.B.; Charleston, L.; Tassorelli, C.; Brevig, T.; Hirman, J.; Cady, R. Patient-reported outcomes, health-related quality of life, and acute medication use in patients with a ≥75% response to eptinezumab: Subgroup pooled analysis of the PROMISE trials. *J. Headache Pain* **2022**, *23*, 23. [CrossRef]
27. Ashina, M.; Goadsby, P.J.; Reuter, U.; Silberstein, S.; Dodick, D.W.; Xue, F.; Zhang, F.; Lima, G.P.d.S.; Cheng, S.; Mikol, D.D. Long-term efficacy and safety of erenumab in migraine prevention: Results from a 5-year, open-label treatment phase of a randomized clinical trial. *Eur. J. Neurol.* **2021**, *28*, 1716–1725. [CrossRef]
28. Goadsby, P.J.; Silberstein, S.D.; Yeung, P.P.; Cohen, J.M.; Ning, X.; Yang, R.; Dodick, D.W. Long-term safety, tolerability, and efficacy of fremanezumab in migraine: A randomized study. *Neurology* **2020**, *95*, e2487–e2499. [CrossRef]
29. Camporeale, A.; Kudrow, D.; Sides, R.; Wang, S.; Van Dycke, A.; Selzler, K.J.; Stauffer, V.L. A phase 3, long-term, open-label safety study of Galcanezumab in patients with migraine. *BMC Neurol.* **2018**, *18*, 188. [CrossRef]
30. Kudrow, D.; Cady, R.K.; Allan, B.; Pederson, S.M.; Hirman, J.; Mehta, L.R.; Schaeffler, B.A. Long-term safety and tolerability of eptinezumab in patients with chronic migraine: A 2-year, open-label, phase 3 trial. *BMC Neurol.* **2021**, *21*, 126. [CrossRef]
31. Moher, D.; Liberati, A.; Tetzlaff, J.; Altman, D.G.; The PRISMA Group. Preferred reporting items for systematic reviews and meta-analyses: The PRISMA statement. *BMJ* **2009**, *339*, b2535. [CrossRef] [PubMed]
32. Autio, H.; Purmonen, T.; Kurki, S.; Mocevic, E.; Korolainen, M.A.; Tuominen, S.; Lassenius, M.I.; Nissilä, M. Erenumab Decreases Headache-Related Sick Leave Days and Health Care Visits: A Retrospective Real-World Study in Working Patients with Migraine. *Neurol. Ther.* **2022**, *11*, 223–235. [CrossRef] [PubMed]
33. Chandler, D.; Szekely, C.; Aggarwal, S.; Cyprien, L.; Bensink, M. Migraine Characteristics, Comorbidities, Healthcare Resource Utilization, and Associated Costs of Early Users of Erenumab in the USA: A Retrospective Cohort Study Using Administrative Claims Data. *Pain Ther.* **2021**, *10*, 1551–1566. [CrossRef] [PubMed]
34. Gladstone, J.; Chhibber, S.; Minhas, J.; Neish, C.S.; Power, G.S.; Lan, Z.; Rochdi, D.; Lanthier-Martel, J.; Bastien, N. Real-world persistence of erenumab for preventive treatment of chronic and episodic migraine: Retrospective real-world study. *Headache* **2022**, *62*, 78–88. [CrossRef]
35. Hines, D.M.; Shah, S.; Multani, J.K.; Wade, R.L.; Buse, D.C.; Bensink, M. Erenumab patient characteristics, medication adherence, and treatment patterns in the United States. *Headache* **2021**, *61*, 590–602. [CrossRef]
36. McAllister, P.; Lamerato, L.; Krasenbaum, L.J.; Cohen, J.M.; Tangirala, K.; Thompson, S.; Driessen, M.; Casciano, J.; Dotiwala, Z.; Mauskop, A. Real-world impact of fremanezumab on migraine symptoms and resource utilization in the United States. *J. Headache Pain* **2021**, *22*, 156. [CrossRef]
37. Tepper, S.J.; Fang, J.; Zhou, L.; Shen, Y.; Vo, P.; Abdrabboh, A.; Glassberg, M.B.; Ferraris, M. Effectiveness of erenumab and onabotulinumtoxinA on acute medication usage and health care resource utilization as migraine prevention in the United States. *J. Manag. Care Spéc. Pharm.* **2021**, *27*, 1157–1170. [CrossRef]
38. Tepper, S.J.; Fang, J.; Vo, P.; Shen, Y.; Zhou, L.; Abdrabboh, A.; Glassberg, M.; Ferraris, M. Impact of erenumab on acute medication usage and health care resource utilization among migraine patients: A US claims database study. *J. Headache Pain* **2021**, *22*, 27. [CrossRef]
39. Varnado, O.J.; Manjelievskaia, J.; Ye, W.; Perry, A.; Schuh, K.; Wenzel, R. Treatment Patterns for Calcitonin Gene-Related Peptide Monoclonal Antibodies Including Galcanezumab versus Conventional Preventive Treatments for Migraine: A Retrospective US Claims Study. *Patient Prefer. Adherence* **2022**, *16*, 821–839. [CrossRef]
40. Alex, A.; Vaughn, C.; Rayhill, M. Safety and Tolerability of 3 CGRP Monoclonal Antibodies in Practice: A Retrospective Cohort Study. *Headache* **2020**, *60*, 2454–2462. [CrossRef]
41. Armanious, M.; Khalil, N.; Lu, Y.; Jimenez-Sanders, R. Erenumab and Onabotulinumtoxin A Combination Therapy for the Prevention of Intractable Chronic Migraine without Aura: A Retrospective Analysis. *J. Pain Palliat. Care Pharmacother.* **2021**, *35*, 1–6. [CrossRef] [PubMed]

42. Baraldi, C.; Castro, F.L.; Cainazzo, M.M.; Pani, L.; Guerzoni, S. Predictors of response to erenumab after 12 months of treatment. *Brain Behav.* **2021**, *11*, e2260. [CrossRef] [PubMed]
43. Belvís, R.; Irimia, P.; Pozo-Rosich, P.; González-Oria, C.; Cano, A.; Viguera, J.; Sánchez, B.; Molina, F.; Beltrán, I.; Oterino, A.; et al. MAB-MIG: Registry of the spanish neurological society of erenumab for migraine prevention. *J. Headache Pain* **2021**, *22*, 74. [CrossRef] [PubMed]
44. Blumenfeld, A.M.; Frishberg, B.M.; Schim, J.D.; Iannone, A.; Schneider, G.; Yedigarova, L.; Adams, A.M. Real-World Evidence for Control of Chronic Migraine Patients Receiving CGRP Monoclonal Antibody Therapy Added to Onabotulinumtoxin A: A Retrospective Chart Review. *Pain Ther.* **2021**, *10*, 809–826. [CrossRef] [PubMed]
45. Cainazzo, M.M.; Baraldi, C.; Ferrari, A.; Castro, F.L.; Pani, L.; Guerzoni, S. Erenumab for the preventive treatment of chronic migraine complicated with medication overuse headache: An observational, retrospective, 12-month real-life study. *Neurol. Sci.* **2021**, *42*, 4193–4202. [CrossRef] [PubMed]
46. Cohen, F.; Armand, C.; Lipton, R.B.; Vollbracht, S. Efficacy and Tolerability of Calcitonin Gene–Related Peptide–Targeted Monoclonal Antibody Medications as Add-on Therapy to Onabotulinumtoxin A in Patients with Chronic Migraine. *Pain Med.* **2021**, *22*, 1857–1863. [CrossRef]
47. Dapkutė, A.; Vainauskienė, J.; Ryliškienė, K. Patient-reported outcomes of migraine treatment with erenumab: Results from a national patient survey. *Neurol. Sci.* **2022**, *43*, 3305–3312. [CrossRef]
48. Dinh, B.B.K.; Aziz, W.H.; Terruzzi, A.; Krieger, D.W. Initial experience with novel CGRP-receptor inhibitor therapy in Migraine in the United Arab Emirates: A retrospective observational study. *BMC Neurol.* **2021**, *21*, 486. [CrossRef]
49. Driessen, M.T.; Cohen, J.M.; Patterson-Lomba, O.; Thompson, S.F.; Seminerio, M.; Carr, K.; Totev, T.I.; Sun, R.; Yim, E.; Mu, F.; et al. Real-world effectiveness of fremanezumab in migraine patients initiating treatment in the United States: Results from a retrospective chart study. *J. Headache Pain* **2022**, *23*, 47. [CrossRef]
50. Eghtesadi, M.; Leroux, E.; Pagé, G. Real-Life Response to Erenumab in a Therapy-Resistant Case Series of Migraine Patients From the Province of Québec, Eastern Canada. *Clin. Drug Investig.* **2021**, *41*, 733–739. [CrossRef]
51. Faust, E.; Pivneva, I.; Yang, K.; Betts, K.A.; Ahmed, Z.; Joshi, S.; Hogan, R.; Blumenfeld, A.; Schim, J.; Feoktistov, A.; et al. Real-World Treatment Profiles, Clinical Outcomes, and Healthcare Resource Utilization of Patients with Migraine Prescribed Erenumab: A Multicenter Chart-Review Study of US Headache Centers. *Neurol. Ther.* **2021**, *10*, 293–306. [CrossRef] [PubMed]
52. Gantenbein, A.R.; Agosti, R.; Gobbi, C.; Flügel, D.; Schankin, C.J.; Viceic, D.; Zecca, C.; Pohl, H. Impact on monthly migraine days of discontinuing anti-CGRP antibodies after one year of treatment—A real-life cohort study. *Cephalalgia* **2021**, *41*, 1181–1186. [CrossRef]
53. Kanaan, S.; Hettie, G.; Loder, E.; Burch, R. Real-world effectiveness and tolerability of erenumab: A retrospective cohort study. *Cephalalgia* **2020**, *40*, 1511–1522. [CrossRef] [PubMed]
54. López-Bravo, A.; Oliveros-Cid, A.; Sevillano-Orte, L. Treatment satisfaction with calcitonin gene-related peptide monoclonal antibodies as a new patient-reported outcome measure: A real-life experience in migraine. *Acta Neurol. Scand.* **2022**, *145*, 669–675. [CrossRef] [PubMed]
55. Maraia, Z.; Ricci, D.; Rocchi, M.B.L.; Moretti, A.; Bufarini, C.; Cavaliere, A.; Peverini, M. Real-Life Analysis with Erenumab: First Target Therapy in the Episodic and Chronic Migraine's Prophylaxis. *J. Clin. Med.* **2021**, *10*, 4425. [CrossRef] [PubMed]
56. Mechtler, L.; Saikali, N.; McVige, J.; Hughes, O.; Traut, A.; Adams, A.M. Real-World Evidence for the Safety and Efficacy of CGRP Monoclonal Antibody Therapy Added to OnabotulinumtoxinA Treatment for Migraine Prevention in Adult Patients With Chronic Migraine. *Front. Neurol.* **2022**, *12*. [CrossRef]
57. Nandyala, A.S.; Suri, H.; Dougherty, C.O.; Ailani, J. A retrospective evaluation of the combination of erenumab and onabotulinum toxin A for the prevention of chronic migraine. *Clin. Neurol. Neurosurg.* **2022**, *215*, 107200. [CrossRef]
58. Ornello, R.; Baraldi, C.; Guerzoni, S.; Lambru, G.; Fuccaro, M.; Raffaelli, B.; Gendolla, A.; Barbanti, P.; Aurilia, C.; Cevoli, S.; et al. Gender Differences in 3-Month Outcomes of Erenumab Treatment—Study on Efficacy and Safety of Treatment With Erenumab in Men. *Front. Neurol.* **2021**, *12*, 774341. [CrossRef]
59. Ornello, R.; Baraldi, C.; Guerzoni, S.; Lambru, G.; Andreou, A.P.; Raffaelli, B.; Gendolla, A.; Barbanti, P.; Aurilia, C.; Egeo, G.; et al. Comparing the relative and absolute effect of erenumab: Is a 50% response enough? Results from the ESTEEMen study. *J. Headache Pain* **2022**, *23*, 38. [CrossRef]
60. Ornello, R.; Frattale, I.; Caponnetto, V.; De Matteis, E.; Pistoia, F.; Sacco, S. Menstrual Headache in Women with Chronic Migraine Treated with Erenumab: An Observational Case Series. *Brain Sci.* **2021**, *11*, 370. [CrossRef]
61. Ornello, R.; Casalena, A.; Frattale, I.; Gabriele, A.; Affaitati, G.; Giamberardino, M.A.; Assetta, M.; Maddestra, M.; Marzoli, F.; Viola, S.; et al. Real-life data on the efficacy and safety of erenumab in the Abruzzo region, central Italy. *J. Headache Pain* **2020**, *21*, 32. [CrossRef] [PubMed]
62. Overeem, L.H.; Peikert, A.; Hofacker, M.D.; Kamm, K.; Ruscheweyh, R.; Gendolla, A.; Raffaelli, B.; Reuter, U.; Neeb, L. Effect of antibody switch in non-responders to a CGRP receptor antibody treatment in migraine: A multi-center retrospective cohort study. *Cephalalgia* **2022**, *42*, 291–301. [CrossRef] [PubMed]
63. Raffaelli, B.; Kalantzis, R.; Mecklenburg, J.; Overeem, L.H.; Neeb, L.; Gendolla, A.; Reuter, U. Erenumab in Chronic Migraine Patients Who Previously Failed Five First-Line Oral Prophylactics and Onabotulinumtoxin A: A Dual-Center Retrospective Observational Study. *Front. Neurol.* **2020**, *11*, 417. [CrossRef] [PubMed]

64. Robblee, J.; Devick, K.L.; Mendez, N.; Potter, J.; Slonaker, J.; Starling, A.J. Real-World Patient Experience With Erenumab for the Preventive Treatment of Migraine. *Headache* **2020**, *60*, 2014–2025. [CrossRef]
65. Ruiz, I.P.; Ferrández, J.S.-R.; Fonfría, A.C.; García, T.M. Early Experiences in Switching between Monoclonal Antibodies in Patients with Nonresponsive Migraine in Spain: A Case Series. *Eur. Neurol.* **2022**, *85*, 132–135. [CrossRef]
66. Scheffler, A.; Schenk, H.; Wurthmann, S.; Nsaka, M.; Kleinschnitz, C.; Glas, M.; Holle, D. CGRP antibody therapy in patients with drug resistant migraine and chronic daily headache: A real-world experience. *J. Headache Pain* **2021**, *22*, 111. [CrossRef]
67. Scheffler, A.; Messel, O.; Wurthmann, S.; Nsaka, M.; Kleinschnitz, C.; Glas, M.; Naegel, S.; Holle, D. Erenumab in highly therapy-refractory migraine patients: First German real-world evidence. *J. Headache Pain* **2020**, *21*, 84. [CrossRef]
68. Sette, L.; Caponnetto, V.; Ornello, R.; Nežádal, T.; Čtrnáctá, D.; Šípková, J.; Matoušová, Z.; Sacco, S. Acute Medication Use in Patients With Migraine Treated With Monoclonal Antibodies Acting on the CGRP Pathway: Results From a Multicenter Study and Proposal of a New Index. *Front. Neurol.* **2022**, *13*, 846717. [CrossRef]
69. Storch, P.; Burow, P.; Möller, B.; Kraya, T.; Heintz, S.; Politz, N.; Naegel, S. Pooled retrospective analysis of 70 mg erenumab in episodic and chronic migraine: A two tertiary headache centers experience during clinical practice. *Acta Neurol. Belg.* **2022**, *122*, 931–937. [CrossRef]
70. Toni, T.; Tamanaha, R.; Newman, B.; Liang, Y.; Lee, J.; Carrazana, E.; Vajjala, V.; Viereck, J.; Liow, K.K. Effectiveness of dual migraine therapy with CGRP inhibitors and onabotulinumtoxin A injections: Case series. *Neurol. Sci.* **2021**, *42*, 5373–5376. [CrossRef]
71. Viudez-Martínez, A.; Pascual-Carrasco, A.; Beltrán-Blasco, I.; Hernandez-Lorido, R.; de Apodaca, R.F.R. Effectiveness and safety of erenumab and galcanezumab in the prevention of chronic and episodic migraine: A retrospective cohort study. *J. Clin. Pharm. Ther.* **2022**, *47*, 814–823. [CrossRef] [PubMed]
72. Barbanti, P.; Aurilia, C.; Egeo, G.; Fofi, L. Erenumab: From scientific evidence to clinical practice—The first Italian real-life data. *Neurol. Sci.* **2019**, *40*, 177–179. [CrossRef] [PubMed]
73. Barbanti, P.; Aurilia, C.; Egeo, G.; Fofi, L.; Cevoli, S.; Colombo, B.; Filippi, M.; Frediani, F.; Bono, F.; Grazzi, L.; et al. Erenumab in the prevention of high-frequency episodic and chronic migraine: Erenumab in Real Life in Italy (EARLY), the first Italian multicenter, prospective real-life study. *Headache* **2021**, *61*, 363–372. [CrossRef] [PubMed]
74. Barbanti, P.; Aurilia, C.; Cevoli, S.; Egeo, G.; Fofi, L.; Messina, R.; Salerno, A.; Torelli, P.; Albanese, M.; Carnevale, A.; et al. Long-term (48 weeks) effectiveness, safety, and tolerability of erenumab in the prevention of high-frequency episodic and chronic migraine in a real world: Results of the EARLY 2 study. *Headache* **2021**, *61*, 1351–1363. [CrossRef]
75. Barbanti, P.; Egeo, G.; Aurilia, C.; D'Onofrio, F.; Albanese, M.; Cetta, I.; Di Fiore, P.; Zucco, M.; Filippi, M.; Bono, F.; et al. Fremanezumab in the prevention of high-frequency episodic and chronic migraine: A 12-week, multicenter, real-life, cohort study (the FRIEND study). *J. Headache Pain* **2022**, *23*, 46. [CrossRef]
76. Becker, W.J.; Spacey, S.; Leroux, E.; Giammarco, R.; Gladstone, J.; Christie, S.; Akaberi, A.; Power, G.S.; Minhas, J.K.; Mancini, J.; et al. A real-world, observational study of erenumab for migraine prevention in Canadian patients. *Headache* **2022**, *62*, 522–529. [CrossRef]
77. Caronna, E.; Gallardo, V.J.; Alpuente, A.; Torres-Ferrus, M.; Pozo-Rosich, P. Anti-CGRP monoclonal antibodies in chronic migraine with medication overuse: Real-life effectiveness and predictors of response at 6 months. *J. Headache Pain* **2021**, *22*, 120. [CrossRef]
78. Cheng, S.; Jenkins, B.; Limberg, N.; Hutton, E. Erenumab in Chronic Migraine: An Australian Experience. *Headache* **2020**, *60*, 2555–2562. [CrossRef]
79. De Luca, C.; Baldacci, F.; Mazzucchi, S.; Lombardo, I.; Curto, L.; Ulivi, M.; Chico, L.; Papa, M.; Siciliano, G.; Gori, S. CGRP Inhibitors and Oxidative Stress Biomarkers in Resistant Migraine: A Real-Life Study with Erenumab, Fremanezumab, and Galcanezumab. *J. Clin. Med.* **2021**, *10*, 4586. [CrossRef]
80. De Matteis, E.; Affaitati, G.; Frattale, I.; Caponnetto, V.; Pistoia, F.; Giamberardino, M.A.; Sacco, S.; Ornello, R. Early outcomes of migraine after erenumab discontinuation: Data from a real-life setting. *Neurol. Sci.* **2021**, *42*, 3297–3303. [CrossRef]
81. Lentsch, S.d.V.; Verhagen, I.E.; Hoek, T.C.V.D.; MaassenVanDenBrink, A.; Terwindt, G.M. Treatment with the monoclonal calcitonin gene-related peptide receptor antibody erenumab: A real-life study. *Eur. J. Neurol.* **2021**, *28*, 4194–4203. [CrossRef] [PubMed]
82. Frattale, I.; Caponnetto, V.; Casalena, A.; Assetta, M.; Maddestra, M.; Marzoli, F.; Affaitati, G.; Giamberardino, M.A.; Viola, S.; Gabriele, A.; et al. Association between response to triptans and response to erenumab: Real-life data. *J. Headache Pain* **2021**, *22*, 1. [CrossRef] [PubMed]
83. Guerzoni, S.; Baraldi, C.; Pensato, U.; Favoni, V.; Castro, F.L.; Cainazzo, M.M.; Cevoli, S.; Pani, L. Chronic migraine evolution after 3 months from erenumab suspension: Real-world-evidence-life data. *Neurol. Sci.* **2022**, *43*, 3823–3830. [CrossRef]
84. Iannone, L.F.; Fattori, D.; Benemei, S.; Chiarugi, A.; Geppetti, P.; De Cesaris, F. Predictors of sustained response and effects of the discontinuation of anti-calcitonin gene related peptide antibodies and reinitiation in resistant chronic migraine. *Eur. J. Neurol.* **2022**, *29*, 1505–1513. [CrossRef] [PubMed]
85. Kwon, S.; Gil, Y.-E.; Lee, M.J. Real-world efficacy of galcanezumab for the treatment of migraine in Korean patients. *Cephalalgia* **2022**, *42*, 705–714. [CrossRef] [PubMed]
86. Lambru, G.; Hill, B.; Murphy, M.; Tylova, I.; Andreou, A.P. A prospective real-world analysis of erenumab in refractory chronic migraine. *J. Headache Pain* **2020**, *21*, 61. [CrossRef]

87. Mahović, D.; Bračić, M.; Jakuš, L.; Cvetkovic, V.V.; Krpan, M. Effectiveness and safety of erenumab in chronic migraine: A Croatian real-world experience. *Clin. Neurol. Neurosurg.* **2022**, *214*, 107169. [CrossRef]
88. Matteo, E.; Favoni, V.; Pascazio, A.; Pensato, U.; Benini, M.; Asioli, G.M.; Merli, E.; Calabrò, C.; Cortelli, P.; Pierangeli, G.; et al. Erenumab in 159 high frequency and chronic migraine patients: Real-life results from the Bologna Headache Center. *Neurol. Sci.* **2020**, *41*, 483–484. [CrossRef]
89. Ornello, R.; Casalena, A.; Frattale, I.; Caponnetto, V.; Gabriele, A.; Affaitati, G.; Giamberardino, M.A.; Assetta, M.; Maddestra, M.; Marzoli, F.; et al. Conversion from chronic to episodic migraine in patients treated with erenumab: Real-life data from an Italian region. *J. Headache Pain* **2020**, *21*, 102. [CrossRef]
90. Pensato, U.; Baraldi, C.; Favoni, V.; Cainazzo, M.M.; Torelli, P.; Querzani, P.; Pascazio, A.; Mascarella, D.; Matteo, E.; Quintana, S.; et al. Real-life assessment of erenumab in refractory chronic migraine with medication overuse headache. *Neurol. Sci.* **2022**, *43*, 1273–1280. [CrossRef]
91. Pensato, U.; Favoni, V.; Pascazio, A.; Benini, M.; Asioli, G.M.; Merli, E.; Calabrò, C.; Cortelli, P.; Pierangeli, G.; Cevoli, S. Erenumab efficacy in highly resistant chronic migraine: A real-life study. *Neurol. Sci.* **2020**, *41*, 457–459. [CrossRef] [PubMed]
92. Raffaelli, B.; Terhart, M.; Mecklenburg, J.; Neeb, L.; Overeem, L.H.; Siebert, A.; Steinicke, M.; Reuter, U. Resumption of migraine preventive treatment with CGRP(-receptor) antibodies after a 3-month drug holiday: A real-world experience. *J. Headache Pain* **2022**, *23*, 40. [CrossRef] [PubMed]
93. Russo, A.; Silvestro, M.; di Clemente, F.S.; Trojsi, F.; Bisecco, A.; Bonavita, S.; Tessitore, A.; Tedeschi, G. Multidimensional assessment of the effects of erenumab in chronic migraine patients with previous unsuccessful preventive treatments: A comprehensive real-world experience. *J. Headache Pain* **2020**, *21*, 69. [CrossRef] [PubMed]
94. Sánchez-Marín, B.; Ledesma, D.H.; Álvarez, A.L.; Ullrich, J.G. Immunotherapy for migraine: The use of erenumab in real life. *Rev. Clin. Esp.* **2021**, *221*, 557–559. [CrossRef]
95. Schoenen, J.; Timmermans, G.; Nonis, R.; Manise, M.; Fumal, A.; Gérard, P. Erenumab for Migraine Prevention in a 1-Year Compassionate Use Program: Efficacy, Tolerability, and Differences Between Clinical Phenotypes. *Front. Neurol.* **2021**, *12*, 805334. [CrossRef]
96. Talbot, J.; Stuckey, R.; Crawford, L.; Weatherby, S.; Mullin, S. Improvements in pain, medication use and quality of life in onabotulinumtoxinA-resistant chronic migraine patients following erenumab treatment—Real world outcomes. *J. Headache Pain* **2021**, *22*, 5. [CrossRef] [PubMed]
97. Torres-Ferrús, M.; Gallardo, V.J.; Alpuente, A.; Caronna, E.; Gine-Cipres, E.; Pozo-Rosich, P. The impact of anti-CGRP monoclonal antibodies in resistant migraine patients: A real-world evidence observational study. *J. Neurol.* **2021**, *268*, 3789–3798. [CrossRef]
98. Tziakouri, A.; Tsangari, H.; Michaelides, C. Assessment of the Effect of Erenumab on Efficacy and Quality-of-Life Parameters in a Cohort of Migraine Patients With Treatment Failure in Cyprus. *Front. Neurol.* **2021**, *12*, 687697. [CrossRef]
99. Vernieri, F.; Altamura, C.; Brunelli, N.; Costa, C.M.; Aurilia, C.; Egeo, G.; Fofi, L.; Favoni, V.; Pierangeli, G.; Lovati, C.; et al. Galcanezumab for the prevention of high frequency episodic and chronic migraine in real life in Italy: A multicenter prospective cohort study (the GARLIT study). *J. Headache Pain* **2021**, *22*, 35. [CrossRef]
100. Vernieri, F.; Altamura, C.; Brunelli, N.; Costa, C.M.; Aurilia, C.; Egeo, G.; Fofi, L.; Favoni, V.; Lovati, C.; Bertuzzo, D.; et al. Rapid response to galcanezumab and predictive factors in chronic migraine patients: A 3-month observational, longitudinal, cohort, multicenter, Italian real-life study. *Eur. J. Neurol.* **2022**, *29*, 1198–1208. [CrossRef]
101. Vernieri, F.; Brunelli, N.; Messina, R.; Costa, C.M.; Colombo, B.; Torelli, P.; Quintana, S.; Cevoli, S.; Favoni, V.; D'Onofrio, F.; et al. Discontinuing monoclonal antibodies targeting CGRP pathway after one-year treatment: An observational longitudinal cohort study. *J. Headache Pain* **2021**, *22*, 154. [CrossRef] [PubMed]
102. Zecca, C.; Cargnin, S.; Schankin, C.; Giannantoni, N.M.; Viana, M.; Maraffi, I.; Riccitelli, G.C.; Sihabdeen, S.; Terrazzino, S.; Gobbi, C. Clinic and genetic predictors in response to erenumab. *Eur. J. Neurol.* **2022**, *29*, 1209–1217. [CrossRef] [PubMed]
103. Cetta, I.; Messina, R.; Zanandrea, L.; Colombo, B.; Filippi, M. Comparison of efficacy and safety of erenumab between over and under 65-year-old refractory migraine patients: A pivotal study. *Neurol. Sci.* **2022**, *43*, 5769–5771. [CrossRef] [PubMed]
104. De Icco, R.; Vaghi, G.; Allena, M.; Ghiotto, N.; Guaschino, E.; Martinelli, D.; Ahmad, L.; Corrado, M.; Bighiani, F.; Tanganelli, F.; et al. Does MIDAS reduction at 3 months predict the outcome of erenumab treatment? A real-world, open-label trial. *J. Headache Pain* **2022**, *23*, 123. [CrossRef]
105. Khalil, M.; Moreno-Ajona, D.; Villar-Martínez, M.D.; Greenwood, F.; Hoffmann, J.; Goadsby, P.J. Erenumab in chronic migraine: Experience from a UK tertiary centre and comparison with other real-world evidence. *Eur. J. Neurol.* **2022**, *29*, 2473–2480. [CrossRef] [PubMed]
106. Alsaadi, T.; Noori, S.; Varakian, R.; Youssef, S.; Almadani, A. Real-world experience of erenumab in patients with chronic or episodic migraine in the UAE. *BMC Neurol.* **2022**, *22*, 221. [CrossRef] [PubMed]
107. Cullum, C.K.; Do, T.P.; Ashina, M.; Bendtsen, L.; Hugger, S.S.; Iljazi, A.; Gusatovic, J.; Snellman, J.; Lopez-Lopez, C.; Ashina, H.; et al. Real-world long-term efficacy and safety of erenumab in adults with chronic migraine: A 52-week, single-center, prospective, observational study. *J. Headache Pain* **2022**, *23*, 61. [CrossRef]
108. Gantenbein, A.R.; Agosti, R.; Kamm, C.P.; Landmann, G.; Meier, N.; Merki-Feld, G.S.; Petersen, J.A.; Pohl, H.; Ryvlin, P.; Schankin, C.J.; et al. Swiss QUality of life and healthcare impact Assessment in a Real-world Erenumab treated migraine population (SQUARE study): Interim results. *J. Headache Pain* **2022**, *23*, 142. [CrossRef]

109. Schenk, H.; Holle, D.; Nsaka, M.; Kleinschnitz, C.; Glas, M.; Scheffler, A. Twelve-month safety, tolerability and susceptibility to adverse events of prophylactic migraine therapy with erenumab: A retrospective real-world study. *J. Headache Pain* **2022**, *23*, 55. [CrossRef]
110. Andreou, A.P.; Fuccaro, M.; Hill, B.; Murphy, M.; Caponnetto, V.; Kilner, R.; Lambru, G. Two-year effectiveness of erenumab in resistant chronic migraine: A prospective real-world analysis. *J. Headache Pain* **2022**, *23*, 139. [CrossRef]
111. Driessen, M.T.; Cohen, J.M.; Thompson, S.F.; Patterson-Lomba, O.; Seminerio, M.J.; Carr, K.; Totev, T.I.; Sun, R.; Yim, E.; Mu, F.; et al. Real-world effectiveness after initiating fremanezumab treatment in US patients with episodic and chronic migraine or difficult-to-treat migraine. *J. Headache Pain* **2022**, *23*, 56. [CrossRef] [PubMed]
112. Altamura, C.; Brunelli, N.; Marcosano, M.; Aurilia, C.; Egeo, G.; Lovati, C.; Favoni, V.; Perrotta, A.; Maestrini, I.; Di Cola, F.S.; et al. Conversion from chronic to episodic migraine in patients treated with galcanezumab in real life in Italy: The 12-month observational, longitudinal, cohort multicenter GARLIT experience. *J. Neurol.* **2022**, *269*, 5848–5857. [CrossRef] [PubMed]
113. Silvestro, M.; Tessitore, A.; Orologio, I.; De Micco, R.; Tartaglione, L.; Trojsi, F.; Tedeschi, G.; Russo, A. Galcanezumab effect on "whole pain burden" and multidimensional outcomes in migraine patients with previous unsuccessful treatments: A real-world experience. *J. Headache Pain* **2022**, *23*, 69. [CrossRef] [PubMed]
114. Fofi, L.; Altamura, C.; Fiorentini, G.; Brunelli, N.; Marcosano, M.; Barbanti, P.; Vernieri, F. Improving distress perception and mutuality in migraine caregivers after 6 months of galcanezumab treatment. *Headache* **2022**, *62*, 1143–1147. [CrossRef]
115. Vernieri, F.; Brunelli, N.; Marcosano, M.; Aurilia, C.; Egeo, G.; Lovati, C.; Favoni, V.; Perrotta, A.; Maestrini, I.; Rao, R.; et al. Maintenance of response and predictive factors of 1-year GalcanezumAb treatment in real-life migraine patients in Italy: The multicenter prospective cohort GARLIT study. *Eur. J. Neurol.* **2022**, *30*, 224–234. [CrossRef]
116. Lentsch, S.d.V.; van der Arend, B.W.H.; Brink, A.M.V.D.; Terwindt, G.M. Blood Pressure in Patients With Migraine Treated With Monoclonal Anti-CGRP (Receptor) Antibodies: A Prospective Follow-up Study. *Neurology* **2022**, *99*, e1897–e1904. [CrossRef]
117. Nowaczewska, M.; Straburzyński, M.; Waliszewska-Prosół, M.; Meder, G.; Janiak-Kiszka, J.; Kaźmierczak, W. Cerebral Blood Flow and Other Predictors of Responsiveness to Erenumab and Fremanezumab in Migraine—A Real-Life Study. *Front. Neurol.* **2022**, *13*, 895476. [CrossRef]
118. Quintana, S.; Russo, M.; Manzoni, G.C.; Torelli, P. Comparison study between erenumab, fremanezumab, and galcanezumab in the preventive treatment of high frequency episodic migraine and chronic migraine. *Neurol. Sci.* **2022**, *43*, 5757–5758. [CrossRef]
119. De Matteis, E.; Caponnetto, V.; Casalena, A.; Frattale, I.; Gabriele, A.; Affaitati, G.; Giamberardino, M.A.; Maddestra, M.; Viola, S.; Pistoia, F.; et al. Cranial autonomic symptoms and response to monoclonal antibodies targeting the Calcitonin gene-related peptide pathway: A real-world study. *Front. Neurol.* **2022**, *13*, 973226. [CrossRef]
120. Cantarelli, L.; Grafiña, D.P.; Perez, A.G.; Gil, S.G.; Nicolás, F.G.; Santana, E.R.; Dávila, M.A.N.; Pérez, S.M.O.; Gómez, G.C.; Reyes, S.P.; et al. Efficacy and Safety of Erenumab, Galcanezumab, and Fremanezumab in the Treatment of Drug-Resistant Chronic Migraine: Experience in Real Clinical Practice. *Ann. Pharmacother.* **2022**. [CrossRef]
121. Barbanti, P.; Egeo, G.; Aurilia, C.; Altamura, C.; D'Onofrio, F.; Finocchi, C.; Albanese, M.; Aguggia, M.; Rao, R.; Zucco, M.; et al. Predictors of response to anti-CGRP monoclonal antibodies: A 24-week, multicenter, prospective study on 864 migraine patients. *J. Headache Pain* **2022**, *23*, 138. [CrossRef] [PubMed]
122. Nsaka, M.; Scheffler, A.; Wurthmann, S.; Schenk, H.; Kleinschnitz, C.; Glas, M.; Holle, D. Real-world evidence following a mandatory treatment break after a 1-year prophylactic treatment with calcitonin gene-related peptide (pathway) monoclonal antibodies. *Brain Behav.* **2022**, *12*, e2662. [CrossRef] [PubMed]
123. Al-Hassany, L.; de Vries, T.; A Carpay, J.; MaassenVanDenBrink, A. Could erectile dysfunction be a side effect of CGRP inhibition? A case report. *Cephalalgia* **2022**, *42*, 257–261. [CrossRef] [PubMed]
124. Albanese, M.; Mercuri, N.B. Could the New Anti-CGRP Monoclonal Antibodies Be Effective in Migraine Aura? Case Reports and Literature Review. *J. Clin. Med.* **2022**, *11*, 1228. [CrossRef] [PubMed]
125. Aradi, S.; Kaiser, E.; Cucchiara, B. Ischemic Stroke Associated With Calcitonin Gene-Related Peptide Inhibitor Therapy for Migraine: A Case Report. *J. Stroke Cerebrovasc. Dis.* **2019**, *28*, 104286. [CrossRef] [PubMed]
126. Arca, K.N.; Starling, A.J. Treatment-Refractory Headache in the Setting of COVID-19 Pneumonia: Migraine or Meningoencephalitis? Case Report. *SN Compr. Clin. Med.* **2020**, *2*, 1200–1203. [CrossRef] [PubMed]
127. Cullum, C.K.; Olsen, M.K.; Kocadag, H.B.; Ashina, M.; Amin, F.M. Extreme ecchymoses in a migraine patient using concomitant treatment with calcitonin gene-related peptide receptor antibodies and fish oil supplements: A case report. *BMC Neurol.* **2021**, *21*, 257. [CrossRef]
128. Evans, R.W. Raynaud's Phenomenon Associated With Calcitonin Gene-Related Peptide Monoclonal Antibody Antagonists. *Headache* **2019**, *59*, 1360–1364. [CrossRef]
129. Fofi, L.; Egeo, G.; Aurilia, C.; Barbanti, P. Erenumab during pregnancy: A case report in a patient with chronic migraine. *Neurol. Sci.* **2021**, *42*, 2145–2146. [CrossRef]
130. Frattale, I.; Ornello, R.; Pistoia, F.; Caponnetto, V.; Colangeli, E.; Sacco, S. Paralytic ileus after planned abdominal surgery in a patient on treatment with erenumab. *Intern. Emerg. Med.* **2021**, *16*, 227–228. [CrossRef]
131. González-Quintanilla, V.; Pérez-Pereda, S.; González-Suárez, A.; Madera, J.; Toriello, M.; Pascual, J. Restless legs-like syndrome as an emergent adverse event of CGRP monoclonal antibodies: A report of two cases. *Cephalalgia* **2021**, *41*, 1272–1275. [CrossRef]

132. Grassini, A.; Marcinnò, A.; Roveta, F.; Gallo, E.; Cermelli, A.; Boschi, S.; Rubino, E.; Rainero, I. Impact of COVID-19 on chronic migraine treated with erenumab: A case report. *Neurol. Sci.* **2021**, *42*, 3079–3081. [CrossRef] [PubMed]
133. Henze, T. Erenumab During Breastfeeding. *Breastfeed. Med.* **2019**, *14*, 513–514. [CrossRef] [PubMed]
134. Iannone, L.F.; Fattori, D.; Geppetti, P.; De Cesaris, F. Galcanezumab effectiveness on comorbid cluster headache and chronic migraine: A prospective case series. *Neurol. Sci.* **2022**, *43*, 697–703. [CrossRef] [PubMed]
135. Iannone, L.F.; Geppetti, P.; Chiarugi, A.; De Cesaris, F. COVID-19 pneumonia during long-term migraine prophylaxis with fremanezumab: A case report. *Intern. Emerg. Med.* **2021**, *16*, 2309–2311. [CrossRef] [PubMed]
136. Kearney, E.; Collins, T.; Sengupta, S. De Novo Visual Aura Onset in a Migraineur on Galcanezumab-Gnlm. *Headache* **2020**, *60*, 1435–1437. [CrossRef]
137. Klager, S.; Khalil, M.; Shulman, K.; Sami, N. Galcanezumab-induced fixed drug eruption. *JAAD Case Rep.* **2021**, *9*, 90–92. [CrossRef] [PubMed]
138. Lehman, L.L.; Bruccoleri, R.; Danehy, A.; Swanson, J.; Mrakotsky, C.; Smith, E.; Orbach, D.B.; Burstein, R. Adverse effects of erenumab on cerebral proliferative angiopathy: A case report. *Cephalalgia* **2021**, *41*, 122–126. [CrossRef]
139. Makarevičius, G.; Ryliškienė, K. Successful treatment of primary headache associated with sexual activity using erenumab: Case report. *Cephalalgia* **2022**, *42*, 680–683. [CrossRef]
140. Mathew, P.G.; Krivitski, D.; Sharon, R. Erenumab-Induced Severe Nausea Leading to Smoking Cessation: A Retrospective Case Series. *Headache* **2020**, *60*, 2563–2569. [CrossRef]
141. Moya, B.; Barranco, R.; García-Moguel, I.; Puerta-Peña, M.; Alonso, L.; Fernández-Crespo, J. First confirmed case of nonimmediate hypersensitivity to fremanezumab during chronic migraine treatment. *Contact Dermat.* **2022**, *86*, 308–310. [CrossRef] [PubMed]
142. Mullin, K.; Kudrow, D.; Croop, R.; Lovegren, M.; Conway, C.M.; Coric, V.; Lipton, R.B. Potential for treatment benefit of small molecule CGRP receptor antagonist plus monoclonal antibody in migraine therapy. *Neurology* **2020**, *94*, e2121–e2125. [CrossRef] [PubMed]
143. Ornello, R.; Frattale, I.; Pistoia, F.; Sacco, S.; Notturno, F. Erenumab Plus Subcutaneous Immunoglobulin in a Patient With Comorbid Chronic Migraine and Myasthenia Gravis. *Headache* **2020**, *60*, 787–788. [CrossRef] [PubMed]
144. Rota, E.; Aguggia, M.; Immovilli, P.; Morelli, N.; Renosio, D.; Barbanera, A. Change in the second exteroceptive suppression period of the temporalis muscle during erenumab treatment. *Naunyn-Schmiedeberg's Arch. Pharmacol.* **2022**, *395*, 607–611. [CrossRef] [PubMed]
145. Rozen, T.D.; A Bhatt, A. Reversible cerebral vasoconstriction syndrome developing after an erenumab injection for migraine prevention. *Cephalalgia* **2022**, *42*, 250–256. [CrossRef]
146. Selvi-Sabater, P.; Carvajal-Sanchez, M.A.; Carrera-Hueso, F.J. Two possible cases of erenumab-induced xerostomia. *J. Clin. Pharm. Ther.* **2022**, *47*, 824–825. [CrossRef]
147. Silvestro, M.; Tessitore, A.; di Clemente, F.S.; Battista, G.; Tedeschi, G.; Russo, A. Additive Interaction Between Onabotulinumtoxin-A and Erenumab in Patients With Refractory Migraine. *Front. Neurol.* **2021**, *12*, 656294. [CrossRef]
148. Silvestro, M.; Tessitore, A.; Di Clemente, F.S.; Tedeschi, G.; Russo, A. Erenumab Efficacy on Comorbid Cluster Headache in Patients With Migraine: A Real-World Case Series. *Headache* **2020**, *60*, 1187–1195. [CrossRef]
149. Spector, A.R.; Kerkow, J.F.; Collins, T.A.; Kerkow, M.J.F. Sleep terrors prodromal for migraine headaches responsive to galcanezumab: A case report. *Headache* **2021**, *61*, 216–217. [CrossRef]
150. Wong, S.M.; Kim, J.Y.; A Maguire, G. Migraine and adult-onset stuttering: A proposed autoimmune phenomenon. *Ann. Clin. Psychiatry* **2021**, *33*, 56–57. [CrossRef]
151. Wurthmann, S.; Nägel, S.; Hadaschik, E.; Schlott, S.; Scheffler, A.; Kleinschnitz, C.; Holle, D. Impaired wound healing in a migraine patient as a possible side effect of calcitonin gene-related peptide receptor antibody treatment: A case report. *Cephalalgia* **2020**, *40*, 1255–1260. [CrossRef] [PubMed]
152. Ziegeler, C.; May, A. Non-Responders to Treatment With Antibodies to the CGRP-Receptor May Profit From a Switch of Antibody Class. *Headache* **2020**, *60*, 469–470. [CrossRef] [PubMed]
153. Primiano, G.; Rollo, E.; Romozzi, M.; Calabresi, P.; Servidei, S.; Vollono, C. Preventive migraine treatment in mitochondrial diseases: A case report of erenumab efficacy and literature review. *Neurol. Sci.* **2022**, *43*, 6955–6959. [CrossRef] [PubMed]
154. Göbel, C.H.; Heinze, A.; Karstedt, S.; Cirkel, A.; Münte, T.F.; Göbel, H. First Report of Symmetrical Drug-related Intertriginous and Flexural Exanthema (SDRIFE or Baboon Syndrome) After Erenumab Application for Migraine Prevention. *Pain Ther.* **2022**, *11*, 1483–1491. [CrossRef] [PubMed]
155. Bonifácio, G.V.; de Carvalho, S.C.; Oliveira, R.; Gil-Gouveia, R. Gestational exposure to erenumab—The outcome of three pregnancies. *Headache* **2022**, *62*, 1218–1221. [CrossRef]
156. Perino, J.; Corand, V.; Laurent, E.; Théophile, H.; Miremont-Salamé, G.; Pariente, A.; Colas, J.; Couffinhal, T.; Salvo, F. Myocardial infarction associated with erenumab: A case report. *Pharmacother. J. Hum. Pharmacol. Drug Ther.* **2022**, *42*, 585–589. [CrossRef]
157. Vig, S.J.; Garza, J.; Tao, Y. The use of erenumab for migraine prophylaxis during pregnancy: A case report and narrative review. *Headache* **2022**, *62*, 1256–1263. [CrossRef]
158. Kawamura, S.; Katsuki, M.; Kashiwagi, K.; Koh, A. Fremanezumab Improved Migraine and Headache Attributed to Glioblastoma. *Cureus* **2022**, *14*, e30064. [CrossRef]
159. Suzuki, S.; Suzuki, K.; Shiina, T.; Kobayashi, S.; Hirata, K. Can calcitonin gene-related peptide monoclonal antibody improve migraine and restless legs syndrome? *J. Neurol. Sci.* **2022**, *443*, 120462. [CrossRef]

160. López-Bravo, A.; Oliveros-Cid, A.; Mínguez-Olaondo, A.; Cuadrado, M.L. Nummular headache responsive to anti-calcitonin gene-related peptide monoclonal antibodies in a patient with migraine. *Headache* **2022**, *62*, 1063–1066. [CrossRef]
161. Ray, J.C.; Allen, P.; Bacsi, A.; Bosco, J.J.; Chen, L.; Eller, M.; Kua, H.; Lim, L.L.; Matharu, M.S.; Monif, M.; et al. Inflammatory complications of CGRP monoclonal antibodies: A case series. *J. Headache Pain* **2021**, *22*, 121. [CrossRef] [PubMed]
162. Cetta, I.; Messina, R.; Colombo, B.; Filippi, M. Myocardial infarction in a patient with migraine and triptan overuse treated with anti-CGRP receptor monoclonal antibody: A case report. *J. Neurol.* **2022**, *269*, 5170–5172. [CrossRef]
163. Dodick, D.W.; Tepper, S.J.; Ailani, J.; Pannacciulli, N.; Navetta, M.S.; Loop, B.; Zhang, F.; Khodavirdi, A.C.; Mann, A.; Abdrabboh, A.; et al. Risk of hypertension in erenumab-treated patients with migraine: Analyses of clinical trial and postmarketing data. *Headache* **2021**, *61*, 1411–1420. [CrossRef] [PubMed]
164. Saely, S.; Croteau, D.; Jawidzik, L.; Brinker, A.; Kortepeter, C.; Saely, B.S. Hypertension: A new safety risk for patients treated with erenumab. *Headache* **2021**, *61*, 202–208. [CrossRef] [PubMed]
165. Gérard, A.O.; Merino, D.; Van Obberghen, E.K.; Rocher, F.; Destere, A.; Lantéri-Minet, M.; Drici, M.-D. Calcitonin gene-related peptide-targeting drugs and Raynaud's phenomenon: A real-world potential safety signal from the WHO pharmacovigilance database. *J. Headache Pain* **2022**, *23*, 53. [CrossRef] [PubMed]
166. Silberstein, S.D.; Reshef, S.; Cohen, J.M.; Gandhi, S.; Seminerio, M.; Ramirez Campos, V.; Kessler, Y.; Thompson, S.F.; Blumenfeld, A. Adverse Events Reported with Therapies Targeting the CGRP Pathway During the First 6 Months Post-launch: A Retrospective Analysis Using the FDA Adverse Events Reporting System. *Adv Ther.* **2022**. [CrossRef]
167. Breen, I.D.; Brumfiel, C.M.; Patel, M.H.; Butterfield, R.J.; VanderPluym, J.H.; Griffing, L.; Pittelkow, M.R.; Mangold, A.R. Evaluation of the Safety of Calcitonin Gene-Related Peptide Antagonists for Migraine Treatment Among Adults With Raynaud Phenomenon. *JAMA Netw. Open* **2021**, *4*, e217934. [CrossRef]
168. Headache Classification Committee of the International Headache Society (IHS). The International Classification of Headache Disorders, 3rd edition. *Cephalalgia* **2018**, *38*, 1–211. [CrossRef]
169. Diener, H.-C.; Tassorelli, C.; Dodick, D.W.; Silberstein, S.D.; Lipton, R.B.; Ashina, M.; Becker, W.J.; Ferrari, M.D.; Goadsby, P.J.; Pozo-Rosich, P.; et al. Guidelines of the International Headache Society for controlled trials of preventive treatment of migraine attacks in episodic migraine in adults. *Cephalalgia* **2020**, *40*, 1026–1044. [CrossRef]
170. Diener, H.-C.; Tassorelli, C.; Dodick, D.W.; Silberstein, S.D.; Lipton, R.B.; Ashina, M.; Becker, W.J.; Ferrari, M.D.; Goadsby, P.J.; Pozo-Rosich, P.; et al. Guidelines of the International Headache Society for controlled trials of acute treatment of migraine attacks in adults: Fourth edition. *Cephalalgia* **2019**, *39*, 687–710. [CrossRef]
171. Tassorelli, C.; Diener, H.-C.; Dodick, D.W.; Silberstein, S.D.; Lipton, R.B.; Ashina, M.; Becker, W.J.; Ferrari, M.D.; Goadsby, P.J.; Pozo-Rosich, P.; et al. Guidelines of the International Headache Society for controlled trials of preventive treatment of chronic migraine in adults. *Cephalalgia* **2018**, *38*, 815–832. [CrossRef] [PubMed]
172. Noseda, R.; Bedussi, F.; Gobbi, C.; Zecca, C.; Ceschi, A. Safety profile of erenumab, galcanezumab and fremanezumab in pregnancy and lactation: Analysis of the WHO pharmacovigilance database. *Cephalalgia* **2021**, *41*, 789–798. [CrossRef] [PubMed]
173. Edvinsson, L.; Chan, K.Y.; Eftekhari, S.; Nilsson, E.; De Vries, R.; Säveland, H.; Dirven, C.M.; Danser, A.J.; MaassenVanDenBrink, A. Effect of the calcitonin gene-related peptide (CGRP) receptor antagonist telcagepant in human cranial arteries. *Cephalalgia* **2010**, *30*, 1233–1240. [CrossRef] [PubMed]
174. Edvinsson, L.; Nilsson, E.; Jansen-Olesen, I. Inhibitory effect of BIBN4096BS, CGRP(8-37), a CGRP antibody and an RNA-Spiegelmer on CGRP induced vasodilatation in the perfused and non-perfused rat middle cerebral artery. *Br. J. Pharmacol.* **2007**, *150*, 633–640. [CrossRef]
175. Benemei, S.; Cortese, F.; Labastida-Ramírez, A.; Marchese, F.; Pellesi, L.; Romoli, M.; Vollesen, A.L.; Lampl, C.; Ashina, M. School of Advanced Studies of the European Headache Federation (EHF-SAS). Triptans and CGRP blockade—Impact on the cranial vasculature. *J. Headache Pain* **2017**, *18*, 103. [CrossRef]

cells

MDPI

Review

From CGRP to PACAP, VIP, and Beyond: Unraveling the Next Chapters in Migraine Treatment

Masaru Tanaka [1], Ágnes Szabó [2,3], Tamás Körtési [1,4,5], Délia Szok [2], János Tajti [2] and László Vécsei [1,2,*]

1 HUN-REN-SZTE Neuroscience Research Group, Hungarian Research Network, University of Szeged (HUN-REN-SZTE), Danube Neuroscience Research Laboratory, Tisza Lajos krt. 113, H-6725 Szeged, Hungary; tanaka.masaru.1@med.u-szeged.hu
2 Department of Neurology, Albert Szent-Györgyi Medical School, University of Szeged, Semmelweis u. 6, H-6725 Szeged, Hungary; szabo.agnes.4@med.u-szeged.hu (Á.S.); szok.delia@med.u-szeged.hu (D.S.); tajti.janos@med.u-szeged.hu (J.T.)
3 Doctoral School of Clinical Medicine, University of Szeged, Korányi fasor 6, H-6720 Szeged, Hungary
4 Faculty of Health Sciences and Social Studies, University of Szeged, Temesvári krt. 31, H-6726 Szeged, Hungary; kortesi.tamas@szte.hu
5 Preventive Health Sciences Research Group, Incubation Competence Centre of the Centre of Excellence for Interdisciplinary Research, Development and Innovation of the University of Szeged, H-6720 Szeged, Hungary
* Correspondence: vecsei.laszlo@med.u-szeged.hu; Tel.: +36-62-545-351

Abstract: Migraine is a neurovascular disorder that can be debilitating for individuals and society. Current research focuses on finding effective analgesics and management strategies for migraines by targeting specific receptors and neuropeptides. Nonetheless, newly approved calcitonin gene-related peptide (CGRP) monoclonal antibodies (mAbs) have a 50% responder rate ranging from 27 to 71.0%, whereas CGRP receptor inhibitors have a 50% responder rate ranging from 56 to 71%. To address the need for novel therapeutic targets, researchers are exploring the potential of another secretin family peptide, pituitary adenylate cyclase-activating polypeptide (PACAP), as a ground-breaking treatment avenue for migraine. Preclinical models have revealed how PACAP affects the trigeminal system, which is implicated in headache disorders. Clinical studies have demonstrated the significance of PACAP in migraine pathophysiology; however, a few clinical trials remain inconclusive: the pituitary adenylate cyclase-activating peptide 1 receptor mAb, AMG 301 showed no benefit for migraine prevention, while the PACAP ligand mAb, Lu AG09222 significantly reduced the number of monthly migraine days over placebo in a phase 2 clinical trial. Meanwhile, another secretin family peptide vasoactive intestinal peptide (VIP) is gaining interest as a potential new target. In light of recent advances in PACAP research, we emphasize the potential of PACAP as a promising target for migraine treatment, highlighting the significance of exploring PACAP as a member of the antimigraine armamentarium, especially for patients who do not respond to or contraindicated to anti-CGRP therapies. By updating our knowledge of PACAP and its unique contribution to migraine pathophysiology, we can pave the way for reinforcing PACAP and other secretin peptides, including VIP, as a novel treatment option for migraines.

Keywords: migraine disorders; headache disorders; nociceptive pain; analgesics; calcitonin gene-related peptide; pituitary adenylate cyclase-activating polypeptide (PACAP); vasoactive intestinal peptide; adrenomedullin; neuropeptides; drug development

Citation: Tanaka, M.; Szabó, Á.; Körtési, T.; Szok, D.; Tajti, J.; Vécsei, L. From CGRP to PACAP, VIP, and Beyond: Unraveling the Next Chapters in Migraine Treatment. *Cells* **2023**, *11*, 2649. https://doi.org/10.3390/cells12222649

Academic Editors: Arpad Dobolyi and Alessandro Castorina

Received: 5 September 2023
Revised: 13 November 2023
Accepted: 14 November 2023
Published: 17 November 2023

1. Introduction

Migraines are neurological disorders causing recurrent, severe headaches and other symptoms like sensitivity to light, sound, smell, or touch, and nausea or vomiting [1]. Their cause is unclear but involves genetic, environmental, and lifestyle factors [2–5]. Triggers vary among individuals and include stress, hormonal changes, certain diets, and sleep disturbances [6–9]. Identifying and managing triggers can be crucial in preventing the onset of migraine attacks and reducing their frequency, duration, and severity [10]. Migraines are complex neurological disorders that can significantly impact individuals' quality of life [11,12]. Comprehending the distinct stages, symptoms, triggers, and treatment options is fundamental for healthcare professionals and researchers, as it facilitates enhanced management and support for individuals affected by migraines [10].

Neuropeptides like calcitonin gene-related peptide (CGRP), pituitary adenylate cyclase-activating polypeptide (PACAP), vasoactive intestinal polypeptide (VIP), islet amyloid polypeptide (IAPP)/amylin, substance P, and adrenomedullin (ADM) [13–18]. The secretin family of peptides, including CGRP, PACAP, ADM, and amylin, control G protein-coupled receptors (GPCR) activity. They share homology, receptor cross-reactivity, and similar biological actions, suggesting they belong to this family (Figure 1) [19]. These neuropeptides play diverse roles in migraine pathogenesis, contributing to our understanding of the disorder's mechanisms [20,21].

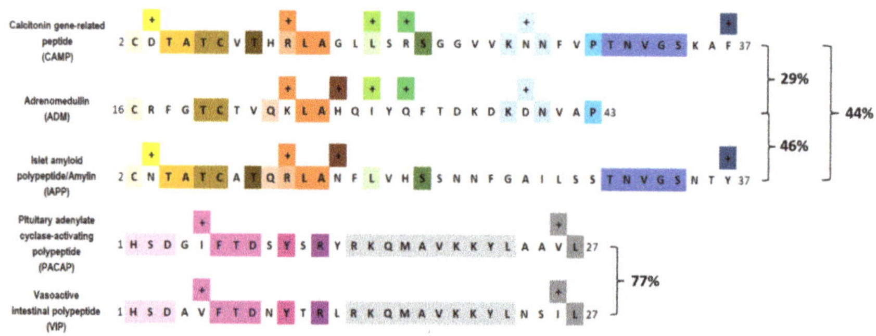

Figure 1. The amino acid sequence alignment analysis of the main secretin family peptides. Those amino acids with matching hues are identical amino acid sequences. The alignment similarity between peptides is displayed as a percentage next to the brackets.

CGRP and PACAP, two neuropeptides, are released during migraine and cluster headache attacks, acting as potent vasodilators that trigger migraine-like symptoms [22–24]. Their expression increases when the trigeminovascular system is activated, contributing to pain signal transmission and the development of mechanical hyperalgesia [25–28]. Despite their similar functions, PACAP and CGRP likely have distinct roles in causing migraine-like symptoms. In rodent models, their pathways seem to operate independently; therefore, elevated levels of these substances in peripheral blood during migraine attacks may serve as prospective biomarkers [29–33]. Different PACAP variants also contribute uniquely to migraine development [34–37]. The neuropeptide VIP, found in the trigeminal nerve, plays a key role in the progression of migraines development. It dilates blood vessels during attacks, influences neurotransmitter release, regulates inflammation and immune responses, and may affect migraine intensity and frequency by modulating pain signal sensitivity [38–42].

New drugs targeting the CGRP signaling pathway have been developed for migraine treatment and prevention. These include monoclonal antibodies (mAbs) directed at either CGRP ligand or receptor and CGRP receptor inhibitors [43–47]. While promising, these treatments have some downsides [48,49]. They have been shown to reduce the frequency of migraine attacks: according to reports from double-blind placebo-controlled clinical trials and open-label trials, the ≥50% responder rate for CGRP mAbs ranges from 27 to 62%, or 44.5 to 71.0%, respectively [50,51]. For CGRP receptor inhibitors, the ≥50% responder rate ranges from 56 to 61%, or 44.5 to 71.0%, in double-blind placebo-controlled clinical trials and open-label trials, respectively [52,53]. The relatively higher effective rates of open-label trials compared to double-blind placebo-controlled clinical trials could be attributed to the possibility that placebo plays a role in real life. Nevertheless, the responder rate varies depending on the type of and the duration of treatment, the response criteria, and the patient characteristics. Additionally, these drugs can be costly with limited insurance coverage [54]. While generally well-tolerated, CGRP-targeting mAbs can cause gastrointestinal disorders like constipation, and gepants can cause fatigue, nausea, dizziness, tiredness, and dry mouth [55].

Humans typically experience migraines, but preclinical research using animal models reveals the interaction of genetic and environmental factors contributing to neurological disorders like migraines [56–67]. These models simulate disease conditions, aiding in identifying pathogenic processes, evaluating symptoms and comorbidities, and discovering interventions [68–74]. The integration of preclinical and clinical research contributes to innovative therapeutics and personalized medicine [75–78]. This review discusses the pathogenesis of migraines and the need for new treatment targets. It highlights the potential of secretin family peptides' ligands and receptors as novel targets. The importance of further research into the roles of PACAP and VIP in migraine pathophysiology is emphasized, along with the development of targeted therapies. The review also considers the pituitary adenylate cyclase-activating peptide 1 receptor and other emerging therapeutic targets, such as PACAP1–38. It explores the similarities between PACAP and VIP, which are involved in sleep regulation and circadian rhythm, suggesting their key roles in migraines.

2. Pituitary Adenylate Cyclase-Activating Peptide and Vasoactive Intestinal Peptide

PACAP is a multi-functional peptide that has therapeutic potential in a variety of pathophysiological conditions and represents a promising avenue for intervention. PACAP is a neuropeptide that plays a crucial role in both neural and endocrine functions [78]. This peptide is widely distributed throughout the body and is involved in diverse physiological processes, including circadian rhythm and immune system regulations, modulation of pain perception, and stress response [79]. PACAP also has neuroprotective effects and has been shown to support nerve cell survival and regeneration in various neurological disorders [80]. GPCRs control the signaling pathways and cause the activation of adenylate cyclase (AC), the release of cyclic AMP, and the activation of protein kinase A (PKA) and calcium channels [81,82]. PACAP is a multi-functional peptide that has therapeutic potential in a variety of pathophysiological conditions and represents a promising avenue for therapeutic intervention [83].

2.1. Background

PACAP was found in ovine hypothalamic extracts in 1989. It is a 38-amino acid peptide hormone that stimulates AC activity in the pituitary gland [84]. Subsequently, it was found to be widely distributed in the central and peripheral nervous systems, as well as in non-neural tissues, including the adrenal gland, pancreas, gut, and reproductive system [85]. PACAP exists in three biologically active forms: PACAP1–38, 6–38-amino acid form of PACAP (PACAP6–38), and PACAP1–27 [86]. PACAP-related peptide (PRP) is also a member of the PACAP family [87]. Radioimmunoassay demonstrated that PACAP1–38 levels were approximately 60 times greater than PACAP1–27 levels and 10 times greater than PRP levels [88].

Since its discovery, PACAP has been extensively studied for its potent neuroprotective effects against a diverse range of neurological disorders, including stroke, traumatic brain injury, Parkinson's disease, and Alzheimer's disease [89,90]. Recent findings suggest that PACAP may also play a key role in the regulation of immune cell function and cytokine production, highlighting its potential as a therapeutic target for immune-mediated diseases such as rheumatoid arthritis, multiple sclerosis, and asthma [91]. Furthermore, PACAP has been implicated in the regulation of energy metabolism, making it a promising therapeutic agent for the treatment of metabolic disorders such as obesity and diabetes [92]. Overall, the growing body of evidence on the multifunctional properties of PACAP highlights its potential as a novel therapeutic target for a wide range of diseases.

VIP is a 28-amino acid polypeptide that was first characterized in 1970. It is secreted by cells throughout the intestinal tract and is widespread in many internal organs and systems [93]. VIP plays important roles in many biological functions, such as stimulation of contractility in the heart, vasodilation, promoting neuroendocrine–immune communication, lowering arterial blood pressure, and anti-inflammatory and immune-modulatory activity [94]. VIP stimulates the secretion of electrolytes and water by the intestinal mucosa and acts as a neurotransmitter, inducing a relaxation effect in some tissues [95]. VIP is also involved in the pathophysiology of various diseases, including osteoarthritis, cancer, and autoimmune disorders [94]. Furthermore, VIP is implicated in the physiological and pathophysiological roles of migraine [96]. In this context, VIP has been studied for its potential therapeutic applications.

2.2. Receptor and Signaling Mechanisms of PACAP and VIP

PACAP plays an important role in a wide range of biological processes such as feeding behavior, stress response, neuroprotection, and regulation of neurotransmitter release. It activates three different GPCRs named PAC1, vasoactive intestinal peptide receptor (VPAC) 1, and VPAC2; these receptors are widely expressed in the central and peripheral nervous systems, endocrine systems, and immune systems [97]. The binding of PACAP to these receptors leads to the activation of multiple signaling mechanisms (Table 1) [98].

Table 1. The secretin family peptides, their receptors, and their binding affinity.

Peptides	Receptors
CGRP	CLR
PACAP1–38	>>PAC1, <VPAC1, <VPAC2
PACAP6–38	?
PACAP1–27	>PAC1, <VPAC1, <VPAC2
PRP	?
VIP	>VPAC1, >VPAC2, <PAC1

CGRP: calcitonin gene-related peptide; PACAP: pituitary adenylate cyclase-activating polypeptide: PRP: PACAP-related peptide; VIP: vasoactive intestinal peptide; CLR: calcitonin receptor-like receptor; PAC1: pituitary adenylate cyclase-activating polypeptide type I; VPAC: vasoactive intestinal peptide receptor; ?: unknown; >>: much higher; >: higher; <: lower.

Activation of the PAC1 receptor by PACAP leads to the activation of the adenylyl cyclase enzyme, which in turn leads to the production of cyclic adenosine monophosphate (cAMP) and the activation of PKA [99]. It also triggers the activation of phospholipase C, which leads to the breakdown of phosphatidyl inositol 4,5-bisphosphate (PIP2) into inositol triphosphate and diacylglycerol (DAG), which activates protein kinase C (PKC) [100]. On the other hand, VPAC1 and VPAC2 receptor activation leads to AC enzyme activation, which leads to the generation of cAMP and the activation of PKA [101]. Also, PACAP signaling turns on calcium signaling, which causes intracellular calcium to be released and calcium/calmodulin-dependent kinase II to be activated [102]. PACAP signaling also activates the mitogen-activated protein kinase (MAPK), extracellular signal-regulated kinase (ERK), and jun N-terminal kinase signaling pathways [103]. These signaling mechanisms contribute to the diverse biological effects of PACAP on cellular functions. The regulation of PACAP gene expression is presented in Figure 2.

PACAP and VIP are neuropeptides that interact specifically with three receptors (VPAC1, VPAC2, and PAC1) from the class II B GPCR family [104]. The similarities between PACAP and VIP in receptor and signaling mechanisms include the following: PACAP and VIP share nearly 70% amino acid sequence identity; PACAP binds with high affinity to all three receptors, while VIP binds with high affinity to VPAC1 and VPAC2 receptors and has a thousand fold lower affinity for the PAC1 receptor compared to PACAP; both PACAP and VIP receptors are preferentially coupled to Gαs, leading to activation of AC, subsequent cAMP production, and activation of PKA; and PKA may in turn activate ERKs, PACAP and VIP receptor-mediated signaling pathways [105–108]. Due to the wide distribution of VIP and PACAP receptors in the body, potential therapeutic applications of drugs targeting these receptors, as well as expected unwanted side effects, are numerous [109]. Designing selective therapeutics targeting these receptors remains challenging due to their structural similarities.

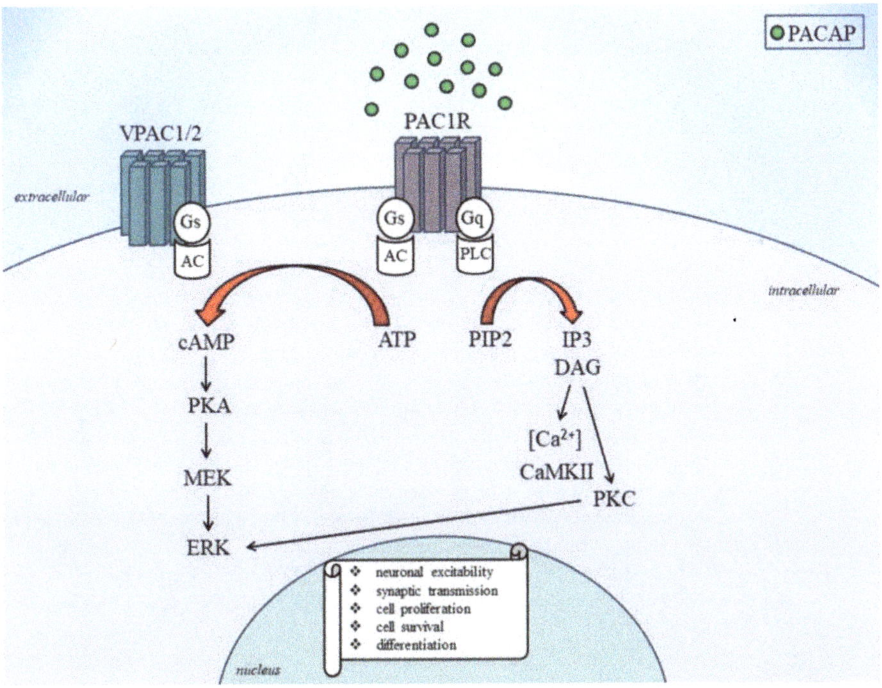

Figure 2. PACAP receptors signaling to ERK activation. AC, adenylate cylase; ATP: adenosine monophosphate; cAMP: cyclic adenosine monophosphate; DAG: diacylglycerol; ERK, extracellular signal-regulated kinase; Gs and Gq: stimulatory G protein; MEK: mitogen-activated protein kinase kinase; PKA: protein kinase A; PKC: protein kinase C; PACAP: pituitary adenylate cyclase-activating polypeptide; PAC1: PACAP 1 receptor; PIP2: phosphatidylinositol bisphosphate; VPAC1: vasoactive intestinal peptide receptor type 1; VPAC2: vasoactive intestinal peptide receptor type 2.

2.3. *Role of PACAP and VIP in Migraine*

PACAP has been strongly associated with the pathophysiology of migraine. PACAP is found in high levels in the trigeminal nerve, which is known to play a critical role in this condition. PACAP is known to increase the sensitivity of the trigeminal nerve, cause dilation of blood vessels in the brain, and trigger inflammation. All these biological effects have been implicated in the development of migraine attacks [110]. Several studies have been conducted to investigate the role of PACAP in migraine. One study showed that PACAP levels in the blood are significantly higher in migraine patients during an attack compared to headache-free controls [111]. This study suggests that PACAP could be used as a potential biomarker for migraine. Another study demonstrated that the venous infusion of PACAP into migraine patients resulted in the development of migraine-like attacks [112]. This finding strongly supports the hypothesis that PACAP plays a crucial role in the pathophysiology of migraine and suggests that blocking PACAP could be a potential therapeutic target for the treatment of migraines. The role of PACAP in migraine pathology is well established, and there is strong evidence that this neuropeptide plays a crucial role in the development of migraine attacks. Further research is needed to better understand the mechanism of action of PACAP and to develop new pharmacological agents that target PACAP for the treatment of migraines.

Both CGRP and PACAP are multifunctional peptides with many roles in the nervous, cardiovascular, respiratory, gastrointestinal, and reproductive systems. They play a role in vasodilation, neurogenic inflammation, and nociception. While CGRP plays an integral role in migraine, PACAP is likely to play a similar but distinct role as CGRP based on similarities and differences observed in both clinical and preclinical studies [113]. In rodent models, the PACAP pathway appears to be independent of the CGRP pathway, suggesting that CGRP and PACAP act in parallel ways that cause a migraine-like symptom [114]. In migraine without aura, the first double-blinded placebo-controlled study reported that 33% of the patients developed delayed migraine attacks after CGRP administration [115]. The studies have identified the involvement of two endogenous neuropeptides, CGRP and PACAP, in the pathogenesis of migraines [116].

VIP has also been implicated in the pathophysiology of migraine [117]. The similarities between PACAP and VIP in their roles in pathogenesis include the following: PACAP and VIP are released in conjunction with migraine and cluster headache attacks [118]; PACAP and VIP are potent vasodilators and can cause migraine-like attacks when infused into people [119]; a 2-h infusion of VIP caused migraine attacks, indicating that VIP plays a significant role in pathophysiology and intravenous administration of PACAP-38 caused headaches in all healthy subjects and migraine-like attacks in 58% of patients with a history of migraine without aura [15,35]; PACAP and VIP receptors are preferentially coupled to Gαs, leading to activation of AC, subsequent cAMP production, and activation of PKA [120]; PKA may in turn activate ERKs [121]; PACAP and VIP receptor-mediated signaling pathways are shown to share activities, including vasodilation, neurogenic inflammation, and nociception in rodents [122]; PACAP and VIP receptors provide a rich set of targets to complement and augment the current CGRP-based migraine therapeutics; VPAC1 receptors play a dominant role in PACAP-induced vasorelaxation in female mice [123]. Also, PG 99-465, a selective VPAC2 receptor antagonist that has been used in a number of physiological studies, has been shown to have significant activity at VPAC1 and PAC1 receptors [124].

2.4. Preclinical Studies

In addition to in vitro systems, a variety of organisms are used in experimental medicine [125–127]. Understanding the effects of endogenous neuropeptides, neurohormones, and metabolites has advanced significantly thanks to the information gathered using laboratory animals [128–133]. Animal models are a crucial tool for bridging the knowledge gap between data- and hypothesis-driven benchwork and its application to clinical bedside management. PACAP has been extensively studied as a neuromodulator in the trigeminal nociceptive pathway [134]. Preclinical studies have shown that PACAP is involved in the transmission of pain signals from the periphery to the central nervous system and is therefore a potential target for the treatment of migraine and other headache disorders [135,136].

In animal models, PACAP has been shown to play a role in trigeminal sensitization, which is the process by which nociceptive signals become amplified and persistent, leading to chronic pain [137]. Studies have also found that PACAP is involved in the activation of inflammatory pathways in the trigeminal nerve, further contributing to pain and inflammation [138]. In addition, PACAP has been implicated in the regulation of blood flow to the brain, which may also play a role in headache pathophysiology [139] and other neurological [26] or neuropsychological conditions [88]. In an experimental model of migraine, intraperitoneal administration of nitroglycerol caused marked photophobia and meningeal vasodilatation, and increased the number of c-fos-positive activated neurons in the TNC in wild-type mice but not in PACAP1–38-deficient mice [140]. In line with this, an increased concentration of PACAP1–38 was detected in the TNC after the activation of the TS in different animal models [141,142].

PAC1 receptor antagonists include PACAP6–38, N-stearyl-[Nle17] neurotensin-(6-11)/VIP-(7-28), deletion mutants of maxadilan, M65, and Max.d.4, and synthesized small-molecule acyl hydrazides, including PG 97-269 [143]. PACAP6–38 has been used as a PAC1 receptor antagonist in many studies, but it has an affinity for VPAC2 receptors [144]. N-stearyl-[Nle17] neurotensin-(6-11)/VIP-(7-28) (SNV) is a chimeric peptide analog that antagonizes the VIP2/PACAP receptor subclass. SNV is a better mitogen for the keratinocytic cell line and can increase AC activity in rat brain membranes 100 times more than VIP1-28 [145,146]. No migraine-related studies have been documented. The maxadilan is a vasodilator peptide derived from the salivary glands of sandflies. Its deletion mutants, M65 and Max.d.4, have been reported to be selective PAC1 receptor antagonists but have not been extensively used due to problems of availability [147,148]. PG 97-269 is a selective VPAC1 receptor antagonist with negligible affinity for the PACAP1 receptor. It did not stimulate AC activity but inhibited competitively the effect of VIP on AC activity in cells expressing the VIP1 receptor [146]. VIP and PACAP-induced vasodilation were partially blocked by PG 97-269, indicating that PACAP and VIP may play a role in migraine pathophysiology and that PG 97-269 may have therapeutic potential for migraine [149] (Table 2). Thus, preclinical studies suggest that concentrating on the PACAP signaling pathways in the trigeminal nociceptive system could be an effective strategy for discovering novel treatments for headache disorders. However, more research is needed to fully understand the mechanisms underlying PACAPs' role in headache pathophysiology and to develop effective and safe PACAP-targeted therapies.

Table 2. Preclinical findings of PACAP receptor antagonists.

Antagonists	Characteristics	Ref.
PACAP6–38	PAC1 receptor antagonist, affinity for VPAC2 receptors	[144]
N-stearyl-[Nle17] neurotensin-(6–11)/VIP-(7–28)	VIP2/PACAP receptor antagonist, mitogen for the keratinocytic cell line and can increase AC activity	[145,146]
Maxadilan mutants	PAC1 receptor antagonists, increased AC activity	[147,148]
PG 97-269	selective PAC1 receptor antagonists	[146]

VIP plays a key role in sensory processing and the modulation of pain pathways in the trigeminal system. In preclinical studies, VIP has been shown to change the activity of nociceptive neurons in the trigeminal ganglion and make the TNC more sensitive, which can cause chronic pain or migraines [150]. In response to noxious stimuli, the trigeminal sensory neurons release VIP. This can activate VIP receptors on nearby neurons and cause the release of a number of signaling molecules involved in pain amplification [151]. VIP-mediated sensitization of trigeminal neurons can lead to hyperexcitability and increased responsiveness to noxious stimuli, which may contribute to the development and maintenance of chronic pain or migraine [152]. Targeting VIP signaling pathways may therefore represent a promising approach for the development of novel therapies for chronic pain or migraine.

2.5. Clinical Studies

A growing body of clinical research suggests that PACAP plays an important role in migraine pathophysiology. Patients with migraines exhibit higher levels of PACAP compared to control groups [153]. PACAP is a neuropeptide recognized for its involvement in the activation of nociceptive pathways, contributing to the development of migraines. The high levels of PACAP in migraineurs have been associated with increased headache severity and frequency, and this has led to the exploration of PACAP as a therapeutic target for treatment [154]. In migraineurs without aura, the development of PACAP1–38-evoked migraine-like attacks was independent of the severity of family load [35,155]. In the same study, 90 min after the injection, the levels of numerous migraine-related molecular markers were increased in the plasma of patients [156]. Magnetic resonance imaging angiography examinations revealed that PACAP1–38-induced headache was associated with prolonged vasodilatation of the middle meningeal artery (MMA) but not the middle cerebral artery (MCA). Sumatriptan, an antimigraine medication, was able to alleviate the headache, which mirrored the contraction of the MMA but not the MCA, indicating that PACAP1–38-induced headaches may originate from extracerebral arteries [157].

An increasing number of clinical studies have shown that targeting PACAP signaling may be a promising therapeutic strategy for migraine treatment. In terms of safety, PACAP has been generally well tolerated in clinical trials [158]. One study found that PACAP induces headaches via sustained vasodilation and that targeting the PACAP pathway may be a promising approach for treatment [159]. AMG 301, a mAb that targets the PAC1 receptor, was administered to patients with episodic or chronic migraines in a randomized, double-blind, placebo-controlled phase 2 study. There was no significant difference between the AMG 301 group and the placebo group, suggesting that AMG 301 was ineffective for prevention [160,161]. On the other hand, the PACAP ligand mAb, Lu AG09222, was shown to reduce the number of monthly migraine days from baseline to weeks 1–4 of treatment statistically significantly more than placebo [162,163]. Additionally, the mAb targeting the PAC1 receptor, LY3451838, is currently undergoing phase 2 clinical trials for adults with treatment-resistant migraine. This trial is in progress, and the results are not yet available [164] (Table 3). Overall, the efficacy and safety of PACAP as a migraine treatment in clinical studies suggest that it is a promising option for patients with this debilitating condition. Further research is needed to fully understand the potential of PACAP as a treatment for migraines, but the current evidence is encouraging.

Table 3. Pituitary adenylate cyclase-activating polypeptide (PACAP) monoclonal antibodies under clinical trials.

ClinicalTrials.gov Identifier	Monoclonal Antibody	Target	Status	Ref.
NCT03238781	AMG 301	receptor	No benefit over placebo for migraine prevention	[160,161]
NCT05133323	Lu AG09222	ligand	No results posted; the press release announced a decrease in the number of migraine days per month	[162,163]
NCT04498910	LY3451838	receptor	No results posted	[164]

VIP infusion has been studied in the context of migraines, with a particular focus on its potential to provoke migraine attacks and its role in pathophysiology. A phase 2 clinical trial investigated the effects of a long-lasting infusion of VIP on headaches, cranial hemodynamics, and autonomic symptoms in episodic migraine patients without aura [165]. The study found that a 2-h infusion of VIP promoted long-lasting cranial vasodilation and delayed headaches in healthy volunteers, resembling the effect of prophylaxis. However, other studies have suggested that VIP infusions may actually provoke migraine attacks. For example, a randomized clinical trial found that a 2-h infusion of VIP caused migraine episodes, suggesting an important role of VIP in migraine pathophysiology [15]. It remains unclear whether the lack of migraine induction can be attributed to the only transient vasodilatory response after a 20-min infusion of VIP. Overall, the search results suggest that VIP infusion may have a role in migraine pathophysiology, but further research is needed to fully understand its effects and potential therapeutic applications.

3. Discussion

This review paper aims to provide insights into the roles of PACAP in migraine by comparing its actions with those of VIP. By analyzing existing studies, this paper hopes to shed light on the pathophysiology of migraines and pave the way towards more effective treatments. The ultimate goal of this review is to explore the potential of developing antimigraine drugs that target the PACAP pathways. Identifying and producing new ways to target the PACAP system may provide an alternative therapeutic option for migraineurs. The authors aim to consolidate the current evidence on the PACAP system's role in migraines and evaluate potential drug targets within the pathway, hoping to pave the way for more extensive research to develop new and effective antimigraine drugs that target the PACAP pathways.

The PACAP system presents a significant challenge when it comes to targeted therapies due to its pleiotropic roles in the body, both physiologically and pathologically [78–82]. PACAP plays crucial roles in various aspects of the body, such as neural development, pain regulation, immune functions, and stress responses. These diverse roles make the PACAP system difficult to target effectively without affecting other physiological functions. Furthermore, PACAP signaling is often dysregulated in pathological conditions such as inflammatory disorders, neurodegenerative diseases, and cancers [91,92]. Conversely, PACAP has been shown to have protective effects in certain diseases, such as ischemic stroke and Alzheimer's disease [89,90]. Thus, finding a balance between targeting the PACAP system to treat diseases while preserving its physiological functions remains a significant challenge in the field of medicine.

The PACAP system has emerged as a potential target for the treatment of migraines, especially after the discovery of the role of CGRP and its receptors in pathophysiology [110–112]. PACAP is a peptide that belongs to the family of CGRP peptides and is highly expressed in the TS. The TS is the neural network that causes migraine pain [137,141,142]. PACAP receptors have been found to be co-localized with CGRP receptors in the TS, suggesting that the two systems could be acting in a synergistic manner to induce migraine pain [113–115]. Therefore, targeting the PACAP system could provide an additional therapeutic approach for the treatment of migraine, and several drugs that inhibit PACAP or its receptors are currently under development.

The present review holds notable significance in shedding light on the critical role of PACAP in comparison with other neuropeptides like CGRP and VIP, which have been extensively studied as potential therapeutic targets for various neurological disorders. The differences in symptomatic manifestation observed in preclinical studies of CGRP, PACAP, and VIP are most likely due to their distinct roles in migraine physiology and pathophysiology [105–109,113–115]. Thus, elucidating the mechanisms of those neuropeptides may not only lead to a better understanding of the etiology of migraine but may also provide a variety of therapeutic targets, potentially supplying a more diverse palette of antimigraine regimens [150]. By thoroughly analyzing the preclinical studies, the review highlights the promising findings that suggest the potential translation of PACAP's therapeutic benefits from laboratory settings to clinical practice. The authors' critical evaluation and systematic compilation of the latest research on PACAP is bound to have a relevant impact on the scientific community and serve as a foundation for further clinical research. Ultimately, the knowledge and insights gained from this review will be instrumental in developing advanced treatments for a range of debilitating neurological conditions.

The difference between those two clinical outcomes of PACAP mAbs could be explained by the fact that mAbs are designed to target specific receptors or ligands with high selectivity. The difference in how mAbs target receptors or ligands can result in different outcomes due to a variety of factors. Initially, mAbs can bind to various receptors or ligands in a variety of ways, which can alter their efficacy and the biological effects that follow [166]. Secondly, mAbs can have a variety of mechanisms of action when interacting with their targets, such as inhibiting cell surface receptors or promoting target cell death [167]. Thirdly, biological and clinical activities can vary greatly depending on the target and antibody design. This includes differences in the efficacy of the treatment, the occurrence of adverse effects, and the overall health of the patient [166]. Fourthly, mAbs exhibit exceptional target selectivity, with the choice of target influencing the antibody's specificity and safety profile. When mAbs interact with their targets, they can perform a variety of actions, such as inhibiting the action of other molecules, killing cells, or altering the immune system's function [166,167]. The choice of target and antibody design is crucial in determining the therapeutic effectiveness of mAbs.

The review also highlights limitations and challenges in PACAP research, such as the complexity of its signaling mechanism, variations in its effects on different cell types, and the limited availability of specific antibodies against PACAP and its receptors. The high cost of producing PACAP analogs and the lack of standardized protocols for their synthesis and purification are also limitations. The scarcity of studies on PACAP and VIP is also a major challenge for this field. It is difficult to establish a general agreement on the preclinical results and their relevance for human trials. Meta-analyses could be helpful in this regard, but they require more studies to be published. Therefore, more clinical investigations are necessary to gather evidence and, hopefully, derive conclusions from the clinical research. These challenges and limitations make it difficult to fully understand the mechanisms of PACAP action and to develop effective therapeutic interventions.

The development of PACAP-based therapeutics for migraines will focus on two main approaches: targeting PACAP ligands and receptors. Studies using animal models of migraines have demonstrated that blocking the PACAP receptor reduces symptoms while inhibiting PACAP signaling reduces pain sensitivity. Currently, clinical trials are underway to assess the safety and effectiveness of various PACAP-based drugs for migraines in humans. PACAP-based therapies may offer an alternative to current treatments by targeting the underlying mechanisms of the disorder and reducing the risk of side effects. In addition, the role of additional secretin family peptides, ADM, and amylin in the pathogenesis of migraine remains to be investigated. Further research in this area could lead to the development of better treatments for migraines.

The future direction of migraine research holds great promise for advancing our understanding of this complex neurological disorder. The combination of preclinical and clinical data, along with computational tools, has provided invaluable insights into various aspects of diseases, including neurological and psychiatric disorders [168–189]. The use of preclinical models and clinical studies has shed light on the underlying mechanisms of migraine. These studies have contributed to the identification of structural and functional changes in the brain that occur in neurological and psychiatric disorders, such as migraine attacks, as well as conditions like depression and other mental health problems [190–206]. Understanding these changes is crucial for identifying biomarkers, developing targeted treatments, and improving diagnosis [207–209].

Migraine is not just a pain disorder, but it is also interrelated to emotional and cognitive domains [210]. This condition is commonly linked with a broad range of psychiatric comorbidities, especially among subjects with migraine with aura or chronic migraine [211]. The comorbidity between neurological and psychiatric disorders likely suggests multiple causes, such as unidirectional causal explanations or shared environmental and/or genetic risk factors, communication with other parts of the body, and their interaction on multiple levels [212–226]. Emotional distress is commonly recognized as a migraine trigger, and being affected by psychiatric disorders is considered an independent modifiable factor of progression toward chronification of migraine and a tendency to overuse medication [227]. Therefore, revealing the mechanisms of comorbidity between migraine and psychiatric disorders may lead to a clue to prevention and management. Many biological and neural aspects of the comorbidity need to be clarified in order to better understand the true nature of the migraine–psychiatric disorder association.

The integration of computational tools in migraine research has allowed for the testing and evaluation of potential treatments. These tools enable researchers to simulate the effects of different interventions, including brain stimulation, and assess their therapeutic efficacy [228–232]. This approach holds promise for the development of novel and more effective treatments. Advanced imaging techniques have played a crucial role in migraine research. Neuroimaging studies have revealed structural and functional brain changes associated with migraine [233–240]. These imaging techniques provide valuable insights into the pathophysiology of the disorder and can help identify unique clinical cases. The use of human brain organoids in migraine research is an emerging area of study. Brain organoids are three-dimensional models that mimic the structure and function of the human brain. They can be used to investigate altered neuronal pathways, protein expression, and metabolic pathways associated with migraines [241–244]. This approach offers a unique opportunity to study the disease in a more physiologically relevant system.

4. Conclusions

PACAP is a neuropeptide that has been linked to the pathophysiology of primary headaches such as migraine. The release of PACAP is associated with this condition and cluster headache attacks, and it has been shown to be a potent vasodilator that dilates cranial arteries and causes migraines when infused into patients. Like CGRP, PACAP is located near sensory nerve fibers and has nociceptive functions. Both peptides are promising targets for migraine therapeutics, and growing evidence supports the involvement of PACAP-related mechanisms in migraines. While CGRP and PACAP share similar functions, the PACAP pathway appears to be independent of the CGRP pathway, suggesting that they act in parallel ways to cause a migraine-like symptom. Therefore, a better understanding of the role of PACAP and other secretin family peptides, including VIP, in migraine pathogenesis could lead to new treatment options for this debilitating condition.

Author Contributions: Conceptualization, M.T. and L.V.; writing—original draft preparation, M.T.; writing—review and editing, M.T., Á.S., T.K., D.S., J.T. and L.V; visualization, Á.S. and T.K.; supervision, J.T. and L.V.; project administration, L.V.; funding acquisition, L.V. All authors have read and agreed to the published version of the manuscript.

Funding: This work was supported by OTKA-138125-K, TUDFO/47138-1/2019-ITM, and the HUN-REN Hungarian Research Network.

Conflicts of Interest: The authors declare no conflict of interest.

Abbreviations

AC	adenylate cyclase
ADM	adrenomedullin
cAMP	cyclic adenosine monophosphate
CGRP	calcitonin gene-related peptide
DAG	diacylglycerol
ERK	extracellular signal-regulated kinase
GPCR	G protein-coupled receptors
IAPP	islet amyloid polypeptide/amylin
mAbs	monoclonal antibodies
MAPK	mitogen-activated protein kinase
MCA	middle cerebral artery
MMA	middle meningeal artery
MEK:	mitogen-activated protein kinase kinase
PACAP	pituitary adenylate cyclase-activating polypeptide
PACAP1–38	38-amino acid form of PACAP
PACAP1–27	27-amino acid form of PACAP
PACAP6–38	6–38-amino acid form of PACAP
PIP2	phosphatidyl inositol 4,5-bisphosphate
PKA	protein kinase A
PKC	activates protein kinase C
PRP	PACAP-related peptide
SNV	N-stearyl-[Nle17] neurotensin-(6–11)/VIP-(7–28)
TNC	trigeminal nucleus caudalis
TS	trigeminovascular system
VPAC	vasoactive intestinal peptide receptor
VIP	vasoactive intestinal polypeptide

References

1. IHS Classification ICHD-3. Available online: https://ichd-3.org/1-migraine/1-2-migraine-with-aura/#:~:text=Description%3A, headache%20and%20associated%20migraine%20symptoms (accessed on 12 October 2023).
2. Amiri, P.; Kazeminasab, S.; Nejadghaderi, S.A.; Mohammadinasab, R.; Pourfathi, H.; Araj-Khodaei, M.; Sullman, M.J.M.; Kolahi, A.A.; Safiri, S. Migraine: A Review on Its History, Global Epidemiology, Risk Factors, and Comorbidities. *Front. Neurol.* **2022**, *12*, 800605. [CrossRef] [PubMed]
3. Ferrari, M.D.; Goadsby, P.J.; Burstein, R.; Kurth, T.; Ayata, C.; Charles, A.; Ashina, M.; van den Maagdenberg, A.M.J.M.; Dodick, D.W. Migraine. *Nat. Rev. Dis. Primers* **2022**, *8*, 2. [CrossRef]
4. Ashina, M. Migraine. *N. Engl. J. Med.* **2020**, *383*, 1866–1876. [CrossRef] [PubMed]
5. Gasparini, C.F.; Sutherland, H.G.; Griffiths, L.R. Studies on the pathophysiology and genetic basis of migraine. *Curr. Genom.* **2013**, *14*, 300–315. [CrossRef] [PubMed]
6. Burstein, R.; Noseda, R.; Borsook, D. Migraine: Multiple processes, complex pathophysiology. *J. Neurosci.* **2015**, *35*, 6619–6629. [CrossRef]
7. Dodick, D.W. A Phase-by-Phase Review of Migraine Pathophysiology. *Headache* **2018**, *58* (Suppl. S1), 4–16. [CrossRef]
8. Schankin, C.J.; Viana, M.; Goadsby, P.J. Persistent and Repetitive Visual Disturbances in Migraine: A Review. *Headache* **2017**, *57*, 1–16. [CrossRef]
9. Khan, J.; Asoom, L.I.A.; Sunni, A.A.; Rafique, N.; Latif, R.; Saif, S.A.; Almandil, N.B.; Almohazey, D.; AbdulAzeez, S.; Borgio, J.F. Genetics, pathophysiology, diagnosis, treatment, management, and prevention of migraine. *Biomed. Pharmacother.* **2021**, *139*, 111557. [CrossRef]
10. Buse, D.C.; Rupnow, M.F.; Lipton, R.B. Assessing and managing all aspects of migraine: Migraine attacks, migraine-related functional impairment, common comorbidities, and quality of life. *Mayo Clin. Proc.* **2009**, *84*, 422–435. [CrossRef]
11. Gupta, J.; Gaurkar, S.S. Migraine: An Underestimated Neurological Condition Affecting Billions. *Cureus* **2022**, *14*, e28347. [CrossRef]
12. Villar-Martinez, M.D.; Goadsby, P.J. Pathophysiology and Therapy of Associated Features of Migraine. *Cells* **2022**, *11*, 2767. [CrossRef]
13. Durham, P.L. Calcitonin gene-related peptide (CGRP) and migraine. *Headache* **2006**, *46* (Suppl. S1), S3–S8. [CrossRef] [PubMed]
14. Waschek, J.A.; Baca, S.M.; Akerman, S. PACAP and migraine headache: Immunomodulation of neural circuits in autonomic ganglia and brain parenchyma. *J. Headache Pain* **2018**, *19*, 23. [CrossRef] [PubMed]
15. Pellesi, L.; Al-Karagholi, M.A.; De Icco, R.; Coskun, H.; Elbahi, F.A.; Lopez-Lopez, C.; Snellman, J.; Hannibal, J.; Amin, F.M.; Ashina, M. Effect of Vasoactive Intestinal Polypeptide on Development of Migraine Headaches: A Randomized Clinical Trial. *JAMA Netw. Open* **2021**, *4*, e2118543. [CrossRef] [PubMed]
16. Ghanizada, H.; Al-Karagholi, M.A.; Walker, C.S.; Arngrim, N.; Rees, T.; Petersen, J.; Siow, A.; Mørch-Rasmussen, M.; Tan, S.; O'Carroll, S.J.; et al. Amylin Analog Pramlintide Induces Migraine-like Attacks in Patients. *Ann. Neurol.* **2021**, *89*, 1157–1171. [CrossRef] [PubMed]
17. May, A.; Goadsby, P.J. Substance P receptor antagonists in the therapy of migraine. *Expert Opin. Investig. Drugs* **2001**, *10*, 673–678. [CrossRef]
18. Petersen, K.A.; Birk, S.; Kitamura, K.; Olesen, J. Effect of adrenomedullin on the cerebral circulation: Relevance to primary headache disorders. *Cephalalgia* **2009**, *29*, 23–30. [CrossRef]
19. Hay, D.L.; Garelja, M.L.; Poyner, D.R.; Walker, C.S. Update on the pharmacology of calcitonin/CGRP family of peptides: IUPHAR Review 25. *Br. J. Pharmacol.* **2018**, *175*, 3–17. [CrossRef]
20. Poyner, D.R.; Hay, D.L. Secretin family (Class B) G protein-coupled receptors—From molecular to clinical perspectives. *Br. J. Pharmacol.* **2012**, *166*, 1–3. [CrossRef]
21. Edvinsson, L.; Grell, A.S.; Warfvinge, K. Expression of the CGRP Family of Neuropeptides and their Receptors in the Trigeminal Ganglion. *J. Mol. Neurosci.* **2020**, *70*, 930–944. [CrossRef]
22. Dux, M.; Vogler, B.; Kuhn, A.; Mackenzie, K.D.; Stratton, J.; Messlinger, K. The Anti-CGRP Antibody Fremanezumab Lowers CGRP Release from Rat Dura Mater and Meningeal Blood Flow. *Cells* **2022**, *11*, 1768. [CrossRef]
23. Pavelic, A.R.; Wöber, C.; Riederer, F.; Zebenholzer, K. Monoclonal Antibodies against Calcitonin Gene-Related Peptide for Migraine Prophylaxis: A Systematic Review of Real-World Data. *Cells* **2023**, *12*, 143. [CrossRef] [PubMed]
24. Körtési, T.; Spekker, E.; Vécsei, L. Exploring the Tryptophan Metabolic Pathways in Migraine-Related Mechanisms. *Cells* **2022**, *11*, 3795. [CrossRef] [PubMed]
25. Iyengar, S.; Johnson, K.W.; Ossipov, M.H.; Aurora, S.K. CGRP and the Trigeminal System in Migraine. *Headache* **2019**, *59*, 659–681. [CrossRef] [PubMed]
26. Edvinsson, L.; Tajti, J.; Szalárdy, L.; Vécsei, L. PACAP and its role in primary headaches. *J. Headache Pain* **2018**, *19*, 21. [CrossRef] [PubMed]
27. Körtési, T.; Tuka, B.; Nyári, A.; Vécsei, L.; Tajti, J. The effect of orofacial complete Freund's adjuvant treatment on the expression of migraine-related molecules. *J. Headache Pain* **2019**, *20*, 43. [CrossRef] [PubMed]
28. Russo, A.F. Calcitonin gene-related peptide (CGRP): A new target for migraine. *Annu. Rev. Pharmacol. Toxicol.* **2015**, *55*, 533–552. [CrossRef]
29. Kaiser, E.A.; Russo, A.F. CGRP and migraine: Could PACAP play a role too? *Neuropeptides* **2013**, *47*, 451–461. [CrossRef]

30. Kuburas, A.; Mason, B.N.; Hing, B.; Wattiez, A.S.; Reis, A.S.; Sowers, L.P.; Moldovan Loomis, C.; Garcia-Martinez, L.F.; Russo, A.F. PACAP Induces Light Aversion in Mice by an Inheritable Mechanism Independent of CGRP. *J. Neurosci.* **2021**, *41*, 4697–4715. [CrossRef]
31. Russo, A.F. CGRP as a neuropeptide in migraine: Lessons from mice. *Br. J. Clin. Pharmacol.* **2015**, *80*, 403–414. [CrossRef]
32. Goadsby, P.J.; Edvinsson, L.; Ekman, R. Release of vasoactive peptides in the extracerebral circulation of humans and the cat during activation of the trigeminovascular system. *Ann. Neurol.* **1988**, *23*, 193–196. [CrossRef] [PubMed]
33. Tuka, B.; Helyes, Z.; Markovics, A.; Bagoly, T.; Szolcsányi, J.; Szabó, N.; Tóth, E.; Kincses, Z.T.; Vécsei, L.; Tajti, J. Alterations in PACAP-38-like immunoreactivity in the plasma during ictal and interictal periods of migraine patients. *Cephalalgia* **2013**, *33*, 1085–1095. [CrossRef] [PubMed]
34. Lassen, L.H.; Haderslev, P.A.; Jacobsen, V.B.; Iversen, H.K.; Sperling, B.; Olesen, J. CGRP may play a causative role in migraine. *Cephalalgia* **2002**, *22*, 54–61. [CrossRef]
35. Schytz, H.W.; Birk, S.; Wienecke, T.; Kruuse, C.; Olesen, J.; Ashina, M. PACAP38 induces migraine-like attacks in patients with migraine without aura. *Brain* **2009**, *132 Pt 1*, 16–25. [CrossRef] [PubMed]
36. Vollesen, A.L.H.; Amin, F.M.; Ashina, M. Targeted Pituitary Adenylate Cyclase-Activating Peptide Therapies for Migraine. *Neurotherapeutics* **2018**, *15*, 371–376. [CrossRef] [PubMed]
37. Ghanizada, H.; Al-Karagholi, M.A.; Arngrim, N.; Olesen, J.; Ashina, M. PACAP27 induces migraine-like attacks in migraine patients. *Cephalalgia* **2020**, *40*, 57–67. [CrossRef] [PubMed]
38. Cernuda-Morollón, E.; Martínez-Camblor, P.; Alvarez, R.; Larrosa, D.; Ramón, C.; Pascual, J. Increased VIP levels in peripheral blood outside migraine attacks as a potential biomarker of cranial parasympathetic activation in chronic migraine. *Cephalalgia* **2015**, *35*, 310–316. [CrossRef] [PubMed]
39. Hery, M.; Faudon, M.; Hery, F. Effect of vasoactive intestinal peptide on serotonin release in the suprachiasmatic area of the rat. Modulation by oestradiol. *Peptides* **1984**, *5*, 313–317. [CrossRef]
40. Pellesi, L.; Al-Karagholi, M.A.; De Icco, R.; Chaudhry, B.A.; Lopez, C.L.; Snellman, J.; Hannibal, J.; Amin, F.M.; Ashina, M. Plasma Levels of CGRP During a 2-h Infusion of VIP in Healthy Volunteers and Patients With Migraine: An Exploratory Study. *Front. Neurol.* **2022**, *13*, 871176. [CrossRef]
41. De la Fuente, M.; Delgado, M.; Gomariz, R.P. VIP modulation of immune cell functions. *Adv. Neuroimmunol.* **1996**, *6*, 75–91. [CrossRef]
42. Hoffmann, J.; Baca, S.M.; Akerman, S. Neurovascular mechanisms of migraine and cluster headache. *J. Cereb. Blood. Flow Metab.* **2019**, *39*, 573–594. [CrossRef]
43. Ocheretyaner, E.R.; Kofman, M.; Quattrocchi, E. Calcitonin gene-related peptide (CGRP) receptor antagonists for the acute treatment of migraines in adults. *Drugs Context.* **2022**, *11*, 2022-3-5. [CrossRef] [PubMed]
44. Wang, X.; Chen, Y.; Song, J.; You, C. Efficacy and Safety of Monoclonal Antibody Against Calcitonin Gene-Related Peptide or Its Receptor for Migraine: A Systematic Review and Network Meta-analysis. *Front. Pharmacol.* **2021**, *12*, 649143. [CrossRef] [PubMed]
45. Berger, A.A.; Winnick, A.; Popovsky, D.; Kaneb, A.; Berardino, K.; Kaye, A.M.; Cornett, E.M.; Kaye, A.D.; Viswanath, O.; Urits, I. Lasmiditan for the Treatment of Migraines With or Without Aura in Adults. *Psychopharmacol. Bull.* **2020**, *50* (Suppl. S1), 163–188. [PubMed]
46. Rissardo, J.P.; Caprara, A.L.F. Gepants for Acute and Preventive Migraine Treatment: A Narrative Review. *Brain Sci.* **2022**, *12*, 1612. [CrossRef] [PubMed]
47. Ibekwe, A.; Perras, C.; Mierzwinski-Urban, M. Monoclonal Antibodies to Prevent Migraine Headaches. In *CADTH Issues in Emerging Health Technologies*; Canadian Agency for Drugs and Technologies in Health: Ottawa, ON, Canada, 2016. Available online: https://www.ncbi.nlm.nih.gov/books/NBK538376/ (accessed on 28 August 2023).
48. Nie, L.; Sun, K.; Gong, Z.; Li, H.; Quinn, J.P.; Wang, M. Src Family Kinases Facilitate the Crosstalk between CGRP and Cytokines in Sensitizing Trigeminal Ganglion via Transmitting CGRP Receptor/PKA Pathway. *Cells* **2022**, *11*, 3498. [CrossRef] [PubMed]
49. Greco, R.; Demartini, C.; Francavilla, M.; Zanaboni, A.M.; Tassorelli, C. Antagonism of CGRP Receptor: Central and Peripheral Mechanisms and Mediators in an Animal Model of Chronic Migraine. *Cells* **2022**, *11*, 3092. [CrossRef]
50. Diener, H.C.; Förderreuther, S.; Gaul, C.; Giese, F.; Hamann, T.; Holle-Lee, D.; Jürgens, T.P.; Kamm, K.; Kraya, T.; Lampl, C.; et al. Prevention of migraine with monoclonal antibodies against CGRP or the CGRP receptor: Addition to the S1 guideline: Therapy of migraine attacks and prevention of migraine. Recommendations of the Germany Society of Neurology and the German Migraine and Headache Society. *Neurol. Res. Pract.* **2020**, *2*, 11. [CrossRef]
51. Raffaelli, B.; De Icco, R.; Corrado, M.; Terhart, M.; Ailani, J. Open-label trials for CGRP-targeted drugs in migraine prevention: A narrative review. *Cephalalgia* **2023**, *43*, 3331024221137091. [CrossRef]
52. FDA Approves Oral CGRP Receptor Antagonist Atogepant for Preventive Migraine Treatment. Available online: https://www.pharmacytimes.com/view/fda-approves-oral-cgrp-receptor-antagonist-atogepant-for-preventive-migraine-treatment (accessed on 13 November 2023).
53. FDA Approves QULIPTA™ (Atogepant), the First and Only Oral CGRP Receptor Antagonist Specifically Developed for the Preventive Treatment of Migraine. Available online: https://news.abbvie.com/news/press-releases/fda-approves-qulipta-atogepant-first-and-only-oral-cgrp-receptor-antagonist-specifically-developed-for-preventive-treatment-migraine.htm (accessed on 13 November 2023).

54. Nguyen, J.L.; Munshi, K.; Peasah, S.K.; Swart, E.C.S.; Kohli, M.; Henderson, R.; Good, C.B. Trends in utilization and costs of migraine medications, 2017–2020. *J. Headache Pain* **2022**, *23*, 111. [CrossRef]
55. Haanes, K.A.; Edvinsson, L.; Sams, A. Understanding side-effects of anti-CGRP and anti-CGRP receptor antibodies. *J. Headache Pain* **2020**, *21*, 26. [CrossRef] [PubMed]
56. Telegdy, G.; Adamik, A.; Tanaka, M.; Schally, A.V. Effects of the LHRH antagonist Cetrorelix on affective and cognitive functions in rats. *Regul. Pept.* **2010**, *159*, 142–147. [CrossRef] [PubMed]
57. Tanaka, M.; Schally, A.V.; Telegdy, G. Neurotransmission of the antidepressant-like effects of the growth hormone-releasing hormone antagonist MZ-4-71. *Behav. Brain Res.* **2012**, *228*, 388–391. [CrossRef] [PubMed]
58. Tanaka, M.; Telegdy, G. Neurotransmissions of antidepressant-like effects of neuromedin U-23 in mice. *Behav. Brain Res.* **2014**, *259*, 196–199. [CrossRef]
59. Tanaka, M.; Csabafi, K.; Telegdy, G. Neurotransmissions of antidepressant-like effects of kisspeptin-13. *Regul. Pept.* **2013**, *180*, 1–4. [CrossRef]
60. Telegdy, G.; Tanaka, M.; Schally, A.V. Effects of the growth hormone-releasing hormone (GH-RH) antagonist on brain functions in mice. *Behav. Brain Res.* **2011**, *224*, 155–158. [CrossRef] [PubMed]
61. Rákosi, K.; Masaru, T.; Zarándia, M.; Telegdy, G.; Tóth, G.K. Short analogs and mimetics of human urocortin 3 display antidepressant effects in vivo. *Peptides* **2014**, *62*, 59–66. [CrossRef]
62. Tran, K.N.; Nguyen, N.P.K.; Nguyen, L.T.H.; Shin, H.-M.; Yang, I.-J. Screening for Neuroprotective and Rapid Antidepressant-like Effects of 20 Essential Oils. *Biomedicines* **2023**, *11*, 1248. [CrossRef]
63. Tanaka, M.; Kádár, K.; Tóth, G.; Telegdy, G. Antidepressant-like effects of urocortin 3 fragments. *Brain Res. Bull.* **2011**, *84*, 414–418. [CrossRef]
64. Baliellas, D.E.M.; Barros, M.P.; Vardaris, C.V.; Guariroba, M.; Poppe, S.C.; Martins, M.F.; Pereira, Á.A.F.; Bondan, E.F. Propentofylline Improves Thiol-Based Antioxidant Defenses and Limits Lipid Peroxidation following Gliotoxic Injury in the Rat Brainstem. *Biomedicines* **2023**, *11*, 1652. [CrossRef]
65. Montanari, M.; Imbriani, P.; Bonsi, P.; Martella, G.; Peppe, A. Beyond the Microbiota: Understanding the Role of the Enteric Nervous System in Parkinson's Disease from Mice to Human. *Biomedicines* **2023**, *11*, 1560. [CrossRef] [PubMed]
66. Garifulin, R.; Davleeva, M.; Izmailov, A.; Fadeev, F.; Markosyan, V.; Shevchenko, R.; Minyazeva, I.; Minekayev, T.; Lavrov, I.; Islamov, R. Evaluation of the Autologous Genetically Enriched Leucoconcentrate on the Lumbar Spinal Cord Morpho-Functional Recovery in a Mini Pig with Thoracic Spine Contusion Injury. *Biomedicines* **2023**, *11*, 1331. [CrossRef] [PubMed]
67. Bueno, C.R.d.S.; Tonin, M.C.C.; Buchaim, D.V.; Barraviera, B.; Ferreira Junior, R.S.; Santos, P.S.d.S.; Reis, C.H.B.; Pastori, C.M.; Pereira, E.d.S.B.M.; Nogueira, D.M.B.; et al. Morphofunctional Improvement of the Facial Nerve and Muscles with Repair Using Heterologous Fibrin Biopolymer and Photobiomodulation. *Pharmaceuticals* **2023**, *16*, 653. [CrossRef] [PubMed]
68. Sojka, A.; Żarowski, M.; Steinborn, B.; Hedzelek, W.; Wiśniewska-Spychała, B.; Dorocka-Bobkowska, B. Temporomandibular disorders in adolescents with headache. *Adv. Clin. Exp. Med.* **2018**, *27*, 193–199. [CrossRef]
69. Polyák, H.; Galla, Z.; Nánási, N.; Cseh, E.K.; Rajda, C.; Veres, G.; Spekker, E.; Szabó, Á.; Klivényi, P.; Tanaka, M.; et al. The Tryptophan-Kynurenine Metabolic System Is Suppressed in Cuprizone-Induced Model of Demyelination Simulating Progressive Multiple Sclerosis. *Biomedicines* **2023**, *11*, 945. [CrossRef]
70. Tanaka, M.; Szabó, Á.; Vécsei, L. Preclinical modeling in depression and anxiety: Current challenges and future research directions. *Adv. Clin. Exp. Med.* **2023**, *32*, 505–509. [CrossRef]
71. Chu, P.-C.; Huang, C.-S.; Chang, P.-K.; Chen, R.-S.; Chen, K.-T.; Hsieh, T.-H.; Liu, H.-L. Weak Ultrasound Contributes to Neuromodulatory Effects in the Rat Motor Cortex. *Int. J. Mol. Sci.* **2023**, *24*, 2578. [CrossRef]
72. Gecse, K.; Édes, A.E.; Nagy, T.; Demeter, A.K.; Virág, D.; Király, M.; Dalmadi Kiss, B.; Ludányi, K.; Környei, Z.; Denes, A.; et al. Citalopram Neuroendocrine Challenge Shows Altered Tryptophan and Kynurenine Metabolism in Migraine. *Cells* **2022**, *11*, 2258. [CrossRef]
73. Nasini, S.; Tidei, S.; Shkodra, A.; De Gregorio, D.; Cambiaghi, M.; Comai, S. Age-Related Effects of Exogenous Melatonin on Anxiety-like Behavior in C57/B6J Mice. *Biomedicines* **2023**, *11*, 1705. [CrossRef]
74. Chen, W.-C.; Wang, T.-S.; Chang, F.-Y.; Chen, P.-A.; Chen, Y.-C. Age, Dose, and Locomotion: Decoding Vulnerability to Ketamine in C57BL/6J and BALB/c Mice. *Biomedicines* **2023**, *11*, 1821. [CrossRef]
75. Statsenko, Y.; Habuza, T.; Smetanina, D.; Simiyu, G.L.; Meribout, S.; King, F.C.; Gelovani, J.G.; Das, K.M.; Gorkom, K.N.-V.; Zaręba, K.; et al. Unraveling Lifelong Brain Morphometric Dynamics: A Protocol for Systematic Review and Meta-Analysis in Healthy Neurodevelopment and Ageing. *Biomedicines* **2023**, *11*, 1999. [CrossRef] [PubMed]
76. Dang, J.; Tao, Q.; Niu, X.; Zhang, M.; Gao, X.; Yang, Z.; Yu, M.; Wang, W.; Han, S.; Cheng, J.; et al. Meta-Analysis of Structural and Functional Brain Abnormalities in Cocaine Addiction. *Front. Psychiatry* **2022**, *13*, 927075. [CrossRef]
77. Balogh, L.; Tanaka, M.; Török, N.; Vécsei, L.; Taguchi, S. Crosstalk between Existential Phenomenological Psychotherapy and Neurological Sciences in Mood and Anxiety Disorders. *Biomedicines* **2021**, *9*, 340. [CrossRef] [PubMed]
78. Rassler, B.; Blinowska, K.; Kaminski, M.; Pfurtscheller, G. Analysis of Respiratory Sinus Arrhythmia and Directed Information Flow between Brain and Body Indicate Different Management Strategies of fMRI-Related Anxiety. *Biomedicines* **2023**, *11*, 1028. [CrossRef]
79. Arimura, A. Perspectives on pituitary adenylate cyclase activating polypeptide (PACAP) in the neuroendocrine, endocrine, and nervous systems. *Jpn. J. Physiol.* **1998**, *48*, 301–331. [CrossRef] [PubMed]

80. Holland, P.R.; Barloese, M.; Fahrenkrug, J. PACAP in hypothalamic regulation of sleep and circadian rhythm: Importance for headache. *J. Headache Pain* **2018**, *19*, 20. [CrossRef]
81. Maugeri, G.; D'Amico, A.G.; Musumeci, G.; Reglodi, D.; D'Agata, V. Effects of PACAP on Schwann Cells: Focus on Nerve Injury. *Int. J. Mol. Sci.* **2020**, *21*, 8233. [CrossRef]
82. Johnson, G.C.; May, V.; Parsons, R.L.; Hammack, S.E. Parallel signaling pathways of pituitary adenylate cyclase activating polypeptide (PACAP) regulate several intrinsic ion channels. *Ann. N. Y. Acad. Sci.* **2019**, *1455*, 105–112. [CrossRef]
83. Clement, A.; Guo, S.; Jansen-Olesen, I.; Christensen, S.L. ATP-Sensitive Potassium Channels in Migraine: Translational Findings and Therapeutic Potential. *Cells* **2022**, *11*, 2406. [CrossRef]
84. Miyata, A.; Arimura, A.; Dahl, R.R.; Minamino, N.; Uehara, A.; Jiang, L.; Culler, M.D.; Coy, D.H. Isolation of a novel 38 residue-hypothalamic polypeptide which stimulates adenylate cyclase in pituitary cells. *Biochem. Biophys. Res. Commun.* **1989**, *164*, 567–574. [CrossRef]
85. Denes, V.; Geck, P.; Mester, A.; Gabriel, R. Pituitary Adenylate Cyclase-Activating Polypeptide: 30 Years in Research Spotlight and 600 Million Years in Service. *J. Clin. Med.* **2019**, *8*, 1488. [CrossRef] [PubMed]
86. Vaudry, D.; Nakamachi, T.; Basille, M.; Wurtz, O.; Fournier, A.; Shoida, S. Hypophysis Adenylate Cyclase Activating Polypeptide. In *Handbook of Biologically Active Peptides*, 2nd ed.; Academic Press: Boston, MA, USA, 2013. Available online: https://www.sciencedirect.com/topics/pharmacology-toxicology-and-pharmaceutical-science/hypophysis-adenylate-cyclase-activating-polypeptide (accessed on 28 August 2023).
87. Tam, J.K.; Lee, L.T.; Chow, B.K. PACAP-related peptide (PRP)—Molecular evolution and potential functions. *Peptides* **2007**, *28*, 1920–1929. [CrossRef]
88. Köves, K.; Szabó, E.; Kántor, O.; Heinzlmann, A.; Szabó, F.; Csáki, Á. Current State of Understanding of the Role of PACAP in the Hypothalamo-Hypophyseal Gonadotropin Functions of Mammals. *Front. Endocrinol.* **2020**, *11*, 88. [CrossRef]
89. Lee, E.H.; Seo, S.R. Neuroprotective roles of pituitary adenylate cyclase-activating polypeptide in neurodegenerative diseases. *BMB Rep.* **2014**, *47*, 369–375. [CrossRef] [PubMed]
90. Sadanandan, N.; Cozene, B.; Park, Y.J.; Farooq, J.; Kingsbury, C.; Wang, Z.J.; Moscatello, A.; Saft, M.; Cho, J.; Gonzales-Portillo, B.; et al. Pituitary Adenylate Cyclase-Activating Polypeptide: A Potent Therapeutic Agent in Oxidative Stress. *Antioxidants* **2021**, *10*, 354. [CrossRef] [PubMed]
91. Waschek, J.A. VIP and PACAP: Neuropeptide modulators of CNS inflammation, injury, and repair. *Br. J. Pharmacol.* **2013**, *169*, 512–523. [CrossRef]
92. Xiao, X.; Qiu, P.; Gong, H.Z.; Chen, X.M.; Sun, Y.; Hong, A.; Ma, Y. PACAP ameliorates hepatic metabolism and inflammation through up-regulating FAIM in obesity. *J. Cell Mol. Med.* **2019**, *23*, 5970–5980. [CrossRef] [PubMed]
93. Gonkowski, S. Vasoactive Intestinal Polypeptide in the Carotid Body—A History of Forty Years of Research. A Mini Review. *Int. J. Mol. Sci.* **2020**, *21*, 4692. [CrossRef]
94. Jiang, W.; Wang, H.; Li, Y.S.; Luo, W. Role of vasoactive intestinal peptide in osteoarthritis. *J. Biomed. Sci.* **2016**, *23*, 63. [CrossRef]
95. Cao, S.G.; Wu, W.C.; Han, Z.; Wang, M.Y. Effects of psychological stress on small intestinal motility and expression of cholecystokinin and vasoactive intestinal polypeptide in plasma and small intestine in mice. *World J. Gastroenterol.* **2005**, *11*, 737–740. [CrossRef]
96. Jacobs, B.; Dussor, G. Neurovascular contributions to migraine: Moving beyond vasodilation. *Neuroscience* **2016**, *338*, 130–144. [CrossRef] [PubMed]
97. Langer, I.; Jeandriens, J.; Couvineau, A.; Sanmukh, S.; Latek, D. Signal Transduction by VIP and PACAP Receptors. *Biomedicines* **2022**, *10*, 406. [CrossRef] [PubMed]
98. Fizanne, L.; Sigaudo-Roussel, D.; Saumet, J.L.; Fromy, B. Evidence for the involvement of VPAC1 and VPAC2 receptors in pressure-induced vasodilatation in rodents. *J. Physiol.* **2004**, *554 Pt 2*, 519–528. [CrossRef]
99. Parsons, R.L.; May, V. PACAP-Induced PAC1 Receptor Internalization and Recruitment of Endosomal Signaling Regulate Cardiac Neuron Excitability. *J. Mol. Neurosci.* **2019**, *68*, 340–347. [CrossRef]
100. Bill, C.A.; Vines, C.M. Phospholipase C. *Adv. Exp. Med. Biol.* **2020**, *1131*, 215–242. [CrossRef] [PubMed]
101. Barloese, M.; Chitgar, M.; Hannibal, J.; Møller, S. Pituitary adenylate cyclase-activating peptide: Potential roles in the pathophysiology and complications of cirrhosis. *Liver Int.* **2020**, *40*, 2578–2589. [CrossRef] [PubMed]
102. Makhinson, M.; Chotiner, J.K.; Watson, J.B.; O'Dell, T.J. Adenylyl cyclase activation modulates activity-dependent changes in synaptic strength and Ca2+/calmodulin-dependent kinase II autophosphorylation. *J. Neurosci.* **1999**, *19*, 2500–2510. [CrossRef]
103. Johnson, G.L.; Lapadat, R. Mitogen-activated protein kinase pathways mediated by ERK, JNK, and p38 protein kinases. *Science* **2002**, *298*, 1911–1912. [CrossRef]
104. Lu, J.; Piper, S.J.; Zhao, P.; Miller, L.J.; Wootten, D.; Sexton, P.M. Targeting VIP and PACAP Receptor Signaling: New Insights into Designing Drugs for the PACAP Subfamily of Receptors. *Int. J. Mol. Sci.* **2022**, *23*, 8069. [CrossRef]
105. Hirabayashi, T.; Nakamachi, T.; Shioda, S. Discovery of PACAP and its receptors in the brain. *J. Headache Pain* **2018**, *19*, 28. [CrossRef]
106. Watanabe, J. Vasoactive Intestinal Polypeptide Receptor 1. In *Methods in Enzymology*; Academic Press: Boston, MA, USA, 2013. Available online: https://www.sciencedirect.com/topics/medicine-and-dentistry/vasoactive-intestinal-polypeptide-receptor-1 (accessed on 28 August 2023).

107. Vasoactive Intestinal Polypeptide Receptor. Autonomic Neuroscience. 2007. Available online: https://www.sciencedirect.com/topics/neuroscience/vasoactive-intestinal-polypeptide-receptor (accessed on 28 August 2023).
108. May, V.; Buttolph, T.R.; Girard, B.M.; Clason, T.A.; Parsons, R.L. PACAP-induced ERK activation in HEK cells expressing PAC1 receptors involves both receptor internalization and PKC signaling. *Am. J. Physiol. Cell Physiol.* **2014**, *306*, C1068–C1079. [CrossRef] [PubMed]
109. Hou, X.; Yang, D.; Yang, G.; Li, M.; Zhang, J.; Zhang, J.; Zhang, Y.; Liu, Y. Therapeutic potential of vasoactive intestinal peptide and its receptor VPAC2 in type 2 diabetes. *Front. Endocrinol.* **2022**, *13*, 984198. [CrossRef] [PubMed]
110. Sundrum, T.; Walker, C.S. Pituitary adenylate cyclase-activating polypeptide receptors in the trigeminovascular system: Implications for migraine. *Br. J. Pharmacol.* **2018**, *175*, 4109–4120. [CrossRef] [PubMed]
111. Liu, J.; Wang, G.; Dan, Y.; Liu, X. CGRP and PACAP-38 play an important role in diagnosing pediatric migraine. *J. Headache Pain* **2022**, *23*, 68. [CrossRef]
112. Schytz, H.W.; Olesen, J.; Ashina, M. The PACAP receptor: A novel target for migraine treatment. *Neurotherapeutics* **2010**, *7*, 191–196. [CrossRef]
113. Kuburas, A.; Russo, A.F. Shared and independent roles of CGRP and PACAP in migraine pathophysiology. *J. Headache Pain* **2023**, *24*, 34. [CrossRef]
114. Ernstsen, C.; Christensen, S.L.; Rasmussen, R.H.; Nielsen, B.S.; Jansen-Olesen, I.; Olesen, J.; Kristensen, D.M. The PACAP pathway is independent of CGRP in mouse models of migraine: Possible new drug target? *Brain* **2022**, *145*, 2450–2460. [CrossRef]
115. Christensen, C.E.; Ashina, M.; Amin, F.M. Calcitonin Gene-Related Peptide (CGRP) and Pituitary Adenylate Cyclase-Activating Polypeptide (PACAP) in Migraine Pathogenesis. *Pharmaceuticals* **2022**, *15*, 1189. [CrossRef]
116. Anapindi, K.D.B.; Yang, N.; Romanova, E.V.; Rubakhin, S.S.; Tipton, A.; Dripps, I.; Sheets, Z.; Sweedler, J.V.; Pradhan, A.A. PACAP and Other Neuropeptide Targets Link Chronic Migraine and Opioid-induced Hyperalgesia in Mouse Models. *Mol. Cell Proteomics* **2019**, *18*, 2447–2458. [CrossRef]
117. Silvestro, M.; Iannone, L.F.; Orologio, I.; Tessitore, A.; Tedeschi, G.; Geppetti, P.; Russo, A. Migraine Treatment: Towards New Pharmacological Targets. *Int. J. Mol. Sci.* **2023**, *24*, 12268. [CrossRef]
118. Pellesi, L.; Chaudhry, B.A.; Vollesen, A.L.H.; Snoer, A.H.; Baumann, K.; Skov, P.S.; Jensen, R.H.; Ashina, M. PACAP38- and VIP-induced cluster headache attacks are not associated with changes of plasma CGRP or markers of mast cell activation. *Cephalalgia* **2022**, *42*, 687–695. [CrossRef]
119. Rasmussen, L.; Deligianni, C.; Christensen, C.E.; Karlsson, W.K.; Al-Khazali, H.M.; Van de Casteele, T.; Granhall, C.; Amin, F.M.; Ashina, M. The effect of Lu AG09222 on PACAP38- and VIP-induced vasodilation, heart rate increase, and headache in healthy subjects: An interventional, randomized, double-blind, parallel-group, placebo-controlled study. *J. Headache Pain* **2023**, *24*, 60. [CrossRef] [PubMed]
120. Vu, J.P.; Germano, P.M.; Pisegna, J.R. Adenylate Cyclase. Adenylate Cyclases Are Enzymes That Catalyze the Conversion of ATP to cAMP and Pyrophosphate. In *From: The Senses: A Comprehensive Reference*; Academic Press: Boston, MA, USA, 2008. Available online: https://www.sciencedirect.com/topics/agricultural-and-biological-sciences/adenylate-cyclase (accessed on 29 August 2023).
121. Roberts, R.E. The extracellular signal-regulated kinase (ERK) pathway: A potential therapeutic target in hypertension. *J. Exp. Pharmacol.* **2012**, *4*, 77–83. [CrossRef]
122. Lund, A.M.; Hannibal, J. Localization of the neuropeptides pituitary adenylate cyclase-activating polypeptide, vasoactive intestinal peptide, and their receptors in the basal brain blood vessels and trigeminal ganglion of the mouse CNS; an immunohistochemical study. *Front. Neuroanat.* **2022**, *16*, 991403. [CrossRef] [PubMed]
123. Ivic, I.; Balasko, M.; Fulop, B.D.; Hashimoto, H.; Toth, G.; Tamas, A.; Juhasz, T.; Koller, A.; Reglodi, D.; Solymár, M. VPAC1 receptors play a dominant role in PACAP-induced vasorelaxation in female mice. *PLoS ONE* **2019**, *14*, e0211433. [CrossRef] [PubMed]
124. VIP and PACAP Receptors—IUPHAR/BPS Guide to PHARMACOLOGY. Available online: https://www.guidetopharmacology.org/GRAC/FamilyDisplayForward?familyId=67 (accessed on 29 August 2023).
125. Datki, Z.; Sinka, R. Translational biomedicine-oriented exploratory research on bioactive rotifer-specific biopolymers. *Adv. Clin. Exp. Med.* **2022**, *31*, 931–935. [CrossRef]
126. Palotai, M.; Telegdy, G.; Tanaka, M.; Bagosi, Z.; Jászberényi, M. Neuropeptide AF induces anxiety-like and antidepressant-like behavior in mice. *Behav. Brain Res.* **2014**, *274*, 264–269. [CrossRef]
127. Lieb, A.; Thaler, G.; Fogli, B.; Trovato, O.; Posch, M.A.; Kaserer, T.; Zangrandi, L. Functional Characterization of Spinocerebellar Ataxia Associated Dynorphin A Mutant Peptides. *Biomedicines* **2021**, *9*, 1882. [CrossRef]
128. Skobeleva, K.; Shalygin, A.; Mikhaylova, E.; Guzhova, I.; Ryazantseva, M.; Kaznacheyeva, E. The STIM1/2-Regulated Calcium Homeostasis Is Impaired in Hippocampal Neurons of the 5xFAD Mouse Model of Alzheimer's Disease. *Int. J. Mol. Sci.* **2022**, *23*, 14810. [CrossRef]
129. Martos, D.; Tuka, B.; Tanaka, M.; Vécsei, L.; Telegdy, G. Memory Enhancement with Kynurenic Acid and Its Mechanisms in Neurotransmission. *Biomedicines* **2022**, *10*, 849. [CrossRef]
130. Tanaka, M.; Szabó, Á.; Spekker, E.; Polyák, H.; Tóth, F.; Vécsei, L. Mitochondrial Impairment: A Common Motif in Neuropsychiatric Presentation? The Link to the Tryptophan–Kynurenine Metabolic System. *Cells* **2022**, *11*, 2607. [CrossRef] [PubMed]

131. Tanaka, M.; Bohár, Z.; Martos, D.; Telegdy, G.; Vécsei, L. Antidepressant-like effects of kynurenic acid in a modified forced swim test. *Pharmacol. Rep.* **2020**, *72*, 449–455. [CrossRef]
132. Tanaka, M.; Telegdy, G. Involvement of adrenergic and serotonergic receptors in antidepressant-like effect of urocortin 3 in a modified forced swimming test in mice. *Brain Res. Bull.* **2008**, *77*, 301–305. [CrossRef] [PubMed]
133. Tanaka, M.; Spekker, E.; Szabó, Á.; Polyák, H.; Vécsei, L. Modelling the neurodevelopmental pathogenesis in neuropsychiatric disorders. Bioactive kynurenines and their analogues as neuroprotective agents-in celebration of 80th birthday of Professor Peter Riederer. *J. Neural. Transm.* **2022**, *129*, 627–642. [CrossRef]
134. Reducha, P.V.; Edvinsson, L.; Haanes, K.A. Could Experimental Inflammation Provide Better Understanding of Migraines? *Cells* **2022**, *11*, 2444. [CrossRef] [PubMed]
135. Ojala, J.; Tooke, K.; Hsiang, H.; Girard, B.M.; May, V.; Vizzard, M.A. PACAP/PAC1 Expression and Function in Micturition Pathways. *J. Mol. Neurosci.* **2019**, *68*, 357–367. [CrossRef] [PubMed]
136. Tamas, A.; Reglodi, D.; Farkas, O.; Kovesdi, E.; Pal, J.; Povlishock, J.T.; Schwarcz, A.; Czeiter, E.; Szanto, Z.; Doczi, T.; et al. Effect of PACAP in central and peripheral nerve injuries. *Int. J. Mol. Sci.* **2012**, *13*, 8430–8448. [CrossRef]
137. Zhang, L.; Zhou, Y.; Yang, L.; Wang, Y.; Xiao, Z. PACAP6-38 improves nitroglycerin-induced central sensitization by modulating synaptic plasticity at the trigeminal nucleus caudalis in a male rat model of chronic migraine. *J. Headache Pain* **2023**, *24*, 66. [CrossRef]
138. Takács-Lovász, K.; Kun, J.; Aczél, T.; Urbán, P.; Gyenesei, A.; Bölcskei, K.; Szőke, É.; Helyes, Z. PACAP-38 Induces Transcriptomic Changes in Rat Trigeminal Ganglion Cells Related to Neuroinflammation and Altered Mitochondrial Function Presumably via PAC1/VPAC2 Receptor-Independent Mechanism. *Int. J. Mol. Sci.* **2022**, *23*, 2120. [CrossRef]
139. Frederiksen, S.D.; Haanes, K.A.; Warfvinge, K.; Edvinsson, L. Perivascular neurotransmitters: Regulation of cerebral blood flow and role in primary headaches. *J. Cereb. Blood Flow Metab.* **2019**, *39*, 610–632. [CrossRef]
140. Markovics, A.; Kormos, V.; Gaszner, B.; Lashgarara, A.; Szoke, E.; Sandor, K.; Szabadfi, K.; Tuka, B.; Tajti, J.; Szolcsanyi, J.; et al. Pituitary adenylate cyclase-activating polypeptide plays a key role in nitroglycerol-induced trigeminovascular activation in mice. *Neurobiol. Dis.* **2012**, *45*, 633–644. [CrossRef] [PubMed]
141. Edvinsson, L. PACAP and its receptors in migraine pathophysiology: Commentary on Walker et al. Br. J. Pharmacol. 171: 1521–1533. *Br. J. Pharmacol.* **2015**, *172*, 4782–4784. [CrossRef] [PubMed]
142. Saposnik, G.; Montalban, X.; Selchen, D.; Terzaghi, M.A.; Bakdache, F.; Montoya, A.; Fruns, M.; Caceres, F.; Oh, J. Therapeutic Inertia in Multiple Sclerosis Care: A Study of Canadian Neurologists. *Front. Neurol.* **2018**, *9*, 781. [CrossRef]
143. Harding, S.D.; Armstrong, J.F.; Faccenda, E.; Southan, C.; Alexander, S.P.H.; Davenport, A.P.; Pawson, A.J.; Spedding, M.; Davies, J.A.; NC-IUPHAR. The IUPHAR/BPS guide to PHARMACOLOGY in 2022: Curating pharmacology for COVID-19, malaria and antibacterials. *Nucl. Acids Res.* **2022**, *50*, D1282–D1294. [CrossRef]
144. Guo, S.; Jansen-Olesen, I.; Olesen, J.; Christensen, S.L. Role of PACAP in migraine: An alternative to CGRP? *Neurobiol. Dis.* **2023**, *176*, 105946. [CrossRef] [PubMed]
145. Granoth, R.; Fridkin, M.; Gozes, I. VIP and the potent analog, stearyl-Nle(17)-VIP, induce proliferation of keratinocytes. *FEBS Lett.* **2000**, *475*, 78–83. [CrossRef]
146. Gourlet, P.; De Neef, P.; Cnudde, J.; Waelbroeck, M.; Robberecht, P. In vitro properties of a high affinity selective antagonist of the VIP1 receptor. *Peptide* **1997**, *18*, 1555–1560. [CrossRef]
147. Beebe, X.; Darczak, D.; Davis-Taber, R.A.; Uchic, M.E.; Scott, V.E.; Jarvis, M.F.; Stewart, A.O. Discovery and SAR of hydrazide antagonists of the pituitary adenylate cyclase-activating polypeptide (PACAP) receptor type 1 (PAC1-R). *Bioorg. Med. Chem. Lett.* **2008**, *18*, 2162–2166. [CrossRef]
148. Laburthe, M.; Couvineau, A.; Tan, V. Class II G protein-coupled receptors for VIP and PACAP: Structure, models of activation and pharmacology. *Peptides* **2007**, *28*, 1631–1639. [CrossRef]
149. Tasma, Z.; Siow, A.; Harris, P.W.R.; Brimble, M.A.; Hay, D.L.; Walker, C.S. Characterisation of agonist signalling profiles and agonist-dependent antagonism at PACAP-responsive receptors: Implications for drug discovery. *Br. J. Pharmacol.* **2022**, *179*, 435–453. [CrossRef]
150. Spekker, E.; Tanaka, M.; Szabó, Á.; Vécsei, L. Neurogenic Inflammation: The Participant in Migraine and Recent Advancements in Translational Research. *Biomedicines* **2022**, *10*, 76. [CrossRef] [PubMed]
151. Pinho-Ribeiro, F.A.; Verri, W.A., Jr.; Chiu, I.M. Nociceptor Sensory Neuron-Immune Interactions in Pain and Inflammation. *Trends Immunol.* **2017**, *38*, 5–19. [CrossRef] [PubMed]
152. Guo, S.; Vollesen, A.L.; Hansen, R.D.; Esserlind, A.L.; Amin, F.M.; Christensen, A.F.; Olesen, J.; Ashina, M. Part I: Pituitary adenylate cyclase-activating polypeptide-38 induced migraine-like attacks in patients with and without familial aggregation of migraine. *Cephalalgia* **2017**, *37*, 125–135. [CrossRef] [PubMed]
153. ClinicalTrials.gov. PACAP Induced Migraine Attacks in Patients with High and Low Genetic Load. Available online: https://classic.clinicaltrials.gov/ct2/show/NCT02158221 (accessed on 29 August 2023).
154. Togha, M.; Ghorbani, Z.; Ramazi, S.; Zavvari, F.; Karimzadeh, F. Evaluation of Serum Levels of Transient Receptor Potential Cation Channel Subfamily V Member 1, Vasoactive Intestinal Polypeptide, and Pituitary Adenylate Cyclase-Activating Polypeptide in Chronic and Episodic Migraine: The Possible Role in Migraine Transformation. *Front. Neurol.* **2021**, *12*, 770980. [CrossRef] [PubMed]

155. Körtési, T.; Tuka, B.; Tajti, J.; Bagoly, T.; Fülöp, F.; Helyes, Z.; Vécsei, L. Kynurenic Acid Inhibits the Electrical Stimulation Induced Elevated Pituitary Adenylate Cyclase-Activating Polypeptide Expression in the TNC. *Front. Neurol.* **2018**, *8*, 745. [CrossRef]
156. Guo, S.; Vollesen, A.L.; Hansen, Y.B.; Frandsen, E.; Andersen, M.R.; Amin, F.M.; Fahrenkrug, J.; Olesen, J.; Ashina, M. Part II: Biochemical changes after pituitary adenylate cyclase-activating polypeptide-38 infusion in migraine patients. *Cephalalgia* **2017**, *37*, 136–147. [CrossRef]
157. Amin, F.M.; Asghar, M.S.; Guo, S.; Hougaard, A.; Hansen, A.E.; Schytz, H.W.; van der Geest, R.J.; de Koning, P.J.; Larsson, H.B.; Olesen, J.; et al. Headache and prolonged dilatation of the middle meningeal artery by PACAP38 in healthy volunteers. *Cephalalgia* **2012**, *32*, 140–149. [CrossRef] [PubMed]
158. Maasz, G.; Zrinyi, Z.; Reglodi, D.; Petrovics, D.; Rivnyak, A.; Kiss, T.; Jungling, A.; Tamas, A.; Pirger, Z. Pituitary adenylate cyclase-activating polypeptide (PACAP) has a neuroprotective function in dopamine-based neurodegeneration in rat and snail parkinsonian models. *Dis. Model Mech.* **2017**, *10*, 127–139. [CrossRef]
159. Rubio-Beltrán, E.; Correnti, E.; Deen, M.; Kamm, K.; Kelderman, T.; Papetti, L.; Vigneri, S.; MaassenVanDenBrink, A.; Edvinsson, L.; European Headache Federation School of Advanced Studies (EHF-SAS). PACAP38 and PAC_1 receptor blockade: A new target for headache? *J. Headache Pain* **2018**, *19*, 64. [CrossRef]
160. Ashina, M.; Doležil, D.; Bonner, J.H.; Zhou, L.; Klatt, J.; Picard, H.; Mikol, D.D. A phase 2, randomized, double-blind, placebo-controlled trial of AMG 301, a pituitary adenylate cyclase-activating polypeptide PAC1 receptor monoclonal antibody for migraine prevention. *Cephalalgia* **2021**, *41*, 33–44. [CrossRef]
161. Study to Evaluate the Efficacy and Safety of AMG 301 in Migraine Prevention 2020. Available online: https://ClinicalTrials.gov/show/NCT03238781 (accessed on 4 September 2023).
162. Lundbeck News Room: Lundbeck Announced the Start of a Phase II Clinical Study to Assess Lu AG09222 for Migraine Prevention. Available online: https://newsroom.lundbeckus.com/news-release/2021/lundbeck-announced-start-of-phase-ii-clinical-study-for-migraine-prevention (accessed on 29 August 2023).
163. A Study With Lu AG09222 in Adults With Migraine Who Have Not Been Helped by Prior Preventive Treatments 2023. Available online: https://classic.clinicaltrials.gov/ct2/show/NCT05133323 (accessed on 4 September 2023).
164. A Study of LY3451838 in Participants with Migraine 2023. Available online: https://ClinicalTrials.gov/show/NCT04498910 (accessed on 4 September 2023).
165. ClinicalTrials.gov. The Effects of a Long-Lasting Infusion of Vasoactive Intestinal Peptide (VIP) in Episodic Migraine Patients. Available online: https://classic.clinicaltrials.gov/ct2/show/NCT04260035 (accessed on 29 August 2023).
166. Terry, J.S.; Anderson, L.B.; Scherman, M.S.; McAlister, C.E.; Perera, R.; Schountz, T.; Geiss, B.J. Development of a SARS-CoV-2 nucleocapsid specific monoclonal antibody. *Virology* **2021**, *558*, 28–37. [CrossRef] [PubMed]
167. Sgro, C. Side-effects of a monoclonal antibody, muromonab CD3/orthoclone OKT3: Bibliographic review. *Toxicology* **1995**, *105*, 23–29. [CrossRef] [PubMed]
168. Tanaka, M.; Szabó, Á.; Vécsei, L. Integrating Armchair, Bench, and Bedside Research for Behavioral Neurology and Neuropsychiatry: Editorial. *Biomedicines* **2022**, *10*, 2999. [CrossRef] [PubMed]
169. Gaebler, A.J.; Finner-Prével, M.; Sudar, F.P.; Langer, F.H.; Keskin, F.; Gebel, A.; Zweerings, J.; Mathiak, K. The Interplay between Vitamin D, Exposure of Anticholinergic Antipsychotics and Cognition in Schizophrenia. *Biomedicines* **2022**, *10*, 1096. [CrossRef] [PubMed]
170. Castillo-Mariqueo, L.; Giménez-Llort, L. Impact of Behavioral Assessment and Re-Test as Functional Trainings That Modify Survival, Anxiety and Functional Profile (Physical Endurance and Motor Learning) of Old Male and Female 3xTg-AD Mice and NTg Mice with Normal Aging. *Biomedicines* **2022**, *10*, 973. [CrossRef]
171. Lee, E.C.; Hong, D.-Y.; Lee, D.-H.; Park, S.-W.; Lee, J.Y.; Jeong, J.H.; Kim, E.-Y.; Chung, H.-M.; Hong, K.-S.; Park, S.-P.; et al. Inflammation and Rho-Associated Protein Kinase-Induced Brain Changes in Vascular Dementia. *Biomedicines* **2022**, *10*, 446. [CrossRef]
172. Simonato, M.; Dall'Acqua, S.; Zilli, C.; Sut, S.; Tenconi, R.; Gallo, N.; Sfriso, P.; Sartori, L.; Cavallin, F.; Fiocco, U.; et al. Tryptophan Metabolites, Cytokines, and Fatty Acid Binding Protein 2 in Myalgic Encephalomyelitis/Chronic Fatigue Syndrome. *Biomedicines* **2021**, *9*, 1724. [CrossRef]
173. Smagin, D.A.; Kovalenko, I.L.; Galyamina, A.G.; Belozertseva, I.V.; Tamkovich, N.V.; Baranov, K.O.; Kudryavtseva, N.N. Chronic Lithium Treatment Affects Anxious Behaviors and theExpression of Serotonergic Genes in Midbrain Raphe Nuclei of Defeated Male Mice. *Biomedicines* **2021**, *9*, 1293. [CrossRef]
174. Tanaka, M.; Tóth, F.; Polyák, H.; Szabó, Á.; Mándi, Y.; Vécsei, L. Immune Influencers in Action: Metabolites and Enzymes of the Tryptophan-Kynurenine Metabolic Pathway. *Biomedicines* **2021**, *9*, 734. [CrossRef]
175. Tanaka, M.; Török, N.; Tóth, F.; Szabó, Á.; Vécsei, L. Co-Players in Chronic Pain: Neuroinflammation and the Tryptophan-Kynurenine Metabolic Pathway. *Biomedicines* **2021**, *9*, 897. [CrossRef]
176. Vila-Merkle, H.; González-Martínez, A.; Campos-Jiménez, R.; Martínez-Ricós, J.; Teruel-Martí, V.; Blasco-Serra, A.; Lloret, A.; Celada, P.; Cervera-Ferri, A. The Oscillatory Profile Induced by the Anxiogenic Drug FG-7142 in the Amygdala–Hippocampal Network Is Reversed by Infralimbic Deep Brain Stimulation: Relevance for Mood Disorders. *Biomedicines* **2021**, *9*, 783. [CrossRef]
177. Santana-Santana, M.; Bayascas, J.-R.; Giménez-Llort, L. Fine-Tuning the PI3K/Akt Signaling Pathway Intensity by Sex and Genotype-Load: Sex-Dependent Homozygotic Threshold for Somatic Growth but Feminization of Anxious Phenotype in Middle-Aged PDK1 K465E Knock-In and Heterozygous Mice. *Biomedicines* **2021**, *9*, 747. [CrossRef]

178. Muntsant, A.; Giménez-Llort, L. Genotype Load Modulates Amyloid Burden and Anxiety-Like Patterns in Male 3xTg-AD Survivors despite Similar Neuro-Immunoendocrine, Synaptic and Cognitive Impairments. *Biomedicines* **2021**, *9*, 715. [CrossRef] [PubMed]
179. Giménez-Llort, L.; Marin-Pardo, D.; Marazuela, P.; Hernández-Guillamón, M. Survival Bias and Crosstalk between Chronological and Behavioral Age: Age- and Genotype-Sensitivity Tests Define Behavioral Signatures in Middle-Aged, Old, and Long-Lived Mice with Normal and AD-Associated Aging. *Biomedicines* **2021**, *9*, 636. [CrossRef] [PubMed]
180. Komatsu, H.; Watanabe, E.; Fukuchi, M. Psychiatric Neural Networks and Precision Therapeutics by Machine Learning. *Biomedicines* **2021**, *9*, 403. [CrossRef] [PubMed]
181. Caruso, G.; Godos, J.; Castellano, S.; Micek, A.; Murabito, P.; Galvano, F.; Ferri, R.; Grosso, G.; Caraci, F. The Therapeutic Potential of Carnosine/Anserine Supplementation against Cognitive Decline: A Systematic Review with Meta-Analysis. *Biomedicines* **2021**, *9*, 253. [CrossRef]
182. Correia, B.S.B.; Nani, J.V.; Waladares Ricardo, R.; Stanisic, D.; Costa, T.B.B.C.; Hayashi, M.A.F.; Tasic, L. Effects of Psychostimulants and Antipsychotics on Serum Lipids in an Animal Model for Schizophrenia. *Biomedicines* **2021**, *9*, 235. [CrossRef] [PubMed]
183. Ikonnikova, A.; Anisimova, A.; Galkin, S.; Gunchenko, A.; Abdukhalikova, Z.; Filippova, M.; Surzhikov, S.; Selyaeva, L.; Shershov, V.; Zasedatelev, A.; et al. Genetic Association Study and Machine Learning to Investigate Differences in Platelet Reactivity in Patients with Acute Ischemic Stroke Treated with Aspirin. *Biomedicines* **2022**, *10*, 2564. [CrossRef]
184. Fan, P.; Miranda, O.; Qi, X.; Kofler, J.; Sweet, R.A.; Wang, L. Unveiling the Enigma: Exploring Risk Factors and Mechanisms for Psychotic Symptoms in Alzheimer's Disease through Electronic Medical Records with Deep Learning Models. *Pharmaceuticals* **2023**, *16*, 911. [CrossRef]
185. Parolini, F.; Goethel, M.; Becker, K.; Fernandes, C.; Fernandes, R.J.; Ervilha, U.F.; Santos, R.; Vilas-Boas, J.P. Breaking Barriers: Artificial Intelligence Interpreting the Interplay between Mental Illness and Pain as Defined by the International Association for the Study of Pain. *Biomedicines* **2023**, *11*, 2042. [CrossRef]
186. Tanaka, M.; Szabó, Á.; Vécsei, L.; Giménez-Llort, L. Emerging Translational Research in Neurological and Psychiatric Diseases: From In Vitro to In Vivo Models. *Int. J. Mol. Sci.* **2023**, *24*, 15739. [CrossRef]
187. Tanaka, M.; Telegdy, G. Antidepressant-like effects of the CRF family peptides, urocortin 1, urocortin 2 and urocortin 3 in a modified forced swimming test in mice. *Brain Res. Bull.* **2008**, *75*, 509–512. [CrossRef] [PubMed]
188. Homberg, J.R.; Kyzar, E.J.; Scattoni, M.L.; Norton, W.H.; Pittman, J.; Gaikwad, S.; Nguyen, M.; Poudel, M.K.; Ullmann, J.F.; Diamond, D.M.; et al. Genetic and environmental modulation of neurodevelopmental disorders: Translational insights from labs to beds. *Brain Res. Bull.* **2016**, *125*, 79–91. [CrossRef] [PubMed]
189. Ren, J.; Xiao, H. Exercise for Mental Well-Being: Exploring Neurobiological Advances and Intervention Effects in Depression. *Life* **2023**, *13*, 1505. [CrossRef] [PubMed]
190. Tanaka, M.; Diano, M.; Battaglia, S. Editorial: Insights into structural and functional organization of the brain: Evidence from neuroimaging and non-invasive brain stimulation techniques. *Front. Psychiatry.* **2023**, *14*, 1225755. [CrossRef] [PubMed]
191. Simon, C.; Soga, T.; Ahemad, N.; Bhuvanendran, S.; Parhar, I. Kisspeptin-10 Rescues Cholinergic Differentiated SHSY-5Y Cells from α-Synuclein-Induced Toxicity In Vitro. *Int. J. Mol. Sci.* **2022**, *23*, 5193. [CrossRef]
192. Okanda Nyatega, C.; Qiang, L.; Jajere Adamu, M.; Bello Kawuwa, H. Altered striatal functional connectivity and structural dysconnectivity in individuals with bipolar disorder: A resting state magnetic resonance imaging study. *Front. Psychiatry.* **2022**, *13*, 1054380. [CrossRef] [PubMed]
193. Liu, N.; Li, Y.; Hong, Y.; Huo, J.; Chang, T.; Wang, H.; Huang, Y.; Li, W.; Zhang, Y. Altered brain activities in mesocorticolimbic pathway in primary dysmenorrhea patients of long-term menstrual pain. *Front. Neurosci.* **2023**, *17*, 1098573. [CrossRef] [PubMed]
194. Du, H.; Yang, B.; Wang, H.; Zeng, Y.; Xin, J.; Li, X. The non-linear correlation between the volume of cerebral white matter lesions and incidence of bipolar disorder: A secondary analysis of data from a cross-sectional study. *Front. Psychiatry* **2023**, *14*, 1149663. [CrossRef]
195. Chen, Y.; Yu, R.; DeSouza, J.F.X.; Shen, Y.; Zhang, H.; Zhu, C.; Huang, P.; Wang, C. Differential responses from the left postcentral gyrus, right middle frontal gyrus, and precuneus to meal ingestion in patients with functional dyspepsia. *Front. Psychiatry* **2023**, *14*, 1184797. [CrossRef]
196. Adamu, M.J.; Qiang, L.; Nyatega, C.O.; Younis, A.; Kawuwa, H.B.; Jabire, A.H.; Saminu, S. Unraveling the pathophysiology of schizophrenia: Insights from structural magnetic resonance imaging studies. *Front. Psychiatry* **2023**, *14*, 1188603. [CrossRef] [PubMed]
197. Battaglia, S.; Nazzi, C.; Thayer, J.F. Fear-induced bradycardia in mental disorders: Foundations, current advances, future perspectives. *Neurosci. Biobehav. Rev.* **2023**, *149*, 105163. [CrossRef] [PubMed]
198. Battaglia, S.; Nazzi, C.; Thayer, J.F. Heart's tale of trauma: Fear-conditioned heart rate changes in post-traumatic stress disorder. *Acta Psychiatr. Scand.* **2023**, *148*, 463–466, Advance online publication. [CrossRef] [PubMed]
199. Battaglia, S.; Di Fazio, C.; Vicario, C.M.; Avenanti, A. Neuropharmacological Modulation of N-methyl-D-aspartate, Noradrenaline and Endocannabinoid Receptors in Fear Extinction Learning: Synaptic Transmission and Plasticity. *Int. J. Mol. Sci.* **2023**, *24*, 5926. [CrossRef]
200. Battaglia, M.R.; Di Fazio, C.; Battaglia, S. Activated Tryptophan-Kynurenine metabolic system in the human brain is associated with learned fear. *Front. Mol. Neurosci.* **2023**, *16*, 1217090. [CrossRef]

201. Battaglia, S.; Garofalo, S.; di Pellegrino, G.; Starita, F. Revaluing the Role of vmPFC in the Acquisition of Pavlovian Threat Conditioning in Humans. *J. Neurosci.* **2020**, *40*, 8491–8500. [CrossRef]
202. Tanaka, M. Crosstalk between Depression, Anxiety, Dementia, and Chronic Pain: Comorbidity in Behavioral Neurology and Neuropsychiatry 2.0. *Biomedicines* **2023**. [CrossRef]
203. Tanaka, M.; Telegdy, G. Antidepressant-Like Effects of Neuropeptide SF (NPSF). *arXiv* **2005**, arXiv:2005.13256. [CrossRef]
204. Battaglia, S.; Harrison, B.J.; Fullana, M.A. Does the human ventromedial prefrontal cortex support fear learning, fear extinction or both? A commentary on subregional contributions. *Mol. Psychiatry* **2022**, *27*, 784–786. [CrossRef]
205. Di Gregorio, F.; Battaglia, S. Advances in EEG-based functional connectivity approaches to the study of the central nervous system in health and disease. *Adv. Clin. Exp. Med.* **2023**, *32*, 607–612. [CrossRef]
206. Tanaka, M.; Vécsei, L. Editorial of Special Issue "Crosstalk between Depression, Anxiety, and Dementia: Comorbidity in Behavioral Neurology and Neuropsychiatry". *Biomedicines* **2021**, *9*, 517. [CrossRef] [PubMed]
207. Tanaka, M.; Vécsei, L. Monitoring the kynurenine system: Concentrations, ratios or what else? *Adv. Clin. Exp. Med.* **2021**, *30*, 775–778. [CrossRef] [PubMed]
208. Tanaka, M.; Vécsei, L. Monitoring the Redox Status in Multiple Sclerosis. *Biomedicines* **2020**, *8*, 406. [CrossRef] [PubMed]
209. Tanaka, M.; Török, N.; Vécsei, L. Are 5-HT_1 receptor agonists effective anti-migraine drugs? *Expert. Opin. Pharmacother.* **2021**, *22*, 1221–1225. [CrossRef]
210. Vuralli, D.; Ayata, C.; Bolay, H. Cognitive dysfunction and migraine. *J. Headache Pain* **2018**, *19*, 109. [CrossRef]
211. Minen, M.T.; Begasse De Dhaem, O.; Kroon Van Diest, A.; Powers, S.; Schwedt, T.J.; Lipton, R.; Silbersweig, D. Migraine and its psychiatric comorbidities. *J. Neurol. Neurosurg. Psychiatry* **2016**, *87*, 741–749. [CrossRef]
212. Tanaka, M.; Chen, C. Editorial: Towards a mechanistic understanding of depression, anxiety, and their comorbidity: Perspectives from cognitive neuroscience. *Front. Behav. Neurosci.* **2023**, *17*, 1268156. [CrossRef]
213. Gonzalez-Escamilla, G.; Dörfel, D.; Becke, M.; Trefz, J.; Bonanno, G.A.; Groppa, S. Associating Flexible Regulation of Emotional Expression With Psychopathological Symptoms. *Front. Behav. Neurosci.* **2022**, *16*, 924305. [CrossRef]
214. Battaglia, S.; Cardellicchio, P.; Di Fazio, C.; Nazzi, C.; Fracasso, A.; Borgomaneri, S. The Influence of Vicarious Fear-Learning in "Infecting" Reactive Action Inhibition. *Front. Behav. Neurosci.* **2022**, *16*, 946263. [CrossRef]
215. Battaglia, S.; Cardellicchio, P.; Di Fazio, C.; Nazzi, C.; Fracasso, A.; Borgomaneri, S. Stopping in (e)motion: Reactive action inhibition when facing valence-independent emotional stimuli. *Front. Behav. Neurosci.* **2022**, *16*, 998714. [CrossRef]
216. Ironside, M.; DeVille, D.C.; Kuplicki, R.T.; Burrows, K.P.; Smith, R.; Teed, A.R.; Paulus, M.P.; Khalsa, S.S. The unique face of comorbid anxiety and depression: Increased interoceptive fearfulness and reactivity. *Front. Behav. Neurosci.* **2023**, *16*, 1083357. [CrossRef] [PubMed]
217. Rajkumar, R.P. Comorbid depression and anxiety: Integration of insights from attachment theory and cognitive neuroscience, and their implications for research and treatment. *Front. Behav. Neurosci.* **2022**, *16*, 1104928. [CrossRef] [PubMed]
218. Vila-Merkle, H.; González-Martínez, A.; Campos-Jiménez, R.; Martínez-Ricós, J.; Teruel-Martí, V.; Lloret, A.; Blasco-Serra, A.; Cervera-Ferri, A. Sex differences in amygdalohippocampal oscillations and neuronal activation in a rodent anxiety model and in response to infralimbic deep brain stimulation. *Front. Behav. Neurosci.* **2023**, *17*, 1122163. [CrossRef] [PubMed]
219. Panov, G.; Panova, P. Obsessive-compulsive symptoms in patient with schizophrenia: The influence of disorganized symptoms, duration of schizophrenia, and drug resistance. *Front. Psychiatry* **2023**, *14*, 1120974. [CrossRef] [PubMed]
220. Hakamata, Y.; Hori, H.; Mizukami, S.; Izawa, S.; Yoshida, F.; Moriguchi, Y.; Hanakawa, T.; Inoue, Y.; Tagaya, H. Blunted diurnal interleukin-6 rhythm is associated with amygdala emotional hyporeactivity and depression: A modulating role of gene-stressor interactions. *Front. Psychiatry* **2023**, *14*, 1196235. [CrossRef] [PubMed]
221. Fraile-Ramos, J.; Garrit, A.; Reig-Vilallonga, J.; Giménez-Llort, L. Hepatic Oxi-Inflammation and Neophobia as Potential Liver–Brain Axis Targets for Alzheimer's Disease and Aging, with Strong Sensitivity to Sex, Isolation, and Obesity. *Cells* **2023**, *12*, 1517. [CrossRef] [PubMed]
222. Sobolewska-Nowak, J.; Wachowska, K.; Nowak, A.; Orzechowska, A.; Szulc, A.; Płaza, O.; Gałecki, P. Exploring the Heart–Mind Connection: Unraveling the Shared Pathways between Depression and Cardiovascular Diseases. *Biomedicines* **2023**, *11*, 1903. [CrossRef]
223. Festa, F.; Medori, S.; Macrì, M. Move Your Body, Boost Your Brain: The Positive Impact of Physical Activity on Cognition across All Age Groups. *Biomedicines* **2023**, *11*, 1765. [CrossRef]
224. Spekker, E.; Bohár, Z.; Fejes-Szabó, A.; Szűcs, M.; Vécsei, L.; Párdutz, Á. Estradiol Treatment Enhances Behavioral and Molecular Changes Induced by Repetitive Trigeminal Activation in a Rat Model of Migraine. *Biomedicines* **2022**, *10*, 3175. [CrossRef]
225. Park, S.Y.; Lee, S.P.; Kim, D.; Kim, W.J. Gut Dysbiosis: A New Avenue for Stroke Prevention and Therapeutics. *Biomedicines* **2023**, *11*, 2352. [CrossRef]
226. Cui, Y.; Zhang, H.; Wang, S.; Lu, J.; He, J.; Liu, L.; Liu, W. Stimulated Parotid Saliva Is a Better Method for Depression Prediction. *Biomedicines* **2022**, *10*, 2220. [CrossRef] [PubMed]
227. Altamura, C.; Corbelli, I.; de Tommaso, M.; Di Lorenzo, C.; Di Lorenzo, G.; Di Renzo, A.; Filippi, M.; Jannini, T.B.; Messina, R.; Parisi, P.; et al. Pathophysiological Bases of Comorbidity in Migraine. *Front. Hum. Neurosci.* **2021**, *15*, 640574. [CrossRef] [PubMed]
228. Schwedt, T.J.; Vargas, B. Neurostimulation for Treatment of Migraine and Cluster Headache. *Pain Med.* **2015**, *16*, 1827–1834. [CrossRef]

229. de Albuquerque, L.L.; Pantovic, M.; Clingo, M.; Fischer, K.; Jalene, S.; Landers, M.; Mari, Z.; Poston, B. A Single Application of Cerebellar Transcranial Direct Current Stimulation Fails to Enhance Motor Skill Acquisition in Parkinson's Disease: A Pilot Study. *Biomedicines* **2023**, *11*, 2219. [CrossRef] [PubMed]
230. Senevirathne, D.K.L.; Mahboob, A.; Zhai, K.; Paul, P.; Kammen, A.; Lee, D.J.; Yousef, M.S.; Chaari, A. Deep Brain Stimulation beyond the Clinic: Navigating the Future of Parkinson's and Alzheimer's Disease Therapy. *Cells* **2023**, *12*, 1478. [CrossRef]
231. Borgomaneri, S.; Battaglia, S.; Sciamanna, G.; Tortora, F.; Laricchiuta, D. Memories are not written in stone: Re-writing fear memories by means of non-invasive brain stimulation and optogenetic manipulations. *Neurosci. Biobehav. Rev.* **2021**, *127*, 334–352. [CrossRef]
232. Adeel, M.; Chen, C.-C.; Lin, B.-S.; Chen, H.-C.; Liou, J.-C.; Li, Y.-T.; Peng, C.-W. Safety of Special Waveform of Transcranial Electrical Stimulation (TES): In Vivo Assessment. *Int. J. Mol. Sci.* **2022**, *23*, 6850. [CrossRef]
233. Battaglia, S.; Schmidt, A.; Hassel, S.; Tanaka, M. Editorial: Case reports in neuroimaging and stimulation. *Front. Psychiatry* **2023**, *14*, 1264669. [CrossRef]
234. Chang, C.H.; Wang, W.L.; Shieh, Y.H.; Peng, H.Y.; Ho, C.S.; Tsai, H.C. Case Report: Low-Frequency Repetitive Transcranial Magnetic Stimulation to Dorsolateral Prefrontal Cortex and Auditory Cortex in a Patient With Tinnitus and Depression. *Front. Psychiatry* **2022**, *13*, 847618. [CrossRef]
235. Zakia, H.; Iskandar, S. Case report: Depressive disorder with peripartum onset camouflages suspected intracranial tuberculoma. *Front. Psychiatry* **2022**, *13*, 932635. [CrossRef]
236. Nyatega, C.O.; Qiang, L.; Adamu, M.J.; Kawuwa, H.B. Gray matter, white matter and cerebrospinal fluid abnormalities in Parkinson's disease: A voxel-based morphometry study. *Front. Psychiatry* **2022**, *13*, 1027907. [CrossRef] [PubMed]
237. Rymaszewska, J.; Wieczorek, T.; Fila-Witecka, K.; Smarzewska, K.; Weiser, A.; Piotrowski, P.; Tabakow, P. Various neuromodulation methods including Deep Brain Stimulation of the medial forebrain bundle combined with psychopharmacotherapy of treatment-resistant depression-Case report. *Front. Psychiatry* **2023**, *13*, 1068054. [CrossRef] [PubMed]
238. Liu, M.; Xie, X.; Xie, J.; Tian, S.; Du, X.; Feng, H.; Zhang, H. Early-onset Alzheimer's disease with depression as the first symptom: A case report with literature review. *Front. Psychiatry* **2023**, *14*, 1192562. [CrossRef] [PubMed]
239. Kim, B.H.; Kim, S.H.; Han, C.; Jeong, H.G.; Lee, M.S.; Kim, J. Antidepressant-induced mania in panic disorder: A single-case study of clinical and functional connectivity characteristics. *Front. Psychiatry* **2023**, *14*, 1205126. [CrossRef]
240. Zhou, J.; Cao, Y.; Deng, G.; Fang, J.; Qiu, C. Transient splenial lesion syndrome in bipolar-II disorder: A case report highlighting reversible brain changes during hypomanic episodes. *Front. Psychiatry* **2023**, *14*, 1219592. [CrossRef]
241. Logan, S.; Arzua, T.; Canfield, S.G.; Seminary, E.R.; Sison, S.L.; Ebert, A.D.; Bai, X. Studying Human Neurological Disorders Using Induced Pluripotent Stem Cells: From 2D Monolayer to 3D Organoid and Blood Brain Barrier Models. *Compr. Physiol.* **2019**, *9*, 565–611. [CrossRef]
242. Gazerani, P. Human Brain Organoids in Migraine Research: Pathogenesis and Drug Development. *Int. J. Mol. Sci.* **2023**, *24*, 3113. [CrossRef]
243. Jalink, P.; Caiazzo, M. Brain Organoids: Filling the Need for a Human Model of Neurological Disorder. *Biology* **2021**, *10*, 740. [CrossRef]
244. Zhang, D.Y.; Song, H.; Ming, G.L. Modeling neurological disorders using brain organoids. *Semin. Cell Dev. Biol.* **2021**, *111*, 4–14. [CrossRef]

cells

MDPI

Article

Glycerol Trinitrate Acts Downstream of Calcitonin Gene-Related Peptide in Trigeminal Nociception—Evidence from Rodent Experiments with Anti-CGRP Antibody Fremanezumab

Nicola Benedicter [1], Birgit Vogler [1], Annette Kuhn [1], Jana Schramm [1], Kimberly D. Mackenzie [2], Jennifer Stratton [2], Mária Dux [3] and Karl Messlinger [1,*]

1 Institute of Physiology and Pathophysiology, Friedrich-Alexander-University, D-91054 Erlangen, Germany; nicolabenedicter@gmail.com (N.B.)
2 Teva Pharmaceuticals, Redwood City, CA 94063, USA; kimberly.mackenzie01@gmail.com (K.D.M.); jennifer_stratton@mac.com (J.S.)
3 Department of Physiology, University of Szeged, H-6720 Szeged, Hungary; dux.maria@med.u-szeged.hu
* Correspondence: karl.messlinger@fau.de

Abstract: Calcitonin gene-related peptide (CGRP) and nitric oxide (NO) have been recognized as important mediators in migraine but their mechanisms of action and interaction have not been fully elucidated. Monoclonal anti-CGRP antibodies like fremanezumab are successful preventives of frequent migraine and can be used to study CGRP actions in preclinical experiments. Fremanezumab (30 mg/kg) or an isotype control monoclonal antibody was subcutaneously injected to Wistar rats of both sexes. One to several days later, glyceroltrinitrate (GTN, 5 mg/kg) mimicking nitric oxide (NO) was intraperitoneally injected, either once or for three consecutive days. The trigeminal ganglia were removed to determine the concentration of CGRP using an enzyme-linked immunosorbent assay (ELISA). In one series of experiments, the animals were trained to reach an attractive sugar solution, the access to which could be limited by mechanical or thermal barriers. Using a semi-automated registration system, the frequency of approaches to the source, the residence time at the source, and the consumed solution were registered. The results were compared with previous data of rats not treated with GTN. The CGRP concentration in the trigeminal ganglia was generally higher in male rats and tended to be increased in animals treated once with GTN, whereas the CGRP concentration decreased after repetitive GTN treatment. No significant difference in CGRP concentration was observed between animals having received fremanezumab or the control antibody. Animals treated with GTN generally spent less time at the source and consumed less sugar solution. Without barriers, there was no significant difference between animals having received fremanezumab or the control antibody. Under mechanical barrier conditions, all behavioral parameters tended to be reduced but animals that had received fremanezumab tended to be more active, partly compensating for the depressive effect of GTN. In conclusion, GTN treatment seems to increase the production of CGRP in the trigeminal ganglion independently of the antibodies applied, but repetitive GTN administration may deplete CGRP stores. GTN treatment generally tends to suppress the animals' activity and increase facial sensitivity, which is partly compensated by fremanezumab through reduced CGRP signaling. If CGRP and NO signaling share the same pathway in sensitizing trigeminal afferents, GTN and NO may act downstream of CGRP to increase facial sensitivity.

Keywords: fremanezumab; monoclonal antibody; calcitonin gene-related peptide; glycerol trinitrate; CGRP release; CGRP concentration; rat; migraine pain

Citation: Benedicter, N.; Vogler, B.; Kuhn, A.; Schramm, J.; Mackenzie, K.D.; Stratton, J.; Dux, M.; Messlinger, K. Glycerol Trinitrate Acts Downstream of Calcitonin Gene-Related Peptide in Trigeminal Nociception—Evidence from Rodent Experiments with Anti-CGRP Antibody Fremanezumab. *Cells* **2024**, *13*, 572. https://doi.org/10.3390/cells13070572

Academic Editor: Arpad Dobolyi

Received: 22 February 2024
Revised: 16 March 2024
Accepted: 20 March 2024
Published: 25 March 2024

1. Introduction

The neuropeptide calcitonin gene-related peptide (CGRP) and nitrogen species such as nitric oxide (NO) are potent vasodilatory agents and important mediators in migraine [1,2].

During spontaneous migraine attacks, elevated concentrations of CGRP have been measured in venous plasma [3–5], saliva [6,7], and tear fluid [8]. Inhibiting CGRP release with triptans and ditans; targeting CGRP using monoclonal antibodies (mAbs) against the CGRP ligand; or blocking CGRP receptors by the CGRP receptor antibody or gepants are effective treatments in migraine [9,10]. Notably, the infusion of CGRP can provoke short-lasting headaches, which are attributed to the vasodilatation of cranial blood vessels and, additionally, delayed migraine-like pulsating pain in migraineurs [11,12].

Similarly, elevated plasma concentrations of nitrogen metabolites have been found in plasma during migraine [13], and the infusion of nitroglycerin (glycerol trinitrate, GTN), which mimics NO, has been shown to cause transient vasodilatory headaches and, in addition, migraine-like pain in migraineurs [14,15]. The inhibition of endogenous NO production and NO-mediated mechanisms have not been used for migraine therapy, partly due to adverse side effects [16]. Due to the similarities of CGRP and GTN administration in headache provocation, there might be an association between endogenous CGRP and NO actions. The question is whether and how these two mediator systems may be linked. GTN-induced migraine-like attacks have been found to be associated with an increase in plasma CGRP levels [17], which responds to treatment with sumatriptan [18,19], suggesting CGRP release provoked by NO mechanisms. However, an increase in plasma CGRP during GTN-induced headache could not be verified by another group [20]; the headache was not prevented by blocking CGRP receptors [21]. Thus, a causal link between CGRP and NO signaling in migraine is still an open question.

To generate models of migraine related to the mentioned clinical experiments, several groups have administered NO donors or GTN in rodents and collected multiple structural, functional, and behavioral data [22–26]. Our group has recently examined the effects of fremanezumab, an anti-CGRP monoclonal antibody (mAb) used for the prevention of chronic and frequent migraine, on meningeal CGRP release and blood flow, CGRP concentration in trigeminal ganglia, and facial sensitivity [27–29]. To compare these data in an animal model of trigeminal nociception, we have repeated some of these experiments after the administration of GTN and found deviating results, which may contribute to the question of how endogenous CGRP and NO signaling may be linked. Our data provide evidence that CGRP signaling is bypassed by NO signaling, which favors the hypothesis that NO may act downstream of CGRP signaling in migraine pathophysiology.

2. Materials and Methods

Animal housing and all experiments were carried out according to the regulations for the care and treatment of laboratory animals of the European Communities Council Directive 1986 (86/609/EEC), amended 2010 (2010/63/EU). The experimental protocols were reviewed by an ethics committee and approved by the District Government of Middle Franconia (54-2532.1-21/12).

2.1. Animals

Fifty-two adult Wistar rats of both sexes (body weight of females: 220–370 g; males: 310–450 g), bred and housed in our animal facility, were used. Groups of 3–4 animals were kept in cages in a 12 h day–night cycle and received pellet food and water ad libitum. Equal numbers of animals were matched according to their sex and weight and allocated to the treatments. The estrus state of the females was not assessed.

2.2. Administration of Monoclonal Antibodies

Administration of mAbs was performed as previously reported [27–29]. In short, either 30 mg/kg of the anti-CGRP mAb fremanezumab, or the isotype control mAb (both Teva Pharmaceuticals, Redwood City, CA, USA) diluted in saline (10 mg/mL), was subcutaneously injected into the neck of the animals under short isoflurane anesthesia. Fremanezumab binds to an epitope that is identical in human α-CGRP, β-CGRP, and rat β-CGRP. There is one similar amino acid change within the epitope for rat α-CGRP, which has a minor effect on overall binding affinity. The operator was blinded to the identity of the antibodies. The animals were individually marked, placed back in their cages, and inspected daily regarding their health state and behavior until they were used for further procedures.

2.3. Administration of GTN and Dissection

On day 1, 3, 10, or 30 after the antibody injection the animals received an intraperitoneal (i.p.) injection of 5 mg/kg GTN (Nitrolingual, containing 1 mg/mL GTN, Pohl-Boskamp, Hohenlockstadt, Germany) or the same volume of synthetic interstitial fluid (SIF) as the vehicle [27] under short isoflurane anesthesia at 8 a.m. (Figure 1A). Four hours later, the animals were deeply anesthetized and sacrificed in a rising CO_2 atmosphere. Another group of animals, which were used for behavioral testing as outlined below, received, on three consecutive days, 5 mg/kg, 2.5 mg/kg and 1.25 mg/kg GTN, namely on days 4–6 and 11–13 after mAb administration, for a total of 6 applications (Figure 1B). With this design, a cumulative sensitizing effect of GTN causing chronic hyperalgesia [30] should be avoided. Animals were sacrificed on day 14 after mAb administration, 10 days after the first, and one day after the last GTN injection. The heads of the animals were separated, skinned, and cut into halves along the sagittal line, as previously described for CGRP release measurements [27]. The trigeminal ganglia were excised from the skull base at a length of 6–7 mm, placed in Eppendorf cups, and frozen at −20 °C until further processing. From each animal, both trigeminal ganglia were harvested, resulting in 104 samples.

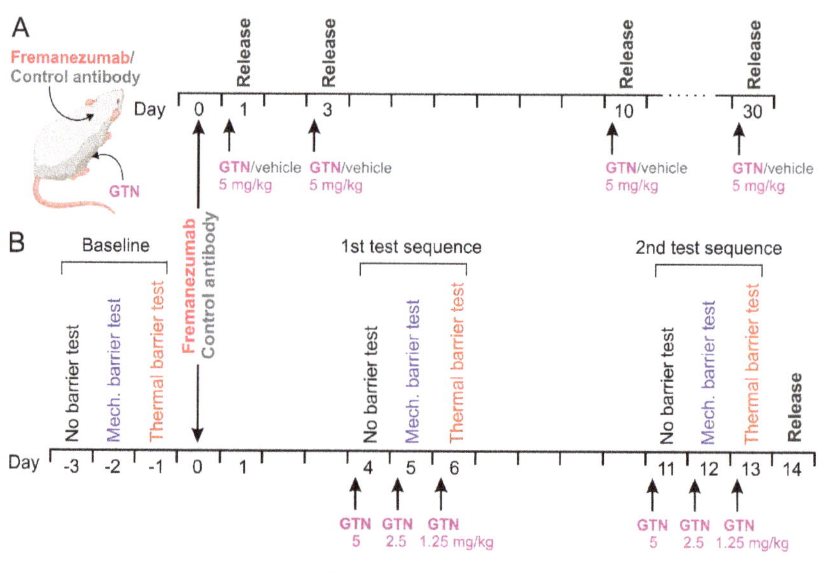

Figure 1. Schematic workflow of the CGRP release experiments (**A**) and the behavioral experiments followed by CGRP release (**B**). Either anti-CGRP mAb fremanezumab or control monoclonal antibody was subcutaneously injected, in B after baseline recordings without barrier, under mechanical and under thermal barrier conditions. Release experiments were performed 4 h after intraperitoneal injections of glycerol trinitrate (GTN) in **A** or after two test sequences according to the baseline recordings and repetitive injections of GTN in **B**.

2.4. Preparation and CGRP Determination

The processing of tissue samples and CGRP measurements were performed as previously reported [29]. In short, after defrosting, the trigeminal ganglia were dipped in an absorbent tissue, weighed, and immersed in 1 mL of 2 M acetic acid, boiled at 95 °C for 10 min, and homogenized. The homogenate was again boiled for 10 min and centrifuged at

2000 rpm. The supernatant was collected, and 100 μL thereof was diluted 4:1 with enzyme-linked immunosorbent assay (ELISA) buffer and processed with an ELISA kit for CGRP with a detection level of 2 pg/mL, according to the instructions of the manufacturer (Bertin Pharma/SPIbio, Montigny le Bretonneux, France). The CGRP concentration (in ng/mg) of each trigeminal ganglion was calculated with regard to the respective ganglion mass.

2.5. Preparation of Animals for Behavioral Tests

The preparation and performance of behavioral tests have previously been reported in detail [28] and are outlined here in short. After priming the animals for an attractive 10% sucrose solution available in standard drinking bottles, the animals were placed in a test cage in which they had to pass with their forehead through an opening; it could be equipped with a mechanical barrier (flexible 0.09 mm steel bristles) or a thermal barrier (circular tube with 50 °C heated water). The opening of the drinking bottle with the sucrose solution could be reached by the animal while its cheeks and forehead touched the steel bristles or the hot tube. The frequency and duration of passing through the opening were recorded automatically with a photo sensor device (Orofacial Stimulation Test System, Ugo Basile, Lugano, Italy) within test periods of 15 min. The volume of consumed sucrose solution during the test periods was determined afterward.

2.6. Sequence of Testing

The test sequence started after 7 days of priming and was always conducted between 12 a.m. and 1 a.m. On the first test day (day-3), the animals could reach the sugar source without a barrier; on the second day (day-2), they had to pass through the mechanical barrier; and on the third day (day-1), they had to pass through the thermal barrier during the 15 min test period (Figure 1B). This provided a baseline measure for the 3 conditions before the animals received fremanezumab or the control mAb on the following day. The experimenter was blinded as to the specific mAb treatment during the whole experiment. On days 4–6 after mAb administration, the animals were injected with GTN as described, followed by the same test sequence 4 h later (Figure 1B). The whole test sequence, including the injection of GTN, was repeated, beginning with day 11 after mAb injection.

2.7. Data Processing and Statistics

Statistical analysis was performed using Statistica software (StatSoft, Release 7, Tulsa, OK, USA). After verification of the normal distribution of data, analysis of variance (factorial ANOVA) with the categorical predictors (factors) mAb, sex, and day after mAb administration, as well as treatment (GTN or saline), was used for a comparison of data. This was extended by Tukey's honest significant difference (HSD) test or the unequal N HSD test. In the case of the behavioral experiments, where the 3 barrier conditions are extremely different situations, data were compared for each condition separately using repeated measures ANOVA for the 3 days of the same tests, with sex and mAb as factors. Sample size calculation was based on the biometric planning for previous measurements [27,28] in accordance with the ethics approval mentioned above. The level of significance was set at $p < 0.05$. Data were displayed as mean ± SEM (standard error of the mean).

3. Results

3.1. CGRP Content of Trigeminal Ganglia

The experiments included 26 female and 26 male rats, either treated with fremanezumab or with the isotype control mAb (13 females and 13 males in each group). The injection of GTN and the final experiment were performed after waiting 1 day in 9 animals; 3 days in 8 animals; 10 days in 13 animals; and 30 days in 10 animals following mAb administration. In further experiments, 12 animals (10 of them also used for behavioral experiments; see below) were treated with GTN on days 4–6 and again on days 11–13 after mAb administration.

3.1.1. General Observations

After mAb administration of either type, none of the animals showed unusual behavior or any other sign of disturbance. Both groups continuously gained weight dependent on the waiting time but without difference between animals treated with fremanezumab ($n = 26$; final body weight 333.6 ± 9.1 g) or the control antibody ($n = 26$; 333.1 ± 7.9 g). As expected, the body weight was significantly different between the sexes (factorial ANOVA, $F_{1,84} = 194.2$, $p < 0.0001$). During the hours after injection of GTN, in particular after repetitive administration, the animals appeared generally less active but showed no signs of spontaneous pain.

3.1.2. CGRP Concentration in Trigeminal Ganglia

In the 104 samples of trigeminal ganglia, factorial ANOVA with the factors mAb (fremanezumab vs. isotype control mAb), sex (female vs. male animals), and time after mAb injection (1, 3, 10, 30 days, and day 14 after repetitive GTN injection) was performed. Surprisingly, ANOVA showed slightly higher ganglion weights in females compared to males ($F_{1,84} = 5.0$, $p < 0.05$) and clearly significant differences between the days after mAb administration ($F_{4,84} = 13.9$, $p < 0.0001$). Calcitonin gene-related peptide concentration was calculated using the measured CGRP content and the ganglion mass, as noted in the Materials and Methods section. The CGRP concentration was significantly different between the sexes (factorial ANOVA, $F_{1,84} = 38.8$, $p < 0.0001$) and between the days after mAb administration ($F_{4,84} = 27.9$, $p < 0.0001$) but not between the two mAbs ($F_{1,84} = 0.1$, $p = 0.74$). The Tukey post hoc test indicated that the difference between days was mainly due to the unusually low ganglion mass of day-30 animals (in both the fremanezumab and the control mAb group) and the significantly lower CGRP concentration in ganglia treated repetitively with GTN. Detailed data are displayed in Table 1.

Table 1. Trigeminal ganglion mass and CGRP concentration (means ± SEM) in ganglion samples comparing different groups of animals treated with fremanezumab or control mAb and GTN.

		Samples (n)	Ganglion Mass (mg)	Difference	CGRP Conc. (ng/mg)	Difference
mAb	Fremanezumab	52	14.7 ± 0.7	n.s.	1.11 ± 0.12	n.s.
	Control antibody	52	13.7 ± 0.8		1.10 ± 0.90	
Sex	Females	52	15.1 ± 0.9	$p < 0.05$	0.87 ± 0.56	$p < 0.0001$
	Males	52	13.4 ± 0.6		1.34 ± 0.13	
Days after mAb administration	Day 1	18	17.7 ± 1.1	$p < 0.0001$	1.22 ± 0.13	$p < 0.0001$
	Day 3	16	15.7 ± 1.6		1.71 ± 0.29	
	Day 10	26	14.7 ± 1.0		1.15 ± 0.13	
	Day 30	20	8.9 ± 0.4		1.25 ± 0.83	
	GTN repetitive	24	14.6 ± 1.0		0.44 ± 0.05	

GTN was applied on the indicated day. "GTN repetitive" animals were treated daily with GTN on days 4–6 and 11–13. Significant differences (p) result from factorial ANOVA comparing two or five variables, respectively; n.s., not significant.

3.1.3. Comparison of CGRP Concentration with Previous Data from Animals Not Treated with GTN

The CGRP concentration of the trigeminal ganglia from the 30 animals sacrificed 1, 3 or 10 days after administration of the mAbs was compared with previous data from 26 rats with a similar distribution of sexes and mAbs [29]. These animals were not injected with GTN, and the 52 trigeminal ganglia were prepared in the same way as described for the present experiments. Data from these previous and present experiments were compared using factorial ANOVA with the factors sex, mAb type, time after mAb injection (1, 3, or 10 days), and treatment (GTN vs. no GTN). The ganglion masses were significantly different between the sexes ($F_{1,88} = 5.3$; $p < 0.05$) but not significant between the mAbs ($F_{1,88} = 2.9$; $p = 0.09$) and the days after mAb administration ($F_{2,88} = 3.1$; $p = 0.05$) or between the treatments ($F_{1,88} = 0.1$; $p = 0.70$). The CGRP concentrations tested with factorial ANOVA were clearly different between the sexes ($F_{1,88} = 50.9$; $p < 0.0001$) but not between the mAbs ($F_{1,88} = 0.16$; $p = 0.69$). The difference between the days was significant ($F_{2,88} = 3.2$; $p < 0.05$), while there was a clear difference between the treatments ($F_{1,88} = 18.9$; $p < 0.0001$), which was mainly due to the male fremanezumab animals (Figure 2).

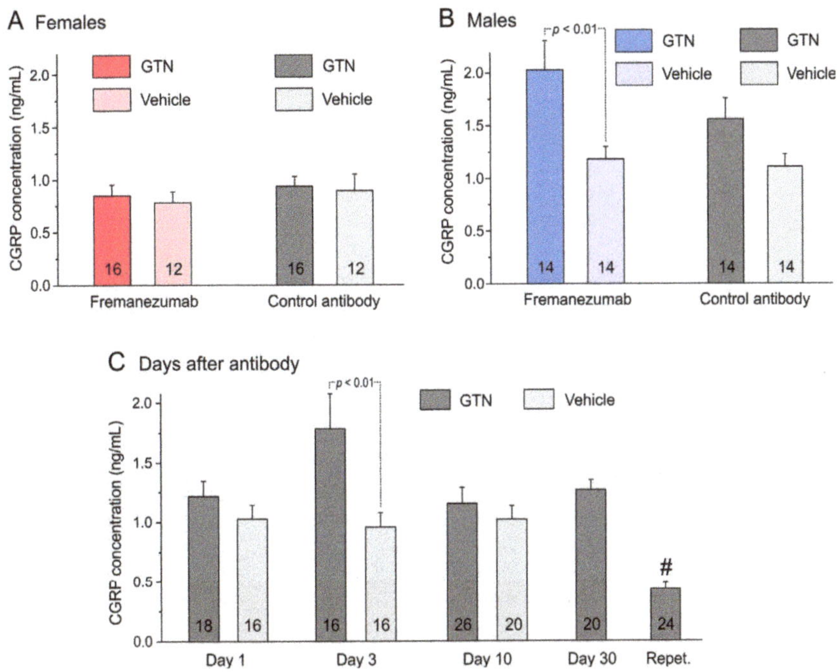

Figure 2. CGRP concentration (means ± SEM) of trigeminal ganglia separated by sexes (**A**,**B**), administration of fremanezumab or control mAb and treatment with GTN or vehicle. The difference between ganglia of GTN and vehicle-treated animals depends largely on males (**B**) with a significant difference between the fremanezumab groups (Tukey's HSD post hoc test following factorial ANOVA, $p < 0.01$). (**C**): The difference between GTN and vehicle-treated groups (sex and mAb groups cumulated) is statistically significant on day 3 after fremanezumab/control mAb treatment (unequal N HSD post hoc test following factorial ANOVA, $p < 0.01$). The low CGRP concentration in the trigeminal ganglion samples of repetitively GTN-treated animals (Repet., #) is significantly different from all other GTN-treated groups (HSD post hoc test following factorial ANOVA, $p < 0.05$–0.0001). Numbers within bars mean number of trigeminal ganglia.

3.2. Behavioral Experiments

Twenty animals (10 females and 10 males) were included in the study. One day after the baseline tests on three consecutive days, namely without a barrier, with a mechanical barrier, and with a thermal barrier, the animals received either fremanezumab or the isotype control mAb (see Figure 1B). On days 4–6 and again on days 11–13 after the mAb administration, the animals were treated with GTN as outlined in the methods and tested again in the same sequence. Data are displayed in Table 2 in detail.

Table 2. Number of approaches (counts), time at the source and consumed volume (means ± SEM) in animals treated with fremanezumab or isotype control mAb and GTN (all groups, $n = 10$).

	Baseline		Day 4–6		Day 11–13	
No barrier	Fremanezumab	Control mAb	Fremanezumab	Control mAb	Fremanezumab	Control mAb
Counts (n)	46.1 ± 4.2	54.6 ± 9.2	58.3 ± 6.8	58.5 ± 10.4	67.6 ± 14.4	48.9 ± 4.6
Time (s)	269.0 ± 21.0	261.7 ± 41.6	287.8 ± 41.5	263.9 ± 53.3	302.1 ± 41.4	294.0 ± 29.2
Volume (mL)	8.3 ± 0.9	7.7 ± 1.3	8.8 ± 1.3	7.4 ± 1.1	10.3 ± 1.1	9.3 ± 1.0
Mech. barrier	Fremanezumab	Control mAb	Fremanezumab	Control mAb	Fremanezumab	Control mAb
Counts (n)	141.5 ± 55.3	44.8 ± 15.1	383.7 ± 152.5	113.9 ± 56.4	553.7 ± 210.6	121.5 ± 56.3
Time (s) *	114.5 ± 33.7	53.3 ± 19.8	162.6 ± 43.8	74.7 ± 30.2	199.2 ± 41.5	106.7 ± 32.6
Volume (mL) *	3.7 ± 1.1	1.9 ± 0.6	5.4 ± 1.3	2.5 ± 1.0	6.1 ± 1.4	2.8 ± 1.0
Therm. barrier	Fremanezumab	Control mAb	Fremanezumab	Control mAb	Fremanezumab	Control mAb
Counts (n)	64.8 ± 9.8	73.3 ± 12.9	66.7 ± 10.3	55.7 ± 11.3	82.9 ± 10.1	62.7 ± 11.5
Time (s)	228.9 ± 38.3	195.6 ± 34.0	230.2 ± 32.5	155.8 ± 41.3	220.3 ± 28.5	175.5 ± 37.8
Volume (mL)	6.8 ± 1.2	5.3 ± 1.0	6.8 ± 0.8	4.5 ± 1.0	6.6 ± 0.8	5.9 ± 1.1

Significant differences (repeated measures ANOVA, $p < 0.05$) between fremanezumab and control mAb (*) are seen under mechanical barrier conditions. Mech., mechanical; Therm., thermal.

3.2.1. Number of Approaches to Source

The number of approaches (counts) to the attractive sugar solution within 15 min was not significantly different between baseline (before mAbs administration) and on days 4–6 and days 11–13 after mAbs administration, i.e., neither without barrier nor with mechanical or thermal barrier. There was also no difference between the two mAbs and sexes.

3.2.2. Time Staying at the Source

The cumulative time spent at the source to consume the sugar solution was not significantly different on the three days under the three barrier conditions. However, solely under the mechanical barrier condition, repeated measures ANOVA indicated a weak difference between the two mAbs ($F_{1,32} = 7.6$, $p < 0.05$) and the sexes ($F_{1,32} = 5.0$, $p < 0.05$), which could not be attributed to a specific group by the post hoc HSD test.

3.2.3. Consumed Volume

The consumed volume of sugar solution was again not significantly different at the three days under the three barrier conditions. Repeated measures ANOVA indicated a difference between the mAbs ($F_{1,32} = 7.2$, $p < 0.05$) under the mechanical barrier, which was not due to a specific group.

3.2.4. Comparison of Behavioral Data with Animals Not Treated with GTN: No Barrier Condition

The behavioral measurements were compared with previous data from 12 rats with an equal distribution of sex and mAbs [28]. Without a barrier, animals treated with GTN tended to approach the attractive source less frequently, stayed for less time at the source, and consumed less sugar solution than animals not treated with GTN (Figure 3A–C). Repeated measures ANOVA with the factors day (baseline vs. days 4 and 11 after mAb injection), treatment (no GTN vs. GTN), sex (male vs. female animals), and mAb (control mAb vs. fremanezumab) indicated no difference between the three days regarding the number of approaches ($F_{2,72} = 1.7$, $p = 0.19$), but the time at the source ($F_{2,72} = 7.8$, $p < 0.001$) and the consumed solution ($F_{2,72} = 16.1$, $p < 0.0001$) increased significantly. This can easily be explained by an increase in body weight of about 45 g on average during the two weeks between the baseline measurements and the second test sequence (see Figure 1B). In addition, the difference between the treatments (GTN vs. no GTN) was highly significant regarding both time at the source ($F_{1,36} = 10.6$, $p < 0.005$) and the consumed volume ($F_{1,36} = 24.1$, $p < 0.0001$), but there were no significant differences between sexes and mAbs regarding any of these measurements.

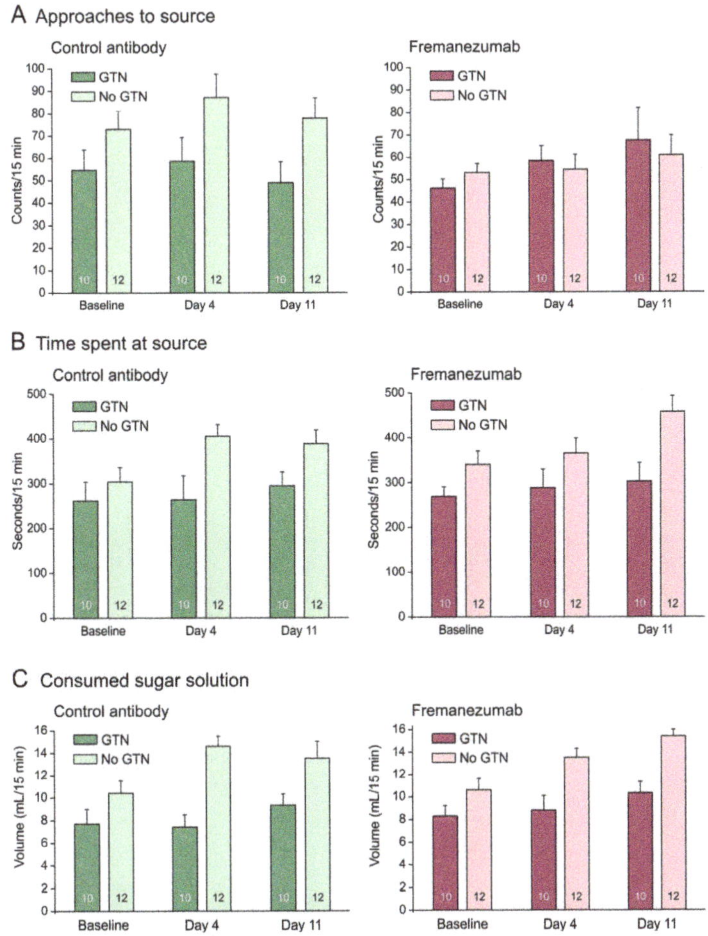

Figure 3. No barrier condition. Behavioral data (means ± SEM) of animals before (baseline) and days after administration of fremanezumab or control mAb and treatment or no treatment with GTN. Female and male data are not presented separately, because there was no sex difference. Animals treated with GTN tended to approach the attractive source less frequently (**A**), stayed less time at the source (**B**) and consumed less sugar solution (**C**) than animals not treated with GTN. There was no difference between fremanezumab and control mAb regarding any of these behavioral data. Numbers within bars mean number of animals.

3.2.5. Comparison of Behavioral Data with Animals Not Treated with GTN: Barrier Condition

With a mechanical barrier, animals treated with GTN tended to approach the attractive source less frequently when they received the control mAb (Figure 4A) but more frequently when they received fremanezumab (Figure 4A–C). GTN animals also tended to stay less time at the source and to consume less sugar solution than animals not treated with GTN (Figure 4B,C). Repeated measures ANOVA with the factors day (baseline vs. days 4 and 11 after mAb injection), treatment (no GTN vs. GTN), sex (male vs. female animals), and mAb (fremanezumab vs. control mAb) indicated differences between the three days regarding the number of approaches ($F_{2,72} = 3.3$, $p < 0.05$), the time spent at the source ($F_{2,72} = 9.8$, $p < 0.0005$), and the consumed solution ($F_{2,72} = 9.9$, $p < 0.0005$). The increase in time at the source and the consumed volume can be explained by the increase in body weight during the two weeks between baseline and day 12. Between the sexes and the treatments, no significant difference appeared. Between the two mAbs, the time spent at the source ($F_{1,36} = 7.6$, $p < 0.01$) and the consumed volume ($F_{1,36} = 7.1$, $p < 0.05$) were significantly different.

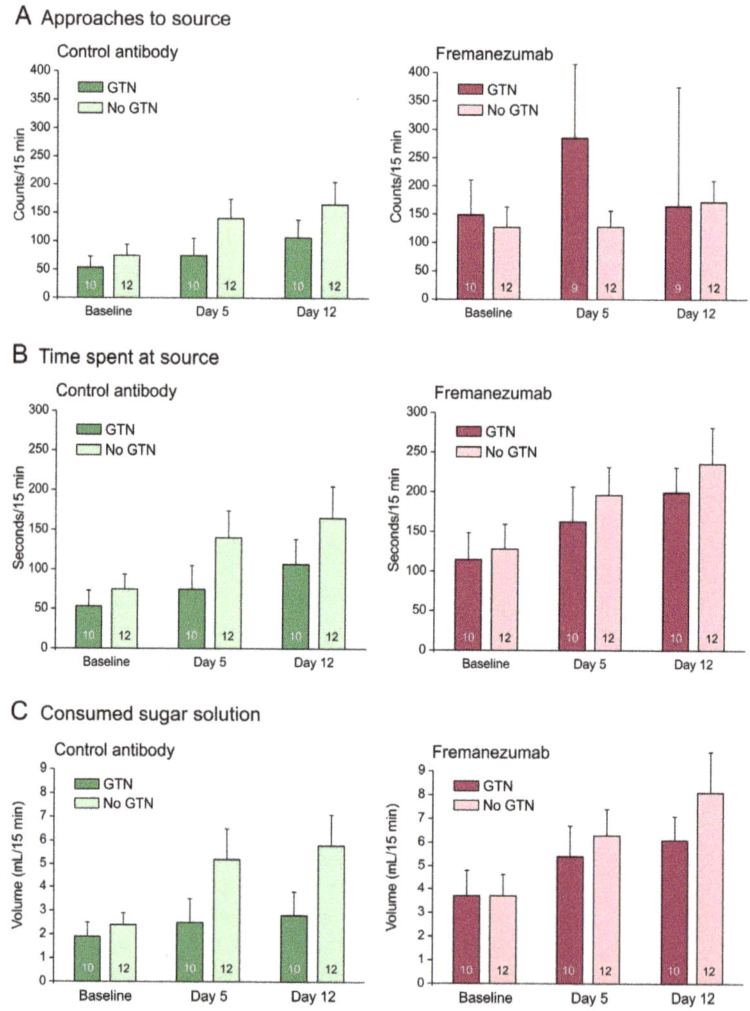

Figure 4. Mechanical barrier condition. Behavioral data of animals before (baseline) and days after administration of fremanezumab or control mAb and treatment or no treatment with GTN (means ± SEM). Female and male animal data are combined because there was no sex difference. Animals treated with GTN tended to approach the attractive source less frequently (**A**), to stay less time at the source (**B**) and to consume less solution (**C**) when they had received control mAb. Animals that received fremanezumab stayed longer at the source and consumed more volume compared to animals that received control mAb. Numbers within bars mean number of animals.

4. Discussion

4.1. Impact of GTN on the CGRP Concentration of Trigeminal Ganglia

In rodent models of migraine or facial pain, mostly 10 mg/kg GTN have been applied as a single dose [31,32], or 5 mg/kg has been administered repetitively every second day [30,33]. Repetitive GTN injection in rats has been reported to induce spinal hyperalgesia and orofacial allodynia along with an increase in CGRP expression [30]. Our results show that the administration of a single dose of 5 mg/kg GTN increased the production of immunologically detectable CGRP in the trigeminal ganglion within four hours. This appears to be a short time for neuropeptide expression and formation, but it is consistent with earlier experiments in our group, wherein anesthetized rats after an i.v. infusion of GTN at 250 μg/kg for two hours and a waiting time of four extra hours [34], or of GTN at 1 mg/kg for two hours and a waiting time of two extra hours [35], was followed by an increased number of CGRP immunoreactive neurons counted in the trigeminal ganglion. In contrast to these results, in our present experiment, a multiple-dose administration of GTN over some days decreased the CGRP concentration in the examined trigeminal ganglia.

The opposite observation of significantly lower CGRP concentrations was made in the trigeminal ganglia of animals repetitively treated with GTN (see Figure 2C). This may have been due to a continuous depletion of CGRP during the six days of treatment. The assumption that GTN treatment causes slow continuous CGRP release from trigeminal afferents is substantiated by our previous finding that the basal (unstimulated) CGRP release from the dura mater was also lower in animals treated with GTN, possibly through draining of the CGRP stores of peptidergic afferents in the dura mater [27]. In contrast, the CGRP release evoked by capsaicin superfusion was higher, indicating a sensitizing effect of GTN on the stimulated CGRP release. However, in animals that received fremanezumab, there was no difference between the vehicle and GTN, pointing to a blocking effect of fremanezumab on the GTN-induced increase in stimulated CGRP release [27]. Different from this result, fremanezumab seems not to influence the proposed GTN-induced increase in CGRP concentration in the trigeminal ganglion, as is indicated by the statistical analysis of animals that received fremanezumab or the control mAb (see Figure 2A,B).

4.2. Sex-Dependent CGRP Concentration of Trigeminal Ganglia

The mean CGRP concentration of ganglia in GTN-treated rats was lower in female compared to male animals (see Table 1 and Figure 2A,B). This confirms our previous experiments in animals not treated with GTN, where both CGRP content and concentration were lower in females [29]. The reason for this sex difference is not clear. It is likely that sex hormones influence the expression or production of CGRP. It has long been reported that in rat dorsal root ganglia, the number of CGRP immunoreactive neurons was lower in females and increased in ovariectomized animals but decreased in estradiol-treated animals [36]. More recent experiments showed that 17β-estradiol decreased CGRP plasma levels and CGRP release from isolated trigeminal ganglia in male rats [37]. However, generally, the CGRP concentration tended to be higher in the trigeminal ganglia of animals treated with GTN compared to non-treated animals, which is prominent after 3 days of mAb administration (see Figure 2C). The reason for this effect could be shrinkage of the ganglion mass (e.g., by fluid loss) or an increase in CGRP expression after GTN treatment. Indeed, post hoc statistical analysis showed that there was a significant interaction between sex and treatment regarding the ganglion weight, which was lower in GTN-treated male, but not female, rats (factorial ANOVA with factors sex and treatment, $F_{1,108} = 11.76$, $p < 0.001$; Tukey's post hoc test, $p < 0.05$ for males but $p = 0.28$ for females). Nevertheless, the results indicate that GTN treatment leads to an increase in the expression of, or at least immunologically detectable, CGRP in the trigeminal ganglion.

4.3. No Impact of Fremanezumab on the CGRP Concentration of Trigeminal Ganglia

We have seen no significant difference in CGRP concentration between the ganglia of rats treated with fremanezumab versus the control antibody, which is consistent with our previous findings, although it contrasts with a decrease in CGRP immunoreactive trigeminal ganglion neurons [29]. The lack of change in CGRP concentration is reminiscent of a recent clinical study in which plasma CGRP levels in patients were determined during the treatment with fremanezumab and four months after discontinuation of the treatment [38]. CGRP plasma concentrations did not differ between during treatment and post-treatment and were on the same level as in healthy controls. Thus, it seems likely that the main therapeutic effect of monoclonal antibodies directed against CGRP depends rather on a decrease in CGRP release during the strong stimulation of trigeminal afferents, as we have demonstrated earlier [27]. On the other hand, long-term effects of anti-CGRP monoclonal antibodies may be based on a decrease in CGRP receptors, which is suggested by a significant reduction in trigeminal ganglion neurons immunoreactive to the CGRP receptor components RAMP1 and CLR [29].

4.4. Limitations Regarding the Calculation of CGRP Concentrations

The CGRP concentration of the tested trigeminal ganglia depends on the amount of CGRP but also the size of the ganglia. We did not remove the dura mater covering each ganglion in order to avoid the destruction of neurons and loss of CGRP during preparation. Therefore, a confounding factor could be the differing amount of connective tissue adhering to the ganglion that influences the weight and hence the CGRP concentration. We nevertheless regard the concentration as a reliable measure for the amount of CGRP, because in our previous study, we saw that there is a high positive correlation between the CGRP content and concentration of each ganglion [29].

4.5. Impact of GTN on Rat Behavior

The behavioral data, in particular the number of approaches to the sugar source, may substantially be influenced by the general activity of animals. The time spent at the source, and particularly the consumed sugar solution, likely depend on the size of the animals, which is reflected by the general increase in these parameters during the 15 days between baseline measurements and day 11/12 after mAb administration (see Figures 2B,C and 3B,C). In addition, GTN treatment seems to have a major impact on these behavioral parameters.

Without a barrier, animals treated with GTN tended to approach the attractive source less frequently, stayed less time at the source, and consumed less sugar solution than animals not treated with GTN (Figure 2A–C). This may be the result of a generally lower activity after GTN, as mentioned above. Alternatively, the GTN effect may be due to a decreased attractiveness of the sugar solution, which is reminiscent of the loss of appetite and nausea during migraine attacks. GTN-induced trigeminal nociception has been used as an animal model for migraine, at least for migraine-related symptoms [24,25], and loss of appetite or nausea is a well-known symptom of migraine. Anti-CGRP antibodies have been found to improve not only the headache severity but also the autonomic symptoms including loss of appetite during migraine attacks [39]. However, because there was no difference between animals that received fremanezumab or the control mAb in the measured parameters of our experiments, the activity factor seems more likely for the decrease in functions.

With a mechanical barrier, like without a barrier, animals treated with GTN tended to stay less time at the source and to consume less sugar solution than animals not treated with GTN (Figure 3B,C), which may have similar reasons as discussed above. However, different to the condition without barrier, animals treated with GTN tended to approach the attractive source more frequently when they received fremanezumab, spent more time at the source, and consumed more sugar solution than animals that received the control mAb (Figure 3A–C). In our previous publication on rats not treated with GTN, we interpreted this observation as a reduced facial sensitivity under the action of the anti-CGRP mAb. Facial hypersensitivity has frequently been observed in animal studies following the administration of GTN [24,40,41]. Therefore, following treatment with GTN, the increase in all three parameters (approaches, time, and intake of solution) in animals that received fremanezumab compared to the control mAb may result from reduced facial hypersensitivity. Reduced incidences of cephalic hyperalgesia/allodynia have also been reported in patients treated with anti-CGRP mAbs like galcanezumab or fremanezumab [42,43].

4.6. Summary of GTN Effects

Taken together, GTN seems to increase the amount of immunologically detectable CGRP in the trigeminal ganglion, but fremanezumab seems not to significantly interact with this GTN effect. Fremanezumab had also no significant effect on the measurable CGRP concentration in trigeminal ganglia without GTN treatment, although in previous experiments, the number of CGRP containing trigeminal ganglion neurons, identified by immunofluorescence, was decreased in animals that received fremanezumab compared to the control mAb [29].

Regarding the behavioral data, GTN treatment generally tended to suppress the activity of the animals, with the final result being that they consumed less of the rewarding sugar solution. However, under the condition of an uncomfortable and irritating barrier, when the animals tried to reach the attractive source more often (but nevertheless consumed less of the solution), fremanezumab seemed to compensate partly for the depressing effect of GTN. The fremanezumab effect may result from its ability to decrease facial sensitivity so that the hypersensitivity caused by GTN is restricted. These results are in line with GTN treatment as an animal model for facial hyperalgesia/allodynia in migraine [24,40,41].

In addition, the results of this study show that the GTN effects are partly independent of manipulating the CGRP signaling system by anti-CGRP mAbs, although the extent of the GTN-induced hypersensitivity seems to be limited by reduced CGRP signaling. The latter is in accordance with previous studies in rodents showing that sumatriptan (lowering CGRP release) and olcegepant (blocking CGRP receptors) alleviate cranial and hind-paw hypersensitivity [44–46]. Performing an extended series of elegant experiments with cell cultures supplemented by measurements of facial sensitivity in mice, it has been recently reported that CGRP acts on glial cells (Schwann cells), which in turn produce NO species that are able to activate adjacent trigeminal afferent neurons [47]. The decisive NO species thereby may be nitroxyl (HNO), a redox sibling of NO formed in the presence of H_2S, which is able to directly activate transient receptor potential channels of the ankyrin type (TRPA1) in trigeminal afferents [48,49]. NO species have recently also been suggested to activate transient potential receptor channels of the vanilloid type (TRPV1) through S-nitrosylation [50]. In any case, the classical intracellular NO-mediated mechanism increasing cyclic guanosine monophosphate [51] is not necessary to explain the activating and sensitizing effects of NO species in pain and migraine generation [52,53].

5. Conclusions

Considering the shared pathways in sensitizing the trigeminal afferent system between CGRP and GTN or NO species, we conclude that GTN and NO act downstream of CGRP to enhance facial sensitivity. The hypothesized NO-induced activation of sensory neurons, as proposed by the previously mentioned group [40], may complete a feedback loop if these neurons release CGRP. It is important to determine in human studies whether this hypothesis holds true for CGRP-NO signaling in migraine pathophysiology. This might explain why the CGRP receptor antagonist olcegepant has been reported not to prevent GTN-induced migraine [21].

Author Contributions: Conceptualization, N.B., M.D. and K.M.; methodology, N.B., B.V., A.K., M.D. and K.M.; investigation, N.B., J.S. (Jana Schramm), B.V. and A.K.; resources, M.D., K.D.M., J.S. (Jennifer Stratton) and K.M.; writing—original draft preparation, M.D. and K.M.; writing—review and editing, N.B., M.D., K.D.M., J.S. (Jennifer Stratton) and K.M.; visualization, M.D. and K.M.; supervision, M.D. and K.M.; project administration, K.M.; funding acquisition, M.D. and K.M. All authors have read and agreed to the published version of the manuscript.

Funding: These studies were part of a sponsored research collaboration with Teva, who provided research support; the anti-CGRP antibody, fremanezumab; and the isotype control antibody. The project was also partly supported by the research grant K119597 project of the Hungarian National Research, Development and Innovation Office, and the Alexander von Humboldt Foundation (Research Group Linkage Program between the Institutes of Physiology of the FAU Erlangen-Nürnberg and the University of Szeged).

Institutional Review Board Statement: These studies were conducted in accordance with the Declaration of Helsinki. The animal study protocol was approved by the District Government of Mittelfranken (54-2532.1-21/12) after review by the Ethics Committee of the District Government of Unterfranken.

Informed Consent Statement: Not applicable.

Data Availability Statement: Data are contained within the article.

Acknowledgments: We thank Rainer Fischer for taking care of the animals.

Conflicts of Interest: Part of this work was performed by the first author (N.B.) for obtaining the degree "Dr. med." at the Friedrich-Alexander-University Erlangen-Nürnberg (FAU). K.D.M. and J.St. were employees of Teva Pharmaceuticals. The other authors declare no conflicts of interest. The funders had no role in the design of the study, in the collection, analyses, or interpretation of data, or in the writing of the manuscript. The submission and publication of the manuscript was authorized by Teva Pharmaceuticals.

References

1. Iyengar, S.; Johnson, K.W.; Ossipov, M.H.; Aurora, S.K. CGRP and the Trigeminal System in Migraine. *Headache* **2019**, *59*, 659–681. [CrossRef]
2. Karsan, N.; Gosalia, H.; Goadsby, P.J. Molecular Mechanisms of Migraine: Nitric Oxide Synthase and Neuropeptides. *Int. J. Mol. Sci.* **2023**, *24*, 11993. [CrossRef]
3. Goadsby, P.J.; Edvinsson, L.; Ekman, R. Vasoactive Peptide Release in the Extracerebral Circulation of Humans during Migraine Headache. *Ann. Neurol.* **1990**, *28*, 183–187. [CrossRef]
4. Edvinsson, L.; Goadsby, P.J. Neuropeptides in Migraine and Cluster Headache. *Cephalalgia Int. J. Headache* **1994**, *14*, 320–327. [CrossRef]
5. Fan, P.-C.; Kuo, P.-H.; Chang, S.-H.; Lee, W.-T.; Wu, R.-M.; Chiou, L.-C. Plasma Calcitonin Gene-Related Peptide in Diagnosing and Predicting Paediatric Migraine. *Cephalalgia Int. J. Headache* **2009**, *29*, 883–890. [CrossRef]
6. Bellamy, J.L.; Cady, R.K.; Durham, P.L. Salivary Levels of CGRP and VIP in Rhinosinusitis and Migraine Patients. *Headache* **2006**, *46*, 24–33. [CrossRef] [PubMed]
7. Alpuente, A.; Gallardo, V.J.; Asskour, L.; Caronna, E.; Torres-Ferrus, M.; Pozo-Rosich, P. Salivary CGRP Can Monitor the Different Migraine Phases: CGRP (in)Dependent Attacks. *Cephalalgia Int. J. Headache* **2022**, *42*, 186–196. [CrossRef] [PubMed]
8. Kamm, K.; Straube, A.; Ruscheweyh, R. Calcitonin Gene-Related Peptide Levels in Tear Fluid Are Elevated in Migraine Patients Compared to Healthy Controls. *Cephalalgia Int. J. Headache* **2019**, *39*, 1535–1543. [CrossRef]
9. Goadsby, P.J.; Holland, P.R.; Martins-Oliveira, M.; Hoffmann, J.; Schankin, C.; Akerman, S. Pathophysiology of Migraine: A Disorder of Sensory Processing. *Physiol. Rev.* **2017**, *97*, 553–622. [CrossRef] [PubMed]
10. de Vries, T.; Villalón, C.M.; MaassenVanDenBrink, A. Pharmacological Treatment of Migraine: CGRP and 5-HT beyond the Triptans. *Pharmacol. Ther.* **2020**, *211*, 107528. [CrossRef] [PubMed]
11. Lassen, L.H.; Haderslev, P.A.; Jacobsen, V.B.; Iversen, H.K.; Sperling, B.; Olesen, J. CGRP May Play a Causative Role in Migraine. *Cephalalgia Int. J. Headache* **2002**, *22*, 54–61. [CrossRef]
12. Hansen, J.M.; Hauge, A.W.; Olesen, J.; Ashina, M. Calcitonin Gene-Related Peptide Triggers Migraine-like Attacks in Patients with Migraine with Aura. *Cephalalgia Int. J. Headache* **2010**, *30*, 1179–1186. [CrossRef] [PubMed]
13. Sarchielli, P.; Alberti, A.; Codini, M.; Floridi, A.; Gallai, V. Nitric Oxide Metabolites, Prostaglandins and Trigeminal Vasoactive Peptides in Internal Jugular Vein Blood during Spontaneous Migraine Attacks. *Cephalalgia Int. J. Headache* **2000**, *20*, 907–918. [CrossRef] [PubMed]
14. Olesen, J.; Thomsen, L.L.; Lassen, L.H.; Olesen, I.J. The Nitric Oxide Hypothesis of Migraine and Other Vascular Headaches. *Cephalalgia Int. J. Headache* **1995**, *15*, 94–100. [CrossRef] [PubMed]
15. Ashina, M.; Bendtsen, L.; Jensen, R.; Olesen, J. Nitric Oxide-Induced Headache in Patients with Chronic Tension-Type Headache. *Brain J. Neurol.* **2000**, *123 Pt 9*, 1830–1837. [CrossRef]
16. Olesen, J. Nitric Oxide-Related Drug Targets in Headache. *Neurotherapeutics* **2010**, *7*, 183–190. [CrossRef] [PubMed]
17. Juhasz, G.; Zsombok, T.; Modos, E.A.; Olajos, S.; Jakab, B.; Nemeth, J.; Szolcsanyi, J.; Vitrai, J.; Bagdy, G. NO-Induced Migraine Attack: Strong Increase in Plasma Calcitonin Gene-Related Peptide (CGRP) Concentration and Negative Correlation with Platelet Serotonin Release. *Pain* **2003**, *106*, 461–470. [CrossRef]

18. Juhasz, G.; Zsombok, T.; Jakab, B.; Nemeth, J.; Szolcsanyi, J.; Bagdy, G. Sumatriptan Causes Parallel Decrease in Plasma Calcitonin Gene-Related Peptide (CGRP) Concentration and Migraine Headache during Nitroglycerin Induced Migraine Attack. *Cephalalgia Int. J. Headache* **2005**, *25*, 179–183. [CrossRef]
19. Vanmolkot, F.; Van der Schueren, B.; de Hoon, J. Sumatriptan Causes Parallel Decrease in Plasma CGRP Concentration and Migraine Headache during Nitroglycerin-Induced Migraine Attack. *Cephalalgia Int. J. Headache* **2006**, *26*, 1037–1038. [CrossRef]
20. Kruuse, C.; Iversen, H.K.; Jansen-Olesen, I.; Edvinsson, L.; Olesen, J. Calcitonin Gene-Related Peptide (CGRP) Levels during Glyceryl Trinitrate (GTN)-Induced Headache in Healthy Volunteers. *Cephalalgia Int. J. Headache* **2010**, *30*, 467–474. [CrossRef]
21. Tvedskov, J.F.; Tfelt-Hansen, P.; Petersen, K.A.; Jensen, L.T.; Olesen, J. CGRP Receptor Antagonist Olcegepant (BIBN4096BS) Does Not Prevent Glyceryl Trinitrate-Induced Migraine. *Cephalalgia Int. J. Headache* **2010**, *30*, 1346–1353. [CrossRef]
22. Tassorelli, C.; Greco, R.; Wang, D.; Sandrini, M.; Sandrini, G.; Nappi, G. Nitroglycerin Induces Hyperalgesia in Rats—A Time-Course Study. *Eur. J. Pharmacol.* **2003**, *464*, 159–162. [CrossRef]
23. Greco, R.; Mangione, A.; Siani, F.; Blandini, F.; Vairetti, M.; Nappi, G.; Sandrini, G.; Buzzi, M.; Tassorelli, C. Effects of CGRP Receptor Antagonism in Nitroglycerin-Induced Hyperalgesia. *Cephalalgia Int. J. Headache* **2013**, *34*, 594–604. [CrossRef]
24. Demartini, C.; Greco, R.; Zanaboni, A.M.; Sances, G.; De Icco, R.; Borsook, D.; Tassorelli, C. Nitroglycerin as a Comparative Experimental Model of Migraine Pain: From Animal to Human and Back. *Prog. Neurobiol.* **2019**, *177*, 15–32. [CrossRef]
25. Sureda-Gibert, P.; Romero-Reyes, M.; Akerman, S. Nitroglycerin as a Model of Migraine: Clinical and Preclinical Review. *Neurobiol. Pain* **2022**, *12*, 100105. [CrossRef]
26. Guo, S.; Ernstsen, C.; Hay-Schmidt, A.; Kristensen, D.M.; Ashina, M.; Olesen, J.; Christensen, S.L. PACAP Signaling Is Not Involved in GTN- and Levcromakalim-Induced Hypersensitivity in Mouse Models of Migraine. *J. Headache Pain* **2022**, *23*, 155. [CrossRef]
27. Dux, M.; Vogler, B.; Kuhn, A.; Mackenzie, K.D.; Stratton, J.; Messlinger, K. The Anti-CGRP Antibody Fremanezumab Lowers CGRP Release from Rat Dura Mater and Meningeal Blood Flow. *Cells* **2022**, *11*, 1768. [CrossRef]
28. Benedicter, N.; Messlinger, K.; Vogler, B.; Mackenzie, K.D.; Stratton, J.; Friedrich, N.; Dux, M. Semi-Automated Recording of Facial Sensitivity in Rat Demonstrates Antinociceptive Effects of the Anti-CGRP Antibody Fremanezumab. *Neurol. Int.* **2023**, *15*, 622–637. [CrossRef]
29. Vogler, B.; Kuhn, A.; Mackenzie, K.D.; Stratton, J.; Dux, M.; Messlinger, K. The Anti-Calcitonin Gene-Related Peptide (Anti-CGRP) Antibody Fremanezumab Reduces Trigeminal Neurons Immunoreactive to CGRP and CGRP Receptor Components in Rats. *Int. J. Mol. Sci.* **2023**, *24*, 13471. [CrossRef]
30. Greco, R.; Demartini, C.; Zanaboni, A.M.; Tassorelli, C. Chronic and Intermittent Administration of Systemic Nitroglycerin in the Rat Induces an Increase in the Gene Expression of CGRP in Central Areas: Potential Contribution to Pain Processing. *J. Headache Pain* **2018**, *19*, 51. [CrossRef]
31. Christensen, S.L.; Ernstsen, C.; Olesen, J.; Kristensen, D.M. No Central Action of CGRP Antagonising Drugs in the GTN Mouse Model of Migraine. *Cephalalgia Int. J. Headache* **2020**, *40*, 924–934. [CrossRef]
32. Andreou, A.P.; Leese, C.; Greco, R.; Demartini, C.; Corrie, E.; Simsek, D.; Zanaboni, A.; Koroleva, K.; Lloyd, J.O.; Lambru, G.; et al. Double-Binding Botulinum Molecule with Reduced Muscle Paralysis: Evaluation in In Vitro and In Vivo Models of Migraine. *Neurotherapeutics* **2021**, *18*, 556–568. [CrossRef] [PubMed]
33. Greco, R.; Demartini, C.; Francavilla, M.; Zanaboni, A.M.; Tassorelli, C. Antagonism of CGRP Receptor: Central and Peripheral Mechanisms and Mediators in an Animal Model of Chronic Migraine. *Cells* **2022**, *11*, 3092. [CrossRef] [PubMed]
34. Dieterle, A.; Fischer, M.J.M.; Link, A.S.; Neuhuber, W.L.; Messlinger, K. Increase in CGRP- and nNOS-Immunoreactive Neurons in the Rat Trigeminal Ganglion after Infusion of an NO Donor. *Cephalalgia Int. J. Headache* **2011**, *31*, 31–42. [CrossRef] [PubMed]
35. Seiler, K.; Nusser, J.I.; Lennerz, J.K.; Neuhuber, W.L.; Messlinger, K. Changes in Calcitonin Gene-Related Peptide (CGRP) Receptor Component and Nitric Oxide Receptor (sGC) Immunoreactivity in Rat Trigeminal Ganglion Following Glyceroltrinitrate Pretreatment. *J. Headache Pain* **2013**, *14*, 74. [CrossRef] [PubMed]
36. Yang, Y.; Ozawa, H.; Lu, H.; Yuri, K.; Hayashi, S.; Nihonyanagi, K.; Kawata, M. Immunocytochemical Analysis of Sex Differences in Calcitonin Gene-Related Peptide in the Rat Dorsal Root Ganglion, with Special Reference to Estrogen and Its Receptor. *Brain Res.* **1998**, *791*, 35–42. [CrossRef] [PubMed]
37. Cetinkaya, A.; Kilinc, E.; Camsari, C.; Ogun, M.N. Effects of Estrogen and Progesterone on the Neurogenic Inflammatory Neuropeptides: Implications for Gender Differences in Migraine. *Exp. Brain Res.* **2020**, *238*, 2625–2639. [CrossRef] [PubMed]
38. Raffaelli, B.; Terhart, M.; Fitzek, M.P.; Lange, K.S.; Mecklenburg, J.; Overeem, L.H.; Siebert, A.; Storch, E.; Reuter, U. Change of CGRP Plasma Concentrations in Migraine after Discontinuation of CGRP-(Receptor) Monoclonal Antibodies. *Pharmaceutics* **2023**, *15*, 293. [CrossRef]
39. Iannone, L.F.; De Cesaris, F.; Ferrari, A.; Benemei, S.; Fattori, D.; Chiarugi, A. Effectiveness of Anti-CGRP Monoclonal Antibodies on Central Symptoms of Migraine. *Cephalalgia Int. J. Headache* **2022**, *42*, 1323–1330. [CrossRef]
40. Harris, H.M.; Carpenter, J.M.; Black, J.R.; Smitherman, T.A.; Sufka, K.J. The Effects of Repeated Nitroglycerin Administrations in Rats; Modeling Migraine-Related Endpoints and Chronification. *J. Neurosci. Methods* **2017**, *284*, 63–70. [CrossRef]
41. Akerman, S.; Romero-Reyes, M.; Karsan, N.; Bose, P.; Hoffmann, J.R.; Holland, P.R.; Goadsby, P.J. Therapeutic Targeting of Nitroglycerin-Mediated Trigeminovascular Neuronal Hypersensitivity Predicts Clinical Outcomes of Migraine Abortives. *Pain* **2021**, *162*, 1567–1577. [CrossRef]

42. Barbanti, P.; Egeo, G.; Aurilia, C.; Altamura, C.; d'Onofrio, F.; Finocchi, C.; Albanese, M.; Aguggia, M.; Rao, R.; Zucco, M.; et al. Predictors of Response to Anti-CGRP Monoclonal Antibodies: A 24-Week, Multicenter, Prospective Study on 864 Migraine Patients. *J. Headache Pain* **2022**, *23*, 138. [CrossRef]
43. Ashina, S.; Melo-Carrillo, A.; Szabo, E.; Borsook, D.; Burstein, R. Pre-Treatment Non-Ictal Cephalic Allodynia Identifies Responders to Prophylactic Treatment of Chronic and Episodic Migraine Patients with Galcanezumab: A Prospective Quantitative Sensory Testing Study (NCT04271202). *Cephalalgia Int. J. Headache* **2023**, *43*, 3331024221147881. [CrossRef]
44. Bates, E.A.; Nikai, T.; Brennan, K.C.; Fu, Y.-H.; Charles, A.C.; Basbaum, A.I.; Ptácek, L.J.; Ahn, A.H. Sumatriptan Alleviates Nitroglycerin-Induced Mechanical and Thermal Allodynia in Mice. *Cephalalgia Int. J. Headache* **2010**, *30*, 170–178. [CrossRef]
45. Akerman, S.; Karsan, N.; Bose, P.; Hoffmann, J.R.; Holland, P.R.; Romero-Reyes, M.; Goadsby, P.J. Nitroglycerine Triggers Triptan-Responsive Cranial Allodynia and Trigeminal Neuronal Hypersensitivity. *Brain J. Neurol.* **2019**, *142*, 103–119. [CrossRef] [PubMed]
46. Ernstsen, C.; Christensen, S.L.; Olesen, J.; Kristensen, D.M. No Additive Effect of Combining Sumatriptan and Olcegepant in the GTN Mouse Model of Migraine. *Cephalalgia Int. J. Headache* **2021**, *41*, 329–339. [CrossRef] [PubMed]
47. De Logu, F.; Nassini, R.; Hegron, A.; Landini, L.; Jensen, D.D.; Latorre, R.; Ding, J.; Marini, M.; Souza Monteiro de Araujo, D.; Ramírez-Garcia, P.; et al. Schwann Cell Endosome CGRP Signals Elicit Periorbital Mechanical Allodynia in Mice. *Nat. Commun.* **2022**, *13*, 646. [CrossRef] [PubMed]
48. Eberhardt, M.; Dux, M.; Namer, B.; Miljkovic, J.; Cordasic, N.; Will, C.; Kichko, T.I.; de la Roche, J.; Fischer, M.; Suárez, S.A.; et al. H_2S and NO Cooperatively Regulate Vascular Tone by Activating a Neuroendocrine HNO-TRPA1-CGRP Signalling Pathway. *Nat. Commun.* **2014**, *5*, 4381. [CrossRef] [PubMed]
49. Dux, M.; Will, C.; Vogler, B.; Filipovic, M.R.; Messlinger, K. Meningeal Blood Flow Is Controlled by H2 S-NO Crosstalk Activating a HNO-TRPA1-CGRP Signalling Pathway. *Br. J. Pharmacol.* **2015**, *173*, 431–445. [CrossRef] [PubMed]
50. Koroleva, K.; Svitko, S.; Ananev, A.; Buglinina, A.; Bogatova, K.; Yakovleva, O.; Nurmieva, D.; Shaidullov, I.; Sitdikova, G. Effects of Nitric Oxide on the Activity of P2X and TRPV1 Receptors in Rat Meningeal Afferents of the Trigeminal Nerve. *Int. J. Mol. Sci.* **2023**, *24*, 7519. [CrossRef] [PubMed]
51. Bryan, N.S.; Bian, K.; Murad, F. Discovery of the Nitric Oxide Signaling Pathway and Targets for Drug Development. *Front. Biosci.* **2009**, *14*, 1–18. [CrossRef] [PubMed]
52. Neeb, L.; Reuter, U. Nitric Oxide in Migraine. *CNS Neurol. Disord. Drug Targets* **2007**, *6*, 258–264. [CrossRef] [PubMed]
53. Spiller, F.; Oliveira Formiga, R.; Fernandes da Silva Coimbra, J.; Alves-Filho, J.C.; Cunha, T.M.; Cunha, F.Q. Targeting Nitric Oxide as a Key Modulator of Sepsis, Arthritis and Pain. *Nitric Oxide Biol. Chem.* **2019**, *89*, 32–40. [CrossRef] [PubMed]

cells

MDPI

Article

Effects of the Dual FAAH/MAGL Inhibitor AKU-005 on Trigeminal Hyperalgesia in Male Rats

Rosaria Greco [1,*], Chiara Demartini [1], Miriam Francavilla [1,2], Anna Maria Zanaboni [1,2], Sara Facchetti [1,2], Michela Palmisani [1,3], Valentina Franco [1,3] and Cristina Tassorelli [1,2]

1 Section of Translational Neurovascular Research, IRCCS Mondino Foundation, Via Mondino 2, 27100 Pavia, Italy; chiara.demartini@mondino.it (C.D.); miriam.francavilla@mondino.it (M.F.); annamaria.zanaboni@unipv.it (A.M.Z.); sara.facchetti@mondino.it (S.F.); michela.palmisani@mondino.it (M.P.); valentina.franco@unipv.it (V.F.); cristina.tassorelli@unipv.it (C.T.)

2 Department of Brain and Behavioral Sciences, University of Pavia, Via Bassi 21, 27100 Pavia, Italy

3 Department of Internal Medicine and Therapeutics, Clinical and Experimental Pharmacology Unit, University of Pavia, Viale Golgi 19, 27100 Pavia, Italy

* Correspondence: rosaria.greco@mondino.it; Tel.: +39-(0382)-380-255

Abstract: The inhibition of endocannabinoid hydrolysis by enzymatic inhibitors may interfere with mechanisms underlying migraine-related pain. The dual FAAH/MAGL inhibitor AKU-005 shows potent inhibitory activity in vitro. Here, we assessed the effect of AKU-005 in a migraine animal model based on nitroglycerin (NTG) administration. Male rats were treated with AKU-005 (0.5 mg/kg, i.p.) or vehicle 3 h after receiving NTG (10 mg/kg, i.p.) or NTG vehicle. One hour later, rats were subjected to the open field test followed by the orofacial formalin test. At the end of the test, we collected serum samples for assessing calcitonin gene-related peptide (CGRP) levels as well as meninges, trigeminal ganglia, and brain areas to assess mRNA levels of CGRP and pro-inflammatory cytokines, and endocannabinoid and related lipid levels. AKU-005 reduced NTG-induced hyperalgesia during the orofacial formalin test but did not influence NTG-induced changes in the open field test. It significantly reduced serum levels of CGRP, CGRP, and pro-inflammatory cytokine mRNA levels in the meninges, trigeminal ganglia, and central areas. Surprisingly, AKU-005 caused no change in endocannabinoids and related lipids in the regions evaluated. The present findings suggest that AKU-005 may have anti-migraine effects by reducing CGRP synthesis and release and the associated inflammatory events. This effect, however, does not seem mediated via an interference with the endocannabinoid pathway.

Keywords: endocannabinoids; FAAH; MAGL; migraine; pain

Citation: Greco, R.; Demartini, C.; Francavilla, M.; Zanaboni, A.M.; Facchetti, S.; Palmisani, M.; Franco, V.; Tassorelli, C. Effects of the Dual FAAH/MAGL Inhibitor AKU-005 on Trigeminal Hyperalgesia in Male Rats. *Cells* **2024**, *13*, 830. https://doi.org/10.3390/cells13100830

Academic Editors: Alessandro Castorina and Alessandra Pacini

Received: 2 April 2024
Revised: 3 May 2024
Accepted: 10 May 2024
Published: 13 May 2024

1. Introduction

The endocannabinoids N-arachidonoylethanolamine (AEA, also known as anandamide) and 2-arachidonoylglycerol (2-AG) are eicosanoid molecules produced from arachidonic acid. These two lipids can activate cannabinoid receptors under various physiological circumstances; they are produced both in the brain and in peripheral tissues, and then secreted by different cell types [1,2]. Endocannabinoids are involved in many physiological and pathophysiological processes, including migraine pain [1,3,4]. In response to neuronal activity, elicited, for instance, by stress, pain, or inflammatory stimulus, 2-AG and AEA are synthesized 'on demand' at postsynaptic sites. They then diffuse to retrogradely activate presynaptic CB1 and other receptors, causing a transient and long-lasting reduction in mediator release, resulting in analgesia [5,6]. The activation of trigeminal CB1 receptors inhibits calcitonin gene-related peptide (CGRP) release [7] and dural vasodilation [8], which are the hallmarks of migraine pathophysiology [9]. Cannabinoids or endocannabinoid system modulators may control pain through neuro-immune interactions in glial cells, which express various endocannabinoid components [10,11]. The involvement of endocannabinoid signaling in the functional coupling of neurons, astrocytes, and microglia supports its contribution to the (neuro)inflammation process [12].

Inhibitors of enzymes involved in endocannabinoid hydrolysis may reduce migraine-related pain by modulating the endocannabinoid system [13,14], although the mechanisms of action underlying this activity are still unclear. In recent years, we have devoted much of our research to inhibiting endocannabinoid degradation in our pre-clinical migraine model based on nitroglycerin (NTG) administration, demonstrating its therapeutic potential. The NTG model is a widely used animal model of migraine that, both in the acute and chronic form, may reproduce some of the features that can be found in migraine patients. Indeed, it produces cephalic and extra-cephalic hypersensitivity, allodynia, and hyperalgesia, as well as the activation of brain areas involved in migraine pain. In addition, the NTG model may reproduce some of the associated symptoms, like photophobia, and comorbid conditions such as anxiety/depression-like behavior [15,16]. This model has also been extensively used to investigate the effects of endocannabinoid system modulators [17–24]. We have shown that under conditions of hyperalgesia, endocannabinoids play a prominent role in reducing inflammatory and pain mediators in central and peripheral regions [22–24].

The piperazine derivative AKU-005 (4-Benzhydrylpiperazin-1-yl) (1H-1,2,4-triazol-4-yl) methadone) has been shown in vitro to be a new potent inhibitor of monoacylglycerol lipase (MAGL) and fatty acid amide hydrolase (FAAH) enzyme activity in several areas of the nervous system involved in migraine pain signaling [25]. Specifically, AKU-005 reduced meningeal afferent excitability [25] via the activation of CB1 receptors.

Interestingly, in mouse brain membrane preparations, AKU-005 can inhibit MAGL activity at sub-nanomolar concentrations (IC50 values range from 0.2 to 1.1 nM) [26], but Della Pietra et al. recently showed in human and rat meningeal samples that the basal FAAH activity is higher than the barely perceptible MAGL activity [27]. The situation is opposite in the TGs and dorsal root ganglia where strong MAGL/low FAAH activity was reported [25,27]. All the abovementioned studies were conducted in vitro, but it is evident that, due to the redundancy and promiscuity of the enzymes involved in the degradation of endocannabinoids, and also because inhibiting these enzymes may have unpredictable effects, even concerning the dose used, it is extremely important to assess their activity in vivo.

In this study, we tested for the first time the pharmacological activity of AKU-005 in vivo in the animal model of migraine based on the administration of NTG to validate its potential effect in migraine treatment.

2. Materials and Methods

2.1. Animals and Drugs

Male Sprague Dawley rats (150–175 g, Charles River Laboratories, Calco, Italy) were housed in pairs in cages at the University of Pavia's animal facility under carefully monitored conditions (temperature 21–22 °C, relative humidity 60–50%, and 12/12 h light cycle) with water and food ad libitum. The Italian Ministry of Health (376/2020-PR) approved all procedures, which were performed in compliance with the guidelines of European Community Directive 2010/63/EU of 22 September 2010.

NTG (Bioindustria L.I.M., Novi Ligure, Italy) was prepared and administered as previously described [23]. NTG was administered intraperitoneally (i.p.) at a dose of 10 mg/kg.

AKU-005 (synthetized by the Department of Chemistry, University of Eastern Finland) was delivered systemically (i.p.) at a dose of 0.5 mg/kg in a volume of 2 mL/kg after being dissolved in tween-80/polyethylene glycol 200/saline (10/10/80; utilized as a vehicle). Considering the lack of pharmacokinetic studies on AKU-005 in vivo, the dose was calculated empirically, based on our previous experience with a similar compound tested in the same experimental paradigm [23].

2.2. Experimental Plan

Rats were randomly allocated the four experimental groups (Table 1) and used in the experimental setting reported in Figure 1.

Table 1. Experimental groups. Each animal received two i.p. injections, indicated as Treatment 1 and 2.

Group Name	Treatment 1	Treatment 2
Control (CT)	NTG vehicle	AKU-005 vehicle
NTG	NTG 10 mg/kg	AKU-005 vehicle
NTG + AKU-005	NTG 10 mg/kg	AKU-005 0.5 mg/kg
AKU-005	NTG vehicle	AKU-005 0.5 mg/kg

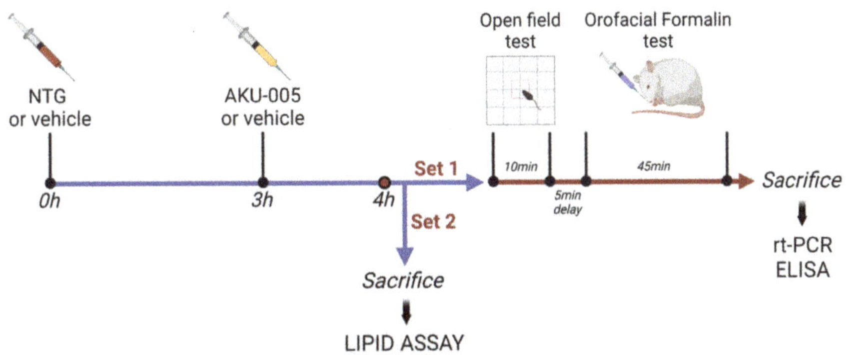

Figure 1. Experimental timeline for the treatment and testing procedures. After treatments, rats belonging to Set 1 underwent the open field test (10 min duration), and after 5 min of delay, they were subjected to the orofacial formalin test. At the end of behavioral testing, the animals were sacrificed and tissue/blood samples collected for gene (by rt-PCR) and protein expression (by ELISA). The animals belonging to Set 2 were immediately sacrificed after treatments, and tissue samples were collected for lipid assay (by means of mass spectrometry). Figure created with BioRender.com (https://www.biorender.com/ accessed on 7 March 2024).

In experimental Set 1, AKU005 (or vehicle) was administered 3 h after NTG (or its vehicle), which is 1 h before the behavioral testing. Animals underwent the open field test and, after a 5 min rest, the orofacial formalin test. At the end of the behavioral testing sessions, rats were sacrificed with a lethal dose of anesthetic (sodium thiopental, 150 mg/Kg, i.p.) followed by decapitation; truncal blood and cranial tissues were collected for ex vivo analysis.

In experimental Set 2, AKU-005 (or vehicle) was administered 3 h after NTG (or its vehicle), and after 1 h, rats were immediately sacrificed by decapitation after exposure to carbon dioxide, in order to collect the samples for lipid assays in specific brain areas.

2.3. Open Field Test

As many physiological and behavioral functions are influenced by endocannabinoid signaling, the open field test was used to assess anxiety, exploratory behaviors, and locomotor ability. After 60 min of acclimatization to the test room, each rat was positioned in the open field arena (Ugo Basile, Gemonio, VA, Italy) and tested for ten minutes as previously reported [24]. The room's light was kept constant from center to corner regions of the arena, thus avoiding possible confounding results since NTG can evoke photophobia. To assess locomotor activity, anxiety, and exploration, we recorded for ten minutes the distance traveled across the arena, the time spent in the center, and rearing behaviors. Furthermore, we looked at spontaneous grooming activity as a sign of heightened nociception. An observer, blinded to the treatment, manually evaluated rearing and grooming behaviors by counting the time the animal stood on its hind legs and the time it spent face- and body-grooming, respectively. As regards the total distance and the time spent in the center, these parameters were assessed by means of the ANY-Maze software (Ugo Basile, application version 4.99g Beta) [24]. This latter was used to conventionally divide the arena into 16 square units to delineate the center of the arena, identified by the 4 central squares, and the 12 surrounding squares represent the periphery.

2.4. Orofacial Formalin Test

On the experimental day, the rats were injected subcutaneously with 50 μL of formalin 1.5% (made of formaldehyde 37% in water and 0.9% saline, v/v) into the right upper lip, just lateral to the nose, and positioned in an observation box. A camera, recording face-rubbing time, was located 50 cm from the box, a 30 × 30 × 30 cm glass chamber with mirrored sides, to provide a clear view of each rat. Face-rubbing was evaluated by a researcher blind to group assignment who counted the seconds the animal spent grooming the injected area with the ipsilateral forepaw or hind paw 0–3 min (Phase I) and 12–45 min (Phase II) after formalin injection. The observation time was divided into 15 blocks of 3 min each. Phase I reflects an acute pain, while Phase II represents the combined effects of afferent input and central sensitization. The orofacial formalin test was utilized in conjunction with the NTG model to induce a state of hyperalgesia that mimics the clinical condition [24].

2.5. Gene Expression

At the end of behavioral testing, rats belonging to Set 1 were euthanized (sodium thiopental, 150 mg/kg, i.p.).

After decapitation, meninges and medulla in toto, cervical spinal cord (CSC, C1–C2), and TG ipsilateral to formalin injection were quickly dissected out, rinsed in cold sterile 0.9% NaCl solution, placed in cryogenic tubes and immediately frozen in liquid nitrogen, and then stored at −80 °C until further processing. Tissue samples were homogenized by means of ceramic beads (PRECELLYS, Bertin Pharma, Montigny le Bretonneux—France) with TRIzol® (Invitrogen, Waltham, MA, USA) to extract the total RNA. All procedures were performed under RNase-free conditions. RNA quality was assessed using a nanodrop spectrophotometer (Euroclone, Pero, MI, Italy) showing that the absorbance ratios (260/280 nm) ranged from 1.9 to 2.0 in all samples, indicating no significant protein (including of blood origin) contamination. The iScript cDNA Synthesis kit (BIO-RAD, Hercules, CA, USA) was used to generate cDNA, and the Fast Eva Green supermix (BIO-RAD, Hercules, CA, USA) was used to assess gene expression. Specifically, we evaluated the gene expression levels of *Calca* (coding for CGRP), tumor necrosis factor alpha (*TNF-alpha*), and interleukin (*IL-6*) using rt-PCR [23,24]. Glyceraldehyde 3-phosphate dehydrogenase (*GAPDH*), the expression of which remained constant in all experimental groups, was used for normalization. The following specific primers were used: *GAPDH* AACCTGCCAAGTATGATGAC (forward), GGAGTTGCTGTTGAAGTCA (reverse); *Calca* (coding for CGRP) CAGTCTCAGCTCCAAGTCATC (forward), TTCCAAGGTTGACCTCAAAG (reverse); *TNF-alpha* CCTCACACTCAGATCATCTTCTC (forward), CGCTTGGTGGTTTGCTAC (reverse); *IL-6* TTCTCTCCGCAAGAGACTTC (forward), and GGTCTGTTGTGGGTGGTATC (reverse). The $2^{-\Delta\Delta Ct} = 2^{-(\Delta Ct\ \text{gene} - \Delta Ct\ \text{housekeeping gene})}$ method was used to investigate the differences in gene expression.

2.6. CGRP Levels

Truncal blood, obtained from the animals belonging to Set 1, was collected in clot activator with gel separator serum tubes and centrifuged for 15 min at 1500× g for CGRP serum evaluations. CGRP levels were assessed using a commercial ELISA kit (α-CGRP: Elabsciences, Houston, TX, USA) and measured using a CLARIOstar microplate reader (BMG LABTECH, Ortenberg, Germany).

2.7. Lipids Levels

To evaluate the levels of 2-AG, AEA, palmitoylethanolamide (PEA), and oleoyl ethanolamide (OEA), the meninges, medulla in toto, CSC, and TG were collected from the animals of Set 2. Samples were weighed before homogenization in methanol with 0.8% formic acid containing cannabidiol-d_3 as an internal standard. Samples were processed and analyzed as thoroughly described in a previous study [23].

Differently to the meningeal samples used for the gene expression analysis, the amount of most of the meninges used for lipid extraction were not enough so that lipids were undetectable. For this reason, the analysis of lipid levels in meningeal samples were not included in the Results section.

A 3200 QTRAP® triple quadrupole mass spectrometer (Applied Biosystems Sciex, Darmstadt, Germany), coupled to an ExionLC 100 integrated high-performance liquid chromatography (HPLC) system (Applied Biosystems Sciex, Darmstadt, Germany), was used for the analysis. The chromatographic column was a monolithic C18 column (Onyx, 100 mm × 3 mm i.d., Phenomenex, Bologna, Italy) maintained at 25 °C. The gradient elution mobile phases were A (water/methanol 98:2 v/v containing 10 mM ammonium formate and 0.1% formic acid), B (methanol/acetonitrile/isopropanol 80:10:10 v/v containing 8 mM ammonium formate and 0.08% formic acid), and C (methanol).

2.8. Statistical Analysis

The face-grooming in Phase II of the orofacial formalin test of a previous study [24] was used to calculate the sample size need for this study. We used GPower (version 3.1.9.4) to perform an a priori power analysis, yielding a statistical power of 0.80 at an alpha level of 0.05. We supposed a difference in the total face-rubbing time between rats injected with NTG (mean 110 ± 39) and those with NTG + AKU-005 of approximately 70 s (mean 40 ± 13). We estimated a sample size of 6 rats in each experimental group with an effect size of 1.95. Because of an intergroup variability, we used a maximum of 9 rats per group. We tested all data for normality using the Kolmogorov–Smirnov normality test. We compared differences among groups using one-way ANOVA followed by post hoc Tukey's multiple comparisons test. A $p < 0.05$ was considered statistically significant. We expressed all data as mean ± SEM and used GraphPad Prism software (version 9.3.1) to perform all statistical analyses.

3. Results

3.1. AKU-005 Effects on the NTG-Induced Behaviors

The systemic administration of NTG reduced locomotor activity and exploratory behavior, expressed as distance traveled and number of rearings, compared with the CT group (Figure 2a,c). It also increased anxiety-like behavior, as indicated by the reduced time spent in the center of the open field, and nociception, as suggested by the increased grooming behavior, compared to the CT group (Figure 2b,d). AKU-005 did not alter the locomotor activity or any other behavior when injected alone, nor it affect the changes induced by NTG administration (Figure 2a–d).

As regards the orofacial formalin test, AKU-005 administration significantly prevented the NTG-induced increase in face-rubbing behavior in phase II of the test (Figure 2f). When AKU-005 was administered alone, without NTG, it did not affect the behavioral response in the test (Figure 2f).

The data show that AKU-005 can counteract the NTG-induced hyperalgesia but does not influence other migraine-like features induced by the NTG challenge.

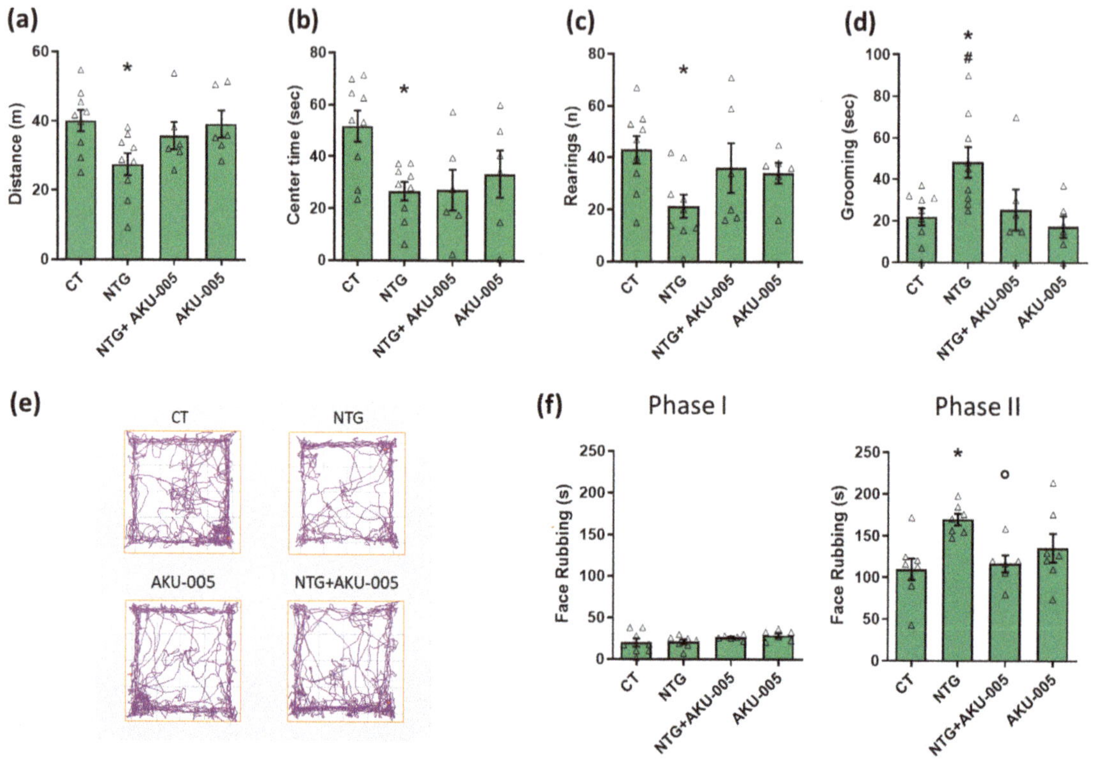

Figure 2. Behavioral testing. Open field test analysis: (**a**) distance (expressed in meters) travelled in the apparatus; (**b**) time spent (expressed in seconds) in the center of the apparatus; (**c**) number of rearings; (**d**) time spent performing grooming behavior (expressed in seconds); (**e**) representative track plots of the experimental groups; and (**f**) face-rubbing behavior (expressed in seconds) of Phase I and II of the orofacial formalin test. Data are expressed as mean ± SEM. One-way ANOVA followed by Tukey's multiple comparisons test; * $p < 0.05$ vs. CT; # $p < 0.05$ vs. AKU-005; ° $p < 0.05$ vs. NTG. Individual subjects' values are represented as triangle; N = 6–9.

3.2. *AKU-005 Effects on CGRP Expression and Serum Levels*

The administration of NTG treatment increased *CGRP* mRNA levels in the meninges, medulla, CSC, and TG ipsilateral to formalin injection as well as CGRP serum levels compared to the CT group (Figure 3a–e). AKU-005 administration significantly attenuated these changes (Figure 3a–e), except for *CGRP* in the CSC (Figure 3b). These findings suggest that the anti-hyperalgesic effects of AKU-005 are associated with a reduction in the CGRP-related pathways, contributing to pain reduction.

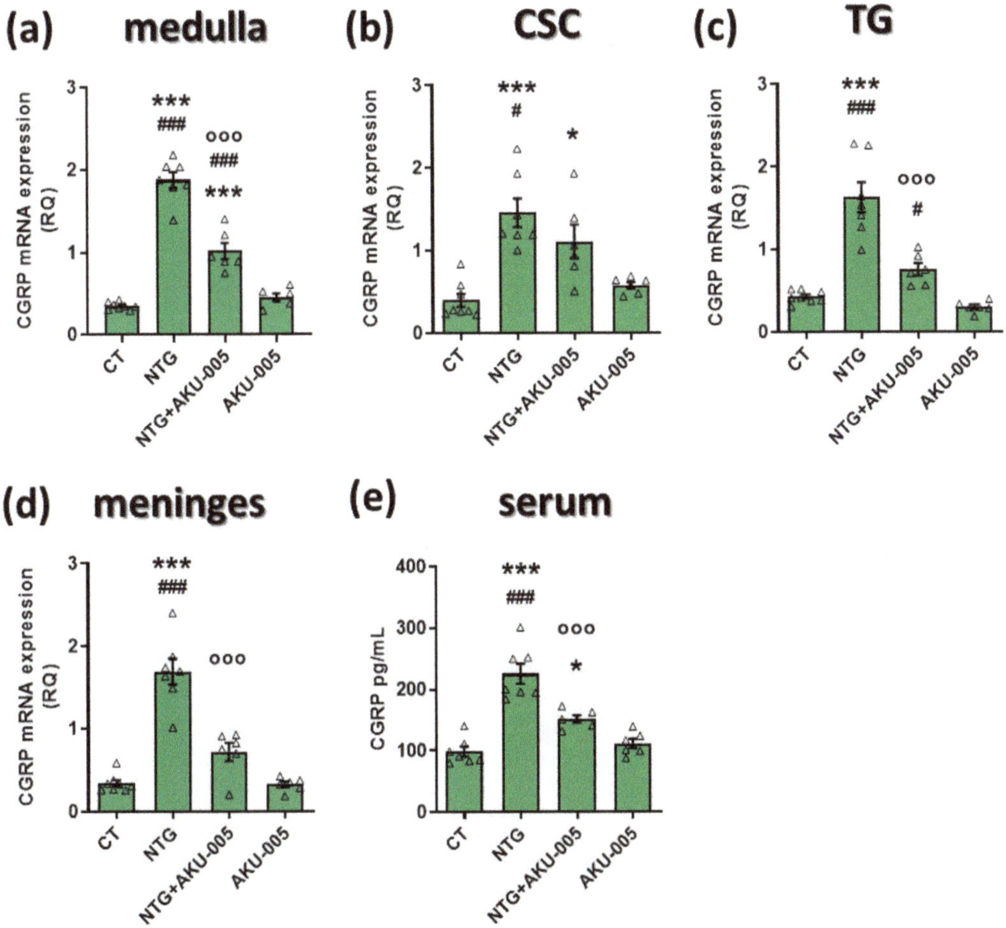

Figure 3. Gene expression and serum protein levels of CGRP. mRNA expression levels of *CGRP* in the (**a**) medulla, (**b**) cervical spinal cord (CSC), (**c**) trigeminal ganglion (TG), and (**d**) meninges; (**e**) CGRP serum levels (pg/mL). Data are expressed as mean ± SEM. One-way ANOVA followed by Tukey's multiple comparisons test; * $p < 0.05$ and *** $p < 0.001$ vs. CT; °°° $p < 0.001$ vs. NTG; # $p < 0.05$ and ### $p < 0.001$ vs. AKU-005. Individual subjects' values are represented as triangle; N = 6–8.

3.3. *AKU-005 Effects on the Gene Expression of Pro-Inflammatory Cytokines*

TNF-alpha and *IL-6* gene expression was significantly higher in the medulla, CSC, and TG ipsilateral to formalin injection in NTG-treated rats. NTG administration also increased the gene expression of cytokines at the meningeal level (Figure 4a–h). AKU-005 treatment prevented an NTG-induced increase in *IL-6* and *TNF-alpha* mRNA levels in all areas, except for *TNF-alpha* in the medulla (Figure 4a).

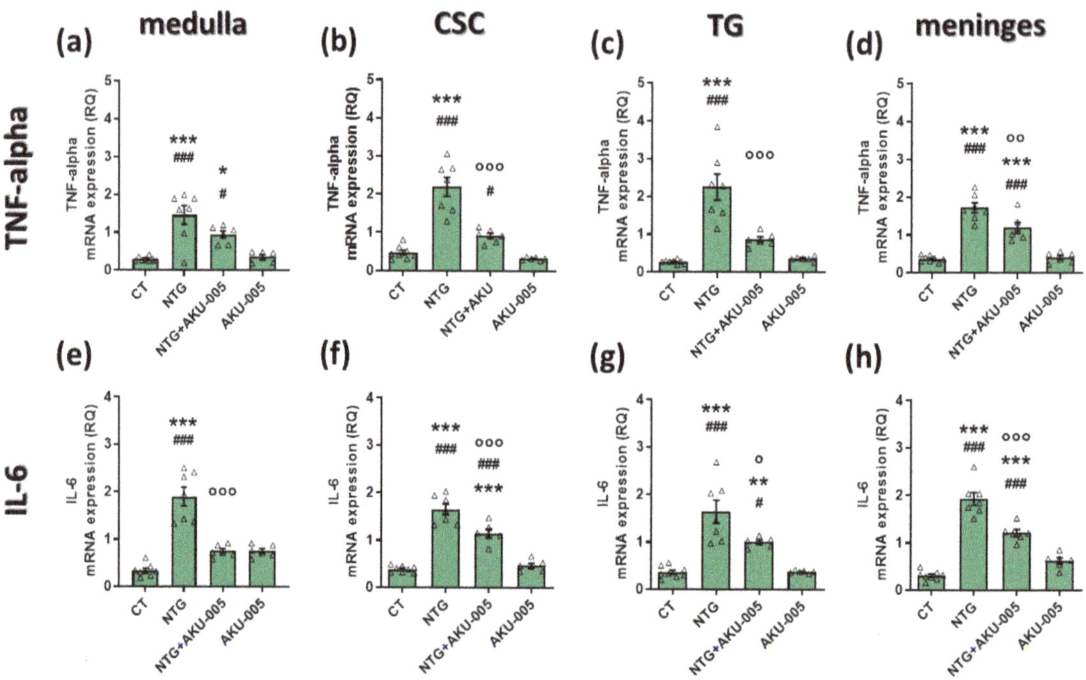

Figure 4. mRNA expression levels of (**a–d**) *TNF-alpha* and (**e–h**) *IL-6* in the medulla, cervical spinal cord (CSC), trigeminal ganglion (TG), and meninges. Data are expressed as mean ± SEM. One-way ANOVA followed by Tukey's multiple comparisons test; * $p < 0.05$, ** $p < 0.01$ and *** $p < 0.001$ vs. CT; ° $p < 0.05$, °° $p < 0.01$ and °°° $p < 0.001$ vs. NTG; # $p < 0.05$ and ### $p < 0.001$ vs. AKU-005. Individual subjects' values are represented as triangle; N = 6–8.

Overall, the data suggest that AKU-005 can counteract the NTG-induced activation of the inflammatory pathway in both peripheral and central areas relevant to migraine pathophysiology.

3.4. AKU-005 Effects on Cranial Lipid Levels

The level of endocannabinoids and related lipids were below the detection threshold in the meninges. In the other areas/tissues, NTG administration did not alter the levels of endocannabinoids and related lipids when compare with the control group, except for a reduction in AEA levels in the TG. The administration of AKU-005 had no effect on such a change (Figure 5). It is reasonable to assume that AKU-005 does not strongly affect the modulation of endocannabinoids and related lipids in these brain regions.

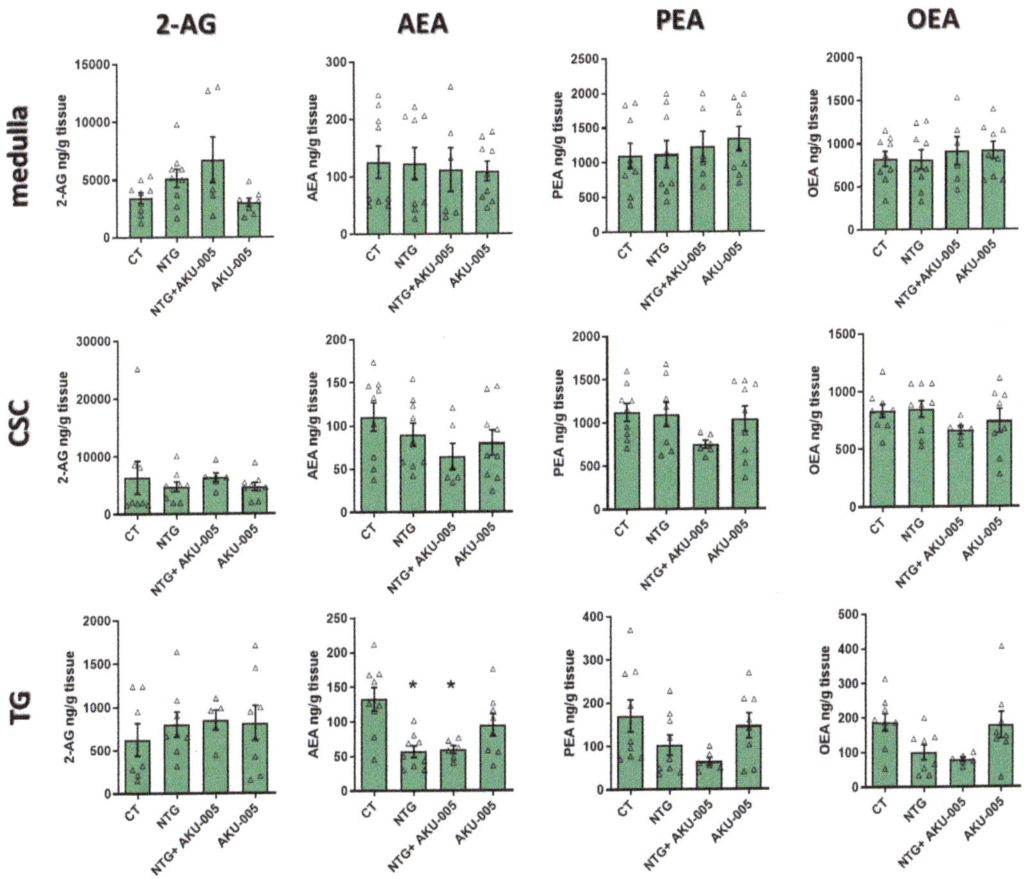

Figure 5. Levels of endocannabinoids and related lipids (expressed as ng/g tissue) in the medulla, cervical spinal cord (CSC), and trigeminal ganglion (TG). Data are expressed as mean ± SEM. One-way ANOVA followed by Tukey's multiple comparisons test; * $p < 0.05$ vs. CT. Individual subjects' values are represented as triangle; N = 6–9. Legend: 2-AG, 2-arachidonoylglycerol; AEA, arachidonoylethanolamine; PEA, palmitoylethanolamide; OEA, oleoyl ethanolamide.

4. Discussion

The local increase of AEA and 2-AG levels synthetized on demand using specific catabolic enzyme inhibitors may represent a means to modulate the endocannabinoid system while limiting the possible adverse side effects.

Here, we tested in vivo the biological activity of the reversible dual AKU-005 FAAH/MAGL inhibitor, which showed potential anti-migraine activity in vitro [27].

Our main findings can be summarized as follows:

(1) The administration of AKU-005 prevented multiple NTG-induced effects: hyperalgesia in the trigeminal region, and CGRP increase in serum and in cranial tissues and brain areas that are relevant to migraine pathophysiology;
(2) The administration of AKU-005 also prevented the NTG-induced activation of the inflammatory response;
(3) These activities were not associated with significant changes in the levels of endocannabinoids or related lipids in cranial tissues and brain areas that are relevant to migraine pathophysiology.

Altogether, these results, obtained in a migraine-specific animal model, suggest that the dual inhibition of FAAH/MAGL activity has a role in the pathophysiology of migraine. The lack of changes in the levels of endocannabinoids and related lipids in areas that are crucial for migraine pathophysiology came as a surprise. Indeed, several papers have demonstrated that the specific inhibition of FAAH or MAGL, the catabolic enzymes for AEA and 2-AG, respectively, counteracts the NTG-induced effects that are relevant to migraine [17,22–24,28]. Furthermore, the in vitro characterization study of AKU-005 by Aaltonen et al. [26] showed that the compound activity is dependent on the elevation of 2-AG and AEA brain levels, acting indirectly on the CB1 receptor. In a previous in vivo study, we also demonstrated that a dual FAAH/MAGL inhibitor counteracted NTG-induced effects via mechanisms that require the activation of CB1 receptors [24].

One possibility to explain why the biological activities of AKU-005 did not impact lipids in our experiments is that AKU-005 may have acted mainly at the level of the trigeminovascular terminals in the meninges, where we were unable to detect endocannabinoids. This may be due to the insufficient amount of meningeal samples needed for lipid extraction, as the trigeminovascular endings are located in the supratentorial dura mater [29].

Along this line of reasoning, we can speculate that AKU-005 induced changes in the lipid levels in the dura mater, which then caused indirect effects on other areas and, consequently, on the behavioral changes observed in the orofacial formalin test.

Similar to the present results, the dual inhibitor JZL195 significantly reduced NTG-induced trigeminal hyperalgesia and pain-associated behavior through cannabinoid receptor-mediated effects [24]. JZL195 also reduced *CGRP* and cytokine gene expression in central and peripheral areas and serum CGRP levels, suggesting that the two dual inhibitors share at least some mechanisms in their biological activity in the NTG paradigm [24]. In this scenario, we can speculate that NTG administration induced the degranulation of meningeal mast cells [30,31], thus contributing to the sensitization and activation of meningeal afferents [32] and causing a localized release of endocannabinoids, which went undetected in our experiments. In agreement, an over-regulation of inducible nitric oxide synthase (*iNOS*) gene expression was reported in the meningeal tissue after NTG administration, confirming an inflammatory state. The iNOS immunoreactivity was mainly expressed within resident meningeal macrophages and was associated with increased IL-1β expression [33]. In line with our data, AKU-005 caused minimal meningeal fiber firing under baseline conditions, while it significantly decreased meningeal firing following KCl stimulation [27], causing an increase in endocannabinoid release. The increase in endocannabinoid levels induced by AKU-005 was mediated by the CB1 receptor [25,27]. In agreement, the enhanced AEA signaling after FAAH inhibition produced significant anti-nociceptive effects in the meninges [34]. In addition, FAAH inhibition was reported to modulate pro-inflammatory activity [35] by interacting with several ion channels expressed in BV2 microglial cells, such as the potassium channel Kv1.5 [36] or the L-type calcium channel [37]. Other endocannabinoid catabolic enzyme inhibitors have shown different effects depending on the dose or timing of administration [38,39]. In addition, it is possible that AKU-005 may had an off-target effect [40,41] which has not yet been identified. For instance, PF3845 and URB597 were able to reduce prostaglandin E_2 (PGE2) production in lipopolysaccharide-stimulated BV2 microglial cells by suppressing the expression of PGE2 biosynthesis rather than the blockade of PGE2 biosynthetic enzymes [42]. It cannot be ruled out that AKU-005, with the dose utilized in vivo, may cause changes in other metabolites that are not closely related to the ECs in the areas investigated [43], thus causing a reduction in nociception through the modulation of CGRP release.

Of course, additional studies on the molecular and cellular characterization of AKU-005 and, specifically, on the assessment of its functional profile in the meninges and/or other brain areas are necessary to further elaborate on this hypothesis; but, still, we feel that the present findings suggest the opportunity to further investigate the high therapeutic potential of this pathway based on the dual inhibition of MAGL and FAAH activity.

Limitations of the Study

Although our animal model is commonly used for studying migraine [15,16], it has some limitations, like any other model. In the NTG human model, only individuals predisposed to migraine are subjected to develop NTG-induced headache, with varying onset times [15,16,44]. Using the animal model, we can mimic only a few aspects of the disease, such as the change in pain sensitivity induced by NTG through the action of CGRP release. In this context, AKU-005 seems to be an ideal candidate as it reduces nociception generation by modulating CGRP release. However, this was a single-dose study, with a dose selected pragmatically due to the lack of informative evidence on AKU-005 in vivo effects. Thus, additional studies are necessary to assess whether the lack of effects of AKU-005 on some of the parameters assessed may be related to an insufficient dose or to tissue-specific mechanisms. The latter seem particularly relevant in pain processing [45]. In this context, a thorough pharmacokinetic profile of the compound and a dose–response assessment will provide crucial information for further studies. Furthermore, we sampled lipid levels only once at a given time point. More information can be derived from a time-course evaluation and from an optimized methodology for obtaining adequate specimens of the meninges.

5. Conclusions

The present findings show for the first time the in vivo potential of the dual MAGL/FAAH inhibitor AKU-005 to counteract several of the NTG-induced effects that are relevant to migraine, probably by modulating the CGRP pathway. Further assessment of the pathways and mediators involved in this biological activity will inform the pathophysiological mechanism of migraine and possibly further characterize a pathway with a high druggable potential.

Author Contributions: Conceptualization, R.G.; Methodology, C.D., M.F., A.M.Z., and V.F.; Validation, R.G., and V.F.; Formal Analysis, R.G.; Investigation, R.G., C.D., M.F., A.M.Z., S.F., M.P., and V.F.; Resources, R.G., C.D., M.F., A.M.Z., M.P., and V.F.; Data Curation, R.G.; Writing—Original Draft Preparation, R.G., and C.D.; Writing—Review and Editing, C.T.; Visualization, R.G., and C.D.; Supervision, R.G., and V.F.; Project Administration, R.G.; Funding Acquisition, C.T. All authors have read and agreed to the published version of the manuscript.

Funding: This research was funded by the Italian Ministry of Health, grant number RC 2022–2024.

Institutional Review Board Statement: The animal study protocol was approved by the Italian Ministry of Health (protocol code 376/2020-PR).

Informed Consent Statement: Not applicable.

Data Availability Statement: The data presented in this study are available in Zenodo repository [doi:10.5281/zenodo.10817751].

Acknowledgments: The authors thank Migraine Research Foundation Impact Award.

Conflicts of Interest: The authors declare no conflict of interest.

References

1. Piomelli, D. The Endocannabinoid System: A Drug Discovery Perspective. *Curr. Opin. Investig. Drugs* **2005**, *6*, 672–679.
2. Finn, D.P.; Haroutounian, S.; Hohmann, A.G.; Krane, E.; Soliman, N.; Rice, A.S.C. Cannabinoids, the Endocannabinoid System, and Pain: A Review of Preclinical Studies. *Pain* **2021**, *162*, S5–S25. [CrossRef]
3. Lo Castro, F.; Baraldi, C.; Pellesi, L.; Guerzoni, S. Clinical Evidence of Cannabinoids in Migraine: A Narrative Review. *J. Clin. Med.* **2022**, *11*, 1479–1488. [CrossRef]
4. Akerman, S.; Holland, P.R.; Lasalandra, M.P.; Goadsby, P.J. Endocannabinoids in the Brainstem Modulate Dural Trigeminovascular Nociceptive Traffic via CB_1 and "Triptan" Receptors: Implications in Migraine. *J. Neurosci.* **2013**, *33*, 14869–14877. [CrossRef]
5. Kano, M.; Ohno-Shosaku, T.; Hashimotodani, Y.; Uchigashima, M.; Watanabe, M. Endocannabinoid-Mediated Control of Synaptic Transmission. *Physiol. Rev.* **2009**, *89*, 309–380. [CrossRef] [PubMed]
6. Di Marzo, V.; Bisogno, T.; De Petrocellis, L. Endocannabinoids and Related Compounds: Walking Back and Forth between Plant Natural Products and Animal Physiology. *Chem. Biol.* **2007**, *14*, 741–756. [CrossRef]

7. Fischer, M.; Messlinger, K. Cannabinoid and Vanilloid Effects of R(+)-Methanandamide in the Hemisected Meningeal Preparation. *Cephalalgia* **2007**, *27*, 422–428. [CrossRef] [PubMed]
8. Akerman, S.; Kaube, H.; Goadsby, P.J. Anandamide Acts as a Vasodilator of Dural Blood Vessels In Vivo by Activating TRPV1 Receptors. *Br. J. Pharmacol.* **2004**, *142*, 1354–1360. [CrossRef]
9. Goadsby, P.J.; Holland, P.R.; Martins-Oliveira, M.; Hoffmann, J.; Schankin, C.; Akerman, S. Pathophysiology of Migraine: A Disorder of Sensory Processing. *Physiol. Rev.* **2017**, *97*, 553–622. [CrossRef]
10. Luongo, L.; Maione, S.; Di Marzo, V. Endocannabinoids and Neuropathic Pain: Focus on Neuron–Glia and Endocannabinoid–Neurotrophin Interactions. *Eur. J. Neurosci.* **2014**, *39*, 401–408. [CrossRef]
11. Cabañero, D.; Ramírez-López, A.; Drews, E.; Schmöle, A.; Otte, D.M.; Wawrzczak-Bargiela, A.; Huerga Encabo, H.; Kummer, S.; Ferrer-Montiel, A.; Przewlocki, R.; et al. Protective Role of Neuronal and Lymphoid Cannabinoid CB2 Receptors in Neuropathic Pain. *Elife* **2020**, *9*, e55582. [CrossRef] [PubMed]
12. Kasatkina, L.A.; Rittchen, S.; Sturm, E.M. Neuroprotective and Immunomodulatory Action of the Endocannabinoid System under Neuroinflammation. *Int. J. Mol. Sci.* **2021**, *22*, 5431. [CrossRef] [PubMed]
13. Leimuranta, P.; Khiroug, L.; Giniatullin, R. Emerging Role of (Endo)Cannabinoids in Migraine. *Front. Pharmacol.* **2018**, *9*, 354945. [CrossRef] [PubMed]
14. Liktor-Busa, E.; Levine, A.A.; Palomino, S.M.; Singh, S.; Wahl, J.; Vanderah, T.W.; Stella, N.; Largent-Milnes, T.M. ABHD6 and MAGL Control 2-AG Levels in the PAG and Allodynia in a CSD-Induced Periorbital Model of Headache. *Front. Pain Res.* **2023**, *4*, 1171188. [CrossRef] [PubMed]
15. Demartini, C.; Greco, R.; Zanaboni, A.M.; Sances, G.; De Icco, R.; Borsook, D.; Tassorelli, C. Nitroglycerin as a Comparative Experimental Model of Migraine Pain: From Animal to Human and Back. *Prog. Neurobiol.* **2019**, *177*, 15–32. [CrossRef] [PubMed]
16. Sureda-Gibert, P.; Romero-Reyes, M.; Akerman, S. Nitroglycerin as a Model of Migraine: Clinical and Preclinical Review. *Neurobiol. Pain* **2022**, *12*, 100105. [CrossRef]
17. Nozaki, C.; Markert, A.; Zimmer, A. Inhibition of FAAH Reduces Nitroglycerin-Induced Migraine-like Pain and Trigeminal Neuronal Hyperactivity in Mice. *Eur. Neuropsychopharmacol.* **2015**, *25*, 1388–1396. [CrossRef]
18. Nagy-Grócz, G.; Bohár, Z.; Fejes-Szabó, A.; Laborc, K.F.; Spekker, E.; Tar, L.; Vécsei, L.; Párdutz, Á. Nitroglycerin Increases Serotonin Transporter Expression in Rat Spinal Cord but Anandamide Modulated This Effect. *J. Chem. Neuroanat.* **2017**, *85*, 13–20. [CrossRef]
19. Nagy-Grócz, G.; Tar, L.; Bohár, Z.; Fejes-Szabó, A.; Laborc, K.F.; Spekker, E.; Vécsei, L.; Párdutz, Á. The Modulatory Effect of Anandamide on Nitroglycerin-Induced Sensitization in the Trigeminal System of the Rat. *Cephalalgia* **2016**, *36*, 849–861. [CrossRef]
20. Peng, L.-M.; Chen, X.-P.; Shi, R.-Z.; Chen, L.; Li, Y.-J.; Yang, T.-L. Involvement of Anandamide Transporter in Calcitonin Gene-Related Peptide Expression Stimulated by Nitroglycerin and Influence of ALDH2 Glu504Lys Polymorphism. *J. Cardiovasc. Pharmacol.* **2014**, *64*, 460–464. [CrossRef]
21. Mansoori, M.; Zarei, M.R.; Chamani, G.; Nazeri, M.; Mohammadi, F.; Alavi, S.S.; Shabani, M. Chronic Migraine Caused a Higher Rate of Tendency to Cannabinoid Agonist Compared to Morphine. *Acta Biomed.* **2020**, *91*, e2020185. [CrossRef] [PubMed]
22. Greco, R.; Demartini, C.; Zanaboni, A.; Casini, I.; De Icco, R.; Reggiani, A.; Misto, A.; Piomelli, D.; Tassorelli, C. Characterization of the Peripheral FAAH Inhibitor, URB937, in Animal Models of Acute and Chronic Migraine. *Neurobiol. Dis.* **2021**, *147*, 105157. [CrossRef]
23. Greco, R.; Francavilla, M.; Demartini, C.; Zanaboni, A.M.; Facchetti, S.; Palmisani, M.; Franco, V.; Tassorelli, C. Activity of FAAH-Inhibitor JZP327A in an Experimental Rat Model of Migraine. *Int. J. Mol. Sci.* **2023**, *24*, 10102. [CrossRef] [PubMed]
24. Greco, R.; Demartini, C.; Francavilla, M.; Zanaboni, A.M.; Tassorelli, C. Dual Inhibition of FAAH and MAGL Counteracts Migraine-like Pain and Behavior in an Animal Model of Migraine. *Cells* **2021**, *10*, 2543. [CrossRef] [PubMed]
25. Della Pietra, A.; Giniatullin, R.; Savinainen, J.R. Distinct Activity of Endocannabinoid-Hydrolyzing Enzymes MAGL and FAAH in Key Regions of Peripheral and Central Nervous System Implicated in Migraine. *Int. J. Mol. Sci.* **2021**, *22*, 1204. [CrossRef] [PubMed]
26. Aaltonen, N.; Savinainen, J.R.; Ribas, C.R.; Rönkkö, J.; Kuusisto, A.; Korhonen, J.; Navia-Paldanius, D.; Häyrinen, J.; Takabe, P.; Käsnänen, H.; et al. Piperazine and Piperidine Triazole Ureas as Ultrapotent and Highly Selective Inhibitors of Monoacylglycerol Lipase. *Chem. Biol.* **2013**, *20*, 379–390. [CrossRef] [PubMed]
27. Della Pietra, A.; Krivoshein, G.; Ivanov, K.; Giniatullina, R.; Jyrkkänen, H.-K.; Leinonen, V.; Lehtonen, M.; van den Maagdenberg, A.M.J.M.; Savinainen, J.; Giniatullin, R. Potent Dual MAGL/FAAH Inhibitor AKU-005 Engages Endocannabinoids to Diminish Meningeal Nociception Implicated in Migraine Pain. *J. Headache Pain.* **2023**, *24*, 38. [CrossRef] [PubMed]
28. Greco, R.; Demartini, C.; Zanaboni, A.M.; Berliocchi, L.; Piomelli, D.; Tassorelli, C. Inhibition of Monoacylglycerol Lipase: Another Signalling Pathway for Potential Therapeutic Targets in Migraine? *Cephalalgia* **2018**, *38*, 1138–1147. [CrossRef]
29. Terrier, L.; Hadjikhani, N.; Velut, S.; Magnain, C.; Amelot, A.; Bernard, F.; Zöllei, L.; Destrieux, C. The Trigeminal System: The Meningovascular Complex—A Review. *J. Anat.* **2021**, *239*, 1–11. [CrossRef]
30. Ghosh, S.; Kinsey, S.G.; Liu, Q.; Hruba, L.; McMahon, L.R.; Grim, T.W.; Merritt, C.R.; Wise, L.E.; Abdullah, R.A.; Selley, D.E.; et al. Full Fatty Acid Amide Hydrolase Inhibition Combined with Partial Monoacylglycerol Lipase Inhibition: Augmented and Sustained Antinociceptive Effects with Reduced Cannabimimetic Side Effects in Mice. *J. Pharmacol. Exp. Ther.* **2015**, *354*, 111–120. [CrossRef]

31. Cabral, G.A.; Marciano-Cabral, F. Cannabinoid Receptors in Microglia of the Central Nervous System: Immune Functional Relevance. *J. Leukoc. Biol.* **2005**, *78*, 1192–1197. [CrossRef] [PubMed]
32. Ferrari, L.F.; Levine, J.D.; Green, P.G. Mechanisms Mediating Nitroglycerin-Induced Delayed-Onset Hyperalgesia in the Rat. *Neuroscience* **2016**, *317*, 121–129. [CrossRef] [PubMed]
33. Reuter, U.; Bolay, H.; Jansen-Olesen, I.; Chiarugi, A.; Del Rio, M.S.; Letourneau, R.; Theoharides, T.C.; Waeber, C.; Moskowitz, M.A. Delayed Inflammation in Rat Meninges: Implications for Migraine Pathophysiology. *Brain* **2001**, *124*, 2490–2502. [CrossRef]
34. Cruz, S.L.; Sánchez-Miranda, E.; Castillo-Arellano, J.I.; Cervantes-Villagrana, R.D.; Ibarra-Sánchez, A.; González-Espinosa, C. Anandamide Inhibits FcεRI-Dependent Degranulation and Cytokine Synthesis in Mast Cells through CB2 and GPR55 Receptor Activation. Possible Involvement of CB2-GPR55 Heteromers. *Int. Immunopharmacol.* **2018**, *64*, 298–307. [CrossRef] [PubMed]
35. Espinosa-Parrilla, J.F.; Martínez-Moreno, M.; Gasull, X.; Mahy, N.; Rodríguez, M.J. The L-Type Voltage-Gated Calcium Channel Modulates Microglial pro-Inflammatory Activity. *Mol. Cell. Neurosci.* **2015**, *64*, 104–115. [CrossRef]
36. Moreno-Galindo, E.G.; Barrio-Echavarría, G.F.; Vásquez, J.C.; Decher, N.; Sachse, F.B.; Tristani-Firouzi, M.; Sánchez-Chapula, J.A.; Navarro-Polanco, R.A. Molecular Basis for a High-Potency Open-Channel Block of Kv1.5 Channel by the Endocannabinoid Anandamide. *Mol. Pharmacol.* **2010**, *77*, 751–758. [CrossRef]
37. Vignali, M.; Benfenati, V.; Caprini, M.; Anderova, M.; Nobile, M.; Ferroni, S. The Endocannabinoid Anandamide Inhibits Potassium Conductance in Rat Cortical Astrocytes. *Glia* **2009**, *57*, 791–806. [CrossRef] [PubMed]
38. Compton, D.R.; Martin, B.R. The Effect of the Enzyme Inhibitor Phenylmethylsulfonyl Fluoride on the Pharmacological Effect of Anandamide in the Mouse Model of Cannabimimetic Activity. *J. Pharmacol. Exp. Ther.* **1997**, *283*, 1138–1143. [PubMed]
39. Okine, B.N.; Norris, L.M.; Woodhams, S.; Burston, J.; Patel, A.; Alexander, S.P.; Barrett, D.A.; Kendall, D.A.; Bennett, A.J.; Chapman, V. Lack of Effect of Chronic Pre-treatment with the FAAH Inhibitor URB597 on Inflammatory Pain Behaviour: Evidence for Plastic Changes in the Endocannabinoid System. *Br. J. Pharmacol.* **2012**, *167*, 627–640. [CrossRef]
40. Liu, J.; Fike, K.R.; Dapper, C.; Klemba, M. Metabolism of Host Lysophosphatidylcholine in *Plasmodium Falciparum*–Infected Erythrocytes. *Proc. Natl. Acad. Sci. USA* **2024**, *121*, e2320262121. [CrossRef]
41. Bosier, B.; Muccioli, G.G.; Lambert, D.M. The FAAH Inhibitor URB597 Efficiently Reduces Tyrosine Hydroxylase Expression through CB_1- and FAAH-independent Mechanisms. *Br. J. Pharmacol.* **2013**, *169*, 794–807. [CrossRef] [PubMed]
42. Tanaka, M.; Yagyu, K.; Sackett, S.; Zhang, Y. Anti-Inflammatory Effects by Pharmacological Inhibition or Knockdown of Fatty Acid Amide Hydrolase in BV2 Microglial Cells. *Cells* **2019**, *8*, 491. [CrossRef]
43. Starowicz, K.; Makuch, W.; Korostynski, M.; Malek, N.; Slezak, M.; Zychowska, M.; Petrosino, S.; De Petrocellis, L.; Cristino, L.; Przewlocka, B.; et al. Full Inhibition of Spinal FAAH Leads to TRPV1-Mediated Analgesic Effects in Neuropathic Rats and Possible Lipoxygenase-Mediated Remodeling of Anandamide Metabolism. *PLoS ONE* **2013**, *8*, e60040. [CrossRef] [PubMed]
44. Pellesi, L. The Human NTG Model of Migraine in Drug Discovery and Development. *Expert. Opin. Drug Discov.* **2023**, *18*, 1077–1085. [CrossRef]
45. Rouzer, C.A.; Marnett, L.J. Endocannabinoid Oxygenation by Cyclooxygenases, Lipoxygenases, and Cytochromes P450: Cross-Talk between the Eicosanoid and Endocannabinoid Signaling Pathways. *Chem. Rev.* **2011**, *111*, 5899–5921. [CrossRef] [PubMed]

MDPI AG
Grosspeteranlage 5
4052 Basel
Switzerland
Tel.: +41 61 683 77 34
www.mdpi.com

Cells Editorial Office
E-mail: cells@mdpi.com
www.mdpi.com/journal/cells

www.ingramcontent.com/pod-product-compliance
Lightning Source LLC
Chambersburg PA
CBHW040804250926
51850CB00007BA/1073
* 9 7 8 3 7 2 5 8 1 7 2 5 2 *